Contemporary Computer-Assisted Approaches to Molecular Structure Elucidation

**New Developments in NMR**

*Editor-in-Chief:*
Professor William S. Price, *University of Western Sydney, Australia*

*Series Editors:*
Professor Bruce Balcom, *University of New Brunswick, Canada*
Professor István Furó, *Industrial NMR Centre at KTH, Sweden*
Professor Masatsune Kainosho, *Nagoya University, Japan*
Professor Maili Liu, *Chinese Academy of Sciences, Wuhan, China*

*Titles in the Series:*
1: Contemporary Computer-Assisted Approaches to Molecular Structure Elucidation

*How to obtain future titles on publication:*
A standing order plan is available for this series. A standing order will bring delivery of each new volume immediately on publication.

*For further information please contact:*
Book Sales Department, Royal Society of Chemistry, Thomas Graham House, Science Park, Milton Road, Cambridge, CB4 0WF, UK
Telephone: +44 (0)1223 420066, Fax: +44 (0)1223 420247, Email: books@rsc.org
Visit our website at http://www.rsc.org/Shop/Books/

# Contemporary Computer-Assisted Approaches to Molecular Structure Elucidation

**Mikhail Elyashberg**
*Advanced Chemistry Development, ACD Limited, Russia*

**Antony Williams**
*Royal Society of Chemistry, Wake Forest, NC, USA*

**Kirill Blinov**
*Advanced Chemistry Development, ACD Limited, Russia*

RSC Publishing

New Developments in NMR No. 1

ISBN: 978-1-84973-432-5
ISSN: 2044-253X

A catalogue record for this book is available from the British Library

Published by The Royal Society of Chemistry,
Thomas Graham House, Science Park, Milton Road,
Cambridge CB4 0WF, UK

Registered Charity Number 207890

For further information see our web site at www.rsc.org

For my wife Natasha, with love and gratitude for patience.

*Mikhail Elyashberg*

For Sharon, Taylor and Tyler. My family, my joy and my loves.

*Antony Williams*

For Tanya, Andrew, Arseniy and Afanasiy.

*Kirill Blinov*

# *Foreword*

Computer-Assisted Structure Elucidation – words that might once have struck terror in the hearts of spectroscopists. But, in the hands of a competent spectroscopist, CASE programs are yet another powerful tool that are complementary to the arsenal of two-dimensional NMR experiments that can be marshaled to solve some of the most daunting chemical structure elucidation problems that are encountered.

The authors have done an excellent job in Chapters 1 and 2 of reviewing the history and carefully delineating the intellectual framework that underlies CASE methods in general before delving into the specifics of these programs. Early experiences with CASE programs were more adversarial than symbiotic, and the early versions of such programs would have been completely incapable of assimilating and utilizing the vast array of 2D NMR data that can now be generated on even sub-milligram samples of materials in a modern NMR lab.

One of the integral facets of CASE methods is the requirement of sifting through and rank-ordering an output file that may contain hundreds, or in some cases as many as tens of thousands of potential solutions to the problem being pursued. The authors provide an excellent treatment of chemical shift prediction in Chapter 3 of this monograph that can be used to sort chemical structures as a function of the agreement between the calculated chemical shift data for the structural possibilities and the measured chemical shifts for the molecule whose structure is sought.

Another critical feature of structure elucidation is the determination of stereochemistry. Chapter 4 shows examples of the application of the Structure Elucidator CASE program to determine stereochemical features of molecules as complex as Taxol and the polyether marine toxin brevetoxin-B. Chapters 5 and 6 deal with CASE systems based on 1D and 2D NMR spectra, respectively. Chapter 7 delves into knowledge bases utilized in conjunction with CASE programs.

New Developments in NMR No. 1
Contemporary Computer-Assisted Approaches to Molecular Structure Elucidation
By Mikhail Elyashberg, Antony Williams and Kirill Blinov
© Royal Society of Chemistry 2012
Published by the Royal Society of Chemistry, www.rsc.org

Next, Chapter 8 addresses the critically important aspect of the preparation and checking of the input data that will be utilized by the CASE program for structure generation. From personal experience, this is perhaps the most time-consuming and demanding aspect of using CASE programs and it is *extremely* important that chemical shifts are identically entered from one experiment to the next.

Chapter 9 addresses algorithmic structure elucidation and highlights the generation of structural fragment information based on $^1$H and $^{13}$C chemical shift information in conjunction with homo- and heteronuclear 2D NMR correlation data. The authors also, very usefully, discuss the ability to input spectroscopist determined fragments and/or forbidden fragment information that can significantly speed the structure generation process. Indeed, for a CASE program to be a useful and synergistic tool for a competent spectro-scopist the program should be capable of assimilating chemical fragment information that the spectroscopist is can quickly deduce.

Beyond "simple" structure elucidation problems, spectroscopists frequently encounter problems where the determination of the structure in question is exacerbated by atypical responses. Examples might include a $^4J_{CH}$ long-range correlation in a HMBC experiment that is of sufficient intensity to be mistaken for a $^3J_{CH}$ correlation. Such correlations can lead to contradictory structural information that can be extremely difficult for even a highly experienced spec-troscopist to work around. These types of problem correlations and the "fuzzy" generation processes developed to address them are covered in Chapter 10.

Chapter 11 considers the application of CASE methods to several challen-ging problems several of which I have had personal and very frustrating experience with. One of these problems, the elucidation of the structure of the bis-indoloquinoline alkaloid cryptotackiine, I worked on periodically for over a decade before the structure was finally elucidated using CASE methods. Severe spectral overlap of the spectra, even at high field, rendered the problem one of multiple simultaneous spectral deconvolutions, the type of problem best dealt with by a computer algorithm. Another of these problems involved the eluci-dation of the structures of multiple degradants of an alkaloid following long term storage. As such, the problems used as illustrations in Chapter 11 show the range of capabilities of CASE programs and their synergy with spectroscopists.

Chapter 12 deals with the the utilization of CASE programs to correct structures in the literature. Complex natural product structures are prone to interpretational errors that can arise as a function of sample size limiting the range of 2D NMR experiments that can be performed on the available sample. Structures reported in the literature are not the only problem structures. Screening libraries obtained from commercial or academic sources also have the potential for incorrectly assigned structures; long term storage of screening libraries with samples stored in DMSO solution can also undergo degradation over time leading to samples that no longer adequately reflect the structure attributed to the sample. Any of these eventualities provide a fertile ground for the application of CASE programs.

Chapter 13 addresses the interesting comparison of human vs. quantum mechanical vs. CASE methods for structure characterization. Human interpretation can involve intuitive deductions that are beyond the range of any currently available CASE program. In contrast, human interpretation will seldom afford all possible structural possibilities consistent with the data. Quantum mechanical methods, while powerful, are too computationally intensive to be amenable to routine applications. CASE methods can routinely elaborate all possible structures. The interplay of these methods should be of considerable interest to a wide range of investigators.

Chapter 14 completes the volume with an evaluation of the performance of CASE programs. As the algorithms become more powerful and knowledge bases accessible by the programs broader, CASE programs will doubtless become more useful to investigator.

Overall, this monograph should be of considerable interest and utility to a wide range of investigators ranging from relative novices to highly experienced spectroscopists who are occasionally challenged by extremely difficult structure problems that even an excellent background in mechanistic chemistry would not predict. The ability of CASE programs to aid spectroscopists in getting "out of the rut" cannot be underestimated. In this reviewer's own experience, there have been several different classes of difficult problems that CASE programs have helped me to successfully navigate in less time than would have been consumed in getting to the answer without them. Indeed, given the difficulty of the elucidation of the structure of the alkaloid cryptotackiine, I might still be pursuing that structure without the intervention of CASE methods!

Gary Martin
Warren, New Jersey

# *Preface*

The structure elucidation of complex chemical compounds remains one of the primary challenges faced by chemists today. Whether the chemicals under study result from a particular synthesis, an impurity or degradant analysis in the pharmaceutical analysis or the isolation of a potentially novel natural product, structure identification is critical to developing an understanding of synthetic routes, metabolic pathways, structure diversity and, in many cases, bringing commercial advantage. There are numerous analytical science approaches that may be utilized to confirm a chemical structure but, since its' discovery in the 1940s nuclear magnetic resonance (NMR) has established itself as the primary structure elucidation tool for chemists. As NMR spectrometers have followed the general trend of miniaturization the instruments have migrated from taking up most of a room and operated by highly skilled spectroscopists to benchtop systems accessed by the majority of chemists using push-button operation and highly automated data processing tools.

The generation of high-quality data is rarely a challenge nowadays. Taking advantage of the many engineering achievements 1D and 2D spectra can now be generated on general sample concentrations in just a few minutes. Even for the most dilute samples the availability of high magnetic field strengths and advanced probe technologies allows high-quality data to be obtained. The analysis of these data remains the primary bottleneck for scientists. The ability to search against reference databases and predict the NMR spectra of hypothetical structures can certainly assist in structure identification and these approaches are now quite common. Nevertheless, the elucidation of a complete unknown, and the full assignment of the spectral features to the structure remains challenging and the potential applications of computers to assist in the process is a natural pursuit.

Computer-Assisted Structure Elucidation, CASE, has been investigated for decades. The improved performance of computers, the development of large databases of assigned NMR spectra and structures and the iterative development of algorithms to perform computer-based analysis has allowed the

New Developments in NMR No. 1
Contemporary Computer-Assisted Approaches to Molecular Structure Elucidation
By Mikhail Elyashberg, Antony Williams and Kirill Blinov
© Royal Society of Chemistry 2012
Published by the Royal Society of Chemistry, www.rsc.org

systems to expand in scope and performance. CASE systems have become a weapon in the armory of a scientist to apply to the elucidation of chemical compounds. It is not limited to small organic molecules but, rather, the elucidation of large complex compounds with over a 100 skeletal atoms and over 1500 mass units has been demonstrated. Many of the problems inherent to human interpretation of the data, specifically bias and an inability to generate all potential structural hypotheses, can be dealt with using CASE systems. Algorithms have been developed to account for the presence of non-standard correlations in the 2D NMR data, can assign relative stereochemistry across compounds with multiple stereocenters and can elucidate complex molecules in just a few seconds for many examples.

This book represents the collective wisdom, knowledge and experiences of almost 90 years of processing, manipulating and interpreting NMR spectroscopy data. Our authors include (1) a scientist who was one of the initial advocates for CASE technologies starting in the late 60s, (2) a scientist who managed multiple spectroscopy laboratories in his career and performed many hundreds of complex structural analyses, and (3) an expert in data-processing and algorithmic development for NMR prediction and structure elucidation. Collectively we have produced a volume which we hope can be consulted as a "getting started guide" for scientists interested in the applications of CASE, can be a reference book for those interested in extending computer-assisted approaches to structure elucidation and will, in any case, document the state of the art in CASE programs as of its completion. Each of us has been involved with the development of a CASE system, Structure Elucidator. This volume will discuss many of our experiences in developing the software over a period of almost two decades to serve the user base in many chemical and pharmaceutical laboratories around the world.

Structure elucidation has, for many years, required a deep knowledge of spectroscopy interpretation skills, not only NMR, and these same skills were often insufficient to remove inherent human bias. We believe that CASE approaches can also be used by non-skilled spectroscopists to interpret complex data and elucidate chemical structures that previously might have been too complex for them to undertake. Software developed with the appropriate warnings can effectively enable chemists through the process and ensure that they do not make common mistakes in the interpretation of data. We also believe that CASE systems offer the opportunity for skilled spectroscopists to embrace a synergistic working relationship with software and offer the opportunity to produce higher quality results and dramatically reduce the number of erroneous reports in the literature and we will effectively demonstrate in this work. The end-goal of this approach will ultimately be the marriage of CASE algorithms to the data generation system and the future of spectroscopic instrumentation is sure to be a report showing a list of rank-ordered structure hypotheses matching the array of experimental data and with the most probably structure highlighted as the answer. This work is already progressing and, as demonstrated by the myriad of examples in this volume, algorithmic approaches can already outperform many of the scientists performing structure elucidation today.

# Contents

New Developments in NMR No. 1
Contemporary Computer-Assisted Approaches to Molecular Structure Elucidation
By Mikhail Elyashberg, Antony Williams and Kirill Blinov
© Royal Society of Chemistry 2012
Published by the Royal Society of Chemistry, www.rsc.org

## Part II Examples of Case Expert Systems

# Acknowledgments

This book represents the efforts of hundreds of researchers during the past half century to develop computer-assisted structure elucidation (CASE). We feel privileged to represent the collective efforts of a group of scientists with the vision, skills and intellect to consider the challenge of developing computer algorithms and approaches to elucidate the structures of chemical compounds. This book also represents many decades of the collective efforts of its authors to develop a state of the art CASE system. We owe a debt of gratitude to those who came before us but also to our immediate colleagues.

The CASE system Structure Elucidator has been developed by the collective efforts of staff at Advanced Chemistry Development (ACD/Labs). This includes our colleagues who have built structure drawing tools, analytical data processing algorithms and databasing systems, who have entered the millions of chemical shifts in the NMR databases, who have designed the graphical user interface elements that knit together the multiple modules necessary to deliver the system, and who have architected the overall vision for an integrated software suite for the analytical sciences. We are proud to have worked with you all on this software package. This book, while authored by us, stands as a testament to their skills and hard work.

We are extremely grateful to Gwen Jones and colleagues at the Royal Society of Chemistry for their assistance with this book. Gwen's ongoing support and focus on bringing this project in on time, and with high quality, kept us on task.

We are honored that Gary Martin chose to contribute a foreword to this volume. He has worked with us to challenge us on the capabilities of the Structure Elucidator software package. Always patient, always encouraging and never settling for second best, Gary has demanded the best of us at all times. We are better for working with him. Many of the examples in this book are taken from works we have published with him and we thank him for his time and effort to support our work.

New Developments in NMR No. 1
Contemporary Computer-Assisted Approaches to Molecular Structure Elucidation
By Mikhail Elyashberg, Antony Williams and Kirill Blinov
© Royal Society of Chemistry 2012
Published by the Royal Society of Chemistry, www.rsc.org

This book is dedicated to our families as they have supported us through this project and allowed us to capture our work, our passion and our obsessions within this volume. We hope that this book will stand the test of time and, in the future, be counted as one of the reference volumes for computer assisted structure elucidation.

*Mikhail Elyashberg and Kirill Blinov*
*Moscow, Russia*

*Antony J. Williams*
*Wake Forest, North Carolina*

Two of us, KB and AJW, take this opportunity to honor our friend and colleague, Mikhail Elyashberg, not only for his guidance and leadership role in developing this volume, but also his mentorship over the years in understanding the depths of CASE systems based on his long historical association with their development. His original idea of applying mathematical logic for spectral analysis was conceived in 1965 (when one of us was still crawling and the other was simply a concept!). In 1966 he submitted 3 articles to Journal of Applied Spectroscopy (Russian): 1) an algorithm for structural-group analysis (with L. A. Gribov), 2) a program realizing the algorithm (with L. A. Moskovkina) 3) an algorithm to determine characteristic spectral features by solving logical equations. All three articles appeared only in 1968 and three other groups (J. Lederberg, M. Munk and S. Sasaki) published their works independently in the same year. His Ph. D thesis issued in 1970 was likely the first Ph. D. devoted to CASE systems in the world (and certainly in the USSR). His book "Molecular spectral analysis and computers" (M. E. Elyashberg, L. A. Gribov, V. V. Serov) was published in 1980, and it was the first monograph on CASE systems in the world literature. Mikhail was the head of the Laboratory of Molecular Spectroscopy (NMR, IR, MS, UV-VIS) in the All-Russian Research Institute of Organic Synthesis for *ca.* 25 years, where thousands of structure elucidations were performed using all available methods. To us, Mikhail is the **father** of Computer Assisted Structure Elucidation and we are honored to have helped bring his vision to fruition. We are humbled to have worked alongside him.

# Part I
# COMPUTER-ASSISTED STRUCTURE
# ELUCIDATION: FUNDAMENTALS

# CHAPTER 1

# General Principles of CASE Systems

## 1.1 Statement of the Problem of Structure Elucidation

The first reports devoted to computer-assisted structure elucidation (CASE) were published by four independent groups of researchers[1–4] in the late 1960s. Prior to describing CASE methods, we will consider the complex nature of the structure elucidation problem.

It is necessary to distinguish two different analytical problems that are associated with the structure elucidation of molecules. The first relates to the identification of a molecule that is assumed to be known already and whose physicochemical characteristics, specifically the associated spectral data, are included into collections of reference data. In this case the solution is found by searching through the reference data and comparing the data measured for the unknown with the available reference data. Most frequently the search is performed using one or more of the mass spectrometry (MS), nuclear magnetic resonance (NMR) or infrared (IR) spectra of the unknown. The second problem is much more challenging and supposes that the unknown is a novel compound synthesized or isolated for the first time. The methods of solving these two problems are quite different. Therefore the first step is the assignment of a given unknown to one of the two mentioned categories. As we will see later in Section 1.6 specific procedures are used for this goal. This book will focus primarily on the structure elucidation of the *new* organic compounds.

The problem of elucidating the structure of a new compound can be divided into the two following sub-problems: (a) establishing the molecular formula, *i.e.* determine the type and number of each of the chemical elements making up the molecule, and (b) determining how the atoms are connected by chemical bonds of different multiplicities in the structure. Modern approaches for the determination of molecular formula are well established and based on

New Developments in NMR No. 1
Contemporary Computer-Assisted Approaches to Molecular Structure Elucidation
By Mikhail Elyashberg, Antony Williams and Kirill Blinov
© Royal Society of Chemistry 2012
Published by the Royal Society of Chemistry, www.rsc.org

high-resolution MS (HRMS). In general, they allow for the unambiguous determination of the molecular formula of an unknown.[5]

Molecular structure determination is such a challenge primarily due to the phenomenon of isomerism. In the 18th century Alexander Humboldt was probably the first who conjectured that there might be chemical substances which are composed of the same set of atoms but have different properties. This was later proven experimentally by Gay-Lussac, Liebig and Wohler[6] at the beginning of the 19th century, and the new term "isomers" was introduced into chemistry by Berzelius[7] in 1830. The following questions then became of interest to chemists: how many isomers can theoretically exist for a given molecule and how can they be exhaustively enumerated? The mathematical challenge of enumerating all isomers corresponding to a specific molecular formula was later realized when the notion of atom valence was defined and the first research into this area was initiated by Cayley[8] in the second half of the 19th century. The computation of the number of isomers from the molecular formula became possible only in the 1960s when computers arrived on the scene and computational chemistry, as a specific area of scientific investigation, was born. Mathematical algorithms and programs were then developed[9-11] that provided a possibility not only to calculate the number of isomers corresponding to a given molecular formula but also to generate the structural formulas for all isomers, which is clearly an important capability. As a result, chemists then had the ability to estimate the magnitude of the number of isomers that could be related to well-known substances and the values were rather unexpected. For example, it turned out that benzene was one of 217 conceivable isomers with the composition of $C_6H_6$ and articles were published in which all of the structural isomers were enumerated and depicted for the first time.[12] The number of isomers produced by the calculations showed that the number is unexpectedly large, even for small molecular formulae.

Figure 1.1 displays the structures associated with a series of modest-sized chemical compounds and the number of potential calculated structural isomers.[13,14]

Figure 1.1 shows that even the simplest of structures can have hundreds of millions, up to trillions, of isomers. For the simple structure with the associated molecular formula of $C_{10}H_{17}Br_2ClO_2$, the number of isomers, $N$, exceeds 50 million and rudimentary inspection suggests that more than 40 million of these could likely exist. It should be noted that the CAS registry[15] contains "only" 56 million known chemical compounds while 45 million are commercially available.

The number of isomers associated with the structures of medium-sized complex organic molecules can be estimated as approximately $10^{20}$–$10^{30}$ isomers (in the order of Avogadro's number). At the same time, the following very important conclusion can be drawn: although the number of possible isomers is huge, those corresponding to a given molecular formula do make up a *countable* (at least in principle) and *finite* set. With this in mind, we can immediately formulate the following general CASE strategy: *to eliminate "superfluous" isomers from the full isomer set by imposing different structural constraints.*

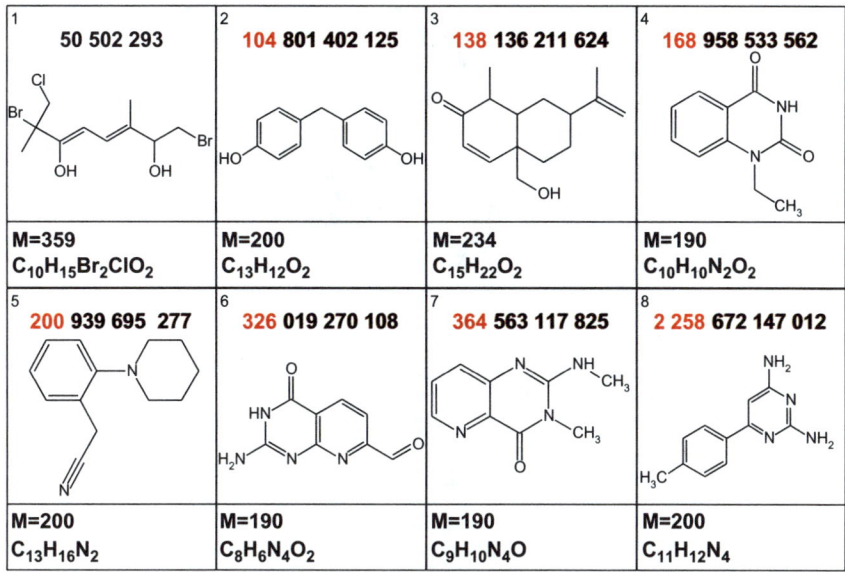

**Figure 1.1** The structures of a series of small molecules and the theoretical numbers of isomers (*N*) associated with the related molecular formulae.

Figuratively, the general CASE strategy is similar to that of a sculptor who removes superfluous material to produce a masterpiece (Figure 1.2).

The more constraints that are imposed, and the more severe they are, then the larger the number of "superfluous" isomers that will be rejected. Since structure identification boils down to the selection of a unique structural formula assigned to an unknown, then a successful result depends on the screening and rejection of $N–1$ structural formulae that do not comply with the experimental data and constraints applied. We will call the set of *n* non-identical isomers ($1 \leq n < N$) selected as a result of imposing a series of constraints the *solution* of the structure elucidation problem. The solution is called *valid* if it contains the correct (genuine) structure and otherwise it is *invalid*. The solution is called *unambiguous* if the response file contains only one structure. It should be noted that an unambiguous solution can be either valid or invalid. If the response file contains no structures ($n = 0$), then the imposed constraints are contradictory and the problem has no solution under the chosen conditions. The conceivable constraints leading either to unique structure or at least to a manageable set of plausible structures are outlined below.

## 1.2 A Molecule as a "Machine" for Coding Structural Information

Molecular structure elucidation is based on the same general cognitive principles that are common to the properties of particles belonging to the atomic and

**Figure 1.2**    The analogy between the CASE strategy and the technique of a sculptor.

subatomic world. In order to obtain information regarding a particular property of a particle it is necessary to stimulate the particle using electromagnetic radiation (or combination of electromagnetic radiation and magnetic field) or stimulate using a stream of particles and then analyze the resulting response signal. In those cases where we want to extract information about the structure of a molecule we excite the system with electromagnetic radiation over a wide frequency range using electrical and magnetic fields and streams of electrons or ions. As a result, we obtain spectral data in the form of ultraviolet-visible (UV-VIS), IR, Raman, NMR or MS or even, when X-rays are used, a 3D structure model of a molecule extracted from the data (Figure 1.3).

In this case the molecule under analysis acts like a specific coding machine (cipher machine) which codes structural information into each kind of spectrum using its own code. The goal of a researcher is to crack these codes and extract the maximum structural information achievable. Figuratively speaking, the CASE problem can be formulated in the following way: create a machine capable of decoding the structural information contained in one or more spectra.

X-ray analysis is the most informative and attractive method as, in principle, it allows one to circumvent the problem of choosing from a huge number of

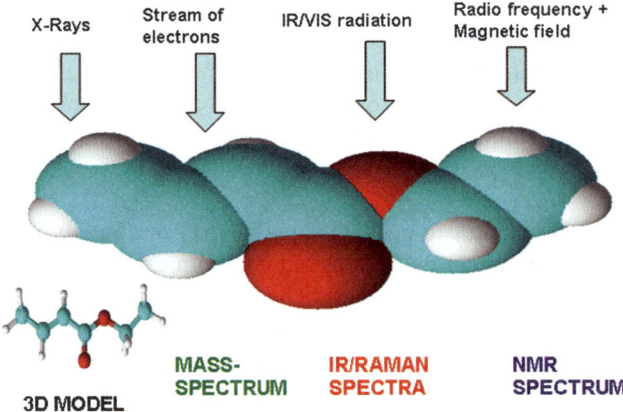

**Figure 1.3** The molecule as a cipher machine for coding structural information.

isomers and leads directly to the determination of a single structure, its 3D model and absolute configuration of the unknown. However, there are a series of reasons why X-ray analysis cannot be used routinely as a universal method for structure elucidation. The reasons are the following. (1) It is not always possible to obtain a crystal of sufficient quality to study. Moreover, it is common that there is no possibility to obtain a crystal at all. (2) To obtain a crystal it is necessary to separate such a large amount of material (10 mg or more) that it is not available in these quantities. In the chemistry of natural products chemists very frequently can only separate 0.1–1 mg of a sample. (3) X-ray analysis is a time-consuming method and at least 1 day can be required to acquire the diffraction data and to perform the necessary mathematical processing. In addition, X-ray crystallography can occasionally lead to misassignments because it does not reveal the positions of hydrogen atoms and it is sometimes difficult to discern between O atoms and NH groups. It can also confuse the identity of atoms within certain functional groups devoid of hydrogen atoms, for example, the assignment of a C atom instead of an N atom (a cyano rather than a diazo group).[16]

Therefore different spectral methods are usually employed at the first stage of the molecular structure elucidation to suggest structural hypotheses, whereas crystallography is used for final structure confirmation and determination of the absolute stereochemistry.

It should be noted that structural information is coded into different types of spectra by codes which are very specific for each kind of spectra. The codes are not only different (*i.e.* they have their own "alphabet"), but they are characterized by different levels of complexity. Mass spectra are composed of lines corresponding to the mass/charge ratios of ions and their associated line intensities. IR and Raman spectra are described by band frequencies and intensities, while half-height linewidths can also carry structural information. NMR spectra include the chemical shifts of the resonance and the associated multiplicity, intensity, coupling constant and linewidth.

Depending on the experimental conditions, MS can produce spectra containing a lot of valuable structural information, but extraction of the details can be very complicated. Nevertheless, MS can usually deliver the mass of the molecular ion and, with the improved resolution and accuracy of recent years, it is now relatively simple to determine the molecular formula for the molecule being studied, which is a key parameter necessary for molecular structure elucidation.

IR and Raman spectra can provide valuable information about the presence or absence of certain functional groups but communicates very little about their environment in the molecule. The richest structural information can be extracted from NMR spectra since the environment of a given magnetically active nucleus ($^1H$, $^{13}C$, $^{15}N$, $^{31}P$, $^{19}F$ *etc.*) can be revealed through chemical shift values and spin–spin couplings with neighboring nuclei. NMR spectra are therefore considered as a primary source of structural information, although maximum efficiency is attained by the combined application of all available spectral methods. An important role in structure elucidation is also played by the researcher who may be able to provide *a priori* information regarding, for example, the sample origin, chemical knowledge, experience and intuition. When used as constraints these data contribute to reducing the set of plausible isomers.

## 1.3   Quantification of Structural Information

The aim of spectral decoding is to extract the maximum amount of structural information from available spectral data. In principle the extracted structural information can be easily quantified[17] if it is possible to count the number of isomers, $N$, corresponding to the molecular formula. Assuming that all conceivable isomers are stable, and let $p_i$ be the probability that the $i$th ($1 \leq i \leq N$) isomer is the genuine structure. Before solving the problem, all isomers can be considered equiprobable and $p_i = 1/N$. Then, the entropy $E_0$, characterizing the initial uncertainty of the solution, can be determined by Shannon's formula[18] as $E_0 = \log_2 N$. The entropy of the valid solution $E_v$ containing $n$ structures will be $E_v = \log_2 n$; obviously, $E_v < E_0$. The quantity of information $I_v$ obtained as a result of the solution of the problem can then be expressed as the difference between the initial and final entropies:

$$I_v = E_0 - E_v = \log_2 N - \log_2 n = \log_2(N/n) \tag{1.1}$$

If $n = 1$ then the structure of the unknown compound is unambiguously determined (assuming that the solution is valid) and the total quantity of structural information $I_0$ obtained as a result of the solution of the problem can be expressed as $I_0 = E_0 = \log_2 N$. Hence, the extraction of new structural information is accompanied by a decrease in the number of isomers that meet constraints imposed by the experimental data (spectra), and the process of structure elucidation reduces to the successive imposition of structural constraints on the number of possible isomers until the extraction of complete structural information is achieved. If complete structural information is taken as 100%, then the fraction of structural information $q$ extracted from the spectra as a result of solving the problem can be estimated by the ratio:

$$q = (I_v/I_o) \cdot 100 = (1 - \log_2 n / \log_2 N) \cdot 100 \tag{1.2}$$

Thus, for each particular problem the most informative method is that which allows for the extraction of the largest fraction of total structural information and the elimination of the maximum number of incorrect structures. The application of this approach allows for selection of the most informative constraints by ranking constraints in the decreasing order of the corresponding amount of structural information.

## 1.4 Straightforward CASE Strategy

At first glance the general strategy for molecular structure elucidation using CASE methods looks rather simple and may create an impression that this procedure can be easily realized using the following straightforward approach to solving the problem: (1) generate, if possible, all isomers from a given molecular formula; and (2) impose different structural constraints one by one until the initial number of isomers is reduced to one, which is the correct structure.

Consider the application of this approach to a simple illustrative example. The molecular formula of $C_6H_{10}O_2$ for an "unknown" compound is determined from the HRMS, and IR, $^1H$ NMR and $^{13}C$ NMR spectra are acquired. The utilization of the IR spectrum of this compound to the process of structure elucidation is initiated (Figure 1.4). An IR spectrum provides information about the typical functional groups via characteristic features (frequencies, intensities and half-height linewidths) of the spectral bands. Careful analysis of the spectral patterns (Figure 1.4) shows that the molecule does not contain alcohol (-OH) or vinyl groups (no absorptions in the range 3100–4000 $cm^{-1}$ and no doublet between 3000 and 3100 $cm^{-1}$). However, alkenic double bond (1660 $cm^{-1}$), carbonyl (1730 $cm^{-1}$) and methyl (1380 $cm^{-1}$) vibrations are present, suggesting the presence of these groups and, in addition, the carbonyl group exists as the ester (1730 $cm^{-1}$). The first step is therefore to find all isomers which meet the constraints produced by considering the sets of functional groups that are present and those that are absent. For the molecular formula $C_6H_{10}O_2$, the number of structural isomers is quite small and is only 4869.

If we impose the constraints obtained from the IR spectrum onto the full set of isomers then the following result is obtained: 4869 → (no O-H) → 1719 → (C=O present) → 384 → (C=C present) → 151 → (CH3 present) → 147 → (O-C=O present) → 26 → (no C=CH2) → 10.

The direct usage of the trivial information regarding the absence of any OH groups reduces the number of candidate structures by a factor of almost three and after the further addition of new constraints only ten candidate structures

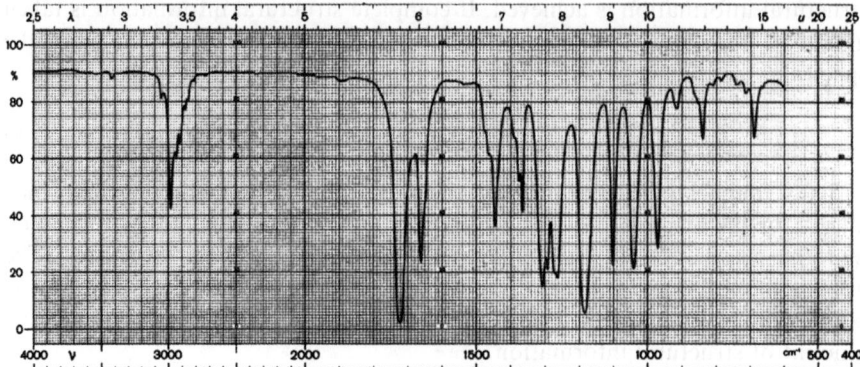

**Figure 1.4**   The IR spectrum of the "unknown" $C_6H_{10}O_2$.

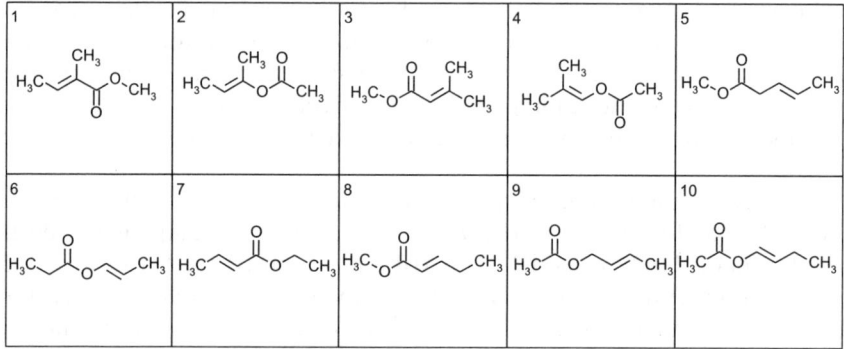

**Figure 1.5**   Candidate structures remaining after the imposition of constraints extracted from the IR spectrum of the "unknown" compound $C_6H_{10}O_2$ on the 4869 conceivable isomers corresponding to this molecular formula.

remain (Figure 1.5). Note that as a result of the application of constraints deduced from the IR spectrum 99.8% of all conceivable isomers were rejected but, according to eqns (1.1) and (1.2), only 73% of the full structural information was extracted.

The structures are quite similar and it is not very simple to distinguish the correct structure using the characteristic features of the IR spectrum only. The $^1$H NMR spectrum of the compound is shown in Figure 1.6.

The three distinct multiplets observed in the 1–4.2 ppm spectral area of the $^1$H NMR spectrum provide unambiguous evidence for the presence of an O-$CH_2$-$CH_3$ fragment, which is absent in all isomers except for structure no. 7 (Figure 1.5). Hence, with this additional constraint, we can complete the structure elucidation of the unknown using the straightforward application of the general CASE strategy. This strategy works in reality and only 3 s execution time for a nominal PC processor running the StrucEluc software (see Part III)

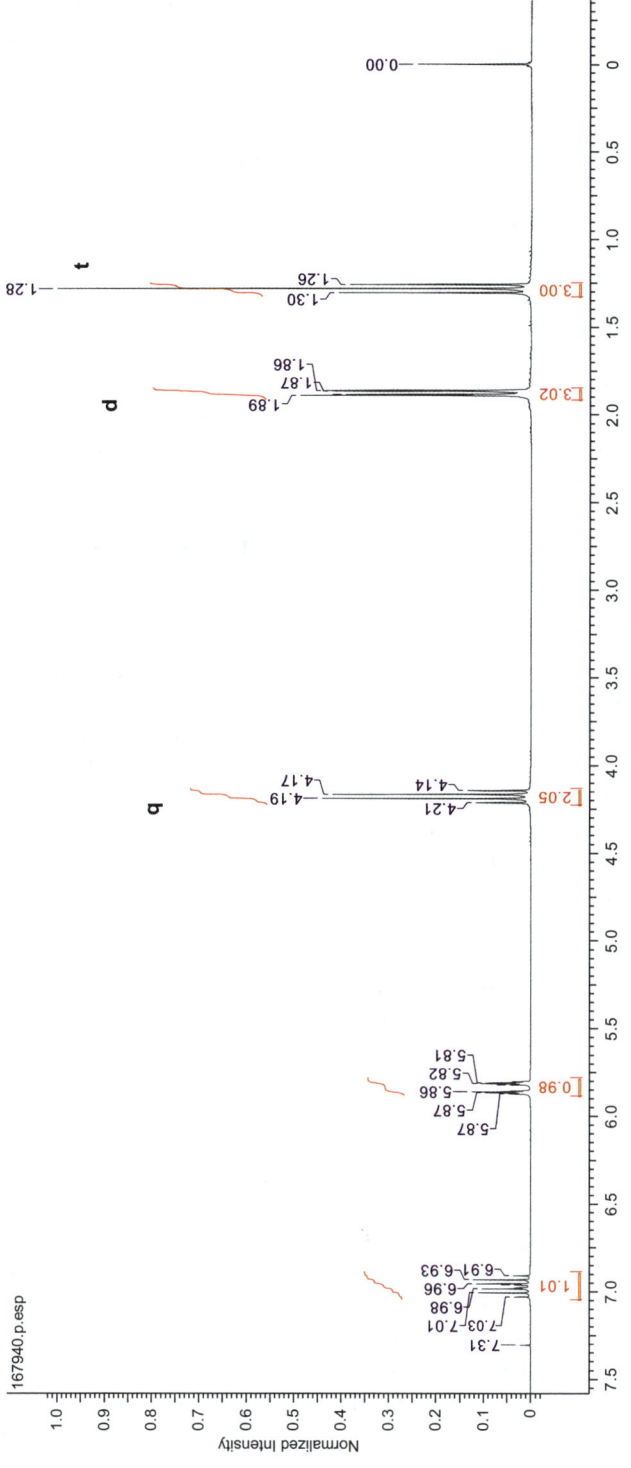

**Figure 1.6** The $^1H$ NMR spectrum of the "unknown" $C_6H_{10}O_2$ acquired on a QE-300 spectrometer in chloroform-d solvent.

were necessary to solve the problem (*i.e.* perform structure generation under the constraints obtained from the IR spectrum). However, such a direct approach is unfortunately justified only in those cases when the molecular formula of the unknown is rather constrained – when the number of heavy atoms does not exceed 10–12. The number of isomers and, consequently, the generation time of the full set of isomers grows at an almost exponential rate with the increase in the number of heavy atoms in the molecular formula and this leads to the so-called "combinatorial explosion". The number of heavy atoms in the molecular formula defines the *problem dimension*. There is only one way to overcome this fundamental difficulty and that is to reduce the dimension of the problem by providing molecular fragments which can encompass ("absorb") a number of the heavy atoms and form "macro-atoms" as discrete units of the structure. In the history of CASE development, optimal methods for the use of fragments for effective structure elucidation took many decades of hard work for a number of research groups. This represents how complex the challenge of developing CASE systems actually is.

## 1.5 Structure Elucidation as an Inverse Problem of Modeling

Figure 1.1 shows only selected examples of the huge number of isomers that can correspond to a given molecular formula. It is only slightly representative of the complexity of the problem of elucidating the structure of an unknown compound. Generally, the compound to be identified is of unknown origin and initially the number of possible structural hypotheses is infinitely large, even if some structural groups may be detected from the spectra. The point at which the number of possible structures starts to be limited, at least in principle, is the determination of the nominal molecular mass. Figure 1.7 illustrates the hierarchical structure of the "chemical world" which can be built from a molecular formula.

A *finite* set of accurate molecular masses corresponding to the nominal mass can be generated and enumerated on the basis of the chemical elements that are allowable. This set makes up only the first series of possibilities. The next series contains all conceivable structural isomers which can be associated with the allowed accurate masses (*i.e.* molecular formulae corresponding to accurate masses), each producing a *finite* number of isomers. The total number of isomers is finite. Finally, each isomer that possesses $n$ stereogenic centers can be considered as a core structure for a stereoisomer family containing a finite number of $\leq 2^n$ conceivable stereoisomers. All three series, or "shells", enumerated from the nominal mass contain both *finite* and *countable* numbers of objects that represent the chemical landscape produced by a given nominal mass. The determination of a single genuine structure and its stereoconfiguration is an enormous challenge for chemists and spectroscopists alike. The challenge is even more difficult if the molecule under investigation possesses conformational flexibility.

Among the manifold problems solved by molecular spectroscopy it is possible to distinguish the two most typical problems encountered. If the molecular structure is known then, in principle, we can predict its spectra and many other

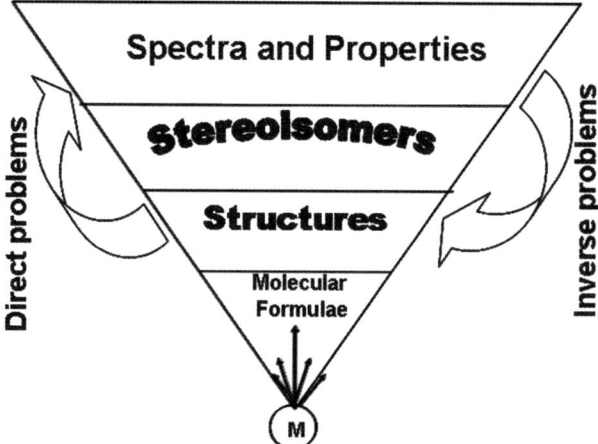

**Figure 1.7** The chemical landscape enumerated from a nominal molecular mass M.

properties with the support of the corresponding theories implemented into mathematical computational algorithms (see Chapter 3). For instance, programs are available which use the quantum mechanical density functional theory (DFT) approximation[19] for the calculation of IR, NMR and UV spectra of suggested organic molecules. A series of empirical and semi-empirical methods utilizing different empirical parameters can be used for fast spectrum modeling (Chapter 3). In so doing, we solve the so-called *direct* or *forward modeling* problem[20–22] (from structure to spectrum, shown by the left-hand arrow in Figure 1.7). The *forward modeling* problems are those that are solved either by direct measurement or by definite computational algorithms, and a unique solution is obtained in these cases. Indeed, both experimental and predicted spectra can be characterized by unique sets of parameters [for instance, frequencies, chemical shifts and band (line) intensities, *etc.*].

To determine the structure of an unknown compound using spectroscopic methods we use experimentally measured spectra and an empirical knowledge of spectrum–structural dependencies. In this case we solve a problem which is related to the class of so-called *inverse problems* which are most frequently *ill-posed*.[20,21] These problems usually do not have unique solutions. To determine a valid solution the introduction of additional information through constraints is necessary. Such direct and inverse problems are the most typical in molecular spectroscopy. Figuratively speaking, there is a cycle of both direct and inverse problems as illustrated in Figure 1.7. Structure determination from spectral data and spectrum prediction, where the structure is known or postulated, is one of the most frequently solved problems of chemists.

Bearing in mind the general strategy of structure elucidation, one can expect that if the available spectrum–structure information is complete enough then the CASE procedure may be performed automatically. Otherwise, additional experimental and theoretical data are necessary to obtain a unique solution. Therefore, before studying the CASE problem in depth, we can already conclude that problems exist that can be solved automatically and there are also those for which an automated solution is impossible. The possibility of obtaining an automated solution is usually revealed only during the process of solving the problem. Therefore, the automated solution of *any* structural problem seems, in general, impossible. This is a very important conclusion associated with structure elucidation and should be remembered when utilizing CASE methods.

## 1.6   A Practical Approach to Solution of the CASE Problem

Modern expert systems have been developed as a result of extensive investigations around the application of artificial intelligence (AI) approaches to molecular structure elucidation. Since the strategy for structure elucidation is based on AI, a computer is expected to mimic the reasoning of a skilled spectroscopist to as large an extent as possible. When a spectrum, or set of spectra (usually NMR, MS, IR and UV), are recorded for an unknown compound a spectroscopist follows two common paths. As discussed earlier these are: (a) application of the *dereplication* process by the examination of existing spectral collections to check for consistency between the experimental data and a previously elucidated structure, and (b) if the compound is *new*, then direct elucidation of the structure by utilizing the spectral data to deduce a structure consistent with all available data.

In general, spectroscopists still prefer the classical approach of elucidation using a brute force approach. However, over the past decade there has been a gradual shift toward the use of computer-based systems. These approaches may include searching a database of tens (or even hundreds) of thousands of spectra and associated chemical structures using specific spectral features to define the search criteria. The intention is to extract a set of structures or substructural fragments that may provide clues to the structure of the unknown or to at least provide core fragments that can speed up the elucidation process. If a search fails to find one or more spectra consistent with the experimental input, the chemists are then forced to perform full interpretation of the available spectral data using their experience of empirical correlations between spectral and structural features. The result of such an analysis is a set of structural fragments whose presence in the analyzed structure conforms to the spectral data and molecular formula. The next stage of the process is the chemist combining individual molecular fragments to provide molecular structures that are consistent with the data available. If more than one hypothetical structure can be suggested using this approach, then the procedure becomes an iterative examination of the data available to determine which is the most likely structure. Alternatively, it may be necessary to acquire additional data to unequivocally confirm the structure of the molecule in question. This process distinguishes the following three main steps in the human-driven process of structure elucidation:

1. Analysis based on substructural moieties. During this phase, the experimental data associated with an unknown compound are utilized to derive empirical spectrum–structure correlations either directly (from 2D NMR data) or with the aid of various databases. Sets of molecular fragments that are consistent with both the experimental spectra and other structural constraints are identified.
2. The second phase of the process involves the construction of hypothetical chemical structures from the substructural fragments identified in step 1. The goal is to produce chemical structures consistent with the experimental spectra, the molecular formula and other imposed structural constraints.
3. The final step in the process is the evaluation of the congruence between the chemical structures generated and the experimental spectra. This task can be accomplished by spectral prediction for the suggested structures and comparison with the experimental data.

These steps serve as the basis for the development of expert systems (ES) for molecular spectroscopy which mimic the reasoning of a spectroscopist. The manifold number of expert systems developed up to now differ in the kinds of spectral data employed and the complexity of the systems, and they may use different algorithms to perform spectral analysis, but the overall work flow of all of the expert systems is fundamentally the same.

These components of the workflow are relatively independent. The workflow can be optimized to deal with different types of spectra. The independent functions are realized using distinctly different mathematical methods. The methods for identification of fragments, the generation of chemical structures and the accurate prediction of spectra are essentially independent areas of computer-based spectroscopy and mathematical chemistry. Truly, the greatest advantages for the derivation of chemical structures from experimental spectroscopic data come from the integration of these independent approaches to produce a single application utilizing the strengths and benefits derived from each area.

Note that the technique of deriving structural information from experimental 2D NMR spectra does not differ significantly from that applied to 1D NMR and vibrational spectra but rather is an extension of the approach. 1D NMR systems are based on spectral fragment libraries that provide a set of possible fragments that are used as structural constraints. Due to the overlap of spectral ranges characteristic of different component fragments, the initial information regarding the chemical structure that can be derived from 1D NMR spectra is ambiguous. The acquisition of 2D NMR spectra allows the establishment of directly bonded atom pairs [*e.g.* $^1$H-$^{13}$C or $^1$H-$^{15}$N *via* heteronuclear single quantum coherence (HSQC)] followed by the arrangement of these heteronuclear pairs into larger structural fragments *via* longer-range correlations across two or more bonds [$^1$H-$^1$H correlation spectroscopy (COSY) and long-range heteronuclear shift correlation data (HMBC)]. The 2D

NMR data provide a set of structural constraints that are also ambiguous by nature (see Section 2.2). Consequently, we can summarize the real strategy for computer-aided structure elucidation by the flow-diagram shown in Figure 1.8.

Structural information should be presented to a CASE system in a form which allows efficient information processing by the associated computer algorithms. Most expert systems deal with organic molecules in a manner that can be accurately described using classical theories of molecular structure.

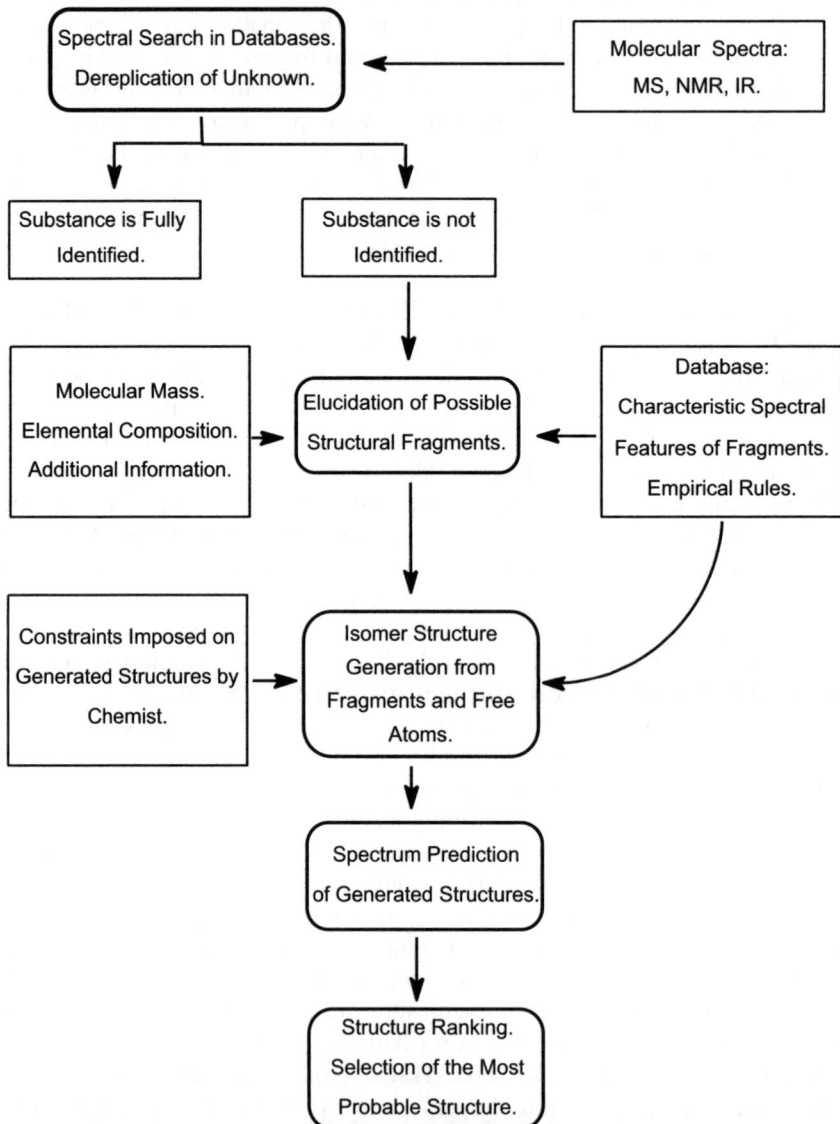

**Figure 1.8**   A flow-diagram of the general CASE strategy.

Mathematically, a structural formula is an object whose abstract properties can be described by graph theory[23,24] which is widely applied to the analysis of molecular structures using methods developed by cheminformatics.[25,26] Some of the elements of cheminformatics necessary for understanding the theoretical background of expert systems are given below.

In the next sections we will briefly consider the main ideas used to realize each of the blocks making up the diagram presented in Figure 1.8. Since spectral searching across databases resembles a manual search in spectral atlases, it is intuitively understandable. Hence we will discuss here those blocks that are integral parts of expert systems.

# 1.7 Detection of Molecular Fragments

To identify structural fragments most expert systems use the characteristic spectral features of either functional groups or atom-centered fragments (ACFs) in IR and NMR, the latter being used most frequently for the structural interpretation and prediction of NMR spectra. The knowledge base under an expert system usually consists of molecular fragments, the range of characteristic features (chemical shifts in NMR spectra and frequencies in IR spectra) and the rules relating the spectra to the structure. In a number of systems[27–29] the libraries of the structures and corresponding spectra are included in the knowledge bases in addition to the molecular fragments. Depending on the type of knowledge base, different methods of structural information retrieval are used.

## 1.7.1 Spectrum–Structure Correlation Method

### 1.7.1.1 A Discrete Model of the "Structure–Spectrum" System

Mathematical algorithms for structural group analysis of unknown compounds were developed on the basis of careful investigation of reasoning approaches used by spectroscopists performing structure elucidation. The detection of potential substructures is the first step in this process, and spectrum–structural correlations in NMR and IR spectra combined with MS data are usually employed for this purpose. This procedure is based on a logical comparison of the spectrum of the molecule with the dependencies already known between the structural elements and their spectral features. These dependencies are, however, frequently rather complicated, are multistage in nature, and include a large number of logical relations that can be rather vague and ambiguous. It is often difficult to retain intermediate conclusions at all stages of analysis. As a consequence, there is a substantial dependence of the results of the spectral data analysis on the experience and intuition of the specialist. This type of investigation should be facilitated using special mathematical techniques in which human arguments could be expressed in an *explicit* form and which allows computer-aided calculation to be performed.

Elyashberg and Gribov[2] have shown that this requirement is satisfied by the principles of symbolic logic, which is the science of discrete, structural simulation of finite discrete systems. The application of symbolic logic is highly fruitful in terms of studying complicated objects of a discrete nature.

Mathematical modeling can be successful if the nature of the object being modeled corresponds to the nature of the mathematical apparatus being applied. It has been shown[9] that a molecule, taken together with the signals observed in its associated spectra, can be considered as a finite discrete system.

A molecule consists of atoms and atomic groups (structural units) and, consequently, is of a discrete nature. A molecular spectrum (IR or NMR) arises due to transfers between discrete energy levels and because of this the spectrum also has a discrete nature. Certain relationships exist between the two sets of elements, one of which is formed from atoms and molecular fragments and another from the spectral bands or lines. The simplest model for these relationships can be presented as a "black box". The substance under investigation can be considered as a "black box", with the atoms, bonds and molecular substructures playing the role of "receivers" of the stimuli (electromagnetic radiation, magnetic field, *etc.*). The output of this "black box" is one or more spectra containing encoded structural information. As mentioned above, in this respect the molecule is considered as a coding "device".

To assist in decoding the structural information, spectroscopic theory has established relationships between observed spectral features and structural elements. These data can be used to provide a physical explanation of the empirical spectrum–structural correlations. For instance, normal coordinate analysis of ketone molecules shows that the absorption bands observed near $1710 \text{ cm}^{-1}$ accounted for stretching vibrations of the C=O bond. Qualitative spectral analysis allows assignment of these connections *directly* on the basis of the concept of characteristic spectral features. The occurrence or absence of characteristic features in a spectrum depending on the presence or absence of certain substructures plays the role of indicating the inter-relation between elements of the system "molecule–spectrum". This indicator is of a discrete type with the number of indicator states being equal to 2, *i.e.* a fragment (spectral feature) is either present or absent. Therefore the complex object, including the molecule and its spectrum, makes up a discrete structural model and symbolic logic can be applied as a mathematical tool for qualitatively describing the mutual relationships between the spectrum and the structure of a molecule. Moreover, experience shows (see Chapter 12) that the ability to apply the main notions of symbolic logic during manual structure elucidation from spectral data helps to avoid the extraction of erroneous structures.

### 1.7.1.2   *Main Notions and Definitions of Symbolic Logic*

A brief description of some basic notions and definitions of propositional calculus[30] is necessary to derive a formal logical concept of molecular structure analysis by characteristic spectral features.

A *proposition* is a statement that may be true or false. A simple proposition is designated by a letter and called an *elementary proposition*. For example, an elementary proposition $A_i$ may be read as: "a molecule contains the $A_i$ fragment". Propositions formed by combining elementary propositions are called *Boolean functions*. For the construction of Boolean functions the following main logical operations are used:

1. The *negation* of the proposition $A_i$ corresponds to the particle "no" and is symbolized by $\bar{A}_i$ ("The molecule does not contain the $A_i$ fragment").
2. The conjunction of the propositions $A_i$ and $A_k$ corresponds to the conjunction "and" and is designed by $A_i \wedge A_k$ ("The molecule contains $A_i$ and $A_k$ fragments").
3. The *disjunction* of $A_i$ and $A_k$ corresponds to the non-excluding "or" and is designated by $A_i \vee A_k$ ("Either the $A_i$ or the $A_k$ fragment, or both together, are present in the molecule").
4. The Boolean function $A_i \rightarrow w_j$ is called an *implication*. It corresponds to the expression "if $A_i$, then $w_j$" ("If the molecule contains the $A_i$ fragment, then the spectral feature $w_j$ is observed in a spectrum").

The following identities of Boolean algebra are valid:

$$a \rightarrow b = \bar{a} \vee b = \bar{b} \rightarrow \bar{a} \tag{1.3}$$

$$\overline{a \wedge b} = \bar{a} \vee \bar{b}, \qquad \overline{a \vee b} = \bar{a} \wedge \bar{b} \tag{1.4}$$

The expressions in eqn (1.4) are referred to as de Morgan's laws.

The problem of structural group analysis can be stated in terms of symbolic logic as indicated below.

## 1.7.1.3 Theoretical Basis of the Structural Group Analysis

In order to ensure appropriate usage of an expert system it is very important to have a clear understanding of the main concepts which the system is based upon. The aim of this section is to give a short description of the formal logical model of structural unit analysis. A logical model for *structural group analysis* (SGA) was first developed[2,31–33] for IR spectroscopy. The basic notion of this model was a *functional group*, since vibrational spectroscopy is most successful in detecting functional groups defined in organic chemistry. When spectrum–structural correlations from $^1$H and $^{13}$C NMR spectroscopy are used for structure elucidation, different atom combinations which are not specified as functional groups can be revealed by their chemical shifts along with "classic" functional groups. Both atomic combinations and functional groups are *discrete units of structure*. We will explain a formal logical model of structural unit analysis that is equally applicable to both NMR and IR spectra.

The problem of structural unit spectral analysis can be formulated in symbolic logic language as follows. Assume that an experimental spectrum (NMR, IR or Raman) of an unknown compound is measured and the relationships between the features observed in the spectrum and the structural elements (for instance, correlation tables) are known. Assume that among the features occurring in the spectrum are those that are characteristic of structural units composing the unknown substance. It is necessary to find all possible combinations of the structural units that may be present in the given sample in accordance with the experimental spectrum.

Consider the following sets:

$\tilde{W} = \{w_j\}, j = 1, 2, ..., m$ is the set of spectral features (chemical shifts, frequencies) observed in the spectrum of the sample;

$A = \{A_i\}, i = 1, 2, ..., N$ is the set of structural units having characteristic features occurring in the set $W$;

$\bar{W} = \{w_j\}, j = m + 1, ..., M$ is the set of frequencies which correspond to some $A_i \in A$, but are not present in the spectrum;

$W = \tilde{W} \cup \bar{W} = \{w_j\}, j = 1, 2, ..., M$ is the set of frequencies which are to be considered in solving a particular problem;

$U = A \cup W$ is the universal set which limits the structural units and the spectral features applicable to the problem under consideration.

We shall stipulate that the elements of these sets are logical variables which are interpreted as the following elementary propositions:

1. $A_i$: molecule contains structural unit $A_i$
2. $\bar{A}_i$: molecule does not contain structural unit $A_i$
3. $w_j$: spectrum contains feature $w_j$
4. $\bar{w}_j$: spectrum does not contain feature $w_j$

Now we have to write all logical connections between the elements of the set $A$ and $W$, in the form of logical functions. In the general case, where a set of $n$ characteristic features is attributed to the group, their relation with group $A_i$ is written as follows:

$$A_i \rightarrow w_1(i) \wedge w_2(i) \wedge ... \wedge w_n(i), \qquad w_\alpha(i) = w_j \subset W \qquad (1.5a)$$

$$\{\bar{w}_1(i) \vee \bar{w}_2(i) \vee ... \vee \bar{w}_n(i)\} \rightarrow \bar{A}_i \qquad (1.5b)$$

Here w(*i*) represents the features which occur in the characteristic ranges $(\Delta w)(i)$, corresponding to the given unit $A_i$.

Eqn (1.5a) indicates that if a molecule contains group $A_i$, then the set of spectral features w$_1$(*i*), w$_2$(*i*), ...w$_n$(*i*) will occur in the spectrum. Indeed, the eqn (1.5a) is the symbolic representation of the experimentally established spectrum–structural correlation. According to eqn (1.5a), the presence of the conjunction of frequencies w$_1$(*i*) ∧ w$_2$(*i*) ... ∧w$_n$(*i*) is a *necessary* condition for the molecule to contain the structural unit $A_i$. Eqn (1.5b), which is equivalent to eqn (1.5a) [see eqns (1.3) and (1.4)], shows that a sufficient condition for the absence of a structural element in the sample under investigation is that the spectrum should not contain at least one of the characteristic features of this group.

If the spectral feature $w_i$ is contained in the sets of the characteristic features of the units $A_1(j)$, $A_2(j)$, ..., $A_g(j)$, $A_\beta(j) \in A$, then

$$w_j \rightarrow \{A_1(j) \vee A_2(j) \vee ... \vee A_\beta(j) \vee ... \vee A_g(j)\}, A_\beta(j) = A_i \in A \qquad (1.6)$$

This expression means that if the feature $w_j$ is detected in the spectrum then the molecule contains at least one of the units, $A_1(j)$, $A_2(j)$,..., $A_g(j)$. Note that the implication (1.6) takes into account the possibility of a signal overlap in the spectra: the feature $w_j$ may appear as a result of overlap of absorption lines associated with any combination of units included into set $A$. Eqns (1.5a) and (1.6) represent the fundamental types of logical relationships between structural units and spectral features in a structural unit analysis.

Note that an implication shows that one proposition follows from another, meaning that the same conclusion can also arise in other ways. Obviously, in principle, the conjunction of the features $w_1(i)$, $w_2(i)$, ..., $w_n(i)$, may be the consequence of the presence of not only group $Ai$, but also due to the presence of some other groups or their combinations in the sample. On the other hand, it may happen that the disjunction, $A_1(j) \vee A_2(j) \vee ... \vee A_g(j)$ is implied not only by the feature, $w_j$, but also by some other spectral attributes. Hence, it follows that the fundamental formulae, eqns (1.5a) and (1.6), exactly represent the nature of the relationships between the structural units and the corresponding characteristic spectral features.

Let $T(A,W)$ denote the Boolean function which describes all the mutual relationships between the spectral features and the structural units related to a given problem. The experimental spectrum as a combination of features may be presented by the Boolean function, $Sp(W)$.

The logical analysis of this situation lies in jointly processing the functions, $T(A,W)$ and $Sp(W)$ so that, as a result, we establish the function, $f(A)$, that enumerates all possible sets (combinations) of structural elements which may be present in the molecule under investigation. Symbolically, this is expressed as:

$$T(A, W) \rightarrow \{Sp(W) \rightarrow f(A)\} \qquad (1.7)$$

Eqn (1.7) is the most general formulation of the qualitative spectral problem in the language of Boolean algebra and is the logical equation for the function $f(A)$. Solution of the logical problem of structural unit analysis lies in calculation of the function $f(A)$. An algorithm for the calculation of the $f(A)$ function[2,34] is based on the concept of designating numbers developed by Ledley.[35] This algorithm was previously used to create a computer-aided method for structural unit analysis[34,36] where logical relationships forming the $T(A,\Omega)$ function (correlation tables) are present in the knowledge base.

It should be emphasized that eqn (1.7) has a high degree of generality and, as such, may be used in describing the methodology of qualitative chemical analysis based on other principles.

The approach described above can be illustrated with the following simple example, in which the elemental composition of an "unknown" compound is $C_{11}H_{12}N_2O_1$, and the IR spectrum measured in the range between 4000 and 400 $cm^{-1}$ contains the following frequencies: 2900, 2770, 2215, 1670, 1590, 1500, 1465, 1355, 830 and 810 $cm^{-1}$. The task is to find a solution for eqn (1.7) in respect to the given molecular formula and the IR spectrum using the STREC program.[37]

The molecular formula and list of frequencies were input into a computer. Using the spectrum–structure correlations collected in the digital library, the program identifies the appropriate elements of the $W$ and $A$ sets and forms all initial hypotheses as logical relations of the type in eqn (1.6). In so doing, only those fragments are selected whose characteristic frequencies are in agreement with the IR spectrum and the empirical formula. The following logical relations (implications) were established by the program:

1. 810 $cm^{-1}$ → 1,3,5-AR ∨ 1,2,4-AR ∨ 1,2,3,4-AR ∨ 1,3 -AR ∨ 1,4-AR
2. 830 $cm^{-1}$ → 1,3,5-AR ∨ 1,2,4-AR ∨ 1,2,3,4-AR ∨ 1,4- AR
3. 1355 $cm^{-1}$ → (C)CH3
4. 1465 $cm^{-1}$ → O-CH3 ∨ (C)CH3 ∨ N-CH3
5. 1500 $cm^{-1}$ → 1,3,5-AR ∨ 1,2,4-AR ∨ 1,2,3,4-AR ∨ 1,3 -AR ∨ 1,4-AR
6. 1590 $cm^{-1}$ → 1,3,5-AR ∨ 1,2,4-AR ∨ 1,2,3,4-AR ∨ 1,3 -AR ∨ 1,4-AR
7. 1670 $cm^{-1}$ → N-C=O ∨ O=C[AR] ∨ O=CH[AR]
8. 2215 $cm^{-1}$ → C≡C ∨ C≡N
9. 2770 $cm^{-1}$ → O-CH$_3$ ∨ O=CH[AR]
10. 2900 $cm^{-1}$ → O-CH$_3$ ∨ (C)CH$_3$ ∨ N-CH$_3$

The right sides of the relations 1, 5 and 6 are identical, so only one of these three implications should be considered during logical reasoning. The same conclusion can be made concerning relations 4 and 10. Implication 3 requires the obligatory presence of the (C)CH$_3$ group in the molecule if a band at 1355 $cm^{-1}$ appears in the spectrum. This implication was generated because an empirical correlation (C)CH$_3$ → [1350–1400 $cm^{-1}$] exists and no other competing fragments were selected from the fragment library. If both implications $A_i → w_j$ and $w_j → A_i$ are true then we obtain an equivalence $A_i ↔ w_j (A_i ≡ w_j)$ which indicates: "$A_i$ only when $w_j$" or "$w_j$ is necessary and sufficient for $A_i$". Since such rigid propositions are rarely true in molecular spectroscopy, in those cases when a "single" implication $ω_j → A_i$ is formed it should be replaced by the implication $A_i → ω_j$ that is true according to spectroscopic knowledge. The latter allows introduction of fragments but in the current case constraint 3 may be simply omitted because the (C)CH$_3$ group is included in both relations 4 and 10.

The computer solution to the logical equation of this problem [the function f($A$)] consists of eight equally probable sets of molecular fragments, each set containing as many or less atoms of all types as exist in the empirical formula. In addition, for any set, the total number of double bonds or rings provided should not exceed that calculated from the empirical formula. The following sets of fragments were found:

1. C≡N O=CH[AR] N-CH₃ 1,3,5-AR
2. C≡N O=CH[AR] N-CH₃ 1,2,4-AR
3. C≡N O=CH[AR] N-CH₃ 1,2,3,4-AR
4. C≡N O=CH[AR] N-CH₃ 1,4-AR
5. C≡N O=CH[AR] (C)CH₃ 1,3,5-AR
6. C≡N O=CH[AR] (C)CH₃ 1,2,4-AR
7. C≡N O=CH[AR] (C)CH₃ 1,2,3,4-AR
8. C≡N O=CH[AR] (C)CH₃ 1,4-AR

Two fragments, C≡N and O=CH[AR], are included in all sets, whereas other fragments are available in different combinations. We have therefore obtained the final result of structural group analysis using the characteristic frequencies of the IR spectrum. The next step is to attempt to build up all possible isomers from each fragment set and to select the most probable structures. All of these procedures, including the structural group analysis, can be executed in the framework of expert systems[37–39] elaborated for the purpose of molecular structure elucidation using 1D NMR and IR spectra. The program generated 523 distinct structures from these fragment sets and selected 53 isomers satisfying the characteristic frequencies observed in the IR spectrum. When the NMR spectral data were added to the input information, the following structure was selected by the system as a solution to the problem:

**1.1**

The approach described above serves as the basis of structural spectrum interpretation in CASE systems such as STREC,[37–39] X-PERT,[40] CHEMICS[41] and EXPIRS.[42] Solution of eqn (1.7) makes it possible to successively produce logically consistent fragment sets that are appropriate for structure generation. However, such an approach has a drawback in that it is too rigid. For instance, a small deviation in a characteristic feature of a fragment in the experimental spectrum from the limits of a given interval leads to the loss of the corresponding fragment from the input set used during structure generation. To circumvent this drawback, methods of fuzzy logic[43] have been proposed[44] where the characteristic ranges are described with the membership function μ which differs from zero outside the corresponding ranges. The approach has been used for the structural interpretation of IR and NMR spectra in the X-PERT[40] system and for the spectral filtering of structures in two generations of the StrucEluc system.[29,45]

## 1.7.2  Methods for Recognition of Large Fragments

To allow for the identification of large molecules, an expert system must have adequate analytical tools for the determination of large fragments. Obviously, this is necessary because large fragments can absorb many skeletal atoms thereby leading to a significant reduction in the size of the problem. Attempts to create libraries containing large fragments and variable ranges of their characteristic spectral features is difficult, again due to the potential for combinatorial explosion, as well as the challenge of adequately predicting the ranges. The problem can, however, be solved using databases containing chemical structures and their associated spectra. This approach relies on the hypothesis that similar spectra (subspectra) correspond to similar structures (substructures). This hypothesis is one of the cornerstones of empirical methods for the structural interpretation of molecular spectra.

Two approaches to the identification of large fragments using NMR, IR and mass spectral databases have been described. In the first approach the algorithm[46–48] is reduced to three main operations:

1. Search for structures whose spectra are similar to a given experimental spectrum and order the resulting set according to decreasing similarity.
2. Select a number of structures from the beginning of the list that, according to the basic hypothesis, have the highest similarity to the spectrum of the analyzed compound.
3. Selection of the largest common fragments that can be plausible constituents of the structure of the unknown compound.

The probability of correct identification of the fragment increases if the database retrieval system selects the data using several types of spectra[46,49] and the selected fragments are tested on the basis of coincidence of the predicted subspectra. This can be most easily performed using $^{13}$C NMR spectra.

The algorithm above requires reliable quantitative criteria for both structural and spectral similarity (a review of structural similarity criteria has been published by Willett and co-workers[50]), as well as methods for revealing the largest common subgraph in a set of chemical graphs.[25] The development of such methods has given rise to a specific area of study in graph theory and numerous studies on the subject have been reported. The second method developed for recognizing large fragments has demonstrated wide applications as a result of the elaboration of large $^{13}$C NMR spectra databases (see, for example, references[51–54]). In these databases, structures are represented by HOSE codes[55] and the spectral resonances are assigned to the corresponding carbon atoms. Such an approach allowed (see, for example, references[28,29,56]) the development of algorithms that can be used to create subspectrum–substructure correlation libraries to allow structure elucidation. First the program compares all subspectra of the library fragments with the experimental $^{13}$C NMR spectrum of the analyzed substance. In this case, the signal intensity (*i.e.* the number of carbon atoms responsible for the observed spectral response) is also considered in addition to the chemical shifts and signal multiplicities of the resonances. As a result, only those substructural fragments whose resonances fall within the limits of allowed deviations with the query subspectra are selected. Subsequently, the most probable fragments are selected by detailed analysis of the spectral information and then ranked in descending order of the number of carbon and skeletal atoms. The largest fragments therefore appear at the beginning of the list thereby helping to optimize the structure generation process. The use of this procedure in an expert system is considered in greater detail in Sections 5.4 and 9.5. When 2D NMR spectra are used the application of $^{1}$H-$^{1}$H COSY and $^{1}$H-$^{1}$H COSY/total correlation spectroscopy (TOCSY) frequently allows for the detection of large spin systems, which is equivalent to the detection of large fragments.

# 1.8 Structure Generation: Descriptive Explanation

This book is focused on addressing the needs of practicing chemists and spectroscopists and we will not consider the strict mathematical algorithms of structure generation. These have been described previously in numerous publications, for example.[10,11,57,58] We will introduce only some of the basic concepts that are necessary to understand how expert systems operate with structural information.

## 1.8.1 Elements of Chemical Informatics

All CASE systems deal with organic molecules that can be accurately described using classical theories of molecular structure. Mathematically, a structural formula is an object whose abstract properties can be described by graph theory[23] that is widely applied to the analysis of molecular structures. Graph theory operates on two initial objects – a *vertex* and the *edge* of a graph. Vertices are depicted as a set of dots in a plane, and edges are represented as lines connecting the vertices. Graph $G$ consists of a non-empty set $V$ containing $p$ vertices and of a given set $X$ whose elements are $q$ pairs of different vertices of $V$ set. Each pair of vertices $x = \{u,v\}$ in a set $X$ is denoted as an edge of $G$ and $x$ is said to connect $u$ and $v$. Vertices $u$ and $v$ are called *adjacent*, and vertex $u$ and edge $x$ are called *incidental*. If edges $x$ and $y$ are incidental to the same vertex, they are called *adjacent*.

A graph is a *multigraph* if it has a pair of vertices that can be connected with more than one edge. Such edges are called *multiple*. An example of a multigraph is displayed below:

A graph is a *labeled* graph with $n$ edges if all of its vertices are labeled with integers from 1 to $n$. The numbers $1, 2,\ldots n$ are called graph *labels*, and vertices are designated as $u_1, u_2, .., u_n$, for example:

The degree or valence of vertex $v$ of graph $G$ is a measure of the number of edges incidental to $v$. In the graph displayed above, the *degree of vertices* $u_1, u_3, u_4, u_5$ is 2 and the degree of vertex $u_2$ equals 4.

A *subgraph* of graph $G$ is a graph whose vertex and edge sets are subsets of those of $G$. A graph is a *connected graph* if any pair of its vertices is connected with a simple edge. A graph is called *acyclic* if it contains no cycles. A connected acyclic graph is called a *tree*.

The *connectivity matrix* $A = [a_{ij}]$ of labeled graph $G$ containing $p$ vertices is called a square matrix of $n$-order whose elements $a_{ij}$ equal $m$, if vertices $u_i$ and $u_j$ are connected with edges of degree $m$, and elements $a_{ij}$ equal 0 if no edge exists between the vertices. For example, for the labeled graph displayed above, the connectivity matrix is represented as:

$$A = \begin{array}{c|ccccc} & 1 & 2 & 3 & 4 & 5 \\ \hline 1 & 0 & 2 & 0 & 0 & 0 \\ 2 & 2 & 0 & 1 & 1 & 0 \\ 3 & 0 & 1 & 0 & 0 & 1 \\ 4 & 0 & 1 & 0 & 0 & 1 \\ 5 & 0 & 0 & 1 & 1 & 0 \end{array}$$

The structural formula of a molecule can be obviously considered as a connected multigraph with labeled vertices. Multigraph vertices are represented by atoms of various types whose chemical notations are vertex labels (all atoms can be numbered). The vertex degree depends on an atom's valence. The number of edges, both ordinary and multiple, branching from a vertex is equal to the number of chemical bonds formed by an atom. The classical theory of molecular structures of organic compounds can therefore be treated in terms of graph theory. It follows that all of the conclusions made on the basis of the theorems of graph theory can be applied to the investigation of topological properties of a structural formula.

Atoms within a molecular formula can be connected with chemical bonds in many different ways to form multiple isomers. For example, 26 mathematically conceivable isomers can be formed for $C_5H_8$. Two of these are shown below:

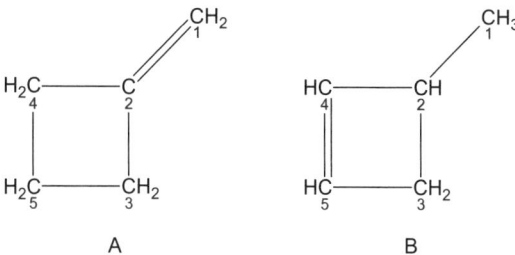

A                                   B

Consider the connectivity matrices $A$ and $B$ for the above structures in accordance with the numbering of skeleton atoms used in the graph example shown above. Hydrogen atoms are usually omitted and the resulting matrices are:

$$A = \begin{array}{c|ccccc} & 1 & 2 & 3 & 4 & 5 \\ \hline 1 & 0 & 2 & 0 & 0 & 0 \\ 2 & 2 & 0 & 1 & 1 & 0 \\ 3 & 0 & 1 & 0 & 0 & 1 \\ 4 & 0 & 1 & 0 & 0 & 1 \\ 5 & 0 & 0 & 1 & 1 & 0 \end{array} \qquad B = \begin{array}{c|ccccc} & 1 & 2 & 3 & 4 & 5 \\ \hline 1 & 0 & 1 & 0 & 0 & 0 \\ 2 & 1 & 0 & 1 & 1 & 0 \\ 3 & 0 & 1 & 0 & 0 & 1 \\ 4 & 0 & 1 & 0 & 0 & 2 \\ 5 & 0 & 0 & 1 & 2 & 0 \end{array}$$

Two different structural formulae therefore correspond to different connectivity matrices. If the atom numbering in structure A is changed then the connectivity matrix $A_1$ is generated for the obtained structure:

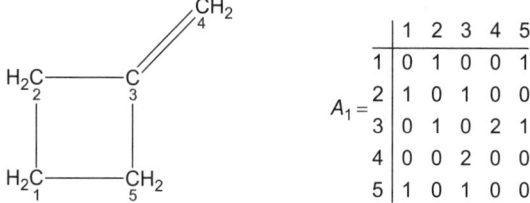

$$A_1 = \begin{array}{c|ccccc} & 1 & 2 & 3 & 4 & 5 \\ \hline 1 & 0 & 1 & 0 & 0 & 1 \\ 2 & 1 & 0 & 1 & 0 & 0 \\ 3 & 0 & 1 & 0 & 2 & 1 \\ 4 & 0 & 0 & 2 & 0 & 0 \\ 5 & 1 & 0 & 1 & 0 & 0 \end{array}$$

This example shows that even when the formula is the same the connectivity matrix can be different. Depending on the order in which the atoms are numbered in a structure, $n!$ connectivity matrices ($n$ is the number of vertices in a graph) can be constructed while a specific connectivity matrix corresponds to only one structural formula. If a molecule contains $k$ types of chemical elements, the number of possible connectivity matrices will be equal to $n_1!,n_2!...n_i!...n_k!$, where $n_i$ is a number of chemical elements of $i$-th type.

The question arises as to whether there is a specific way of atom numbering in a structure that will result in only one connectivity matrix specific to that structure. The answer is positive: it is possible, using the appropriate algorithm, to describe a structure that will differ from other matrices by the presence of a unique specific property. A matrix having this property is called a *canonical* matrix. The notion of matrix canonicity is important for the development of algorithms associated with the generation of structural formulae. To avoid redundancy within the generated structure set, the matrices associated with the generated structures should be *canonicalized*. Different criteria are used during this process including maximum or minimum *lexicographical* codes with regard to the lower or upper triangle of the connectivity matrix or the whole matrix.

For example, the following matrix $A_{can}$ is canonical for structure A. The new canonicalized atom numbering is shown in the picture below:

We will explain the mentioned criterion by constructing lexicographical codes for squared $5 \times 5$ matrices $A$ and $A_{can}$. Let us imagine that each matrix represents a number which is cut into 5 parts and the first part containing 5 number positions is placed in the first row of a matrix, the next 5 number positions in the second row and so on. As a result we will obtain the following two numbers (lexicographic matrix presentations) corresponding to matrices $A$ and $A_{can}$:

$$\#A = 02000\ 20110\ 01001\ 02001\ 00110$$
$$\#A_{can} = 02101\ 20000\ 10010\ 00101\ 10010$$

Because a bond multiplicity value may vary between 0 and 3 in classical structures, the numbers $\#A$ and $\#A_{can}$ can be considered as numbers presented in *quaternary* notation. If we convert these numbers into decimal notation we will find that $\#A^d = 141308787295252$ while $\#A^d_{can} = 159979214488836$. So, in our example, it is obvious that $\#A^d < \#A_{can}$. Calculations show that the $\#A_{can}$ code corresponds to a lexicographically maximal matrix matching structure A, *i.e.* matrix $\#A_{can}$ is canonical and can be considered as a unique representation of structure A.

The general concept behind the algorithm of canonical graph generation assumes that all possible graphs are constructed and each newly built graph is checked against its canonicity. At that time only canonical graphs are considered as newly built. In this way all non-isomorphic (non-identical) graphs for a given set of vertices are obtained.

## 1.8.2   The Generation of Structural Formulae of Isomers

The first chemical structure generators[41,59–66] were developed in the decade covering the 1970s. A number of these could be used as standalone structure generators,[59–61] while others, DENDRAL,[10] CHEMICS,[62] DARC[66] and STREC,[38] were components within their associated expert systems. The merits of these generators have been discussed previously in a series of reviews and monographs.[9,11,34,67–69]

Attention was applied to how to improve the performance of structure-generating software and strategies to more efficiently use structural constraints which can lead to limiting the generated structure sets. As the number of atoms in a molecular formula increases, the number of mathematically possible isomers, as mentioned above, grows in an approximately exponential manner. Examples of the number of possible isomers for a given composition $C_iH_j$ are given in Table 1.1.

Significant efforts have been invested in using structural constraints to as high a degree as possible *during* the structure generation process. This proactive approach is far more efficient, since the imposition of constraints after structure generation is inefficient and leads to significant increases in overall calculation time. Different principles of structure generation have been proposed and discussed in a number of published reports.[70–78] During the elaboration of new expert systems utilizing 2D NMR data, new structure generators were derived from existing approaches and modified to account for the information available from 2D NMR data.

The first work utilizing 2D NMR spectral data (INADEQUATE) for structure generation was published by Lindley *et al.*[79] in 1983. Since then a number of expert systems have been reported (see Chapter 6 and Part III) that utilize 2D NMR data. These systems represent a groundbreaking period in the history of CASE development and have heralded the possibility of elucidating structures with ever-increasing complexity.

In parallel with efficient algorithm development, computer performance was developed along the usual Moore's Law trajectory,[80] *i.e.* it was approximately doubled every 2 years. These parallel developments have facilitated a significant increase in the complexity of the problems which can be solved. Currently, a structural framework containing up to 15–20 skeletal (non-hydrogen) atoms

**Table 1.1**   Isomers of $C_iH_j$ composition.

| C \ H | 0 | 2 | 4 | 6 | 8 | 10 | 12 | 14 | 16 | 18 | 20 | 22 |
|---|---|---|---|---|---|---|---|---|---|---|---|---|
| 0 | – | 1 | | | | | | | | | | |
| 1 | – | – | 1 | | | | | | | | | |
| 2 | 1 | 1 | 1 | 1 | | | | | | | | |
| 3 | 1 | 2 | 3 | 2 | 1 | | | | | | | |
| 4 | 3 | 7 | 11 | 9 | 5 | 2 | | | | | | |
| 5 | 6 | 21 | 40 | 40 | 26 | 10 | 3 | | | | | |
| 6 | 19 | 85 | 185 | 217 | 159 | 77 | 25 | 5 | | | | |
| 7 | 50 | 356 | 920 | 1230 | 1031 | 575 | 222 | 56 | 9 | | | |
| 8 | 204 | 1804 | 5308 | 7982 | 7437 | 4679 | 2082 | 654 | 139 | 18 | | |
| 9 | 832 | 10064 | 33860 | 56437 | 57771 | 40139 | 19983 | 7244 | 1902 | 338 | 35 | |
| 10 | 4330 | 64352 | 241297 | 439373 | 488125 | 369067 | 201578 | 81909 | 24938 | 5568 | 852 | 75 |

can be enumerated in many cases, depending on the molecular formula composition. This is dependent on the valence of the atoms, the number of free bonds in the fragments (if used), the degree of unsaturation of the molecule and several other factors. It is interesting to compare the performance of modern structure generators with that of the first prototypes which were capable of generating a skeletal framework of 7–8 atoms in the 1970s. For instance, it took 3420 s to generate 4347 isomers for a composition of $C_{15}H_{32}$ using a program running on an archaic M-20 computer.[61] A modern structure generator, GENM, implemented in the Structure Elucidator system[81] and installed on a standard personal computer (2.8 GHz Pentium IV) can solve this problem in a fraction of a second (0.14 s), an improvement in performance of four orders of magnitude! Similarly, it took just over 47 h for the CHEMICS[82] system to generate 37 491 structures from a molecular formula of $C_6H_9NO$. The GENM program[75,81] generated 35 759 structures in 0.1 s ($\sim 10^6$ times faster!), while, in addition, correctly removing isomorphic structures. The role of the degree of unsaturation is significant as shown by the following example: for $C_{20}H_{42}$ the GENM program produced 366 319 structures in 32 s, while for $C_{20}H_{40}$ 12 662 282 isomers were generated in 9 min, *i.e.* 35 times longer.

### 1.8.3 Typical Structural Constraints Used in Structure Generators

The basic requirements for structure generators were formulated[9–11] in the 1970s. If the generation is carried out from the molecular formula then the primary expectation is that structure generation must be both exhaustive and non-redundant. In practice this process utilizes both atoms and fragments in the presence of additional structural constraints. The secondary requirement is therefore that, irrespective of the generation algorithm utilized, the software requires flexible tools to allow the imposition of structural constraints. These constraints usually include:

1. Fragments identified at the stage of structural spectrum interpretation. It is usually assumed (it is an attribute of a "classic" generator) that the individual fragments have no overlapping atoms and that the number of free valence bonds for the terminal atoms associated with the fragments are fixed. These fragments have been termed "macro-atoms"[10] and in the structure generation process they are utilized together with free atoms to produce either larger fragments or final structures. The sets of discrete units of structure (DUS) are composed of both atoms and macro-atoms and the total number of DUS sets correlates with the complexity of the problem. Some programs allow partial overlap of the macro-atoms.[60,77] This does result, however, in the algorithm being more complex and a consequent increase in the generation time. Algorithms for structure generation have been developed to allow structure generation from fragments with at least one pair of overlapping atoms.

2. Obligatory and forbidden fragments. In software developed for structure generation the ability to impose the inclusion or require the exclusion of a fragment in the structure generation process is implemented through the use of a GOODLIST or BADLIST function. These structural constraints allow the user to impose direct or intuitive influence over the results, since each structure must meet the requirements of retaining or excluding the listed fragments. The GOODLIST may include fragments found by the program during the spectral interpretation stage, as well as those suggested by the chemist from independent experiments (*e.g.* IR, UV and MS) and additional information including the origin of the sample and simple intuition. These fragments are allowed to overlap with each other, as well as with the DUS structures. For the majority of CASE systems the global constraints also include Bredt's rule[83] as a filter for all generated structures.

3. Constraints associated with both ring size and bond multiplicities. These constraints, and the ability to set minimum and maximum allowed values, make it possible to formulate general requirements for generated structures with consideration given to the DUS set and *a priori* information, such as the presence of multiple bonds in macro-atoms and rings, and the degree of unsaturation and so on.

4. Unlikely fragments. These fragments are those that are *unlikely* to be experienced in chemical structures and include, for instance, small cycles containing triple bonds or/and cumulated double bonds, certain strained fused structures with small cycles and fragments of the -O-O-O- type *etc.* Unlikely fragments of these types are usually included into a permanent BADLIST for the expert system and provide a rigid filter during or after structure generation.

As described later a significant effort has been made to use all constraints in as an efficient manner as possible to ensure that structures which do not comply with the given constraints are not generated. Since the structure elucidation of large molecules remains a primary challenge for CASE systems, the generation stage remains a bottleneck for all expert systems, despite the advances to date. The likelihood of success improves with the number and severity of the imposed structural constraints. As shown later, the application of constraints inferred from rich 2D NMR data allows for the elucidation of complex natural product structures containing more than 100 skeletal atoms when using modern expert systems (see Chapter 6 and Part III).

## 1.8.4 Structure Verification

As pointed out in Section 1.6, the final step in the structure elucidation process is the evaluation of the congruence between the chemical structures generated and the experimental spectra. All constraints related to the given problem are taken into account during structure generation and, at first glance, it could be supposed that all structures included into the output file must match the experimental spectra and additional information. Structure generation is performed from a set of discrete structural units – both fragments and free atoms – which are combined according to valence rules. As a result, certain substructures can be generated that contradict the experimental spectra. For instance, if an unknown is a polycyclic aliphatic hydrocarbon, then double and triple bonds may be generated from free (not encompassed by fragments) carbon atoms. Therefore it is necessary to verify whether or not these structures contradict the NMR and IR spectra. To this end, it is necessary to predict the spectra of candidate structures and compare the predicted spectra with those available experimentally.

It is possible to distinguish two levels of spectrum prediction: *estimated* and *strict*.

The *estimated* level of spectrum prediction is based on characteristic spectral features and a predicted spectrum can be represented as a set of spectral ranges, each of which is inherent to some substructure or some atom existing in the generated molecule.

In our example of a polycyclic aliphatic hydrocarbon, the generated structures containing C=C and C≡C bonds will be rejected because the IR spectra of these compounds should contain bands in the range of 1600–1680 $cm^{-1}$ (C=C) or/and 2100–2260 $cm^{-1}$ (C≡C), while the $^1$H NMR and $^{13}$C NMR have to show resonances in the 5–8 ppm and 70–150 ppm ranges correspondingly. In the case of aliphatic hydrocarbons these spectral intervals are empty, and this is the reason for the rejection of contradictory structures. This procedure of spectral estimation is used during the filtering of generated structures using libraries of spectrum–structure correlations containing fragments accompanied with their characteristic spectral ranges in different spectra. A simple example of structure spectral filtering was presented in Section 1.4. Structures that contain fragments contradicting the experimental spectra are removed from the output file. Note that the characteristic ranges may be rather wide, and the structure is most frequently considered as acceptable if at least one experimental feature falls in a postulated range.

If the solution to the problem of structure elucidation is not unambiguous after filtering, *i.e.* the number of isomers, $n$, in the output file is $n > 1$, it means that although all spectrum–structural constraints and correlations are exhaustively involved in solving the problem, the imposition of additional rigid constraints is necessary. In such cases the extraction of missing structural information can be achieved by *strict* spectrum prediction for candidate structures followed by comparison between the predicted and experimental spectra. In this case it is necessary to check whether *each* predicted spectral feature (chemical shift, vibrational band, *etc.*) can be related to the definite experimental one. As any spectrum simulation will contain errors, meaningful decisions regarding the coincidence or non-coincidence between experimental and predicted spectral features require the development of special algorithms.

In the process of choosing the most effective approach for performing spectrum prediction, the possibilities of predicting MS, IR and NMR spectra have been investigated.[67] The most appropriate method which could be applied to the rejection of incorrect candidate structures has to meet the following requirements:

1. The method must be *automatic* and *fast* enough to provide calculations for a *great* number of *large structures* of arbitrary complexity in a *reasonable time*.
2. The spectrum prediction *accuracy* must be high enough to allow the distinguishing of close structures which are expected in the output file.
3. The predicted spectra must be presented in such a form which provides easy and *automatic comparison* of predicted and experimental spectral features.

Our investigations have showed that only *empirical* methods of NMR spectral prediction (see Chapter 3) meet all of the above mentioned requirements. As discussed in further detail later (see Section 3.3.4), selection of the genuine structure can be achieved on the basis of combining empirical and quantum mechanical methods of NMR chemical shift prediction. Different approaches used for spectrum prediction will be discussed in detail in Chapter 3.

# References

1. J. Lederberg, G. L. Sutherland, B. G. Buchanan, E. A. Feigenbaum, A. V. Robertson, A. M. Duffield and C. Djerassi, *J. Am. Chem. Soc.*, 1968, **91**, 2973.
2. M. E. Elyashberg and L. A. Gribov, *Zh. Prikl. Spectrosk.*, 1968, **8**, 296–300.
3. D. B. Nelson, M. E. Munk, K. B. Gasli and D. L. Horald, *J. Org. Chem.*, 1969, **34**, 3800.
4. S. I. Sasaki, H. Abe, T. Ouki, M. Sakamoto and S. I. Ochia, *Anal. Chem.*, 1968, **40**, 2220.
5. T. A. Gillespie and B. E. Winger, *Mass Spectrom. Rev.*, 2011, **30**, 479.
6. Z. Slanina, *Theoretical Aspects of Isomerism Phenomenon in Chemistry*, Mir, Moscow, 1984.
7. J. J. Berzelius, *Ann. Phys. Chem.*, 1830, **19**, 305.
8. A. Cayley, *Phil. Mag.*, 1874, **47**, 444.
9. M. E. Elyashberg, L. A. Gribov and V. V. Serov, *Molecular Spectral Analysis and Computer*, Nauka, Moscow, 1980.
10. R. K. Lindsay, B. G. Buchanan, E. A. Feigenbaum and J. Lederberg, *Applications of Artificial Intelligence for Organic Chemistry: The DENDRAL Project*, McGraw-Hill, New York, 1980.
11. N. A. B. Gray, *Computer-Assisted Structure Elucidation*, Wiley, New York, 1986.
12. V. V. Serov, M. E. Elyashberg and L. A. Gribov, *Izv. TSKhA*, 1977, **1**, 215.
13. M. E. Elyashberg, K. A. Blinov, A. J. Williams, S. G. Molodtsov and G. E. Martin, *J. Chem. Inf. Model.*, 2006, **46**, 1643–1656.

14. M. E. Elyashberg, A. J. Williams and K. A. Blinov, in *Modern NMR Approaches for the Structure Elucidation of Natural Products*, ed. G. E. Martin and A. J. Williams, Royal Society of Chemistry, in preparation.
15. http://www.cas.org/expertise/cascontent/registry/regsys.html
16. K. C. Nicolaou and S. A. Snyder, *Angew. Chem. Int. Ed.*, 2005, **44**, 1012–1044.
17. M. E. Elyashberg, Y. Z. Karasev, E. R. Martirosian, H. Thiele and H. Somberg, *Anal. Chim. Acta*, 1997, **348**, 443–463.
18. C. E. Shannon, *Bell Syst. Tech. J.*, 1948, **27**, 379–423.
19. A. Bagno and G. Saielli, *Theor. Chem. Acc.*, 2007, **117**, 603–619.
20. A. M. Denisov, *Elements of the Theory of Inverse Problems*, VSP BV, Amsterdam, 1999.
21. A. Tarantola, *Inverse Problem Theory and Methods for Model Parameter Estimation*, SIAM, Philadelphia, 2005.
22. A. Tarantola, *Nat. Phys.*, 2006, **2**, 492–494.
23. F. Harary, *Graph Theory*, Addison-Wesley, Reading, MA, 1971.
24. F. Harary and E. M. Palmer, *Graphical Enumeration*, Academic Press, New York, 1973.
25. A. R. Leach and V. J. Gillet, *An Introduction to Chemoinformatics*, Kluwer Academic Publishers, Dordrecht, 2003.
26. J. Gasteiger and T. Engel, *Chemoinformatics*, Wiley-VCH, Weinheim, 2003.
27. K. A. Blinov, M. E. Elyashberg, S. G. Molodtsov, A. J. Williams and E. R. Martirosian, *Fresenius' J. Anal. Chem.*, 2001, **369**, 709.
28. M. Will, W. Fachinger and J. R. Richert, *J. Chem. Inf. Comput. Sci.*, 1996, **36**, 221.
29. M. E. Elyashberg, K. A. Blinov and E. R. Martirosian, *Lab. Autom. Inf. Manage.*, 1999, **34**, 15.
30. D. Goldrey, *Propositional and Predicate Calculus: A Model of Argument*, Springer, London, 2005.
31. L. A. Gribov and M. E. Elyashberg, *J. Mol. Struct.*, 1970, **5**, 179–198.
32. L. A. Gribov, M. E. Elyashberg and L. A. Moscovkina, *J. Mol. Struct.*, 1971, **9**, 357–371.
33. M. E. Elyashberg, in *Encyclopedia of Computational Chemistry*, ed. P. v. R. Schleyer, N. L. Allinger, T. Clark, J. Gasteiger, P. A. Kollman, H. F. Schaefer III and P. R. Schreiner, John Wiley & Sons, Chichester, 1998, pp. 1307–1312.
34. L. A. Gribov and M. E. Elyashberg, *Crit. Rev. Anal. Chem.*, 1979, **8**, 111–220.
35. R. S. Ledley, *Digital Computer and Control Engineering*, McGraw-Hill, New York, 1960.
36. M. E. Elyashberg and L. A. Moskovkina, *J. Appl. Spectrosc.*, 1968, **6**, 595–597.
37. M. E. Elyashberg, V. V. Serov, E. R. Martirosian, L. A. Zlatina, Y. Z. Karasev, V. N. Koldashov and Y. Y. Yampolskiy, *J. Mol. Struct.*, 1991, **76**, 191–203.

38. L. A. Gribov, M. E. Elyashberg and V. V. Serov, *Anal. Chim. Acta, Comp. Techn. Optimiz.*, 1977, **95**, 75–96.
39. L. A. Gribov, M. E. Elyashberg, V. N. Koldashov and I. V. Pletnjov, *Anal. Chim. Acta*, 1983, **148**, 159–170.
40. M. E. Elyashberg, E. R. Martirosian, Y. Z. Karasev, H. Thiele and H. Somberg, *Anal. Chim. Acta*, 1997, **337**, 265–286.
41. K. Funatsu, Y. Susuta and S. Sasaki, *Anal. Chim. Acta*, 1989, **220**, 155–169.
42. G. N. Andreev, O. K. Argirov and P. N. Penchev, *Anal. Chim. Acta*, 1993, **284**, 31.
43. D. H. Rouvray, *Fuzzy Logics in Chemistry*, Academic Press, San Diego, 1997.
44. V. V. Serov, L. A. Gribov and M. E. Elyashberg, *J. Mol. Struct.*, 1985, **129**, 183–214.
45. M. E. Elyashberg, K. A. Blinov, S. G. Molodtsov, A. J. Williams and G. E. Martin, *J. Chem. Inf. Comput. Sci.*, 2004, **44**, 771–792.
46. K. S. Lebedev and D. J. Cabrol-Bass, *Chem. Inf. Comput. Sci.*, 1998, **38**, 410.
47. K. Varmuza, P. N. Penchev and H. J. Scsibrany, *Chem. Inf. Comput. Sci.*, 1998, **38**, 420.
48. L. Chen and W. J. Robien, *Chem. Inf. Comput. Sci.*, 1994, **34**, 934.
49. I. I. Strokov and K. S. Lebedev, *J. Chem. Inf. Comput. Sci.*, 1999, **39**, 659.
50. P. Willett, J. M. Barnard and G. M. Downs, *J. Chem. Inf. Comput. Sci.*, 1998, **38**, 983.
51. *Advanced Chemistry Development. ACD/NMR Predictors. Prediction suite includes $^1H$, $^{13}C$, $^{15}N$, $^{19}F$, $^{31}P$ NMR prediction*, 2011.
52. W. Robien, *Nachr. Chem. Tech. Lab.*, 1998, **46**, 74.
53. *Specinfo, Chemical Concepts GmbH*, D-69442 Weinheim, Germany.
54. W. Robien, CSEARCH, http://felix.orc.univie.ac.at/~wr/csearch_server_info.html
55. W. Bremser, *Anal. Chim. Acta, Comp. Techn. Optimiz.*, 1978, **2**, 355–365.
56. M. Carabedian, I. Dagane and J. E. Dubois, *Anal. Chem.*, 1988, **60**, 2186.
57. S. G. Molodtsov, *MATCH Commun. Math. Chem.*, 1994, **30**, 203.
58. T. Wieland, A. Kerber and R. Laue, *J. Chem. Inf. Comput. Sci.*, 1996, **36**, 413.
59. L. M. Masinter, N. S. Sridharan, J. Lederberg and D. H. Smith, *J. Am. Chem. Soc.*, 1974, **96**, 7702.
60. R. E. Carhart, D. H. Smith, N. A. B. Gray, J. G. Nourse and C. Djerassi, *J. Org. Chem.*, 1981, **46**, 1709.
61. V. V. Raznikov and V. L. Tal'roze, *Zh. Strukt. Khim.*, 1970, **11**, 357.
62. Y. Kudo and S. Sasaki, *J. Chem. Inf. Comput. Sci.*, 1976, **16**, 43.
63. V. V. Serov, M. E. Elyashberg and L. A. Gribov, *Dokl. Akad. Nauk USSR*, 1975, **224**, 109.
64. C. A. Shelley and M. E. Munk, *Anal. Chim. Acta*, 1981, **133**, 507.
65. C. A. Shelley, T. R. Hays, M. E. Munk and R. V. Roman, *Anal. Chim. Acta*, 1978, **103**, 121.

66. J.-E. Dubois, M. Carabedian and B. Ancian, *C. R. Acad. Sci.*, 1980, **290**, 369.
67. M. E. Elyashberg, *Russ. Chem. Rev.*, 1999, **68**, 525–547.
68. M. E. Munk, *J. Chem. Inf. Comput. Sci.*, 1998, **38**, 997–1009.
69. M. Badertscher, K. Bischofberger and E. Pretsch, *Trends Anal. Chem.*, 1997, **16**, 234.
70. C. Peng, S. Yuan, C. Zheng and Y. Hui, *J. Chem. Inf. Comput. Sci.*, 1994, **34**, 805–813.
71. C. Steinbeck, *J. Chem. Inf. Comput. Sci.*, 2001, **41**, 1500.
72. Y. Han and C. Steinbeck, *J. Chem. Inf. Comput. Sci.*, 2004, **44**, 489–498.
73. J. Meiler and M. Will, *J. Chem. Inf. Comput. Sci.*, 2001, **41**, 1535.
74. C. Benecke, R. Grund, R. Hohberger, A. Kerber, R. Laue and T. Wieland, *Anal. Chim. Acta*, 1995, **314**, 141.
75. S. G. Molodtsov, *MATCH Commun. Math. Chem.*, 1998, **37**, 157.
76. A. Korytko, K.-P. Schulz, M. S. Madison and M. E. Munk, *J. Chem. Inf. Comput. Sci.*, 2003, **43**, 1434.
77. B. D. Christie and M. E. Munk, *J. Chem. Inf. Comput. Sci.*, 1988, **28**, 87.
78. J.-M. Nuzillard, W. Naanaa and S. J. Pimont, *Chem. Inf. Comput. Sci.*, 1995, **35**, 1068.
79. M. Lindley, J. N. Shoolery, D. Smith and C. Djerassi, *Org. Magn. Reson.*, 1983, **21**, 405.
80. G. E. Moore, *Electronics*, 1965, **38**(8), 114.
81. K. A. Blinov, D. Carlson, M. E. Elyashberg, G. E. Martin, E. R. Martirosian, S. G. Molodtsov and A. J. Williams, *Magn. Reson. Chem.*, 2003, **41**, 359–372.
82. K. Funatsu, N. Miyabayashi and S. Sasaki, *J. Chem. Inf. Comput. Sci.*, 1988, **28**, 18.
83. *Bredt's Rule*, http://www.iupac.org/goldbook/B00732.pdf

CHAPTER 2

# Cognitive Peculiarities of the Structure Elucidation Problem

The history of the development of CASE systems to date has convincingly demonstrated the point of view suggested 40 years ago[1,2] that the process of molecular structure elucidation is reduced to the logical inference of the most probable structural hypothesis from a set of statements reflecting the inter-relation between a spectrum and a structure. This methodology was *implicitly* used for a long time before computer methods appeared. Independent of computer-based methods, the path to a target structure is the same and CASE expert systems mimic the approaches of a human expert. The main advantages of CASE systems are as follows:

1. All statements regarding the interrelation between spectra and a structure ("axioms") are expressed explicitly.
2. All logical consequences (structures) following from the system of "axioms" are completely deduced without any exclusions.
3. The process of computer-based structure elucidation is very fast and provides a tremendous saving in both time and labor for the scientist.
4. If the chemist has several alternative sets of "axioms" related to a given structural problem, then an expert system allows for the rapid generation of all structures from each of the sets and identification of the most probable structure by comparison of the solutions obtained.

We assume that the reader is familiar with the basic principles of molecular spectroscopy (MS, 1D and 2D NMR, IR, *etc.*) and has some experience with its application to structure elucidation. In this Chapter we will describe the main kinds of statements used during the process of structure elucidation. These can be conventionally divided into several categories.

New Developments in NMR No. 1
Contemporary Computer-Assisted Approaches to Molecular Structure Elucidation
By Mikhail Elyashberg, Antony Williams and Kirill Blinov
© Royal Society of Chemistry 2012
Published by the Royal Society of Chemistry, www.rsc.org

# 2.1   Axioms and Hypotheses Based on Characteristic Spectral Features

In Chapter 1 we showed that structural group analysis of organic molecules using characteristic spectral features (chemical shifts in NMR, frequencies in IR) could be formalized on the basis of symbolic logics. It turned out that interrelations between definite substructures and their characteristic features were described by logical functions. As a result of joint logical analysis of these interrelations with an experimental spectrum [solution of logical eqn (1.7)], the algorithm produces all of the combinations of substructures that can be present in the unknown structure. Here we will see that application of logical calculus for solving the spectrum–structural problem is not only a mathematical technique, but an aid for more deeply understanding the way of expert reasoning during the structure elucidation of an unknown compound.

In accordance with the definition, we refer to "axioms" as those statements that can be considered true based on prior experience. To elucidate the structure of a new unknown compound, the chemist usually uses spectrum–structure correlations that are established as a result of the efforts of several generations of spectroscopists. In so doing, the chemist implies that each correlation is meaningful and *true*, although experience suggests a probabilistic origin of these correlations. We will see that statements reflecting the existence of characteristic spectral features can be considered as the basic axioms of structure elucidation theory, the latter possessing all traits of the *axiomatic theory*. The general form of typical axioms belonging to this category can be formulated as follows: *If a molecule contains a fragment $A_i$ then the characteristic features of fragment $A_i$ are observed in certain spectrum ranges $[X_1],[X_2],...[X_m]$ which are characteristic for this fragment.*

According to eqn (1.5), this axiom can be presented as a logical function:

$$A_i \rightarrow [X_1] \wedge [X_2] \wedge ... \wedge [X_m] \tag{2.1}$$

For example, if a molecule contains a $CH_2$ group, then a vibrational band around 1450 $cm^{-1}$ is observed in the IR spectrum. If a molecule contains a $CH_3$ group, then two bands at approx. 1450 and 1380 $cm^{-1}$ appear. These axioms can be presented formally in the following way:

$$CH_2 \rightarrow [1450 \ cm^{-1}]; \ CH_3 \rightarrow [1380] \wedge [1450 \ cm^{-1}]$$

Analogously, for characteristic [13]C NMR chemical shifts the following implications are also exemplar axioms:

$$(C)_2C{=}O \rightarrow [200 \ ppm], \ (C)_2C{=}S \rightarrow [200 \ ppm].$$

When characteristic spectral features are used for the detection of fragments that can be present in a molecule under investigation then the chemist usually forms statements for which a typical "template" is as follows: *If a spectral feature is observed in a spectrum range $[X_j]$ then the molecule contains at least* one fragment of the set $A_i(X_j)$, $A_k(X_j)$, ... $A_l(X_j)$, where $A_i$, $A_k$, ... $A_l$ are fragments for which the spectral feature observed in the range $[X_j]$ is characteristic, and the fragments form a finite set.

This assertion is symbolized as:

$$[X_j] \rightarrow A_i(X_j) \vee A_k(X_j) \vee ... \vee A_l(X_j) \qquad (2.2)$$

It is necessary to underline that this statement is a *hypothesis*, not an axiom, because: (1) the feature $X_j$ can be produced by some fragment which is not known as yet, and (2) the feature $X_j$ can appear due to some intramolecular interaction of known fragments. Therefore, if an absorption band is observed at 1450 cm$^{-1}$ in an IR spectrum, then the molecule can contain either $CH_2$ or $CH_3$ groups, both of them (band overlap at 1450 cm$^{-1}$ is allowed) or the 1450 cm$^{-1}$ band can be present as a result of the presence of another unrelated functional group. This statement can be expressed formally using the symbol for logical disjunction ($\vee$):1450 cm$^{-1}$ $\rightarrow$ $CH_2$ $\vee$ $CH_3$ $\vee$ $\alpha$, where $\alpha$ is a "sham fragment" denoting an unknown cause of the feature origin. For our $^{13}$C NMR examples, we may obviously formulate the following hypothesis: 200 ppm $\rightarrow$ $(C)_2C=O$ $\vee$ $(C)_2C=S$. It is very important to have in mind that if $A_i \rightarrow X_j$ is true, then in accord with formal logic the inverse implication:

$$X_j \rightarrow A_i$$

can be true or not true. In other words, the presence of a signal $x_j \subset [X_j]$ belonging to the range of a characteristic spectral feature in a spectrum does not imply the presence of a corresponding fragment. According to symbolic logic rules (see eqn (1.3)), a true implication equivalent to expression $A_i \rightarrow X_j$ is:

$$\bar{X}_j \rightarrow \bar{A}_i.$$

This implication means that if the characteristic spectral feature $X_j$ does *not* occur in a spectrum, then the corresponding fragment $A_i$ is absent from the molecule under investigation. The latter statement can be considered as another equivalent formulation of the basic axiom $A_i \rightarrow X_j$ and it is in good agreement with chemical common sense.

During the process of structure elucidation, to make a conclusion about the presence or the absence of some fragment, its characteristic features in all available spectra are taken into account. In the general case, variable $w_j$ in the eqns (2.3) denotes the characteristic features of the fragment $A_i$ assigned in spectra of different kinds (for instance, IR, $^1$H NMR and $^{13}$C NMR):

$$A_i \rightarrow w_1(i) \wedge w_2(i) \wedge ... \wedge w_n(i), \qquad (2.3a)$$
$$\{\bar{w}_1(i) \vee \bar{w}_2(i) \vee ... \vee \bar{w}_n(i)\} \rightarrow \bar{A}_i \qquad (2.3b)$$

Eqn (2.3b) is equivalent to eqn (2.3a) and obtained from (2.3a) by application of the de Morgan laws (1.3) and (1.4) to conjunction $w_1(i) \wedge w_2(i) \wedge ... \wedge w_n(i)$. Word interpretation of eqn (2.3b) can be presented as the sentence: "if at least one characteristic feature $w_1$, $w_2$, ...$w_n$ is not observed in the experimental spectra, then the fragment $A_i$ is absent from the structure of unknown". It is necessary to underline that the "template" presented by eqn (2.3b) allows chemists to obtain the most reliable conclusions during the structural analysis of an unknown compound. The reason is that we use an *axiom* (2.3a) in the equivalent form (2.3b) for rejecting the presence of a fragment, whereas, as mentioned, the inverse implication $X_j \rightarrow A_i$ is only a hypothesis and it can be true or not true. In agreement with the philosophy of science, truthfulness of a hypothesis is verified by practice. In our case the truthfulness of a suggested hypothesis is determined *a posteriori* when the genuine structure is established.

For instance, the following characteristic spectral ranges are characteristic for the aldehyde group: IR [1720 cm$^{-1}$], $^{13}$C NMR [200 ppm] and $^{1}$H NMR [9.5 ppm]. Then the corresponding logical relation is:

$$HCO \rightarrow \left[1720\,cm^{-1}\right]_{IR} \wedge [200\,ppm]_{13C} \wedge [9.5\,ppm]_{1H}$$

and the equivalent representation of the axiom can be written as:

$$[\overline{1720}\,cm^{-1}]_{IR} \vee [\overline{200}\,ppm]_{13C} \vee [\overline{9.5}\,ppm]_{1H} \rightarrow \overline{HCO}$$

It is evident that absence of any IR band in the range [1720 cm$^{-1}$] or the absence of any resonances in the corresponding intervals of $^{13}$C or $^{1}$H NMR spectra leads to the conclusion that the molecule under investigation can not contain the aldehyde group.

When a newly synthesized or separated large molecule has to be identified, the first stage is commonly to characterize an unknown by the structural groups that may exist in the molecule. The investigation of experts' manner of reasoning leads to the conclusion that the structural group spectral analysis, SGA (both manual and computerized), is based on the following assumptions completing the ones mentioned above:

- Each fragment $A_i \in \mathbf{A}$ has definite ranges for its characteristic spectral features if the fragment's *permissible environment* in a molecule is strictly defined.
- The analyzed molecule must be *additive* or at least contain some closed self-contained groups. *Additive molecules produce additive spectra.*
- Only $A_i \in \mathbf{A}$ fragments can be involved in the solution of the problem. All other conceivable fragments, $A_x \notin \mathbf{A}$, will not be considered, but they may be involved in the game during the structure generation.
- While solving the problem *all of the spectral features* (chemical shifts, frequencies) of the experimental spectra are considered as potential particular attributes of the structural elements, $A_i \in \mathbf{A}$, and may be interpreted only in terms of the $A_i$ fragments.

- The chemical composition and degree of unsaturation, U, for the fragments selected during the SGA must be consistent with the molecular empirical formula.
- While solving the problem every possible fragment set resulting from the structural group analysis must provide an interpretation of all spectral features perceived as characteristic in the experimental spectra.

All these "axioms" are used (in the explicit or implicit form) by experts in the process of SGA. The expert system, based on the mathematical model of SGA, uses them only in the explicit form because these "axioms" are put in the system algorithms. Consequently, all those "real world" contradictions which the chemist may encounter in the process of molecular structure elucidation are typical of the expert system as well.

To overcome these contradictions the system needs input of a chemist who has diverse aids to do it, working in the interactive mode.

Let us consider the most typical contradictions and their possible symptoms which emerge when SGA is performed.

*1. The additivity of a molecule is violated.*
The spectral feature $w_j$ belonging to $A_i$ ($A_i \in A$) is observed outside the characteristic range of the $A_i$ fragment which actually *exists* in the molecule. As a result the $A_i$ fragment will be missed.

If, in addition, the analytical signal $w_j$ shifts in the frequency interval of fragment $A_k$($A_k \in A$), which actually does not exist in the sample at hand, a wrong fragment $A_k$ will be involved in the game (for instance, an ester group instead of ketone one; see example in Chapter 12).

*2. The sample at hand contains an unknown fragment $A_x$ that does not belong to the A set ($A_x \notin A$).*
If the $w_x$ feature of $A_x$ fragment is observed just within the characteristic interval of some $A_k$ fragment which does not actually exist in the sample at hand, then a wrong $A_k$ fragment will be involved in the game.

*3. The experimental spectrum contains a non-characteristic feature $w_x$ caused by some interactions between different structural elements of the molecule (typical in IR) or steric factors (typical in NMR).*
If the $w_x$ feature falls into the characteristic interval of the $A_k$ fragment which actually does not exist in the molecule under investigation (the $w_x$ feature mimics a characteristic attribute of this fragment), then a wrong $A_k$ fragment will be involved in the game.

In addition to these principal contradictions, a wrong result of SGA can be caused by too narrow characteristic intervals of some fragments included in the knowledge base. In this case, a fragment of such type can be missed during the SGA process.

All logically possible sets of fragments should be checked by expert (tedious work) or by the expert system for conformity with the empirical formula of the molecule and its degree of unsaturation, **U**, calculated from the empirical formula. This check allows rejection of a great number of fragment sets which exceed the empirical formula and the corresponding **U** value. However, a situation may occur, when all fragment sets are rejected. This means that the system faces a contradiction between the empirical formula and chemical compositions of all fragment sets. The most probable cause of this contradiction can be explained with the following example taken from IR spectroscopic practice.

Let the molecule under investigation contain [according to the empirical formula and double bond equivalent (DBE)] only one phenyl ring and the IR spectrum shows the presence of intense bands $w_j$ and $w_k$ in the characteristic intervals of mono- and para-substituted benzene correspondingly. Then the logically true implications will be displayed:

$$w_j \rightarrow \textbf{1-Ar}$$

$$w_k \rightarrow \textbf{1,4-Ar}$$

This means that both **1-Ar** and **1,4-Ar** fragments should be present in the molecule, which is impossible. Obviously, this contradiction can be solved by the chemist who can use additional information to make the right choice or, if he/she fails to do it, both fragments must be introduced into the reasoning by means of a logical relation such as:

$$(w_j \wedge w_k) \rightarrow (\textbf{1-Ar} \vee \textbf{1,4-Ar})$$

All possibilities will be verified and plausible structures will be selected.

As a matter of fact, the strategies of solving spectrum–structure problems should be considered as the methods of overcoming different difficult situations which appear when solving a problem, some of which were discussed above in this subsection.

## 2.2   Axioms and Hypotheses of 2D NMR Spectroscopy

2D NMR spectroscopy is a method which, in principle, is capable of inferring a molecular structure from the available spectral data *ab initio* without using any spectrum–structure correlations and additional suppositions. In some cases the 2D NMR data provides sufficient structural information to suggest a manageable set of plausible structures. This is a fairly common situation for small molecules with a lot of protons contained within the molecule. In practice the structure elucidation of large molecules by the *ab initio* application of 2D NMR data only (without 1D NMR spectrum–structure correlations) is generally impossible. The 1D and 2D NMR data are usually combined synergistically to obtain solutions to real analytical problems in the study of natural products.

Experience has shown[3–7] that the size of a molecule is not a crucial obstacle for a CASE system based on 2D NMR data. The number of hydrogen atoms responsible for the propagation of structural information across the molecular skeleton and the number of skeletal heteroatoms are the most influential factors. An abundance of hydrogen atoms and a small number of heteroatoms generally eases the structure elucidation process rather markedly. To date we have failed to determine any specific dependence between molecular composition and the number of plausible structures deduced by an expert system because the different modes for solving a problem are chosen according to the nature of the specific problem (see Section 2.3). Moreover, the complexity of the problem is associated with many factors which cannot be identified before attempts are made to solve the problem. For instance, the complexity of the problem depends on whether the heavy atoms and their attached hydrogen atoms are distributed "evenly" around the molecular skeleton. If at least one "silent" fragment (*i.e.* having no attached hydrogens) is present in a molecule then it can interrupt a chain of heteronuclear multiple-bond correlation (HMBC) and COSY correlations. As a result, the number of structural hypotheses will increase dramatically as reported, for example, in the cryptolepine family.[6]

When 2D NMR data are used to elucidate a molecular structure then the chemist (or an expert system which mimics the manner of the chemist's reasoning) deduces conceivable structures from the molecular formula and a set of hypotheses matching the data from 2D NMR spectroscopy. When we deal with a *new* natural product, we must interpret a *new* 2D NMR spectrum or spectra. In this case we have no possibility to rely on "axioms" valid for the given spectrum–structure matrix, therefore hypotheses which are considered as the most plausible are formed. These hypotheses are based on the general regularities which are the significant *axioms of 2D NMR spectroscopy*. We will attempt to express these axioms in an explicit form and classify them.

There are, of course, various forms of 2D NMR spectroscopy, the most important and common of these being homonuclear $^1$H-$^1$H and heteronuclear $^1$H-$^{13}$C spectroscopy. Although heteronuclear interactions of the nature X1-X2 (X1 and X2 are magnetically active nuclei but not $^1$H nor $^{13}$C) are possible, such spectra are rare and, except for labeled materials, very difficult to acquire in general.

A necessary condition for the application of 2D NMR data to computer-assisted structure elucidation is the chemical shift assignment of all proton-bearing carbon nuclei, (*i.e.* all $CH_n$ groups where $n = 1$–3). This information is extracted from the HSQC [or, alternatively, the heteronuclear multiple quantum correlation (HMQC)] data using the following axiom:

- *If a peak ($\delta C$-i, $\delta H$-i) is observed in the spectrum then the hydrogen atom H-i with chemical shift $\delta H$-i is attached to the carbon atom C-i having chemical shift $\delta C$-i.*

Using notations of symbolic logic we can present this axiom as follows:

$$\delta(C\text{-}i, H\text{-}i) \rightarrow \text{bond}\{(C\text{-}i) - (H\text{-}i)\}$$

The main sources of structural information are COSY and heteronuclear multiple bond correlation (HMBC) correlations which allow the elucidation of the backbone of a molecule. We refer to "standard" correlations[8] as those that satisfy the following axioms reflecting the experience of NMR spectroscopists:

- *If a peak (H-i, H-k) is observed in a COSY spectrum, then a molecule contains the chemical bond (C-i)–(C-k).*

$$\delta(H\text{-}i, H\text{-}k) \rightarrow \text{bond}\{(C\text{-}i)-(C\text{-}k)\}$$

- *If a peak ($\delta$H-i, $\delta$C-k) is observed in a HMBC spectrum, then atoms C-i and C-k are separated in the structure by one or two chemical bonds:*

$$(\text{C-}i)-(\text{C-}k) \text{ or } (\text{C-}i)-(\text{X})-(\text{C-}k), \text{ X=C, O, N...}$$
$$\delta(C\text{-}i, H\text{-}i) \rightarrow \text{bond}\{(\text{C-}i)-(\text{C-}k)\} \vee \text{bonds}\{(\text{C-}i)-(\text{X})-(\text{C-}k)\}$$

The standard COSY and HMBC correlations are characterized by coupling constants $^3J_{HH}$ (COSY) and $^3J_{CH}$ (HMBC) so the topological distance between interacting nuclei is of three bonds.

By analogy, the main axiom associated with employing the nuclear Overhauser effect (NOE) for the purpose of structure elucidation can be formulated in the following manner:

- *If a peak ($\delta$H-i, $\delta$H-k) is observed in a nuclear Overhauser effect/enhancement spectroscopy (NOESY) [rotating-frame Overhauser enhancement spectroscopy (ROESY)] spectrum, then the distance between the atoms H-i and H-k through space is less than 5Å.*

$$\delta(H\text{-}i, H\text{-}k) \rightarrow \{\text{dist}[(H\text{-}i, H\text{-}k)] < 5\text{Å}\}$$

It is important to note, that there is a principal difference between logical interpretations of 1D and 2D NMR axioms. For instance, for COSY there exists the second equivalent form of the main axiom:

- *If a molecule does not contain the chemical bond (C-i)–(C-k), then no peak (H-i, H-k) is observed in a COSY spectrum.*

$$\overline{\text{bond}}\{(C\text{-}i)-(C\text{-}k)\} \rightarrow \overline{\delta}(H\text{-}i, H\text{-}k)$$

Note that according to the property of logical implication, the logical expression

$$\overline{\delta}(H\text{-}i, H\text{-}k) \rightarrow \overline{\text{bond}}\{(C\text{-}i)-(C\text{-}k)\}$$

produced from the true COSY axiom can have any value of truthfulness. In this case, the interpretation of logical implication allows us to conclude that the absence of a peak $\delta(H\text{-}i, H\text{-}k)$ in a COSY spectrum says *nothing* about existence of the chemical bond $(C\text{-}i)–(C\text{-}k)$ in the molecule: the bond may exist and may not exist. Consequently, the expert system does not use the absence of some 2D NMR peak $\delta(H\text{-}i, H\text{-}k)$ for rejection of structures containing the bond $(C\text{-}i)–(C\text{-}k)$. Analogous conclusions are also applicable to HMBC and NOESY spectra. Note that formal conclusions following from symbolic presentation of main 2D NMR axioms fully correspond to conclusions which are usually made by experts during structural interpretation of 2D NMR data.

While it is known that the listed axioms hold in the overwhelming majority of cases, there are many exceptions and these correlations are referred to as *non-standard correlations* (NSCs).[8] The NSCs are those for which the topological distance between interacting nuclei exceeds three bonds, *i.e.* $n > 3$ for coupling constants $^nJ_{HH}$ and $^nJ_{CH}$. Since standard correlations and NSCs are not easily distinguished, the existence of NSCs is the main hurdle to logically inferring the molecular structure from the 2D NMR data. If the 2D NMR data contain both undistinguishable standard correlations and NSCs, then the total set of "axioms" derived from the 2D NMR data will contain contradictions.

This means that the correct structure cannot be inferred from these axioms and, in this case, the structural problem either has no solution or the solution will be incorrect: the set of suggested structures will not contain the genuine structure. Numerous examples of such situations will be considered in the next Chapters.

Unfortunately, as yet, there are no routine NMR techniques which distinguish between 2D NMR signals belonging to standard correlations and NSCs. In some fortunate cases, the application of time-consuming INADEQUATE and 1,1-ADEQUATE experiments,[9] as well as heteronuclear two bond correlation (H2BC)[10] experiments, is expected to help to resolve contradictions, but these techniques are also based on their own axioms which can be violated.

## 2.3 Structural Hypotheses Necessary for the Assembly of Structures

When chemical shifts in 1D and 2D NMR spectra are assigned and all 2D correlations are transformed into *connectivities* between heavy atoms in the skeletal framework, then feasible molecular structures should be assembled from "strict fragments" (suggested on the basis of the 1D NMR, 2D COSY and IR spectra, as well as those postulated by the researcher) and "fuzzy fragments" determined from the 2D HMBC data. To assemble the structures it is necessary to make a series of responsible decisions, which are equivalent to constructing a set of axiomatic hypotheses. At least the following choices should be made:

- Allowable chemical composition(s): CH, CHO, CHNO, CHNOS, CHNOCl, *etc.* The choice is made on the basis of chemical considerations and other additional information that may be available (sample origin, molecular ion cluster, *etc.*). It is evident that if the possibility of some chemical element that really exists in an unknown is declined by the researcher the genuine molecule will never be identified.

- Possible molecular formula (or formulae) as selected from a set of possible accurate molecular masses. The suggestion of a correct molecular formula is crucial for CASE systems and is highly desirable in order to perform dereplication.
- Possible valences of each atom having variable valence: N(3 or 5), S(2 or 4 or 6) or P(3 or 5). If $^{15}$N and $^{31}$P spectra are not available, then, in principle, all of the admissible valences of these atoms should be tried. Obviously, it is practically impossible to perform such a complete search manually. The application of a CASE system allows, in principle, the verification of all conceivable valence combinations, and an example is reported in Section 12.4.
- Hybridization of each carbon atom: *sp*, *sp²*, *sp³* or *not defined*.
- Possible neighborhoods with heteroatoms for each carbon atom: *fb* (forbidden), *ob* (obligatory) or *not defined*. An example of a typical challenge is: does C($\delta = 103$ ppm) indicate a carbon in the *sp²* hybridization state or in the *sp³* hybridization state but connected with two oxygens by ordinary bonds?
- Number of hydrogen atoms attached to carbons that are the nearest neighbors to a given carbon (determined, if possible, from the signal multiplicity in the $^1$H NMR spectrum). This decision may be rather risky and therefore such constraints should be used only with great caution, and in those cases where no signal overlap occurs and signal multiplicity can be reliably determined as in the case of methyl group resonances that are typically singlets or doublets.
- Maximum allowed bond multiplicity: 1, 2 or 3. The main challenges are related to a carbon being a neighbor of heteroatoms, as well as to the presence or absence of a triple bond. Strictly speaking, the latter can be solved reliably only based on IR or Raman spectra; an IR spectrum allows the decision to be made on the presence or absence of terminal -C≡CH and -C≡N groups, whereas a central –C≡C– group is usually revealed as a strong band in a Raman spectrum.
- List of fragments that can be assumed to be present in a molecule according to chemical considerations or based on a fragment search using the $^{13}$C NMR spectrum to search the fragment database. The chemical considerations usually arise from careful analysis of the NMR spectra related to known structures that have the same origin and similar spectra. The presence of the most significant functional groups (C=O, OH, NH, C≡N, C≡C, C≡CH *etc.*) can be suggested from both IR and Raman spectra when the corresponding assumptions are not contradicted by the NMR data and molecular formula of the unknown. Within an expert system such as Structure Elucidator (StrucEluc; see below), a list of obligatory fragments can be automatically offered for consideration by the chemist with them making the final decision in regards to inclusion.

- List of fragments which are forbidden within the given structural problem. These include fragments unlikely in organic chemistry: for example, a triple bond in small cycles or an O-O-O connectivity, *etc.* Additionally, substructures which are uncommon in the chemistry of natural products (for instance, a four-membered cycle). IR and Raman spectra can also hint at the specification of forbidden fragments, and the axiom $\bar{X}_j \rightarrow \bar{A}_i$ is usually a rather reliable basis for making a particular decision. For example, if no characteristic absorption bands are observed in the region 3100–3700 cm$^{-1}$, then an alcohol group will be absent from the unknown. This structural constraint, which can be obtained very simply, leads to the rejection of a huge number of conceivable structures containing the alcohol group (see an illustrative example in Chapter 1).

It should be evident that at least one poor decision based on the points listed above would likely lead to a failure to elucidate the correct structure. We will see examples of this below.

## 2.4 Properties of Information Used for the Structure Elucidation

If we generalize all axioms and hypotheses forming the partial axiomatic theory of a given molecular structure elucidation then we will arrive at the following properties which should be logically analyzed:

- Information is *fuzzy by nature*, *i.e.* there are either two or three bonds between pairs of H-$i$ and C-$k$ atoms associated with a 2D peak $(i,k)$ in the HMBC spectrum.
- Not all possible correlations are observed in the 2D NMR spectra, *i.e.* information is *incomplete*.
- The presence of NSCs frequently results in *contradictory* information.
- The number of NSCs and their lengths are *unknown* and signal overlap leads to the appearance of *ambiguous* correlations. Information is otherwise *uncertain*.
- Information can be *false* if a mistaken hypothesis is suggested.
- Information contained within the "structural axioms" reflects the opinion of the researcher and the information is, therefore, *subjective*, and typically based on biosynthetic arguments.

The incompleteness of the 2D NMR data is illustrated by Figures 2.1 and 2.2. The Figures show the relationship between the number of observed and theoretically possible correlations in both the COSY and HMBC spectra of natural products which were used as problems for challenging the StrucEluc system. The red line corresponds to a ratio of one.

Taking into consideration the information properties above we can assume that the human expert is frequently unable to search *all plausible structural*

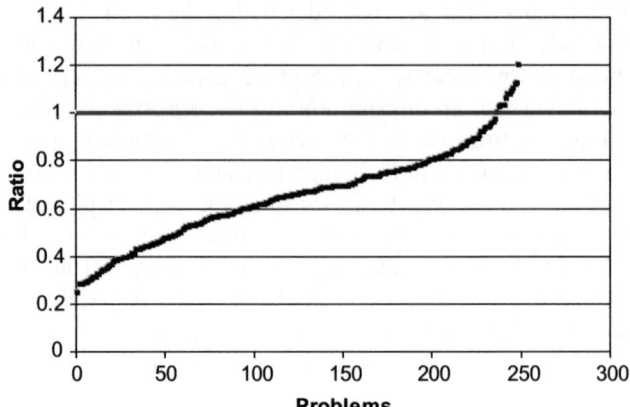

**Figure 2.1**  The ratio of the number of HMBC correlations *vs.* the number of theoretically possible correlations across the 250 natural products used as
problems to challenge the StrucEluc system.[5] The problems are ordered in
ascending order of the ratio.

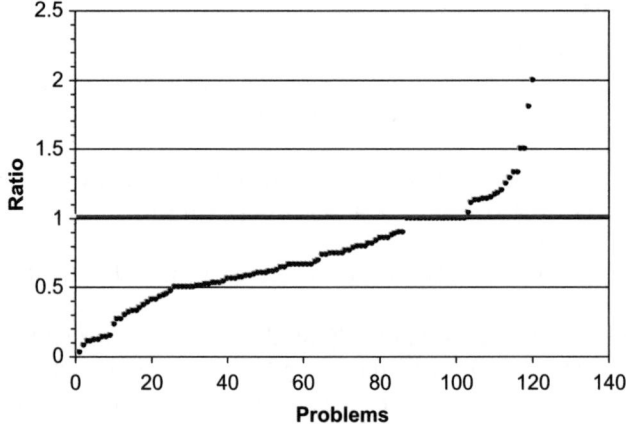

**Figure 2.2**  The ratio of the number of experimental COSY correlations *vs.* the
number of theoretically possible correlations across the 120 natural products used as problems to challenge the StrucEluc system.[5] The problems
are ordered in ascending order of the ratio.

*hypotheses.* Therefore it is not surprising that different researchers arrive at
different structures from the same experimental data and, as a result, articles
revising previously reported chemical structures are quite common (see Chapter
12). Considering the potential errors that can combine in the decision-making
process associated with structure elucidation, it is actually quite surprising that
chemists are so capable of processing such intricate levels of spectrum–structure
information and successfully extracting very complex structures at all. To assist

the chemist to logically process the initial information, a computer program that would be capable of systematically generating and verifying all possible structural hypotheses from ambiguous information would be of value. The expert system StrucEluc,[3–7] which will be described in Part III of this book, comprises a software program and series of algorithms which were specifically developed to process *fuzzy, contradictory, incomplete, uncertain, subjective* and even *false* spectrum–structural information.

# References

1. L. A. Gribov, M. E. Elyashberg and L. A. Moscovkina, *J. Mol. Struct.*, 1971, **9**, 357–371.
2. M. E. Elyashberg, L. A. Gribov and V. V. Serov, *Molecular Spectral Analysis and Computer (in Russian)*, Nauka, Moscow, 1980.
3. M. E. Elyashberg, K. A. Blinov, A. J. Williams, S. G. Molodtsov and E. R. Martirosian, *J. Nat. Prod.*, 2002, **65**, 693–703.
4. M. E. Elyashberg, K. A. Blinov, S. G. Molodtsov, A. J. Williams and G. E. Martin, *J. Chem. Inf. Comput. Sci.*, 2004, **44**, 771–792.
5. M. E. Elyashberg, K. A. Blinov, A. J. Williams, S. G. Molodtsov and G. E. Martin, *J. Chem. Inf. Model.*, 2006, **46**, 1643–1656.
6. K. A. Blinov, D. Carlson, M. E. Elyashberg, G. E. Martin, E. R. Martirosian, S. G. Molodtsov and A. J. Williams, *Magn. Reson. Chem.*, 2003, **41**, 359–372.
7. *Advanced Chemistry Development. ACD/NMR Predictors. Prediction suite includes $^1H$, $^{13}C$, $^{15}N$, $^{19}F$, $^{31}P$ NMR prediction*, 2011.
8. S. G. Molodtsov, M. E. Elyashberg, K. A. Blinov, A. J. Williams, G. M. Martin and B. Lefebvre, *J. Chem. Inf. Comput. Sci.*, 2004, **44**, 1737–1751.
9. S. Berger and S. Braun, *200 and More NMR Experiments: A Practical Course*, Wiley, New York, 2004.
10. N. T. Nyberg, J. O. Duus and O. W. Sorensen, *J. Am. Chem. Soc.*, 2005, **127**, 6154–6155.

# CHAPTER 3

# Methods of NMR Spectrum Prediction and Structure Verification

There are two general categories of problems in molecular spectroscopy amenable to software tools and solutions. The problems that can be addressed algorithmically are those of structure verification and structure elucidation.

Frequently chemists have proposed chemical structures based on their knowledge of the origin or a sample or based on their expectations of the potential product(s) of a particular chemical reaction. Forearmed with expectations of the proposed structure(s), the general workflow is therefore to acquire and then examine the spectral data in terms of the consistency between the proposed structure and the experimental data. This workflow requires experience in spectral interpretation, experimental access to the necessary data and, where appropriate, access to software tools for spectral prediction and comparison. These tools also play a prominent role in CASE expert systems (ES).

Depending on the rigorous nature of the structural constraints imposed by the experimental data, the output file of an expert system may contain tens, hundreds or even tens of thousands of structural formulae. A correct structure cannot easily be distinguished by taking into account changes in the characteristic spectral features of the functional groups and fragments existing in the probable structures. Therefore the selection of the most probable structure is carried out by comparing experimental to predicted spectra, and this step is generally at the conclusion of the expert system workflow.

In the initial development of CASE systems an early posed question was what kinds of molecular spectra would be most amenable to prediction for distinguishing the "best" structural hypothesis among other competing structures? The appropriate methods of spectrum prediction would need to meet, at least, the

New Developments in NMR No. 1
Contemporary Computer-Assisted Approaches to Molecular Structure Elucidation
By Mikhail Elyashberg, Antony Williams and Kirill Blinov
© Royal Society of Chemistry 2012
Published by the Royal Society of Chemistry, www.rsc.org

following requirements: (a) the speed of the spectrum calculation must be fast enough to be applied to large structural files; (b) the calculations should be as automated as possible and, preferably, fully automated; (c) the application of the methods must not depend on molecular size; and (d) the methods must be of high enough accuracy to discriminate between similar candidate structures.

It quickly emerged that only empirical methods of NMR prediction had the ability to address the goals. MS spectral prediction cannot provide theoretical spectra of arbitrary chemical structures, although there are many variables in the mass spectral conditions that can lead to various fragmentation pathways, and fragmentation prediction is used more as a support filter for structure identification. Examination of the potential of IR spectrum prediction shows[1] that semi-empirical methods based on the valence-optical scheme,[2] as well quantum chemical methods, do not satisfy the requirements (a)–(c) and, consequently, they should be declined. In addition, it is challenging to compare experimental IR spectra with calculated IR spectra when band intensities are taken into account; IR spectra of similar isomers frequently differ only in the shape and intensity of some of the absorption bands. In the 1970s the first programs for empirical $^{13}$C NMR chemical shift prediction based on additive rules[3] appeared, and the results obtained looked very promising. These methods of spectral prediction, in contrast to MS and IR spectra, supported the requirements outlined in points (a)–(c) and are amenable to further improvements in accuracy and the speed of chemical shift calculation. During the last two decades quantum mechanical (QM) methods of NMR chemical shift calculation were shown to be capable of predicting the NMR spectra of mid-sized molecules with an accuracy which is sufficient to select the preferable structures among those suggested by the researcher. This improvement was possible due to the development of density functional theory (DFT) methods for nuclear magnetic shielding calculations. These methods are, however, rather time consuming and hence satisfy only the requirement (d). The practical aspects of applying QM methods to solving CASE problems will be discussed in more detail below (see Section 3.3 and Chapters 12 and 13).

As a result of the efforts of many research groups, a series of programs for NMR prediction, both commercial and freely accessible, are now available to the chemistry community. $^{1}$H and $^{13}$C spectra are clearly the primary nuclei of interest and are most utilized by chemists for structure verification. In our experience[4] $^{1}$H NMR is used with at least a 20:1 ratio over $^{13}$C spectroscopy. The development of NMR prediction tools have thus focused on these nuclei. NMR prediction software has been incrementally improved over the past two decades and these tools can now provide reasonable to excellent accuracy in the quality of NMR prediction (*vide infra*). Spectroscopists also regularly make use of other nuclei during their structure elucidation efforts and the predictions of chemical shifts and, in some cases, coupling constants for other nuclei have also been pursued. For example, chemical shift databases and prediction algorithms extracted from these data are available for the following nuclei: $^{11}$B/$^{10}$B[5,6], $^{15}$N, $^{19}$F, $^{29}$Si and $^{31}$P.[5,7–9]

In this Chapter we will discuss software tools that have been developed to allow the prediction of 1D NMR spectra, as well as both homo- and heteronuclear 2D spectra, thereby improving throughput of structure verification.

# 3.1    Methods of $^{13}$C NMR Chemical Shift Prediction

Since carbon generally forms the foundation of a molecular skeleton, $^{13}$C NMR spectroscopy is one of the most powerful methods for elucidation of the structures of organic molecules and for testing structural hypotheses. $^{13}$C NMR spectra can be simulated most easily, since, in the presence of $^1$H broadband decoupling, and assuming the absence of other magnetically active nuclei, the spectra consist of single resonances and each isochronal carbon atom is ascribed to a signal with a specific chemical shift. The multiplicity (C, CH, CH$_2$ or CH$_3$) of an individual resonance can be determined using a myriad of approaches including distortionless enhancement by polarization transfer (DEPT)[10] or multiplicity-edited HSQC.[11] Even in the presence of strong overlap, as long as the overlap is only in one frequency domain, heteronuclear 2D NMR experiments still allow for the determination of resonance multiplicity. The structure verification process compares a calculated chemical shift or spectrum, either 1D or 2D, with the corresponding experimental data and the structural hypothesis is either accepted, rejected or can be revised on the basis of visual inspection or calculated mean or standard deviations.[12] Alternatively, it may be obvious from this analysis that additional homo- or heteronuclear correlation data must be acquired to verify the structure or to test a revised structure. The two most widely used procedures for predicting NMR spectra are the construction of empirical models[3,12–16] and the application of prediction algorithms extracted from data collected within spectral databases.[8,9,17–20] These approaches are provided in a series of commercially available programs.[8,9,21,22] Certain applications[8,23,24] use both approaches simultaneously. During the last decade algorithms for $^{13}$C NMR chemical shift prediction using artificial neural networks (ANNs)[25] have also been developed.[26–38]

## 3.1.1    Additive Rules-based Methods

Grant and Paul[39] suggested the first additive linear model for calculating the chemical shifts of carbon atoms in aliphatic hydrocarbons using increments accounting for environmental effects up to four atoms away from the prediction center. Following their groundbreaking work, linear models were extended to cyclohexanes, alcohols, amines and chlorine-containing compounds. The derived formulae and tables of increments were collected in a number of monographs (see, for example[40,41]). The approach provides reasonable predictions for the narrow classes of structures used for model derivation. The validation of the approach stimulated the development of software for more general structural formulae using an approach of selecting the corresponding additive model for each carbon atom and thereby calculating the chemical shift. This general approach was of limited utility since chemical shifts could be predicted only for sp$^3$-hybridized carbon atoms and the increment values could not be modified.

Fürst and Pretsch[12] removed these restrictive constraints. They used large databases containing structures and their associated assigned $^{13}$C NMR spectra to construct linear models that could be applied to many classes of chemical compounds. The models contain both configuration- and conformation-dependent parameters and take into consideration the configuration of C=C bonds, as well as the presence of axial and equatorial substituents in cyclohexyl rings. The software developed also allows for the modification of parameters, both reference values and increments, and the input of new additivity rules. A test of the program on a set of 170 000 experimentally determined chemical shifts demonstrated that the rules allowed for the calculation of chemical shifts for over 97% of carbon nuclei for which computations were performed. The overall standard deviation of predicted versus experimental chemical shifts was ±5.5 ppm. This program has been incrementally improved and has been commercialized.[21] Currently, the $^{13}$C NMR prediction tool is based on 4000 parameters and it has been reported[22] that over 95% of the shifts can be predicted with a mean deviation of ±3.8 ppm. With these statistics the approach outlined here would be quite adequate for the filtration of an output structure set created by an expert system. As we will see later (Section 3.1.4), recent studies have shown[38,42] that the accuracy of incremental methods for $^{13}$C chemical shift prediction can be improved by a factor of two to provide an average mean deviation of 1.6–1.8 ppm.

## 3.1.2 Fragment-based Methods

In a different approach, databases containing chemical structures and the assigned carbon chemical shifts form the foundation data set to allow for the derivation of prediction algorithms. For every carbon atom associated with each chemical structure contained in the database, atom-centered fragments (ACF) with a prescribed number of concentric layers are generated according to a hierarchical ordering of spherical environments or HOSE code.[43] These fragments, and their corresponding chemical shifts, are stored as an ordered list for use in the prediction algorithms. To predict the spectrum for a candidate structure, the program selects all possible ACFs present in a structure, performs a search for their analogues in the database, and, after statistical processing, ascribes the chemical shifts taken from the reference fragments to the carbon atoms being predicted. If an ACF is not found in the database then the program interpolates using the most similar structural environments available. The results obtained using this approach are generally in good agreement with the experimental data when using a large database containing a diverse set of structures (see, for example[17]). Frequently, the difference between individual predicted and experimental chemical shifts lies within ±1 ppm – limits delivering high prediction accuracy.

Despite the impressive accuracy reported for this approach, it does suffer some drawbacks. The absence of stereochemical information in databases containing only connection tables and not including stereochemistry can strongly affect the prediction quality. It has been shown[44] that neglecting stereochemical effects can increase the deviations between predicted and experimental chemical shifts by more than 10 ppm. This problem was circumvented to some extent by Schutz *et al.*[45] who introduced 3D descriptors to modify the HOSE code. It should be noted that the spectral properties of the reference fragments used to derive the prediction may appear to be unrelated in certain cases but this is simply the nature of the approach. There remain a number of limitations in the algorithm including the fact that the effects of an atom equidistant from another atom can be different in different compounds and this is ignored at the encoding stage. The prediction of a spectrum using this type of approach can take a relatively long time, depending on the complexity of the molecule and the type of computer used. In the early 1980s predictions took tens of seconds to minutes but with modern computers the spectra of even complex organic molecules can generally be completed in less than 10 s.

A method has been described,[24] that is a combination of both of the aforementioned approaches. In the first stage, the environments describing the carbon atoms in the reference and the target (verified) structures are represented as vectors and the Euclidean distance between these vectors is calculated. After repeated application of this procedure, a test set of carbon atoms whose chemical environment is similar to that of the carbon atoms in the target molecule is generated. If the degree of similarity of the chemical environment for the target atom is high then the chemical shift characteristics describing its "twin" in the reference structure is ascribed to the atom in the target structure. Otherwise, a linear model constructed using the test set and descriptors for the topological, geometric and electronic properties of the environment is introduced. Models are constructed using multiple linear regression (MLR), partial least squares (PLS) and principal component analysis (PCA)[46,47] methods. A test of this combined approach on a set of > 38 500 atoms has shown that the mean deviation is ±1.69 ppm, whereas the straightforward database retrieval procedure gave a value of ±1.85 ppm.

A number of commercially available [13]C chemical shift prediction software packages based on the fragment database approach have become available in recent years. The most popular products to date are those of ACD/Labs,[8] Chemical Concepts[9] and Sadtler.[5,6] The authors are familiar with the ACD/Labs product suite and will use these products as examples for future discussion.

The [13]C NMR shift prediction program, ACD/CNMR,[8] has an internal data file containing over 2 540 000 experimental [13]C chemical shifts and over 105 000 coupling constants characterizing interaction with different nuclei. Each $\delta_C$ value is pre-assigned to a specific carbon nucleus in its unique environment. The data file has been generated from the fully assigned spectra of over 200 000 chemical structures. The program uses several different algorithms for estimating $\delta_C$ for those fragments not represented in the internal database of experimental values. Both the algorithms and the details of their derivation constitute proprietary knowledge and have not been described in the literature, although the general concepts are described below.

When a new chemical structure is drawn in the structure-drawing interface of ACD/CNMR, the program automatically splits the structure into a set of unique fragments that are then compared to the structural fragments from the internal database.

- If a fragment from the drawn structure *coincides* with a fragment contained within the database, the program will use this experimental $\delta_C$ value as part of the final set of chemical shifts for the structure. For such $\delta_C$ values the program will not show confidence intervals in the table of chemical shifts. The program utilizes a reference structure of up to 16 spheres in radius for a particular carbon atom. As a result the size of the fragment is defined by the size of the largest fragment common to both the predicted and the reference structure, the fragments being centered on the given carbon atom.
- If some fragments from the structure cannot be found in the internal database, then the program will search for the most similar fragments in the database. First, the program composes sets of fragments from the database that are structurally similar to each of the fragments generated from the analyzed structure. Second, the program estimates $\delta_C$ values for the fragments from the structure using secondary algorithms and compares them to the estimated $\delta_C$ values of fragments selected from the database. This allows the program to narrow down to a set of similar fragments from the database. Third, the program calculates the average values ($\delta_{Av}$) of the experimental data and produces estimated $\delta_C$ values after application of the second criterion described above. The resulting $\delta_C$ value is calculated using both the estimated $\delta_C$ value of the given fragment and the average $\delta_{Av}$ values. The obtained $\delta_C$ values are used to compose the final set of chemical shifts for the structure. After composing the final lists of chemical shifts, ACD/CNMR generates the exact number, location, intensities, and assignment of the spectral lines associated with the structure.

The array of chemical shift, coupling constants, and linewidth parameters describing an NMR spectrum are influenced by many external factors including solvent, concentration, temperature, relaxation times, concentration of paramagnetic impurities, shimming, and observation frequency, to cite just a few. Many of these parameters are simply too complex to take account of during a prediction, but certainly solvent dependence can be accounted for to a certain extent. ACD/CNMR Predictor provides the ability to perform solvent-specific predictions. The user can select from a list of common NMR solvents and predict a solvent-specific NMR spectrum. The stereochemistry of a particular structure is crucial in determining the molecular properties and when the stereochemistry of an atom is included in the submitted chemical structure the information is utilized during the prediction process.

The efficiency and influence of detailed stereochemistry coding of structures included into the database was demonstrated by the CAST [13]C NMR predictor.[48] This database contains about 1500 structures including terpenoids, steroids, macrolides, polyketides, polyethers, and their synthetic intermediates.

Since many basic chemistry research programs are focused primarily on the development of new and novel chemical structures, it is possible that specific fragments are not yet described in the literature and consequently are not contained within the databases used as the foundation of the algorithms. The ideal situation would be to allow a scientist to not only capture and catalog their own structures and assignments but to use this information directly in the prediction algorithms. To this end, ACD/CNMR allows the chemist to create a *user database* of structures and assignments. In this case, the program again splits the structures into unique fragments. As new chemical shifts are assigned to atoms, the program treats these data as an update to its internal database. If a spectrum is predicted for a new chemical structure while the user database is present, the program performs all of the same actions described above, but pays primary attention to the data that have been entered in the *user database*. This form of user database training can have a dramatic impact on the ability of an organization to predict NMR spectra for a diverse array of compounds containing structural moieties that have been represented at a fairly minimal level of 1–2 structures in the training database. With this capability not only are the legacy data generated as a result of the structure elucidation process available for reference through searching, but also it forms the foundation for a large number of organization scientists to improve or benefit from their own predictions.

### 3.1.3   Artificial Neural Networks (ANNs)

Beginning in the early 1990s, the attention of chemists was drawn to the possibilities of promising new mathematical tools developing in computer-based chemistry, *viz.*, ANNs. There was a rapid increase in the number of studies on the application of ANNs to the interpretation, classification, and prediction of spectral data, including $^{13}C$ NMR chemical shift prediction.

A neural network can be considered as a simplified computer model of the human brain consisting of several layers of neurons that send signals to other neurons as a function of the input signals received. A flowchart representing a neural net is presented in Figure 3.1. Typically a neural network consists of an input layer containing a certain number of *input* neurons, hidden layer(s) containing hidden neurons and an output layer. Such networks have a "black box" nature and possess the common ability to construct empirical models of the systems for which theoretical dependencies between the input and output are extremely complicated or even unknown. Models are obtained as a result of training the network. In the course of training, the network is represented in the form of input–output pairs related by a simulated transformation. A network trained using these examples is able to predict the output signals from input examples not presented originally in the training set. The training procedure may be time-consuming (tens of hours), but a network, once trained, generates the result almost instantaneously. For instance, a network can be trained to generate structural information (output) retrieved from a spectrum (input) or to predict a spectrum (output) from structural information (input). The theory of ANNs and examples of their application in chemistry and spectroscopy is described in a monograph by Zupan and Gasteiger.[25]

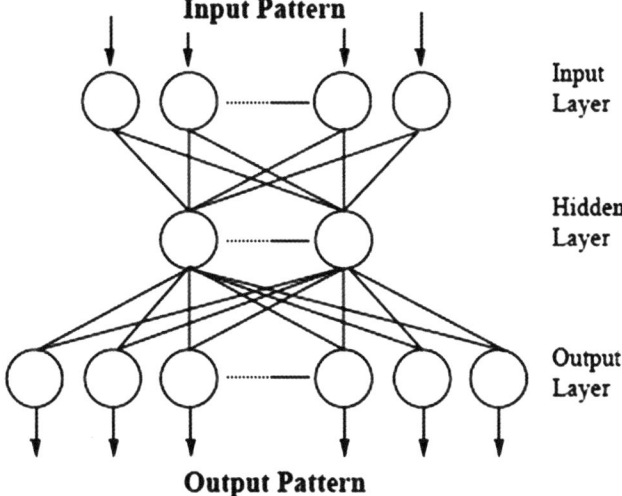

**Figure 3.1** Schematic diagram of a neural network.

In earlier works, attempts to apply ANNs for the prediction of $^{13}$C NMR chemical shifts has been undertaken with regard to some chemical classes of compounds including methyl-substituted cyclohexanes,[30] mono-substituted benzenes,[31] monosaccharides,[34] ribonucleosides,[35] and sp$^2$- and sp$^3$-hybridized α-atoms in acyclic alkenes,[36,49] *etc.* as examples.

ANNs are trained to predict the spectra of compounds *via* a training set including encoded structures and their associated spectra. In the course of training, the network uses reference structures as input information and the output signals are compared with the $^{13}$C NMR spectra of these structures. The training process is complete if the deviations of the predicted spectra from the reference set are less than a chosen threshold. The results of computational experiments performed by Ivanciuc *et al.*[49] will be considered as an example.

The authors[49] trained a network to predict the chemical shifts of sp$^3$-hybridized carbon atoms in the α-position with respect to the double bond in acyclic alkenes. In parallel with the training, a MLR-based model (see above) was developed. Testing of both models showed that the standard deviations were within the range of 1.3–1.5 ppm for the ANN and 3 ppm for the MLR model. This work indicated that ANNs show a better performance than linear models in terms of spectral prediction.

Until recently ANNs were applied to predict the spectra only of compounds belonging to those classes of molecules on which the networks were trained. In these cases the structures included in the output answer files created by the expert system may contain a number of unexpected fragments.

As shown earlier, $^{13}$C NMR spectral prediction approaches relying on structural databases produced predicted spectra for diverse structures with a precision sufficient for practical needs. The main shortcoming of this approach, however, is the speed of prediction. For instance, ACD/CNMR Predictor takes about 2 to 6 s to calculate the $^{13}$C NMR spectra of the typical natural products shown below when both structures are absent from the database: dimethyl-winkleridine (**3.1**, $C_{22}H_{35}NO_6$) and azadirachtin A (**3.2**, $C_{35}H_{44}O_{16}$) (Pentium IV, 2.8 GHz).

**3.1**

**3.2**

This speed of calculation is quite acceptable for single calculations or even for batch predictions when the number of structures is not large, for example, 10–100 structures. However, if the intent is to apply a spectral-prediction program to a large structural file containing large numbers of hypothetical structures, then achieving the same high calculation precision becomes a very time-consuming issue and the application of faster methods is desirable. As mentioned earlier ANNs could be used for this purpose but the programs described are applicable only to specific classes of organic compounds. The ideal of high speed and high precision was pursued by Meiler *et al.*[26] in their development of "C_SHIFT" program.

## 3.1.3.1 ANN Algorithms of C_SHIFT Program

In order to calculate the chemical shift values of carbon atoms with ANNs it is necessary to describe the atom types and chemical environments of all atoms numerically. In doing so, an optimum number of carbon atom descriptors must be found. The number must be as small as possible for computational reasons while at the same time the descriptors must clearly distinguish differences in molecular structures. In the study by Meiler *et al.*[26] 28 descriptors were determined for 28 different atom types (see Table 3.1). The number of representative atom types identified in a database containing approx. 40 000 molecules taken from[9] is indicated in Table 3.1. The atom types are derived from element number, hybridization state, and number of attached hydrogen atoms. Nine different atom types for the more than 525 000 carbon atoms available for the prediction of $^{13}$C NMR chemical shifts were extracted in total.

**Table 3.1** Atom types (1–28) and "sum-types" (29–30) defined to describe the atom environments and their frequencies in the data set.

| ID | Atom Type | Frequency of atoms of this type |
|---|---|---|
| 1 | >C< | 19527 |
| 2 | >CH- | 49556 |
| 3 | -CH$_2$- | 116175 |
| 4 | -CH3 | 73724 |
| 5 | =C< | 50711 |
| 6 | =CH-/=CH$_2$ | 27556 |
| 7 | ≡C-/≡CH/=C= | 3793 |
| 8 | >C- (aryl) | 68416 |
| 9 | >CH (aryl) | 117107 |
| 10 | >N- | 9876 |
| 11 | -NH- | 9115 |
| 12 | -NH$_2$ | 3521 |
| 13 | =N-/=NH | 7427 |
| 14 | ≡N | 2053 |
| 15 | -NO$_2$ | 2688 |
| 16 | >N (aryl) | 3743 |
| 17 | -O- | 31641 |
| 18 | -OH | 20626 |
| 19 | =O | 39259 |
| 20 | >P-/-PH-/-PH$_2$ | 383 |
| 21 | >PO- | 1053 |
| 22 | -S-/-SH | 4146 |
| 23 | =S | 1214 |
| 24 | >SO$_2$ | 1789 |
| 25 | -F | 3613 |
| 26 | -Cl | 9585 |
| 27 | -Br | 2718 |
| 28 | -I | 603 |
| 29 | Sum of all hydrogen atoms bound in this sphere | |
| 30 | Sum of all ring closures in this sphere | |

Two additional descriptors were introduced in order to obtain a more complete description. One descriptor holds the number of all hydrogen atoms located in the individual sphere (type 29 in Table 3.1). A second descriptor identifies the number of ring closures (type 30) in order to identify the influence on the $^{13}C$ NMR chemical shift caused by the formation of rings.

The chemical environments of the carbon atoms are described by sorting the atoms into spheres as shown in Figure 3.2. The occurrence of every atom type in each sphere is counted. As shown, for example, by position 2 in Figure 3.2, five spheres (I to V) are formed for every carbon. All atoms outside the five-sphere radius are projected into an additional "sum sphere" ($\Sigma$). Consequently, 30 numbers for the atom types listed in Table 3.1 are necessary to describe one sphere, and 180 numbers are necessary for the complete description of the environment of an individual carbon. This description is based on the number, kind, and distances of the substituents, and is not sufficient so the descriptors are extended by a so-called "$\pi$-contact area" to take conjugated $\pi$-electronic systems into consideration. For example, to describe the environment of carbon C-2 shown in Figure 3.2, the double bonds in the side-chain must not be considered as a $\pi$-contact, since they do not belong to a conjugated $\pi$-electronic system that includes one neighbor of the C-2 carbon under consideration. Otherwise, two $sp^3$-hybridized carbon atoms would have to be considered. The C-6 and C-9 carbon centers are neighbors of the conjugated $\pi$-electronic system (C3=C4−C5=O), as is the example carbon C-2. The molecular structure descriptors are extended by a second set of 30 numbers for each sphere using the atom types and sum parameters described in Table 3.1. In contrast to the general count associated with the first set, only the atoms in $\pi$-contact with the carbon under examination are considered now. Consequently, for each sphere

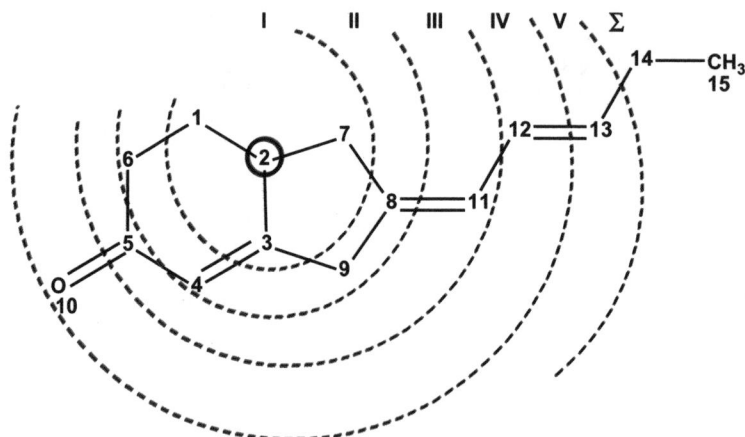

**Figure 3.2**   A schematic representation of the spherical division of the chemical environment of a carbon atom in a molecule. Carbon 2 of the substituted 1,2,3,6,7,7a-hexahydroinden-5-one is the center of attention for this schematic.

two sets of descriptors are used to encode the chemical environment of a given carbon atom.

The $^{13}$C NMR chemical shifts associated with the 40 000 molecules were utilized for a series of calculations and the assignments were related to the nine carbon atom types listed in Table 3.1. All chemical shifts were estimated in either CDCl$_3$ or CCl$_4$ and referred to tetramethylsilane (TMS) as the internal standard. Stereochemical information was not available for the molecules in the database. The environment of every individual carbon atom was encoded with the descriptors from Table 3.1. Three hundred and sixty descriptors were used as the *input vector* for the neural network with the individual $^{13}$C NMR chemical shift value representing the output. For *each of the nine* atom types representing a carbon atom, an individual neural network was constructed and trained using the back-propagation of errors approach.[47] The set of molecules was randomly subdivided into three sets. A set of 90% of the data was used for training the neural networks. A second data set contained 7% of the available data was used for monitoring. The training and monitoring data sets were used simultaneously. A third data set consisting of the remaining 3% of randomly selected molecules was used as an independent testing set for the trained networks.

The number of input and output units for the neural network was fixed while the number of hidden neurons was determined experimentally. The best results were achieved with 5–20 hidden neurons, depending on the carbon atom type. A mean deviation of ±1.79 ppm and a standard deviation of ±2.10 ppm resulted for the more than 15 000 carbons in the test data sets. The accuracy of these calculations was much better than has been attained until the past decade by prediction algorithms based on fixed increments, especially if complex structures are investigated. The largest deviations were obtained for the sp-hybridized carbons (with a standard deviation of ±3.8 ppm for type 7) and for carbons with sp$^2$ hybridization (standard deviations of ±3.7 and ±3.6 ppm for types 5 and 6), respectively.

All of the effects influencing chemical shifts through space were not considered. Therefore, comparable carbons in different stereoisomers and conformers are not distinguishable through their different spatial environments. It is known that one observes an increase in the uncertainty of prediction for sp$^3$-hybridized carbons with an increasing number of non-hydrogen substituents (more spheres should be taken into account). Here, the standard deviation increases from ±1.3 ppm for methyl group carbons up to ±3.4 ppm for the quaternary carbons.

To test the capability of the method for the prediction of $^{13}$C NMR chemical shifts, the authors[26] used the molecule epothilone-A (**3.3**) as a test example. This molecule was not present in either the database or the training set. Experimentally determined $^{13}$C NMR chemical shifts[50] were compared with the values calculated by use of the neural networks[51] and the values predicted using a HOSE code-based approach. Only experimental chemical shifts measured in DMSO were available. As a result, the agreements between the experimental and computed values were not as precise as would be expected for predictions in a non-polar solvent such as chloroform. This restriction, however, affected both approaches. The standard deviations were ±2.5 ppm for the neural networks and ±2.4 ppm for the HOSE code approach.

**3.3**

While the results are of comparable quality, the authors,[26] however, found that with the neural network approach the chemical shifts were computed 100 to 1000 times faster compared to the HOSE code-based determination. The chemical shifts determined with the increments show a larger deviation (±3.8 ppm), as expected. All three methods have difficulties with the conjugated π-system in the heterocyclic ring. Some of the relatively large deviations observed for the results obtained with the neural networks were explained by stereochemical influences. The work[26] showed that neural networks trained by a chemical shift database with significant structural diversity could be successfully used for screening the results of structure generation in CASE systems.

### 3.1.3.2 Improvement of the C_SHIFT Program

The approach described above was improved in the next report devoted to the C_SHIFT program.[27] Improvements were made to account for deviations obtained with conjugated systems, tertiary, and quaternary carbon atoms, small ring systems, and double-bonded carbon atoms. To address the shortcomings for these environments the description of the atom environment was modified. The number of spheres considered in the code was increased from 5 plus the sum-sphere to 7 plus the sum-sphere. This modification ensured that the influence of a conjugated π-electron system on the chemical shift was addressed.

To improve the prediction of tertiary and quaternary carbon shifts, the individual substituents of a carbon atom were handled separately; an input vector was computed and input neurons were provided for each substituent of the carbon. As a result, possible interactions between these substituents were described more precisely than in the original code. A further modification was suggested by the problems that occur in predicting chemical shifts for small ring systems, conjugated π-electron systems, and highly substituted carbon atoms. It was realized that the important interactions and influences of the inner sphere atoms were not properly detected by the sum parameters (29 and 30). Inner atoms were therefore coded individually by introducing eight parameters: the number of valence electrons, period, electronegativity, van der Waals radius, hybridization, bond type to previous atom, number of bonded hydrogen atoms, and ring closures. In addition, the number of atom classes was enlarged from 28 to 30 to include azides -N=N=N and sulfoxides >S=O. The resulting code for a carbon substituent was therefore lengthened to 424 descriptors. Overall, a set of 1.3 million descriptors was constructed out of the SPECINFO database. The same methods as used in the previous article[26] were used for neural network training.

The trained neural networks were implemented in a newer version of the C_SHIFT software.[26] Only a slightly better average deviation of ±1.6 ppm was obtained in comparison to the former neural network (average deviation = ±1.8 ppm), but marked improvements were obtained only for some structural fragments. This is explained by the fact that NN algorithms are of a "black box" nature and it is difficult to predict, *a priori*, the results of introducing new parameters into the atom coding.

In general ANNs deliver results that are qualitatively similar to HOSE code predictions using the complete SPECINFO database, but offer the benefit of being database-independent and about 1000 times faster[27] than HOSE code predictions.

Meiler *et al.*[52,53] have also compared nine methods for predicting the $^{13}$C NMR chemical shifts of small organic molecules and have used the example of the drug Taxol (**3.4**). These methods included ANNs (two versions of C_SHIFT[26]), incremental methods (CHEMDRAW,[21] SpecTool[22]), QM calculation (GAUSSIAN,[54] COSMOS[55]) and HOSE code fragment-based prediction tools (SPECINFO,[9] ACD/CNMR,[8] BIO-RAD Predict-It NMR[56]).

**3.4**

In order to provide a fair comparison of the quality of each prediction method for an unknown and new class of organic substances, it would be necessary to remove all Taxol derivatives from program internal databases as used by SPECINFO, ACD/CNMR, and Predict-It NMR, and indirectly from C_SHIFT, CHEMDRAW, and SPECTOOL while training these methods.

The accuracy achieved by all tested methods of $^{13}$C chemical shift calculation is presented in Table 3.2.

Table 3.2 shows that both the standard and mean deviation of the ACD/CNMR method fall in the same range as the SPECINFO HOSE code prediction and also with the neural network approach used by C_SHIFT. This comparison shows that the HOSE code and neural network approaches, at least for this example, are comparable in performance. The Table shows that incremental methods such as those used in CHEMDRAW and SPECTOOL supply the worst results for this fairly complex structure (standard deviations ±5.6 ppm and ±5.3 ppm, respectively). COSMOS and GAUSSIAN, both quantum chemical programs, also lead to rather large deviations (standard deviations ±4.6 and ±5.1 ppm, respectively), which can be explained mainly by incorrect representation of the 3D structure rather than by the prediction method itself.[27]

### 3.1.3.3   NN vs. Fragmental Approaches: Merits and Demerits

It should be noted that the HOSE code methods relying on fragment databases are readily capable of incorporating new structures while neural networks would need to be retrained dynamically to adjust the prediction algorithms to novel chemical classes. In contrast, quantum chemical methods do not need any adjustment for the prediction of new classes of substances.

The important advantage of the HOSE code-based approach is the large diversity of structures used for the spectrum prediction. Broad chemical diversity can provide a higher probability of obtaining the smallest deviation for the $^{13}$C NMR spectrum calculated for an expert system output file of generated structures.

A particularly useful feature of these software programs includes the ability to review the details of how a predicted spectrum is generated. For example, in the ACD/Labs NMR prediction programs[8] a "chemical shift calculation protocol" can be displayed. When a prediction is performed on a chemical structure that is absent from the database then the different structures used to produce the predicted spectrum are indicated. This calculation protocol allows comparison of the environments of the carbon atoms for which the chemical shifts are calculated with the environments and chemical shifts of the related structures used for prediction. The analysis of the spectrum–structural information presented in the protocol dialog box allows the chemist to validate the accuracy and reliability of an NMR prediction.

**Table 3.2** Comparison of accuracy of $^{13}C$ chemical shift prediction for Taxol performed by different methods.

| Type of deviation | NNs | | Increments | | HOSE code predictions | | | QM | |
| --- | --- | --- | --- | --- | --- | --- | --- | --- | --- |
| | C_SHIFT (new) | C_SHIFT (old) | CS CHEM DRAW PRO | SPECTOOL | SPECINFO | PREDICT-IT NMR 1.3 | ACD/CNMR 6.12 | COSMOS 4.5 | GAUSSIAN 98 |
| Std dev. | **1.3** | 3.1 | 5.6 | 5.3 | **1.0** | 3.7 | **1.3** | 4.3 | 5.1 |
| Aver. dev. | **1.0** | 2.1 | 3.5 | 3.3 | **0.8** | 2.3 | **0.9** | 3.3 | 3.9 |
| Max. dev. | **3.4** | 11.6 | 19.0 | 17.0 | **3.0** | 11.5 | **4.4** | 14.5 | 23.6 |

For example, for the carbon nuclei at 144.9 and 90.5 ppm indicated with arrows in Structure **3.5** the chemical shift assignment has been performed.[57]

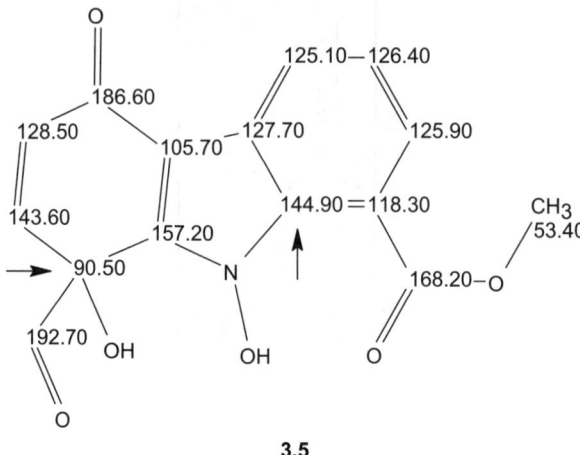

**3.5**

The ACD/CNMR program predicted values of 131.0 and 74.3, respectively. The cause of the large discrepancy between the experimental and predicted shifts is clarified by the calculation protocol. The protocol produced by the program for the carbon atom at 144.9 ppm is presented in Figure 3.3.

The protocol indicates that 3482 hits were selected from the database for prediction of the chemical shift of the carbon nucleus under consideration. The histogram indicates the distribution of hits with chemical shifts assigned to representative carbon nuclei used for prediction. In the left part of the Figure an example structure from the set of hit structures is shown. The fragment centered on the carbon atom of interest is colored red. The histogram shows that the majority of hits have chemical shifts near 135 ppm for this atom, and the calculated value is $\sim 131$ ppm. The unusually large chemical shift can probably be explained by the neighborhood of the N-OH group, which rarely occurs in molecules of natural products.

When the protocol related to the carbon atom with an experimental shift of $\delta_C = 90.5$ ppm was requested, the program displayed the following message (Figure 3.4).

It follows from this message, that no structure was found that contains a carbon-centered fragment suitable for chemical shift calculation for the given atom. Therefore, the chemical shift calculation was performed using an incremental approach that delivered $\delta_C = 74.26$ ppm and the consequent discrepancy relative to the assigned shift.

As mentioned, the neural networks function as a "black box" and do not provide any access to the details associated with how the chemical shift for a given carbon atom is calculated. However, as stated previously, neural networks are significantly faster for $^{13}$C NMR chemical shift calculation and this makes them more effective in the prediction of assignments for large files of

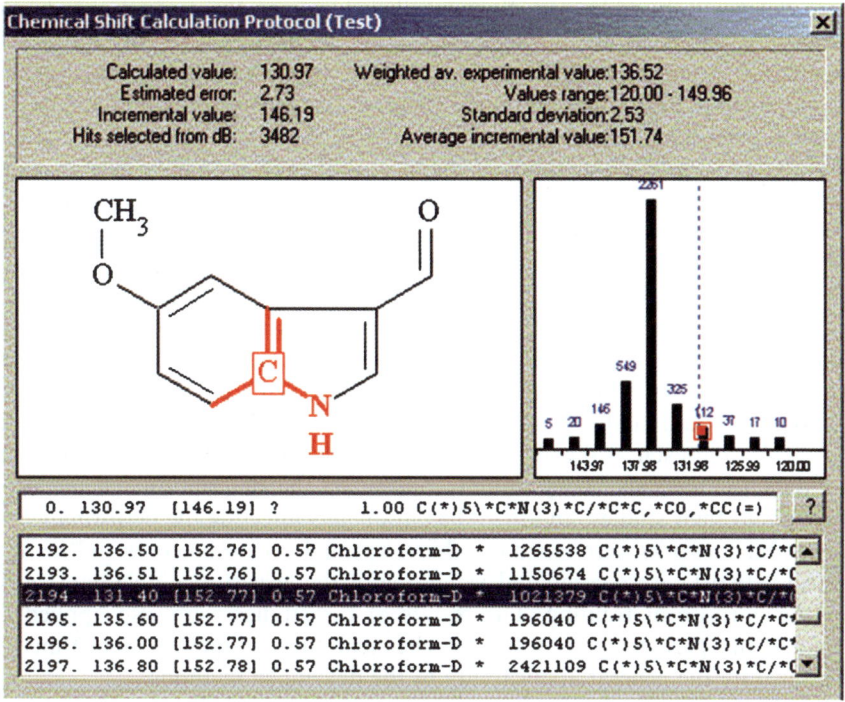

**Figure 3.3** An example of a chemical shift calculation protocol.

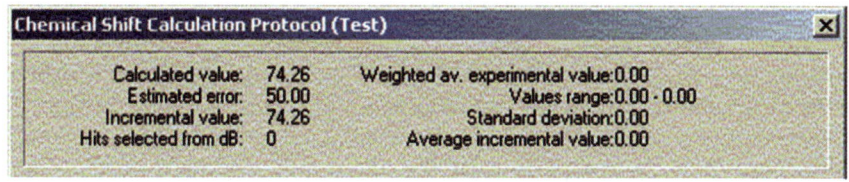

**Figure 3.4** A chemical shift calculation protocol which is displayed when no structures are found to support chemical shift prediction.

chemical structures. Nevertheless, as we will see later (Part III), a rational combination of different approaches – an incremental approach, an approach based on a fragment library, and ANN methods – provides an optimal strategy for identification of the correct structure, even in large output files containing thousands of structures.

## 3.1.4 A New Fast and Accurate Algorithm for $^{13}C$ Chemical Shift Prediction

Despite the intensive research invested in the field of neural networks (NN) and PLS regression applications to $^{13}C$ chemical shift prediction, a number of issues were not yet fully addressed. Successfully addressing these questions should allow researchers to create faster and more accurate algorithms for NMR prediction. Some important problems have been resolved in the further development of algorithms and software programs incorporated into the ACD/ NMR predictor software program. In so doing, more sophisticated algorithms of NN- and PLS-based methods of chemical shift prediction have been created, and the performances of both approaches were comprehensively compared.

The results of chemical shift prediction performed by PLS and NN depend both on the methods of encoding chemical structures into a numerical input and on the associated computational approaches. To provide for more transparent comparison between PLS and NN it is necessary to separate these two parts of the algorithms. These two parts were separately evaluated and compared by Smurnyy *et al.*,[37] and further a robust, fast PLS algorithm for prediction was additionally optimized and examined by Blinov *et al.*[38] Smurnyy *et al.*[37] addressed each of the abovementioned issues using [13]C chemical shifts as the data for analysis. They initially optimized the description scheme used throughout their studies, and then the parameters affecting the performance of the neural networks were tuned to achieve optimal performance. Both PLS and neural networks were compared using one of the largest available databases[8] for the purpose of training. The results of the studies were also applied to examine the performance of the [1]H chemical shift prediction algorithms to demonstrate the general applicability of the developed methods.

### 3.1.4.1 Databases

The carbon and proton chemical shift databases used in work[37] were comprised of approximately 2 million [13]C chemical shift values (note that Meiler *et al.*[26] used approx. 525 000 [13]C chemical shifts). Care was taken to avoid overlap between the datasets used for NN training and algorithm comparison. The *training dataset* was compiled using experimental data published from the early 1990s until 2004. A comparison of the algorithms was performed using a completely independent database compiled from data originally published in 2005 and consisting of approx. 118 000 [13]C chemical shifts.

### 3.1.4.2 Data Encoding

To encode a chemical structure into a numerical representation an atom-based scheme was used as shown in Figure 3.5.

The environment surrounding an atom is divided into spheres, each including all atoms and separated from the center by a definite number of covalent bonds. Typically, it accounted for the nearest six spheres or less. Every atom within a sphere is classified into one of the predefined atomic classes described in Table 3.3. The scheme was inspired by the works of Meiler *et al.*,[26,27] but Smurnyy *et al.*[37] have added extra features, such as a more

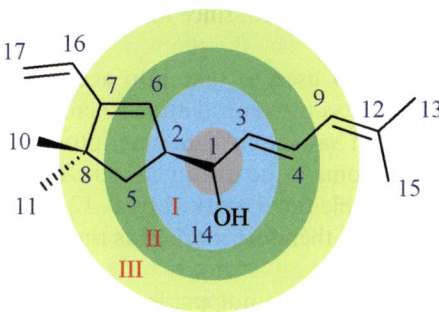

**Figure 3.5** The encoding of the atom environment. The three nearest spheres are indicated in cyan, green and yellow circles, and the Roman numerals denoting the number of each sphere are shown. Atom no. 1 is the central atom. The blue numbers are assigned arbitrarily and serve as references.

**Table 3.3** Atomic classes used to classify atoms. The adjective "aromatic" means that the atom is within an aromatic system. The symbol "(n-)X" designates that $n$ single bonds are attached to the X atom.

| | |
|---|---|
| Carbon | $sp^3$ (C, CH, CH$_2$, CH$_3$), $sp^2$ (C, CH, CH$_2$), sp (C, =C=), aromatic (C, CH), carbonyl. |
| Heteroatoms | (3-)N, (2-)NH, NH$_2$, =NH, N(sp), N(V), aromatic N, (2-)O, OH, =O, (3-)P, P(V), (2-)S, =S, S(IV), S(VI), F, Cl, Br, I. |
| Exotic elements | Si, Ge, Sn, (2-)Se, =Se, (2-)Te, =Te, B, As(III), As(V), Sb(III). |
| Solvent | CHCl$_3$, CH$_2$Cl$_2$, C$_6$H$_6$, (CH$_3$)$_2$SO, dioxane, CH$_3$OH, CH$_3$NO$_2$, tetrahydrofuran, cyclohexane, (CH$_3$)$_2$CO, CH$_3$CN, (CH$_3$)$_2$CONH$_2$, pyridine, CF$_3$COOH, CH$_3$COOH, C$_6$H$_5$NO$_2$, C$_6$H$_5$CH$_3$, CCl$_4$, CS$_2$, H$_2$O, other/unknown. |
| Stereochemistry | Stereoconfiguration relative to double bond; stereoconfiguration relative to 3–6-membered rings. |
| Other parameters | Formal positive charge on atoms; formal negative charge on atoms; involvement into the same π-conjugated system with the central atom. |

complete list of non-typical atoms and the ability to take 20 frequently used solvents into account in the prediction algorithm.

Some additional parameters were introduced to improve chemical shift prediction for atoms included in conjugated systems. For the purpose of this work a conjugated system was defined as atoms forming conjugated double bonds plus all of their immediate neighbors. If a central atom participates in a system, then all of the other atoms in the system are marked with a special flag. For example, in the molecule shown in Figure 3.5 the central atom marked as no. 1 is part of a conjugated system. As a result, atoms 3, 4, 9 and 12 (the diene system) and their neighbors 1 (the center), 13 and 15 will be marked. Note that atoms 6, 7, 16 and 17 are not marked, since they are separated from the central atom by two σ-bonds.

Additional flags were also used to take into account double-bond stereochemistry. If we use atom 4 as the central atom, then atoms 13 and 15 both lie in the third sphere and both have equivalent descriptors. The addition of stereo descriptors allows these atoms to be distinguished. Atom 15 lies on the same side as the atom of the double bond marked by 9–12, whereas atom 13 lies on the opposite side. Atom 15 is therefore marked as the Z atom, whereas atom 13 is marked as the E atom.

Stereochemical descriptors were not implemented systematically throughout the system. However, a separate flag is set for atoms which are one bond away from a 3–6-membered aliphatic ring. These atoms are classified as located either on the same or opposite side of a ring in comparison with position of central atom. In the molecule shown in Figure 3.5 atom 10 lies on the same side of the five-membered ring made up of atoms 2, 6, 7, 8 and 5, whereas atom 11 lies on the opposite side relative to the central atom 1. This method obviously can only be used for relatively rigid rings and is inapplicable to the stereochemistry of flexible systems such as large rings and chains.

"Non-additivity" of substituents influence on chemical shifts was also taken into account. Generally, the total effect from several substituents does not equal the sum of effects of each substituent alone. Special factors, so-called "cross-increments", take into account this feature and improve quality of prediction. These refer to pairs of atoms; for each two atoms separated by a definite small number of bonds an independent identifier is generated and stored. In this study,[37] pairs separated by 1–3 bonds, or by 1–4 bonds in conjugated systems were considered, with both atoms located within the first 3–4 spheres. For example, for the central atom 1, the following pairs of atoms: 3–14, 3–2 and 2–14 (in the first sphere) and 3–4, 2–6, 2–5, 3–6, 3–5, 2–4, 5–6, 4–14, 5–14 and 6–14 in the second sphere should be taken into account. Atom pairs 4–5 and 4–6 are ignored, since the distance between the atoms (4 bonds) is too long in this case. The distances between atoms, as well as sphere numbers, are also used to describe cross-increments. For example, the atoms of the pair 3–4 are both CH ($sp^2$ hybridized) and lie in the first and second spheres correspondingly and are separated by one bond. The atoms in the pair 3–6 are also both CH ($sp^2$ hybridized), lie in the first and second spheres and are separated by three bonds. Obviously, the fact that these atom pairs are not equal and the distance between atoms in a pair allows distinguishing described types of cross-increment.

### 3.1.4.3  Computational Experiments

Neural network[58] and regression[59] algorithms were implemented into the corresponding computer programs and were executed on PC computers operating with clock speeds of 1.8–3.2 GHz and with 1–6 GBytes of memory installed.

All of the neural networks tested in work[37] include an input layer (with different number of input nodes), 1–4 hidden layers, and a single output neuron. Logistic or hyperbolic tangent activation functions were used. Networks were trained using the standard back-propagation algorithm.[58] Neural network input and output values were assumed to be within the interval of [0;1]. For the network inputs $x^i$ the corresponding values of an average of $a_{aver} = x^i_{aver}$ and the associated standard deviations were calculated using all available training patterns in which the input has a non-zero value. The values for the chemical shifts were calculated using all available training patterns and subsequently used for reverse transformation from the [0;1] output to a chemical shift value in ppm delivered to the user.

### 3.1.4.4 Optimization of the Neural Network Performance: Small Data Set

Since the performance of neural network algorithms depends on a large set of factors, the authors[37] first performed a set of test calculations to try and establish an optimal parameter set using a smaller test dataset. For this purpose, 32 000 structures from the main training database were randomly selected. The issue of whether the neural network would benefit from the inclusion of cross-increments was a particular question of interest. As described earlier, these are the parameters that take into account the influence of atom pairs separated by one or more bonds on the chemical shift values predicted for a given carbon atom.

Using a MATLAB implementation of the neural network algorithms, eighteen test runs were set up while varying a few parameters in order to investigate the variation in performance as a result of changes in the different parameters.

The following parameters were examined:

- The number of cross-increments varied as: (1) none, (2) pairs of atoms separated only by one bond or (3) pairs of atoms separated by up to two bonds.
- The geometry of the neural network: (1) 50 hidden neurons in one layer, (2) 30 hidden neurons in *one* layer or (3) *two* hidden layers of 20 and 10 neurons.
- The transfer function:[25] either the logistic function $f(x) = (1 + e^{-x})^{-1}$ or the hyperbolic tangent function (*tanh*).

It was determined that taking into account a reasonable number of cross-increments led to a minimum mean error. At the same time, the arrangement of the neurons into layers appeared to play a somewhat secondary role, since the difference between the networks with the same total number of neurons in the hidden layers but with different geometries is negligible.

These results defined a reference point for further studies. When scaling up the calculations to approx. 2 000 000 $^{13}C$ chemical shifts, the authors[37] mainly utilized networks with 1–3 hidden layers and with a logistic transfer function.

## 3.1.4.5   Choosing the Optimal Number of Sub-databases

Although using a large database is likely to be of value for analysis purposes, the researchers chose to split it into a number of smaller subsets. The most appropriate strategy for splitting a large database is based on the nature of the central atom type. All carbons can be classified according to hybridization, the number of attached protons or both. Also, some groups are so specific and abundant, for example a carbonyl group, that they might deserve a separate class.

Four different splitting strategies were examined:

a. The database was used as a whole.
b. The carbon atoms were classified *only* according to hybridization: three aliphatic classes ($sp^3$, $sp^2$ and sp) and one for all carbons within aromatic rings to give a total of four different classes.
c. The hybridization and the *number of attached protons* were both taken into account. This led to ten classifications: four $sp^3$ (C, CH, $CH_2$, $CH_3$), three $sp^2$ (C, CH, $CH_2$), sp and two aromatic (C, CH).
d. The presence or absence of a heteroatom within one bond of a central atom was additionally taken into account. The fifteen resulting classes include: aliphatic $sp^3$ (seven classes: C, C(het), CH, CH(het), $CH_2$, $CH_2$(het), $CH_3$), aliphatic $sp^2$ (four classes: C, CH, $CH_2$, CO), aliphatic sp, and three aromatic classes: C, C(het), CH. The symbol (het) denotes a heteroatom, oxygen or nitrogen, in this case, nearby.

Some classes were merged into one sub-class, for example, aliphatic $sp^3$ $CH_3$ and $CH_3$(het) because a smaller class was too small or not diverse enough to reliably teach a network, especially one with a higher number of neurons.

It appears that more detailed classifications bring more flexibility. For example, one can introduce specific description schemes for certain atomic types – chemical shifts of the methyl group might highly depend on stereochemistry and it is possible to introduce stereodescriptors specifically for this atomic type. Smaller databases are also easier to handle in terms of computer memory requirements. However, restricting the training set to very similar compounds prevents a neural network from making generalizations and decreases the quality of the results.

The results of the computational experiments using a NN approach are presented in Figure 3.6.

These results suggest that for a large enough database, approx. 50–100 thousand compounds, the errors are only slightly dependent on the classification scheme. For the largest database utilized here, over 207 000 compounds, the difference between the best classification (strategy b, four classes), giving an error of 1.65 ppm, and the worst (strategy a, one class), giving an error of 1.75 ppm, was only 0.10 ppm. This difference is not practically important for carbon NMR shifts. The results obtained allowed the use of the one class strategy in ACD\CNMR Predictor.

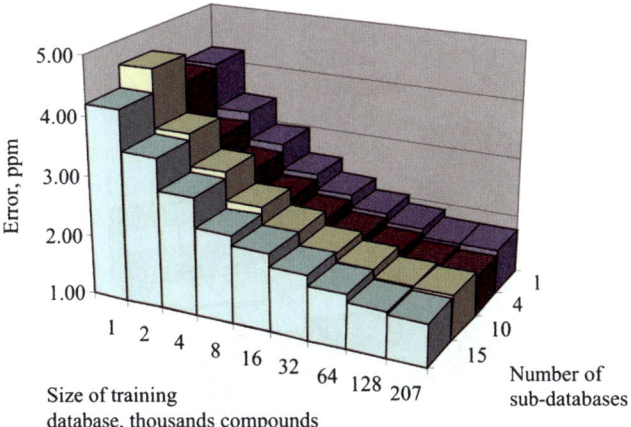

**Figure 3.6** The dependence of the mean error derived from the test set for a number of sub-databases relative to the size of the training set. The results were obtained with neural networks containing 30 neurons in one hidden layer. The splitting schemes leading to one, four, ten and fifteen different classes are described in the text.

## *3.1.4.6 Choosing the Optimal Number of Cross-Increments: Small Dataset*

Using cross-increments is one of the few ways to boost the performance of a regression-based scheme. Although the approach is less popular with neural networks, it was illustrated above that the latter can also benefit from explicitly specified cross-increments. At the same time, providing too many increments and cross-increments leads to an unnecessarily detailed description of a structure and, consequently, to neural net overfitting.[25] Using a small part of the whole database, 16 000 compounds with 212 000 chemical shifts, Smurnyy *et al.*[37] systematically changed a few parameters related to increments and cross-increments in order to elucidate an optimal scheme. Since it is possible to vary the number of atomic spheres under consideration, values from 3 to 7 were tried. Due to limited computational resources it was impossible to generate all possible cross-increments between atoms up to the 6th and 7th spheres. Hence cross-increments were created for atoms located no further than the 3rd sphere (see Figure 3.5). The maximum number of bonds between the atoms in the cross-increments was varied from 0 (no cross-increments) to 4.

For these tests the main training database was split into ten sub-databases (splitting strategy c) and a **PLS** routine was used for carrying out computational experiments.

Figure 3.7 summarizes the findings. The best result was obtained from a combination of six spheres and three bonds to provide a mean error of 2.27 ppm. Practically no difference is observed between the two largest sets of increments with up to three or four bonds between the atoms, and this suggests

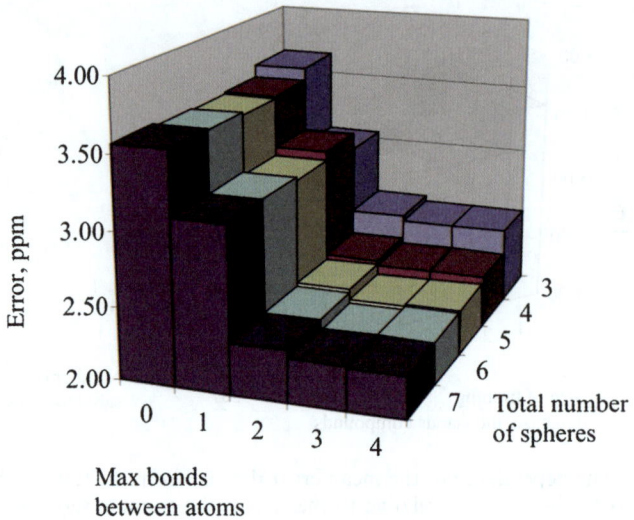

**Figure 3.7**    The *mean error* of a set as a function of the *maximum* number of bonds
between the atoms in the cross-increments and the number of spheres used
to describe an atom's neighborhood.

that further refinement might not be necessary and can lead to overfitting.
Moreover, to speed up calculations a distance of two bonds can be chosen. The
same applies to the total number of spheres – four are enough.

This makes perfect chemical sense, since there are few electrostatic interactions or conjugation, inductive or mesomeric effects in aromatic systems that
span across more than five to six covalent bonds.

### 3.1.4.7    Final Optimization of Parameters: Large Dataset

For evaluation of the performance of the neural networks the following
parameters were varied:

    a. The size of the network while varying the number of hidden layers,
    b. The quantity of neurons in the hidden layers,
    c. The conditions of cross-increment selection.

All other parameters, such as the transfer function, the number of sub-
databases, *etc.* were taken as optimized during the previous steps. An independent test set of 118 000 individual $^{13}$C chemical shifts was used, and calculations were performed with 15 individual sub-databases as described above
(splitting strategy **d**). All cross-increments were constructed from atoms separated by not more than one covalent bond and located within 1–3 spheres from
the central atom.

As shown in Figure 3.8, the performance of neural networks gradually
improves as more hidden neurons and cross-increments are added. Not

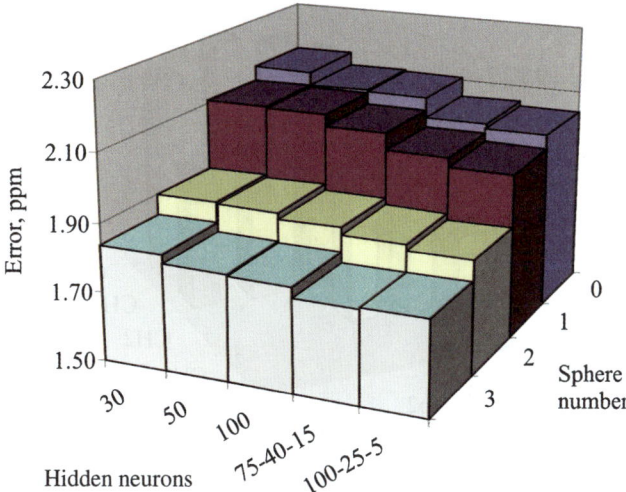

**Figure 3.8**   The mean $^{13}$C chemical shift error as a function of the neural network geometry and the maximum number of spheres used for creating cross-increments.

surprisingly, the best result, providing a mean error of 1.77 ppm, was achieved with the maximum number of cross increments (when distances between paired atoms were up to three bonds) and the largest neural network.

The neural network configuration was further tuned by adding more cross-increments up to the 3$^{rd}$ sphere with no more than two covalent bonds between the atoms and using stereo information in the atomic descriptors (see above). With three layers of 100, 25 and 5 hidden neurons the network produced a mean error of 1.59 ppm, with a largest error value of 85 ppm and 0.6% of the chemical shifts predicted with an error of more than 10 ppm (see Table 3.4).

Values for the mean error determined using NN and presented in Figure 3.9 as a function of the number of substituents (*i.e.* CH$_2$ *vs.* CH for the same hybridization state) suggest that generally the errors are higher for more highly substituted atoms. This would appear reasonable because the cross-influence between the different substituents enhances the non-linear effects.

## 3.1.4.8   Performance of a PLS Algorithm for $^{13}$C Chemical Shift Prediction: Large Dataset

A similar approach was used to evaluate the performance of the PLS algorithm. The same database split of 15 sub-databases was used. For the cross-increments, Smurnyy *et al.*[37] used a configuration of six spheres and cross-increments were taken from three first spheres with atoms separated by not more than three covalent bonds. This setup was shown to be the most effective in previous studies (see above). The results obtained were of a similar quality.

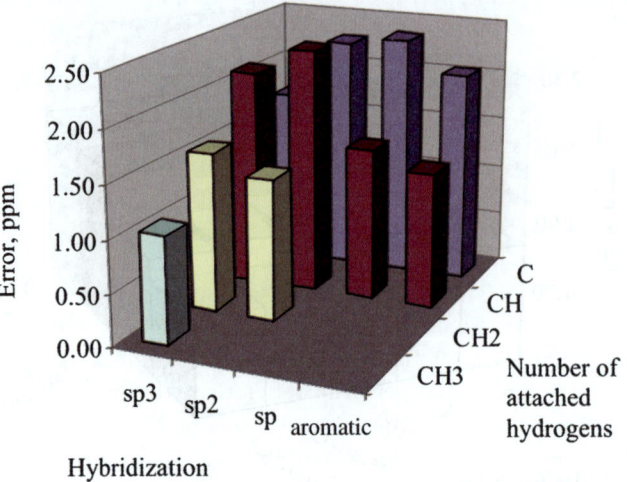

**Figure 3.9** The dependence of the mean error (ppm) on the number of substituents and the hybridization of the central atom. The data shown are for $^{13}C$ chemical shift predictions with the most efficient neural network configurations.

**Table 3.4** The best results for the prediction of both $^{13}C$ chemical shifts by HOSE codes, PLS and NN algorithms. The specific set of parameters for each method is described in the text. Outliers are defined as those shifts predicted with an error of more than 10 ppm.

| Prediction method | Mean error (ppm) | Standard deviation (ppm) | Maximum error (ppm) | Outliers (%) |
|---|---|---|---|---|
| HOSE | 1.80 | 3.05 | 58.0 | 2.8 |
| PLS | 1.71 | 2.61 | 51.6 | 0.7 |
| NN | 1.59 | 2.45 | 85.8 | 0.6 |

The mean $^{13}C$ chemical shift error is slightly higher (1.71 ppm), but other benchmark parameters were close to those obtained for the neural network (see Table 3.3). The number of severe outliers is somewhat lower, with only 0.7% of the centers producing errors of greater than 10 ppm. The maximum error was 51.6 ppm.

## 3.1.4.9  Comparison of HOSE, PLS and NN Algorithms for $^{13}C$ Chemical Shift Prediction

The results of the computational experiments performed with ACD/CNMR Predictor employing HOSE code-based, NN and PLS algorithms produced the statistical data necessary for a comparison of performance of all three methods. The corresponding data are collected in Table 3.4.

The comparison shows that both methods, NN and PLS, can provide results of similar quality after being properly optimized. A neural network, in general, seems to perform better with atoms whose chemical shifts are closer to an average value for the corresponding atomic type. Linear regression can more easily handle exotic fragments (such as in the compound $CI_4$ with a chemical shift of $-292.5$ ppm), since even the most unusual combination of substituents can easily be assigned an appropriate incremental value leading to a more accurate prediction. At the same time, the values of weights for most of the neurons in a network are affected by all of the structures present during the training process and the impact of unusual structures is masked by the majority of the more regular structures. These differences are highlighted by the lower value of the mean error for the neural network. It is interesting to note that both PLS and NN methods perform better than the implementation of the database-based HOSE code approach (see Section 3.1.2), which was evaluated with the same independent test set of 118 000 individual $^{13}C$ chemical shifts. For a given atom the algorithm retrieves few structures from a database which have chemically similar nuclei. The predicted value is the weighted average of chemical shifts contained within the database structures. The approach fails with structures which are under-represented in the database and this raises the standard deviation and the maximum error.

## 3.1.4.10  Speed of $^{13}C$ Chemical Shift Prediction

As mentioned above (Chapter 1), modern CASE expert systems are based on the utilization of 2D NMR data and this allows the identification of newly isolated materials such as natural products or synthesized organic molecules. These systems are capable of elucidating large molecules containing 100 or more skeletal atoms.[60] Since the initial structural information extracted from 2D NMR data is fuzzy by nature (see Chapter 2), the number of structures that are consistent with the spectral data can usually be rather large (up to tens of thousands[61]). As a result, the selection of the most probable structure from a large output file containing many molecules requires an approach whereby the expert systems can utilize both *accurate* and *fast* approaches for NMR chemical shift prediction.

To estimate the speed of $^{13}C$ chemical shift calculations compared in study,[37] the expert system StrucEluc[62] was employed since the ACD/CNMR predictor was already implemented into this software. The prediction speed was estimated by performing the spectral prediction of thousands of candidate structures generated by the program. It was found that the average speed of the $^{13}C$ chemical shift prediction by PLS is approx. 10 000 shifts per second on a 2.8 GHz PC computer, whereas the neural network based algorithm was approx. 2.5–3 times slower. The combination of this high speed of prediction with an appropriate accuracy for prediction with an average deviation of 1.71 ppm makes the PLS approach a powerful tool for computer-aided structure elucidation.

## 3.1.4.11  *Performance Validation of NN-based* $^{13}C$ *NMR Prediction Using a Publicly Available Data Source*

In spite of the fact that the test dataset used for validation of the improved ANN algorithm[37] had no overlap with the training set, it was of interest to determine to what degree the accuracy is dependent on the size, composition and diversity of the structural file used as a testing set. A resource is available on the Internet that has met the above criteria of size and quality to serve as a fair and reliable validation set. This resource is a database called NMRShiftDB[63] and has been created as a collaborative effort by chemists and spectroscopists submitting data to the open access database. The work[42] was devoted to an analysis of the performance of the ANN-based ACD/CNMR predictor using the NMRShiftDB database as the validation set. At the time of the study[42] the database contained approx. 20 000 structures with approx. 215 000 assigned carbon chemical shifts. Datasets entered by contributors are sent to registered reviewers for evaluation, who are presented with the newly entered spectrum, together with a shift prediction and a color-coded table of deviations. A significant part of NMRShiftDB was initially assembled from in-house databases from collaborating institutions and have been entered unchecked. This called for external checks of the data based on independent databases and resources. On the basis of a cursory examination of the structural diversity within the database, these data represent a statistically relevant set to use in an evaluation of predictive accuracy and is the first large dataset available from an independent source which could be used for this purpose.

The NMRShiftDB file was downloaded and the structures and shifts were imported into the format of ACD/Labs. As a first step, an analysis of the degree of overlap between the structures in the training set within the ACD/CNMR Predictor and the validation set made up of NMRShiftDB was undertaken. The presence or absence of a chemical shift in the NMRShiftDB was determined in the following manner. For a specific carbon atom existing in a structure contained within the NMRShiftDB, the HOSE code was determined and then the atom with the same HOSE code was searched in the ACD/CNMR database. If such an atom was identified, then the corresponding chemical shift was deemed to be present in both databases and therefore excluded. Otherwise, the shift was included into the dataset for which the given chemical shift was absent from NMRShiftDB.

It was determined that 57% of the carbon chemical shifts in the NMRShiftDB were already contained within the ACD/Labs database. The NMRShiftDB database was stripped of replicate chemical shifts used as the basis of the prediction algorithms in ACD/CNMR Predictor.

The results of algorithm validation using the NMRShiftDB test dataset are shown in Table 3.5.

It was revealed that from the total number of chemical shifts collected in the NMRShiftDB (214 136) 92 927 shifts were not contained within the ACD/CNMR database and consequently their HOSE codes were new for the NN ACD/CNMR Predictor. At the same time 121 209 chemical shifts made up the

**Table 3.5** General Results of ACD/CNMR Neural Network Predictor validation.

| Calculation method: | Shift Count | Average Error (ppm) | Standard Error (ppm) | Maximum Error (ppm) | % <1 ppm | % <2 ppm | % <3 ppm | % >3 ppm | % >10 ppm |
|---|---|---|---|---|---|---|---|---|---|
| 1. Whole dataset | 214136 | 1.59 | 2.76 | 153.53 | 50 | 75 | 86 | 13 | 0.7 |
| 2. Absent in data file | 92927 | 1.74 | 3.22 | 133.19 | 49 | 72 | 84 | 16 | 1 |
| 3. Present in data file | 121209 | 1.47 | 2.35 | 153.53 | 51 | 76 | 88 | 12 | 0.4 |

overlap of the two databases. The average, standard and maximum errors are displayed for the whole database and two subsets, as well as the percentages of chemical shifts predicted with errors of d < 1 ppm, d < 2 ppm, d < 3 ppm, d > 3 ppm and d > 10 ppm for the entire database and both subsets.

Table 3.5 shows that the mean error calculated for the entire dataset is 1.59 ppm (standard deviation = 2.76 ppm). This value can vary by ± 9% depending on whether dataset 2 or 3 is used for the purpose of validating the performance of the algorithms. Since the NMRShiftDB library is composed of compounds analyzed by chemists working in varied areas of organic chemistry it is appropriate to assume that it is a representative set for general chemistry. Assuming this to be true then validation of our $^{13}$C chemical shift prediction method using a database of *ca.* 214,000 chemical shifts leads to an average deviation which is reliable within ±10%.

The percentage of chemical shift deviations depending on the different ranges of the values calculated for the three datasets presented in Table 3.5 are shown in Figure 3.10.

Approx. 50% of chemical shifts were calculated with mean errors of < 1 ppm in all three cases. This observation allows us to conclude that, independent of the presence or absence of the predicted chemical shifts in the test database, half of all of the predicted chemical shifts are calculated with an error of < 1 ppm. Errors < 2 ppm and < 3 ppm encompass 72–76 and 84–88% of all calculated shifts respectively. Chemical shifts predicted with relatively low accuracy (d > 3 ppm) make up only 12–16% of the entire number of shifts. This could correspond to errors in the database and some have already been identified.[42,64] However, this distribution also correlates with the distribution of problems solved with the aid of the Structure Elucidator software and with chemical shift deviations calculated for genuine structures.[61] Thus the deviations calculated by ACD/CNMR Predictor based on HOSE codes were > 3 ppm for 18% of all of the problems solved with Structure Elucidator and, in each case, a newly identified natural product was elucidated. The percentage of shifts predicted with errors > 10 ppm varies from 0.4% ("present") to 1% ("absent") and are mainly associated with the presence of outliers in the NMRShiftDB database. These outliers can be from differences in predicted *versus* experimental data or

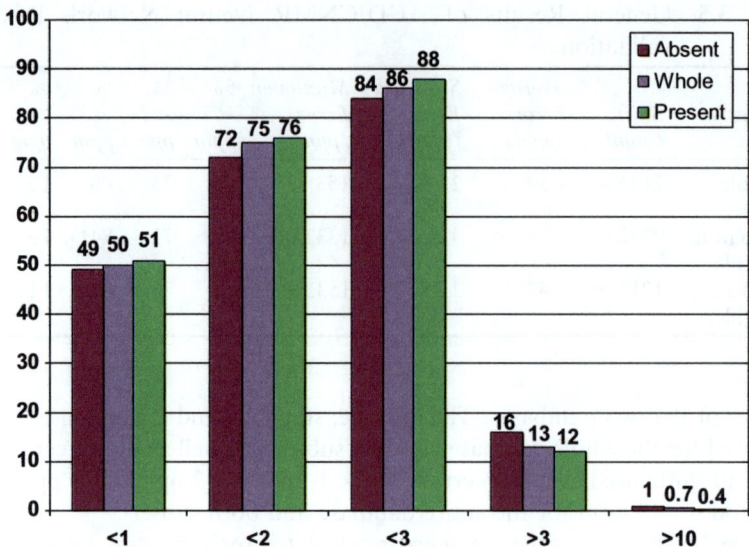

**Figure 3.10**  The percentage of chemical shift deviations depending on the different ranges of values as calculated for the three datasets presented in Table 3.5.

experimental data in error as a result of an incorrect structure representation or mis-assignment. These errors are distributed as follows: d > 10 ppm (1040; 0.5%), d > 25 ppm (141; 0.07%) and d > 50 ppm (31; 0.01%). Examination of the structure giving rise to the maximum error of ∼150 ppm provides evidence that NMRShiftDB either contains some erroneously drawn structures or poorly assigned chemical shifts.[64] This is unavoidable in large databases of this nature. Examples of structures (**3.6** and **3.7**) for which obviously erroneous chemical shift assignments were detected (marked by the circled shifts) as a result of the prediction accuracy validation and are shown below:

When shifts with differences of > 25 ppm were removed from the entire database as well as the two subsets, the average errors decreased only slightly: from 1.59 to 1.56 ppm for the entire DB, from 1.74 to 1.70 ppm for the "absent" subset and from 1.47 to 1.46 ppm for the "present" subset. Consequently, the conclusion regarding the predictive ability of the ACD/CNMR neural network prediction algorithm remains.

The data presented in Table 3.5 and Figure 3.10 describe the average values characterizing the results of the validation study. The accuracy of $^{13}$C NMR spectrum prediction is known to depend on chemical classes to which a given carbon atom belongs, on the number of representatives of a given class in the training sets, and on the influence of stereochemical factors. Therefore, the average prediction accuracy was compared for quaternary carbons, methine, ethyl and methyl groups in aliphatic substructures, as well as for alkenes, alkynes and aromatic compounds. In so doing, Blinov *et al.*[42] have examined how the errors of chemical shift prediction are associated with the presence or absence of a given shift in the validation set. The results of the calculations are graphically represented in Figure 3.11.

A striking observation is that the highest and almost equal accuracy is achieved for the methyl group (d = 1.30–1.07 ppm) and the aromatic protonated carbon atoms (d = 1.30–1.24 ppm) and the errors associated with the aromatic carbons are almost the same for the entire database and both of its subsets. The next in the series is the -CH2- group for which the error varies between 1.41 and 1.71 ppm. At first glance one can expect that the accuracy of chemical shift prediction for >CH- groups should be higher than for

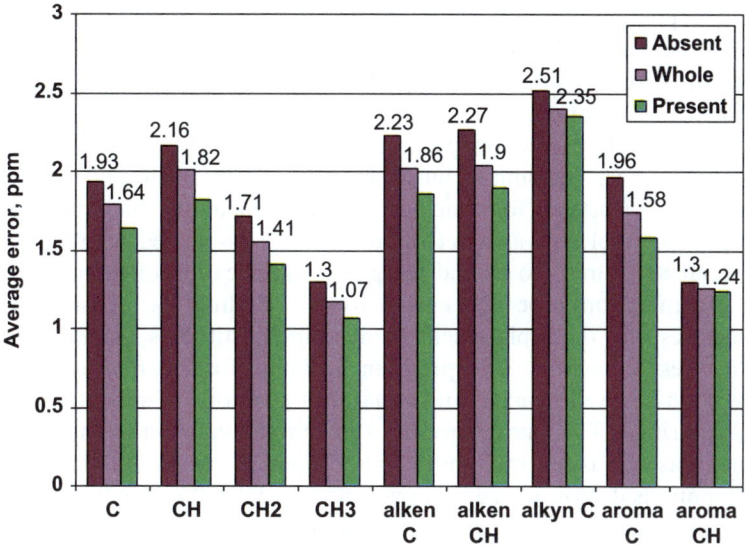

**Figure 3.11** The comparison of chemical shift prediction accuracy for different types of atoms depending on their assignment to one of the three testing sets.

quaternary carbons, since the latter have four non-hydrogen neighbors whose influence on chemical shifts is complex and difficult to take into account. However, the reverse was observed: 1.64–1.93 ppm for quaternary carbons and 1.82–2.16 ppm for >CH- (see Figure 3.11). The accuracy of the chemical shift prediction for =CH and quaternary =C groups in olefins is almost the same if it is considered separately within the entire set and both subsets. The error varies from 1.86 to 2.27 ppm when going from the "present" subset to "absent". The relatively large errors account for the difficulties associated with attempting to allow for stereochemical factors playing an important role in the chemical shift prediction of double bond carbons. The lowest accuracy of shift prediction (d = 2.35–2.51 ppm) was observed for alkynes, which can be explained by a very reduced dataset for such compounds in the training set.

Thus the accuracy of $^{13}$C NMR spectrum calculation by the NN-based ACD/CNMR Predictor was also corroborated using an independent dataset represented by NMRShiftDB, a publicly available internet data source.

## 3.2   Prediction of $^1$H NMR spectra

### 3.2.1   Incremental Approach: Linear Models

As noted earlier, $^{13}$C NMR spectra are both relatively simple to predict as well as to compare with data, since carbon NMR spectra consist almost exclusively of singlets or clearly defined multiplets due to heteronuclear coupling with magnetically active nuclides. The prediction of $^1$H NMR spectra, in contrast, is much more complex because they contain both first-order and higher-order multiplets. The relative complexity of proton spectra accounts for the greater difficulty in comparing the experimental and predicted spectra in an automated fashion. The latter observation probably explains why there are a limited number of software programs for proton NMR relative to carbon chemical shift NMR prediction.

A proton NMR prediction program was developed[65] to predict chemical shifts on the basis of simple linear models.[41] The software automatically determines substructures for which certain rules are applicable, then considers the rest of the molecule as a set of substituents associated with each of the selected substructures. To extend the general application of the program different extrapolation procedures were used, including the division of large substructures and the replacement of missing substituents by closely related substructures if possible. The program accounts for *cis/trans* isomerism of double bonds as well as axial and equatorial positions of substituents in six-membered rings. The linear models were based on a set of approx. 1300 increments as well as several hundred base values. A test on a modest size set of 200 compounds demonstrated that over the range 0.9–9.8 ppm 92% of the 583 different chemical shifts of protons bonded to carbons could be predicted. The mean deviation between predicted and measured values was found to be ±0.08 ppm with a standard deviation of ±0.19 ppm, which is a very impressive performance and providing a small deviation.

Another study[66] evaluating the quality of the proton NMR prediction program using a large test set determined that the standard deviation of chemical shifts was ±0.3 ppm. Using the ASSEMBLE structure generator,[67] 24 300 isomers with 111 600 different proton environments were generated. The chemical shifts could be calculated for 89% of the cases, which can be considered satisfactory. The mean time taken to predict the chemical shifts for one structure on a Macintosh Quadra 650 computer was 0.30 s. When comparing predicted and experimental spectra, the data were input as a list of chemical shifts associated with the CH, $CH_2$, and $CH_3$ groups. The program compares each experimental chemical shift with the corresponding chemical shift in the predicted spectrum. The standard deviations between experimental and calculated spectra are then calculated to rank the generated isomers.

The proton NMR prediction program described above has been integrated into the CambridgeSoft ChemDraw[21] structure editor. The data set associated with the $^1$H-NMR prediction algorithms currently contains 700 base values and approx. 2000 increments. It was reported[22] that chemical shifts of approx. 90% of all $CH_n$ groups could be predicted with a mean deviation of ±0.2– 0.3 ppm. No definition of structural diversity or complexity for this comparison was given but in our experience this proton NMR prediction offers relatively poor performance for complex natural product structures.

The algorithm that was developed for the fast and accurate $^{13}$C chemical shift prediction using a PLS model (Section 3.1.4) was also implemented into the ACD/HNMR Predictor software for calculation of $^1$H NMR chemical shifts. The proton chemical shift database used for creation of models was comprised of approx. 1.2 million $^1$H chemical shift values. The performance of the algorithm was examined using a completely independent database compiled from data originally published in 2005 and consisting of approx. 115 000 $^1$H chemical shifts. As a result, both the speed and accuracy of the chemical shift calculations has been dramatically improved in comparison with previous versions. The ACD/HNMR Predictor based on a PLS approach is now capable of predicting $^1$H spectra with an average deviation of 0.18 ppm and a standard deviation of ±0.26 ppm, while the average calculation speed is 7000–9000 $^1$H chemical shifts per second. Its accuracy and speed fit the requirements which were previously suggested regarding spectrum prediction methods that need to be incorporated into CASE systems.

## 3.2.2 Structural Database Approach: ACD/HNMR Predictor

The most advanced $^1$H NMR prediction algorithm currently available appears to be that embodied in the ACD/HNMR Predictor[8] program. ACD/HNMR Predictor is designed to predict a $^1$H NMR spectrum from a chemical structure. The program has an internal database containing almost 1 760 000 experimental $^1$H chemical shifts and over 620 000 coupling constants. The algorithm used by the program to quantify spin–spin interactions is based on parameters determined for more than 3000 structural fragments. Each $\delta_H$ value in the database is assigned to a particular proton nucleus and each coupling constant is assigned to a pair of interacting nuclei. The internal database of assigned $\delta_H$ and $J_{HH}$ values has been extracted from the analysis of 211 000 experimental $^1H$ NMR spectra. The $\delta_H$ and coupling constant values for proton nuclei in fragments not contained in the database are calculated using proprietary algorithms.

$^1H$ chemical shift calculations can be performed as a function of the spectrometer frequency using an optional parameter setting. Calculated chemical shifts and coupling constants are provided with 95% confidence intervals to provide a measure of the reliability of the calculated values. The calculated spectrum accounts for second-order interactions and long-range coupling constants, thereby allowing for the simulation of spectra with strongly coupled spin systems containing up to eight magnetically inequivalent nuclei. For magnetically equivalent nuclei a larger number of interacting spins can be taken into account. The algorithm recognizes *cis/trans* isomers for alkenes, *syn-anti* isomers of amides, oximes, hydrazones, and nitrosamines, and axial–equatorial isomers of ring systems. The chemical shifts of protons belonging to OH, NH, and SH groups can also be predicted. The program takes into account solvent effects in the following way: if a solvent is specified the program selects for the predictions only those reference structures whose $^1H$ NMR spectra were recorded in the specified solvent in the database. After input of a structure for prediction, the program generates a set of unique fragments that are compared with the fragments contained in the internal database. If any generated structural fragment coincides with a fragment from the database then its experimental $\delta_H$ values are included into the result set of chemical shifts generated for the structure. If some fragments are not identified in the database then the chemical shifts are extrapolated using parameters associated with similar fragments. Once again a "chemical shift protocol" dialog window, similar to that shown in Figure 3.2, allows inspection of the calculation associated with each chemical shift and can be valuable in probing any discrepancies observed.

After composing the final lists of chemical shifts and coupling constants the program composes and diagonalizes the spin Hamiltonian matrix to calculate the number of lines, their positions, and intensities in order to assign the $\delta_H$ values to the hydrogen atoms of a considered structure. The program generates a spectral plot including the positions of the lines and the integral. In the process, the program takes into account the spectrometer frequency, which provides a visual display to allow comparison of the calculated spectral pattern with experimental spectrum. The calculations of the chemical shifts, spectral contours, and integration curve can take from several seconds to tens of seconds on a standard Pentium-based computer. An example of the experimental and calculated $^1H$ NMR spectra of **3.8** (see example in Section 1.4) is shown in Figure 3.12, which shows good agreement between the signal positions and the multiplicities in both spectra.

**3.8**

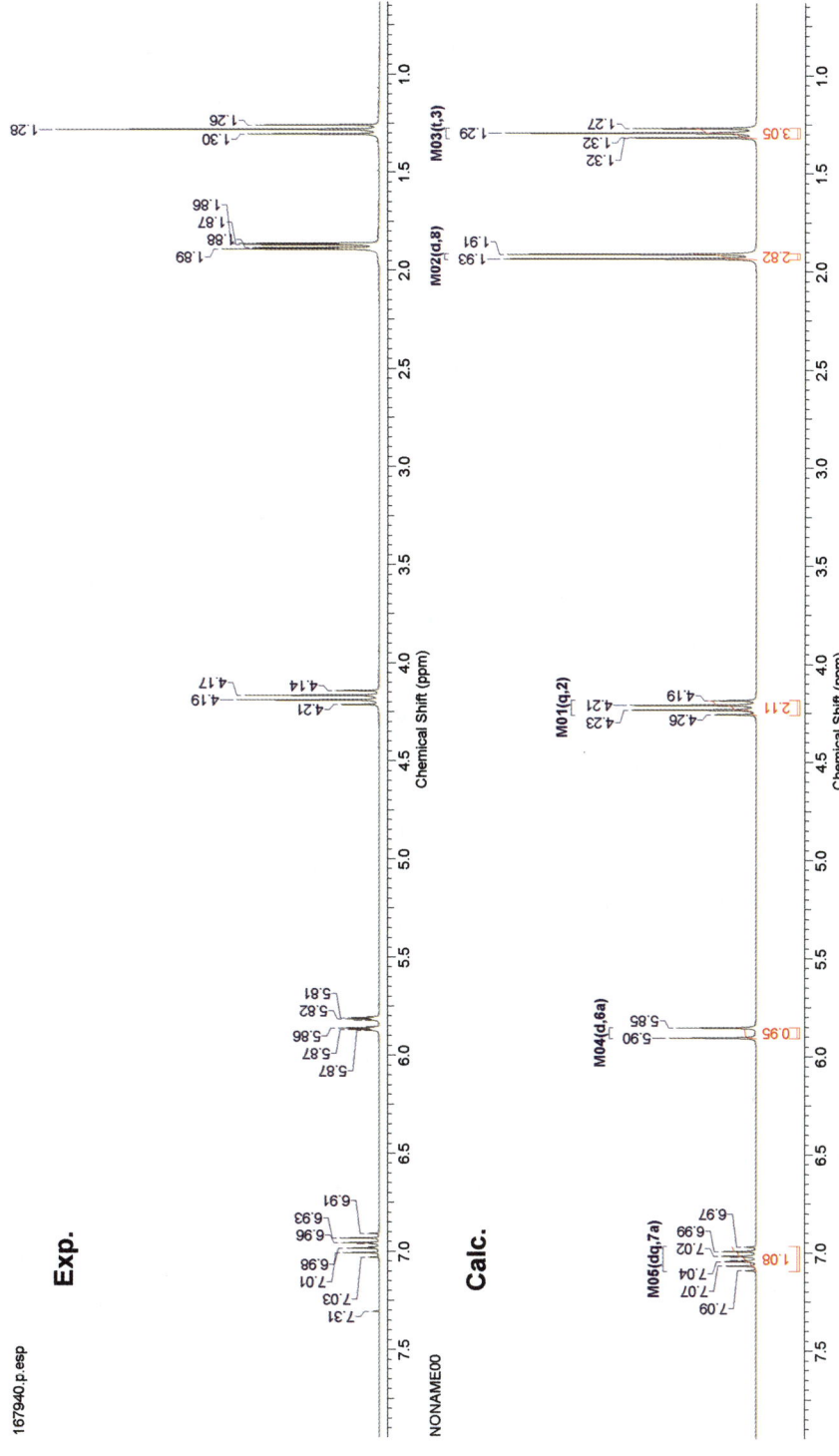

**Figure 3.12** A comparison of the experimental and calculated $^1$H NMR spectra of Structure **3.8**.

The ACD/HNMR Predictor program was developed to aid in the interpretation of experimental spectra and to test structural hypotheses. A preliminary ranking of the structures in the output file created by an expert system can be performed using a fast incremental model of $^{13}$C NMR spectrum prediction. In this case only 20–50 structures placed at the beginning of a rank-ordered final list requires additional confirmation to help select the "best" structure. $^{1}$H NMR predictions are fast enough that the calculation time for hundreds of structures will take only several seconds. The comparison of calculated and experimental $^{1}$H NMR spectra can further aid in the determination of the most probable structures (see Sections 5.4.2 and 13.5). A combination of both $^{1}$H and $^{13}$C NMR prediction is of value in analyzing the output files associated with problems solved using expert systems.

### 3.2.3   NN and PLS Algorithms for $^{1}$H Chemical Shift Prediction

As described above, $^{1}$H NMR chemical shift prediction based on increment rules and structural databases are mostly commonly utilized by chemists and spectroscopists. Predictions based on increment rules were, generally, fast enough but were frequently not sufficiently precise and, as a result, were generally favored by synthetic chemists for *estimates* rather than being relied on by spectroscopists. Employing databases allowed the calculation of $^{1}$H spectra with greater accuracy but with the corresponding cost of greater calculation time. The desire to combine the advantages of both approaches has led researchers[68–70] to investigate the possibilities of applying neural networks to $^{1}$H chemical shift prediction.

The training of neural networks with a small data set of proton chemical shifts extracted from 120 structures has been described by Aires-de-Sousa *et al.*[68] This neural network was used for the fast estimation of NMR chemical shifts associated with $CH_n$ protons. The relationship between protons associated with a structure and the corresponding $^{1}$H NMR chemical shifts was established by counter propagation neural networks (CPG NN).[58] This approach used descriptors for hydrogen atoms as the input and the chemical shift of the corresponding proton as output. With this approach topological, physicochemical, and geometric descriptors were employed. Four different classes of protons were treated separately: protons belonging to aromatic systems, protons belonging to $\pi$ non-aromatic systems, protons belonging to rigid aliphatic substructures, and protons belonging to non-rigid aliphatic substructures. Some descriptors were common to all four classes while others were specific for a single class. When applied to the prediction of 259 chemical shifts associated with a small set of 31 molecules of an independent test set, the mean absolute error obtained for the whole prediction set was ±0.25 ppm.

Binev *et al.*[69,70] employed exactly the same data described in the previous work but extended the approach. They used *ensembles* of feed-forward neural networks[71] (FFNNs) and the optimization of several factors including the selection of variables and the architecture of the networks. In particular, ensembles of neural networks were employed because different nets, trained in order to be consistent with the training data, can give different predictions for new examples. Ensembles of neural networks offer a possible solution to this problem by making predictions on the basis of "votes" from individual networks.[71] They yield more stable and often more accurate predictors than individual networks.

The training set consisted of 744 $^1$H NMR chemical shifts for protons associated with 120 molecular structures. Prediction set A contained 259 chemical shifts from 31 molecules and a larger prediction set B included 952 experimental chemical shifts obtained from 100 structures. The collection was restricted to $CH_n$ protons and to compounds containing elements C, H, N, O, S, F, Cl, Br, or I. As in the previously cited work,[68] the chemical shifts of protons bonded to heteroatoms were not included, and the same four classes of protons were defined.

In order to be submitted to a neural network, each proton was represented by a fixed number of descriptors. For hydrogen atoms belonging to aromatic, rigid aliphatic and non-rigid aliphatic classes, three kinds of descriptors were used: physicochemical, geometric, and topological. Physicochemical descriptors were based on empirical values calculated by the software package PETRA.[47] For example, the partial atomic charge of the proton, the effective polarizability of the proton, the average of partial atomic charges of atoms in the second sphere, the maximum partial atomic charges of atoms in the second sphere, the minimum effective polarizability of atoms in the second sphere, and the average of $\sigma$ electronegativities of atoms in the second sphere were all calculated.

Geometric descriptors were based on the 3D molecular structure generated by the CORINA software package.[72] Topological descriptors were calculated from the connection table and included, for example, the number of carbon atoms in the second sphere centered on the proton and the number of oxygen atoms in the third sphere. In all, 92 descriptors were used for aromatic protons, 119 for non-rigid aliphatic protons, and 174 for rigid aliphatic protons.

Descriptors for the *gem*, *cis*, and *trans* positions relative to a proton were used for non-aromatic π protons. For example, one descriptor was the charge of the atom at the *cis* position, while another descriptor was the charge of the atom at the *trans* position, rather than the average of the charges of atoms three bonds away from the proton.

Binev *et al.*[69] concluded that ensembles of neural networks demonstrated higher robustness and were expected to give better predictions. With an optimized system of networks a global average mean error for an independent test set of 952 cases was ±0.29 ppm and for 90% of the cases the value was ±0.20 ppm. These results were achieved with relatively small training sets, possibly due to the use of physicochemical and topological descriptors that generalize atom types to inherent physicochemical properties.

Since the amount of data available to a system is constantly increasing, neural networks should be able to easily incorporate new data that becomes available after the initial training, thereby expanding their applicability and improving their accuracy. In the next report by Binev *et al.*,[70] two different approaches were explored to incorporate newly available experimental data into previously trained ensembles of FFNNs for the prediction of $^1$H NMR chemical shifts of organic compounds.

One approach used newly available data as additional memory for the previously trained FFNNs, now organized as *associative neural networks*.[73] Associative neural networks (ASNNs) adjust the prediction obtained from an ensemble of NNs on the basis of errors for the $k$ nearest neighbors (KNN) in the memory.

The second approach investigated in this study was based on retraining FFNNs with new data. The same network architectures, the same selected variables, and ensembles with the same sizes were used as these were found to be optimal in the studies with smaller training sets.[69] The networks were trained with the new and old data combined.

For this investigation a data set obtained from Chemical Concepts GmbH containing 5631 experimental chemical shifts and the corresponding nuclei from 482 structures was employed. Stereochemistry was not assigned and only data from spectra obtained in $CDCl_3$ were considered.

The investigation showed that incorporation of new data into the ANN, previously trained with small data sets, resulted in a significant improvement of the predictions. A mean absolute error of $\pm0.19$ ppm was obtained for the entire prediction set. Predictions obtained using the FFNN and ChemDraw were comparable and slightly less accurate than the other two methods. While this investigation cannot be considered as a fully comprehensive and definitive comparison of the methods, the results do suggest that an approach based on ASNN is indeed promising. In terms of prediction speed the entire procedure from structure connection tables to final prediction values of the chemical shifts using the ASNNs took approx. 1 s per structure. On average each structure had 17 hydrogen atoms attached to the skeletal carbon atoms. The speed was measured with a non-optimized system containing several software components running on a PC equipped with a 1.6 GHz Celeron CPU.

Another important feature of the ASNN approach is the possibility to report which protons were used to generate the prediction. The result obtained by submitting a particular proton associated with a particular structure to an ASNN is the prediction of the chemical shift as well as the KNNs in memory whose experimental chemical shifts were utilized in the prediction. This can be considered as a partial explanation of the details of a prediction. The predictions were also clearly improved by comparing the networks trained with small data sets relative to when the neural networks were retrained with more data. An optimal strategy of retraining the FFNNs was both suggested and experimentally confirmed by the authors.[70]

The authors[70] concluded that the amount of available experimental data played a decisive role in the quality of the predictions obtained by neural networks. Incorporation of new experimental data into the *memory* of associative neural networks leads to a significant improvement in the predictions even without retraining of the networks. Improved results could also be achieved after retraining the networks with new data, but this required either careful selection of the training set or initiating the training with the weights of the networks trained based on the initial smaller data sets.

## 3.2.4 ACD/Labs NN and PLS Algorithms for ¹H Chemical Shift Prediction

The experience obtained from the ¹³C-related analysis[37,38,42] was employed to the prediction of ¹H chemical shifts on the basis of the ACD/Labs NN algorithm. Over 1.2 million ¹H chemical shifts (compared with the ∼ 1000 chemical shifts used by Binev *et al.*[69,70]) were used as a training set, and the same test dataset was used as for the ¹³C analysis, a total of approx. 115 000 ¹H chemical shifts.

The whole training database was split into nine sub-datasets, namely the aliphatic $sp^3$ [five classes: CH, CH(het), $CH_2$, $CH_2$(het), $CH_3$], aliphatic $sp^2$, aromatic $sp^2$, aliphatic sp and protons attached to heteroatoms. The neighborhood of an atom was described in exactly the same manner as described earlier for the ¹³C studies. All additional flags specified in Section 3.1.4.2 (stereo configuration, Z/E conjugated system, *etc.*) were included.

The neural network used for the calculation had 100, 25 and 5 hidden neurons arranged in three layers. The six nearest spheres were used; cross-increments were constructed from atoms in the first 3 layers separated by not more than one covalent bond. PLS regression was performed within the same six-sphere vicinity, although with more cross-increments: within the three nearest spheres with atoms separated by three covalent bonds or less. The best results for the ¹H chemical shifts predictions are presented in Table 3.6.

Table 3.6 shows that the neural network and PLS approaches performed in a remarkably similar manner – most of the NN and PLS configurations result in a mean error of approx. 0.2 ppm. In order to reduce the error to that experienced in experimental determinations further optimization such as the detailed description of the 3D geometry might be necessary.

**Table 3.6** The best results for the prediction of ¹H chemical shifts by HOSE codes, PLS and NN algorithms. The specific set of parameters for each method is described in the text. Outliers are defined as those shifts predicted with an error of more than 1.0 ppm.

| Prediction method | Mean error (ppm) | Standard deviation (ppm) | Maximum error (ppm) | Outliers (%) |
|---|---|---|---|---|
| HOSE | 0.19 | 0.30 | 3.94 | 1.3 |
| PLS | 0.18 | 0.26 | 2.72 | 0.7 |
| NN | 0.18 | 0.26 | 3.71 | 0.8 |

## 3.3   Empirical and DFT GIAO QM Methods of $^{13}$C Chemical Shifts Prediction: Competitors or Collaborators?

The prediction of $^{13}$C chemical shifts using QM methods has been a focus for many researchers. An extensive literature is devoted to the development and application of different QM approaches. The discussion of all QM approaches is beyond the scope of this Chapter and we will limit our focus to the gauge-including atomic orbital (GIAO) approximation of the DFT approach which has been increasingly applied to NMR spectral calculations. The reason for the wide applicability of the GIAO DFT calculations is the potential possibility to provide a high enough accuracy to solve many problems for both organic and analytical chemistry at relatively low computational costs. Almost all practicing chemists use different modifications, most frequently the B3LYP functional variant of this method,[74] in which a QM prediction seems necessary. The important advantage of this approach is that it takes into account electron correlation effects. Additional consideration is given to electron correlation by perturbation theory. The calculation of shielding constants at the MP2 level (a perturbation theory of the second order) is available, but these computations are too time-consuming and this is the reason why they are applied only to small molecules. There is an $N^3$ dependence in computational time for the DFT approach and $N^5$ for MP2, where N is the number of basis functions. The Hartree–Fock method is not very computationally expensive, but it also does not take into account electron correlation effects, therefore much larger errors in predicted chemical shifts are expected to be associated with the Hartree–Fock method than with the B3LYP method when electronegative and heavy atoms are present in a molecule.

During the last decade, many publications devoted to the $^{13}$C chemical shift prediction of organic molecules using the GIAO approach have been published. It is possible to distinguish the following goals of these works:

- Search for the most successful combinations of density functionals and basis sets (calculation protocols) capable of providing a prediction of geometry and chemical shifts for sets of organic molecules characterized by structural diversity (for example, see references[75–77]).
- Search for appropriate calculation protocols leading to acceptable predicted chemical shift values for a given compound or class of compounds (for example, see references[78–80]).
- Detailed investigation of the structural and electronic properties for a single molecule or a series of selected molecules (for example, see references[81–83]).
- Selecting the most probable structural hypothesis in the process of molecular structure elucidation (for example, see references[84–93]) and, once the genuine structure is determined, choose its preferable stereochemical configuration.

There are a lot of examples demonstrating that successfully chosen GIAO calculation protocols lead to close coincidence between the predicted chemical and experimental shifts. It is rather common that the functionals and basis sets selected for geometry optimization differ from those used for the chemical shift calculation which hampers guessing the best protocol. Attempts have been made to select an optimum protocol that fits for the purpose of $^{13}$C calculation for both rigid and flexible molecules. For example, Cimino *et al.*[75] tested approx. 50 protocols and concluded that the best prediction of the experimental $^{13}$C values is obtained at the mPW1PW91 level using the 6-31G(d,p) basis set both for the geometry optimization and chemical shift calculation.

Nevertheless, the search for new approaches leading to improved calculation accuracy continues. Recently, for example, Sarotti *et al.*[94] suggested using a multi-standard method (MSTD) for GIAO-based $^{13}$C chemical shift calculations. When the MSTD approach is employed, two reference compounds should be used: (a) methanol – for predicting the chemical shifts of sp$^3$-hybridized carbon atoms, and (b) benzene – for sp- and sp$^2$-hybridized carbon atoms. The authors[94] concluded that the mPW1PW91/6-31G(d) protocol constituted a level of theory that provides maximal reliability and mean average error (MAE) values around 1.5 ppm at minimal computational cost when applying the MSTD approach. This approach looks attractive, and requires further investigation and testing.

Accessibility to programs performing QM calculations encouraged non-specialists in quantum chemistry to use them for the interpretation of different experimental data. Some authors[95] treat the GIAO chemical shift calculation as an almost routine method that can be easily utilized by organic chemists. However, the scattering of observed chemical shift MAE values found by different researchers is evidence that such generalities are not borne out in practice. Theoreticians developing QM-based methods of chemical shift calculations[96] note that "using to full advantage these (GIAO) interpretative potentialities requires perhaps a larger dose of theoretical experience". Experienced researchers also comment that "since the quality of the results obtained depends on the functional and basis set used, their choice must be made wisely and with great attention". The creation of an expert system capable of helping organic chemists to select the appropriate protocol applied to a specific molecular structure could be useful.

The results of QM $^{13}$C NMR shift predictions performed on organic molecules of different chemical compositions and different classes have been published in many articles. As far as we know, the results have not yet been generalized and QM computational errors determined for a large enough structural set were not compared with those obtained from the empirical methods. It is worthy to note that the empirical methods of NMR shift prediction are either almost not mentioned at all in the articles devoted to QM-based computations of chemical shifts or the accuracy attained using a QM approach is commented on without taking into account the latest achievements in the field of empirical methods.

Meanwhile, examples of the application of empirical methods for molecular structure elucidation and the determination of relative stereochemistry, in parallel with QM methods, have been considered.[97-99] The examples show that QM calculations, which are far more computationally expensive in comparison with empirical ones, are frequently used in those cases when empirical shift prediction allows one either to rapidly and reliably find the correct solution of a problem or suggest 1–3 structural hypotheses to be finally discerned by determining additional experimental data and theoretical considerations.

In this connection it would be worthy to cite the following quotation from Dirac's recollections:[100] "The engineering training which I received did teach me to tolerate approximations... If I had not had this engineering training, I should not have had any success with the kind of work that I did later on... Engineers were concerned only with getting equations which were useful for describing nature. They did not very much mind how the equations were obtained. Once they got them they proceeded to use them with their slide rules, and get results which were necessary for their work. And that led me of course to the view that this outlook was really the best outlook to have". We suggest that Dirac's comment should be taken into account when choosing an appropriate method for $^{13}C$ chemical shift prediction. It is quite probable that in many cases an "engineering outlook" represented by empirical methods can be successfully utilized without the additional work associated with the application of QM calculations. Speaking figuratively, it is possible to say that the empirical methods supply practicing chemists with a predictive tool that works automatically like an "engineering slide rule".

The necessity of developing "engineering approaches" to improve the accuracy of NMR chemical shift prediction was also recognized by theoretical chemists who suggested procedures for scaling non-empirically predicted chemical shifts[84] or scaling calculated isotropic tensors of magnetic shielding.[101] Aliev et al.[102] suggested an universal equation for scaling $^{13}C$ chemical shifts calculated with the GIAO B3LYP/6-311 + G(2d,p)//B3LYP/6-31G(d) protocol, which markedly reduces MAE values. Scaling procedures empirically take into account different effects (electron correlation, relativistic effects, interaction with solvent, etc.) influencing calculation accuracy. Reducing prediction errors is the main purpose of the scaling procedures. The MSTD approach mentioned above was also developed having in mind the same goal. One may say the non-empirical methods are indeed "semi-empirical" ones.[95,101] The theoreticians conclude that: "the choice of empirically scaled parameters could be mainly determined by an 'aesthetic drive', i.e. owing to the wish to consider apparently smaller values of the medium average error".[75]

Elyashberg et al.[103] made an attempt to compare the accuracy of $^{13}C$ chemical shift prediction attained by QM and empirical methods for a large number of organic molecules. For this goal, the authors extracted data from over 100 articles in the literature data associated with QM calculations published by different research groups over the last decade and compared the results with those obtained for the same structures using ACD/Labs HOSE code and ANN-based algorithmic approaches. The results of this investigation will be discussed in this Section.

### 3.3.1 Data selection and Processing

For the computational experiments 205 structures have been found for which both assigned experimental and QM-calculated $^{13}C$ chemical shifts were published in the literature. Most of the data were obtained from the *Journal of Molecular Structure, Magnetic Resonance in Chemistry*, and other related journals. Only examples in which the $^{13}C$ experimental spectra were of high quality were chosen for analysis. At the selection stage, it was observed that some authors (for example, see reference[104]) used for the evaluation of QM methods experimental spectra which differed significantly from available reference spectra. In such cases the reference experimental $^{13}C$ NMR spectra which are present in the ACD/Labs database or in the Aldrich spectral atlas[105] were used.

Almost 50% of the structures contained 10 or less carbon atoms and $\sim 85\%$ of the structures contained less than 20 carbon atoms. This distribution reflects the fact that QM chemical shift calculations were applied mostly to molecules of small and modest sizes and the calculations are applicable to molecules with 20–30 carbon atoms, a common situation for natural products. Moreover, $^{13}C$ NMR prediction for a molecule of the size and complexity of Taxol has been reported recently.[102] The compounds were of high enough structural diversity and included the heavy atoms N, O, S, P, Cl, Br and F.

All structures in the test set were input into the ACD/Structure Elucidator software.[106] Carbon atoms were associated with both experimental and QM-calculated $^{13}C$ chemical shifts according to the assignment performed in the corresponding articles. If the QM chemical shifts of a structure were computed using several different protocols, then the best approximation was chosen. In the Structure Elucidator software the structure set under test was included into a user database (UDB), in which all of the results from the calculations could be stored. For all of the structures, $^{13}C$ chemical shifts were calculated using ACD/CNMR Predictor,[8] which is capable of applying all available algorithms: HOSE code-based, ANN and additive rules (increments). Before performing the HOSE-based calculations the program checked whether a given structure was present in the relevant database (207 000 entries) employed for spectrum prediction. If a structure was detected in the database, then it was excluded from the spectrum prediction process. For each of the 205 structures the following values were estimated and stored in the UDB relative to the HOSE, NN, increments and QM methods of prediction:

- The experimental and predicted shifts for each individual carbon atom;
- The differences $\delta_{exp} - \delta_{calc}$ (with their signs) between the experimental and calculated chemical shifts for each carbon atom;
- MAE;

- Standard error (standard deviation, SD);
- Maximum absolute error (maximum deviation, $d_{max}$);
- The regression parameters from linear regression (r, $R^2$, SE, slope $a$, intersect $b$, *etc.*).

For every structure plot showing the $\delta_{calc} = \delta_{exp}$ line (45° line) and linear regression lines for QM, HOSE and NN shift predictions were generated. Utilizing the UDB allowed access to a routine which automatically produces electronic tables containing comprehensive statistical and descriptive information related both to each structure and to the full structural set. The obtained statistical data and plots were carefully analyzed.

### 3.3.2   Statistical Comparison of Methods

The quantitative parameters characterizing the accuracy of the empirical and QM methods of $^{13}$C NMR chemical shift prediction for the set of structures under examination are presented in Table 3.7.

The Table shows that for the given test set of molecules the MAE value obtained for the HOSE-based prediction approach is less than half the value calculated when QM methods were utilized. MAE(NN) is less than MAE(QM) by a factor of 1.7. An analogous trend is observed for MAE(Inc) – the fastest method of chemical shift prediction based on additive rules,[37] which, while not the most accurate, also exceed the QM methods in average precision.

Figures 3.13 and 3.14 show a plot of the MAE and maximal deviations $d_{max}$ values found by the HOSE, NN and QM methods determined for every structure.

Visual assessment allows us to conclude that the majority of MAE values calculated by all three methods are less than 4 ppm, whereas deviations exceeding 4 ppm were shown mainly for the QM predictions. In this case the QM predictions also produce large deviations with values larger than those delivered by the empirical methods. The average values of the maximum deviations ($d_{max}$) are 4.75, 5.15 and 7.40 ppm for HOSE, ANN and QM approaches respectively.

**Table 3.7**   Average Statistical Parameters Calculated for the Test Set of Molecules. The total number of chemical shifts was 2531. MAE is calculated by summation of absolute errors found for each carbon atom divided by the total number of shifts.

| *Method* | *MAE (ppm)* | *SD (ppm)* | $d_{max}$ *(ppm)* |
|---|---|---|---|
| HOSE | 1.58 | 2.55 | 18.9 |
| NN | 1.91 | 2.79 | 21.7 |
| Inc | 2.15 | 3.12 | 22.2 |
| QM | 3.29 | 4.98 | 28.3 |

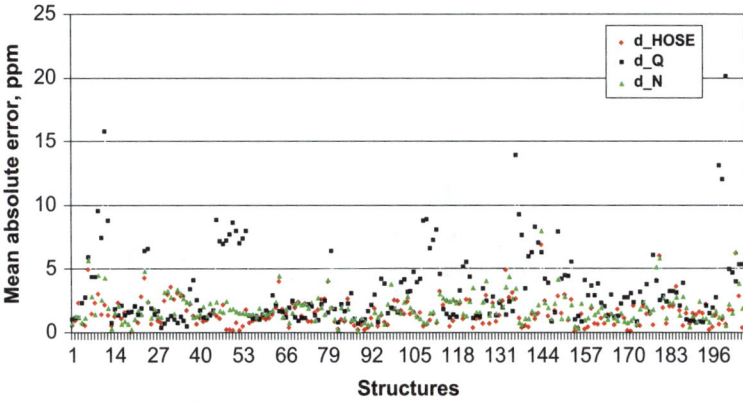

**Figure 3.13**   MAEs calculated by the QM, HOSE and ANN methods.

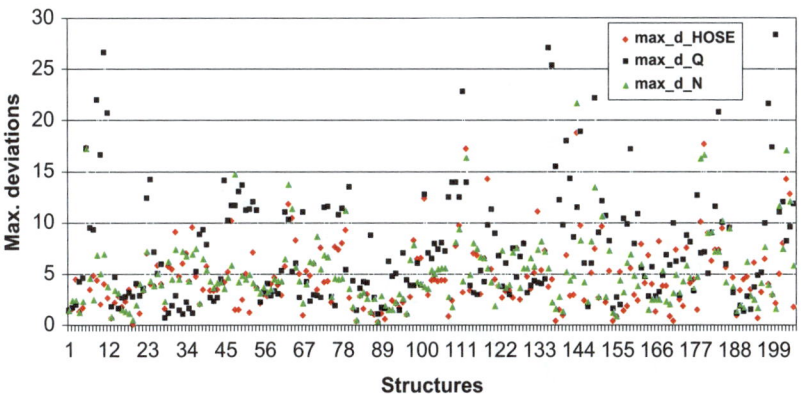

**Figure 3.14**   Maximum deviations ($d_{max}$) calculated by QM, HOSE and ANN methods.

Figure 3.15 shows a comparison of the errors associated with all prediction methods.

The histogram shows that 60–70% of the MAE values provided by the empirical methods are less than 2 ppm and 90% were less than 3 ppm. The corresponding percentages related to the QM methods are 45% and 60% respectively.

The results of a linear regression calculation performed for 2531 experimental and predicted $^{13}$C chemical shifts are presented in Figures 3.16–3.18.

Comparison of the plots and statistical parameters calculated for the examined methods shows that all three models are characterized by acceptable quality. However, both visual inspection and comparison of the linear regression statistical terms shows that the quality gradually decreases in the following order: HOSE > NN > QM, with the QM-based predictions showing the

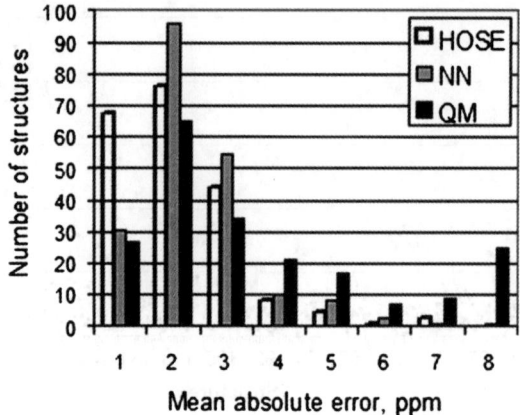

**Figure 3.15**  A comparison plot of the mean absolute errors established for HOSE, ANN and QM methods. The final black column shows that the MAE(QM) exceeds 8 ppm for 25 structures.

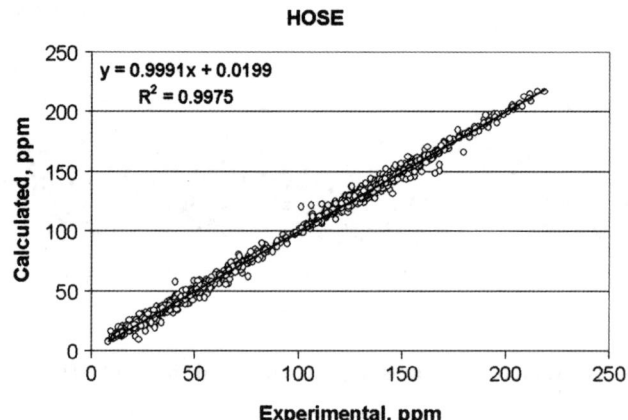

**Figure 3.16**  A linear regression plot showing the dependence of HOSE-based predicted chemical shifts *versus* experimental shifts. The linear regression equation is $\delta_{calc} = 0.9991\delta_{exp} + 0.0199$, $R^2 = 0.9975$.

poorest performance. The HOSE plot practically coincides with the 45°-grade line ($\delta_{calc} = \delta_{exp}$) and is almost coincident with the $\delta_{calc}$ axis zero point, whereas the QM plot is shifted up by 1 ppm, admittedly a small but notable difference. Larger scattering is observed in the QM plot in the interval 100–200 ppm, indicating a decrease in the prediction accuracy. As mentioned earlier, Aliev *et al.*[102] suggested a universal equation $\delta_{scalc} = 0.95\delta_{calc} + 0.3$ for scaling the $^{13}C$ chemical shifts calculated using a GIAO protocol B3LYP/6-311 + G(2d,p)// B3LYP/6-31G(d) (SHIFTS//GEOMETRY). The potential application of this equation to the >2500 chemical shifts calculated by *different* protocols to

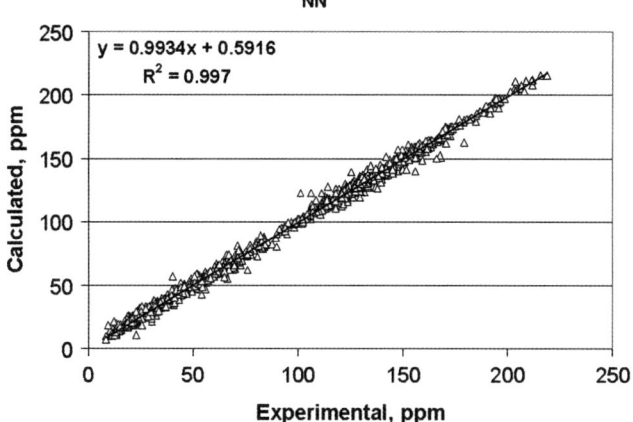

**Figure 3.17** A linear regression plot showing the dependence of NN-based predicted chemical shifts *versus* experimental shifts. The linear regression equation is $\delta_{calc} = 0.9934\delta_{exp} + 0.5916$, $R^2 = 0.9970$.

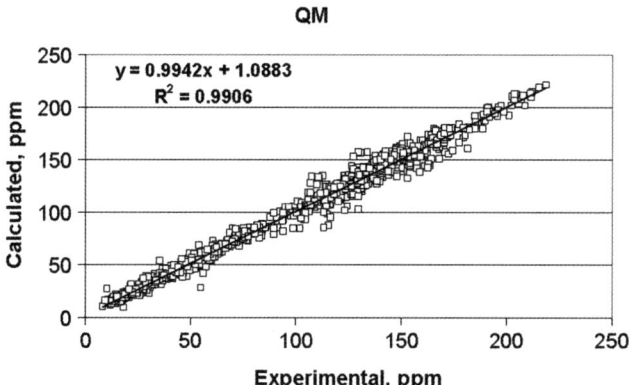

**Figure 3.18** A linear regression plot showing the dependence of QM-based predicted chemical shifts *versus* experimental shifts. The linear regression equation is $\delta_{calc} = 0.9942\delta_{exp} + 1.0883$, $R^2 = 0.9906$.

improve the average MAE value was investigated. When scaling was applied the MAE *increased* from 3.29 ppm to 4.77 ppm and the error distribution shifted to the side of the positive axis; the scaled chemical shifts, in general, were now underestimated especially in the region 100–200 ppm. The suggested scaling equation may thus only be valid when a specific protocol is used.

The results were investigated in more detail specifically examining the calculated MAE values for the various hybridization states: $CH_3$, $CH_2$, CH and quaternary carbons. To extract statistical significance from the analyzed parameters, atom types for which there were less than 50 representatives in the dataset were excluded from consideration. Following this process produced an

**Table 3.8**   The MAE values corresponding to ring carbon atoms of different hybridization states. The symbols C(ar) and CH(ar) denote atoms belonging to aromatic rings.

|                    | $sp^3$ |        |      |       | $sp^2$ |        |       |       |
|--------------------|--------|--------|------|-------|--------|--------|-------|-------|
|                    | $CH_3$ | $CH_2$ | $CH$ | $C_q$ | $=CH$  | $CH(ar)$ | $C(ar)$ | $C_q$ |
| Count[a]           | 273    | 459    | 278  | 99    | 59     | 586    | 405   | 188   |
| HOSE               | 1.51   | 1.46   | 1.97 | 1.34  | 1.90   | 1.20   | 2.05  | 1.79  |
| NN                 | 1.61   | 1.79   | 2.40 | 1.87  | 2.61   | 1.51   | 2.20  | 2.46  |
| QM                 | 2.35   | 1.66   | 2.61 | 2.65  | 2.91   | 3.64   | 4.72  | 5.18  |

[a]Total number of shifts used is 2347 out of a total of 2531.

atom set belonging only to cyclic structures (Table 3.8). This observation is accounted for by the fact that almost all compounds examined by QM chemical shift predictions were related to ring systems, mainly to natural products. The atom lists presented in Table 3.8 are ordered according to both the number of attached hydrogen atoms and the type of hybridization (the ordering also approximately corresponds to increasing chemical shifts) to ease investigation of patterns in the values obtained by QM and empirical methods.

The histogram presented in Figure 3.19 allows visual comparison of the MAE values associated with different atom types. It is evident that the accuracy associated with the empirical methods is essentially independent of the carbon atom type. This implies approximately equal reliability for the calculated shifts across the full chemical shift scale represented (0–200 ppm). In contrast, there is a dependence between the MAE values and the atom types observed for QM-calculated points. A maximum MAE(QM) value of 5.18 ppm is observed for non-aromatic $=C_q$ atoms which can be explained by the influence of substituents attached to quaternary $sp^2$-hybridized carbons. However, it is also likely that the different number of shifts for the non-aromatic and aromatic rings [188 for $=Cq$ and 405 for $=C(ar)$] leads to the observed difference. It has been noted[75] that the GIAO approximation of DFT-based predictions frequently either overestimates or underestimates the predicted chemical shifts for $sp^2$-hybridized carbon atoms depending on the calculation protocol used. This observation is in accord with the data presented here (Figure 3.19) for a large number of shifts ($\sim 1240$). Figure 3.19 also clearly shows that MAE(QM) values increase by a factor of 2 along the chosen plot order of $CH_3$ to $=C_q$ carbon.

It was interesting to learn how the carbon atoms within the test set are distributed as a function of the differences between the experimental and calculated chemical shifts ($\delta_{exp} - \delta_{calc}$). The corresponding distribution plots computed for a deviation interval of $\pm 10$ ppm with a summation step of 0.5 ppm are presented in Figure 3.20. The Figure shows that the distribution corresponding to HOSE-based calculations is a near-normal distribution in nature and characterized by the sharpest peak. The error distribution for the NN approach is represented by a broad bell-shaped curve whose maximum is markedly shifted down relative to the maximum of the HOSE code distribution

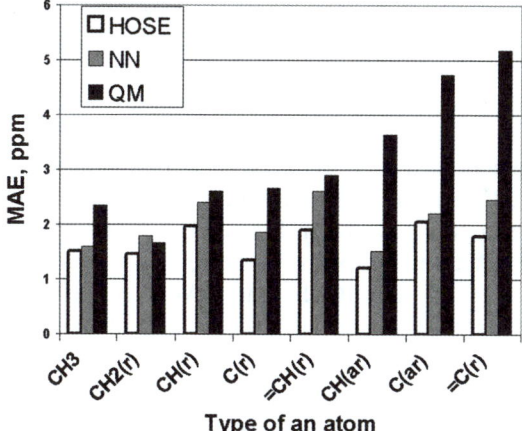

**Figure 3.19** A histogram of the MAE values associated with the corresponding ring carbon atoms in different hybridization states. The symbols C(ar) and CH(ar) denote atoms belonging to aromatic rings.

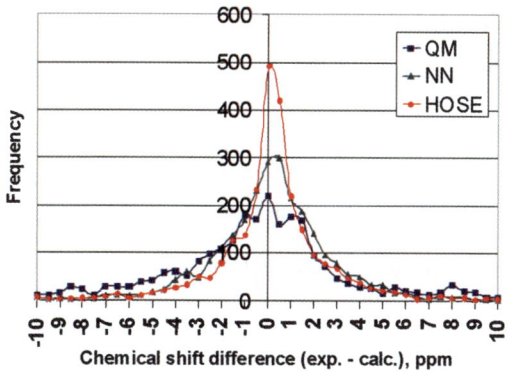

**Figure 3.20** The atom distributions with associated arithmetical differences between experimental and calculated chemical shifts ($\delta_{exp} - \delta_{calc}$).

curve. The shape associated with the QM distribution appears to be far from normal in nature. It has two additional maxima at $\pm 1$ ppm and the negative wing abates markedly slower than the positive one. This observation confirms the fact that the QM approach has a tendency to overestimate calculated chemical shifts when some frequently employed calculation protocols are used.[75]

## 3.3.3 Outliers and Unusual Structures

It was interesting to consider the structures for which the [13]C chemical shift prediction by QM and/or empirical methods produced large MAE values.

MAE values of close to 5 ppm are not rare cases for QM-based calculations (see Figure 3.3), and the structures for which MAE values $> 5$ ppm were obtained at least by one of the methods were examined. Typical structure–outliers with their corresponding MAE values and maximum errors $d_{max}$ were thoroughly analyzed. Analysis of the data showed that some large MAE values associated with the QM predictions related to the presence of: halogen atoms, heteroatoms carrying unshared electron pairs and high molecular flexibility. The contributions from these factors have been discussed in many works devoted to QM chemical shift prediction (for example, see references[75,78,107,108]). Figures 3.21 and 3.22 show plots of the HOSE- and QM-calculated $^{13}C$ chemical shifts *versus* experimental shifts for all atoms included in the structures related to outliers, 274 shifts in total.

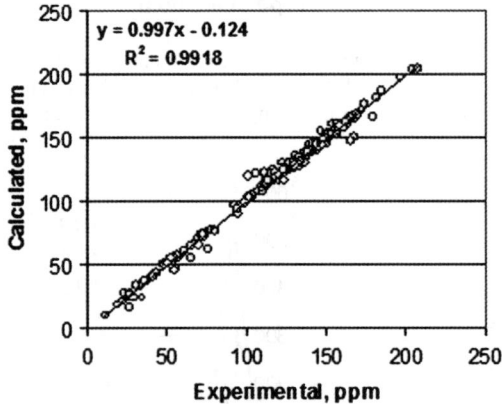

**Figure 3.21**   A linear regression plot of HOSE-based predicted $^{13}C$ chemical shifts *versus* experimental shifts for atoms included in the structures of outliers.

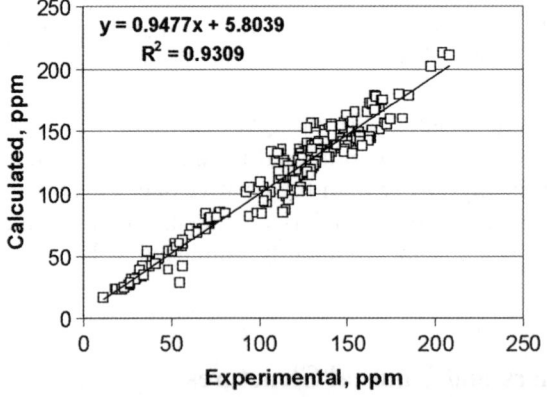

**Figure 3.22**   A linear regression plot of QM-based predicted $^{13}C$ chemical shifts *versus* experiment shifts for atoms included in the structures of outliers.

A comparison of the data presented in Figures 3.21 and 3.22 shows that HOSE-calculated chemical shifts are close to the experimental values (regression statistics: $\delta_{calc} = 0.997\delta_{exp} - 0.124$, $R^2 = 0.992$), whereas the QM-calculated shifts are markedly scattered and the intercept is equal to 5.8 ppm (regression statistics: $\delta_{calc} = 0.948\delta_{exp} + 5.804$, $R^2 = 0.931$).

As we see, the accuracy of QM methods for $^{13}C$ chemical shift prediction, in principle, is enough to distinguish between several possible structural hypotheses. However, the calculations are too slow to allow application of this approach to the selection of the most probable structure in a large output file from a CASE system. They, nevertheless, can play an important role in some of those cases occurring during structure elucidation aided by expert systems. These questions will be considered below in Chapter 12.

### 3.3.4 Synergistic Interaction between Empirical and Non-Empirical Methods

We can conclude that, in principle, both DFT GIAO and empirical calculations can be performed with sufficient accuracy to solve practical problems in organic chemistry. Nevertheless, for the examined structural set the average accuracy of QM methods is 1.5–2 times lower than the accuracy of empirical methods (see Table 3.7).

In regards to prediction speed, molecular size and the level of automation for QM approaches are inferior to empirical based approaches and these limitations, probably, are unlikely to be overcome in the near future. *Accuracy* is therefore the main criterion where QM methods have the potential to complement empirical methods and, in theory, maybe even surpass them.

Empirical methods are known to suffer from at least one principal drawback: if the database created for HOSE prediction, or the training set for the neural net algorithm, do not contain specific atoms representing the atom environments existing in the molecule under investigation, then the empirical methods can fail to predict the chemical shift of such atoms with sufficient accuracy. In these situations QM methods can compensate for the lack of representative data. However, the problem of accuracy should be solved to allow QM methods to be considered as a real analytical tool. We believe that current advances in QM, HOSE and NN $^{13}C$ NMR chemical shift prediction allow for the creation of an efficient strategy for jointly utilizing both empirical and non-empirical methods to solve actual analytical problems.

The most important task requiring the application of chemical shift prediction is that of complete structure elucidation, including the definition of stereochemistry. Empirical methods have been successfully used in this field for many years. Considering the growing capabilities of non-empirical approaches, it is possible to suggest the following strategy for a combined approach using both methods and, in theory, deliver a synergistic effect.

Experience accumulated over the last decade shows[61] that, in the overwhelming majority of cases, empirical methods allow for the successful sorting of structures using MAE(HOSE) values and determination of the most probable structure. Recall that the most probable structure is that which satisfies all constraints imposed by both the 1D and 2D NMR spectra and has the minimal MAE(HOSE) value. This structure fully satisfies the partial axiomatic theory formulated regarding the given spectrum–structural problem. If the MAE(NN) value is also minimal for the preferred structure, then this is considered as additional support for the selection made.

If questionable structures ranked first contain some substructure which seems "exotic" in nature, then it is possible to perform a preliminary search of this substructure in the database used for $^{13}C$ chemical shift prediction. Once it is identified that such a fragment is not contained within the database then a QM calculation could be applied to a rationally selected fragment from the molecule and could be used to deliver reliable chemical shifts which could then be merged in an appropriate fashion with the shifts which were calculated by HOSE and NN methods for the rest of the molecule. Of course, the shifts would be tagged appropriately to label their underlying prediction algorithm. This approach could also be used when the calculation protocol facility of the HOSE-based shift predictor informs the user that it is impossible to predict the chemical shifts for some atoms due to absence of related structures in the database. There are already publications where fragmental QM chemical shift calculations were utilized to select or confirm a structural hypothesis.[90,109]

There is reason to hope that as QM methods for NMR spectrum prediction are improved and the choice of the appropriate calculation protocol becomes a user-independent procedure, these methods will be more readily available for solving different spectrum–structural problems. A reasonable combination of QM and empirical approaches should provide a synergistic effect and will make both approaches more powerful and amenable to be used for practical purposes.

# 3.4   Heteronuclear NMR Spectrum Prediction

## 3.4.1   Phosphorus-31 NMR Spectrum Prediction

While the most prevalent nuclei for study by NMR spectroscopy are $^{1}H$ and $^{13}C$, there are, of course, many other nuclei that are amenable to study by NMR. Among these, the $^{31}P$ nucleus is one of the most commonly studied; it has 100% natural abundance and is a spin-1/2 nuclide. Small structural changes around the phosphorus atom can result in significant changes in the $^{31}P$ NMR spectrum and the chemical shift for a $^{31}P$ nucleus is very useful in determining its chemical environment. If a database of phosphorus-containing structures and associated chemical shifts, either available through a commercial source or produced by the user, is available, then these data can greatly speed up the process of identifying at least the class of a compound. As in the case of both $^{1}H$ and $^{13}C$ NMR prediction, there is considerable value in the prediction of $^{31}P$ NMR spectra to aid in structure identification.

A number of commercial databases and NMR prediction software modules are available[6–9,56] for $^{31}$P. As an example, ACD/PNMR[8] is accompanied by a content database that has been validated by NMR experts and assembled into a database that currently contains over 28 200 unique structures, including 34 700 experimental chemical shifts and 28 000 coupling constants. The database includes references, solvent, and NMR frequency when available.

The ACD/PNMR software program predicts $^{31}$P NMR spectra using modified HOSE code, NN and additive rules based algorithms that are similar to those described earlier for ACD/CNMR and ACD/HNMR. The accuracy of the program has been validated against numerous experimental data[7] and Figure 3.23 shows examples of predicted *versus* experimental chemical shifts. It is obvious that for these examples that the predicted chemical shifts are close enough to the experimental values to demonstrate the value of such predictions

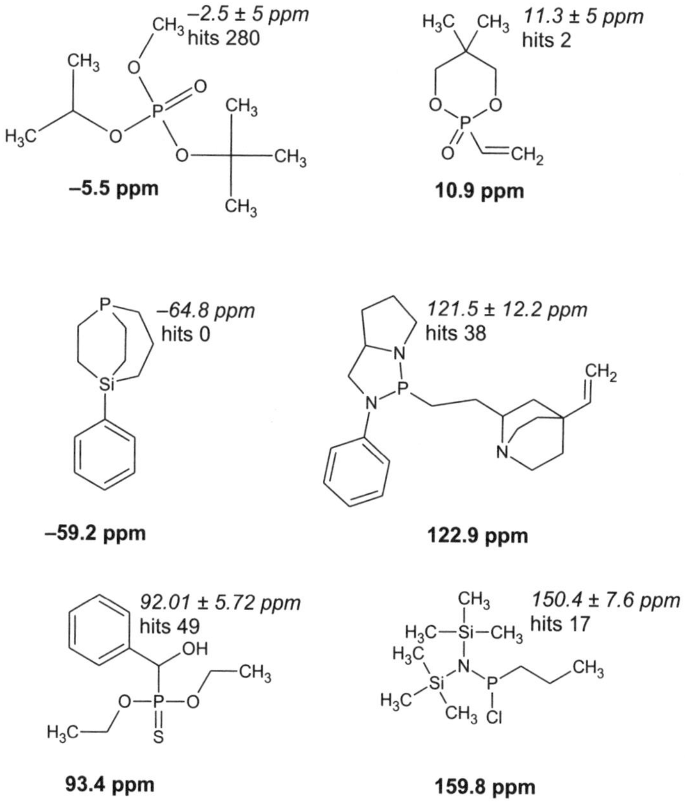

**Figure 3.23** A comparison of $^{31}$P NMR HOSE-calculated chemical shifts (italic) and experimental chemical shifts (bold). Also included are the numbers of database hits used to calculate the chemical shifts, as well as the confidence limits of the predictions. When no similar structures are available, a rules-based prediction is used.

and can provide valuable information to allow an investigator to accept or reject a structural hypotheses being considered by a chemist.

The program also allows a user to create a UDB to store chemical structures, [31]P chemical shifts, couplings, and associated data fields. This database can be used to improve the accuracy of the predictions, as well as providing a knowledge repository for future reference to help prevent duplication of effort in the analysis of phosphorus-containing compounds. In parallel to the capability described earlier for [1]H and [13]C NMR, it is possible to examine the details of how the [31]P NMR shifts were calculated using a chemical shift calculation protocol as shown in Figure 3.2.

The [31]P database may be used as an aid to a spectroscopist by allowing the search of one or more databases for a particular class of phosphorus-containing compound *via* structure, substructure, or similarity of structure searches. It is also possible to search one or more databases for a specific chemical shift or range of shifts to help in determining fragments with similar chemical shifts to help in the identification of an unknown.

## 3.4.2  Fluorine-19 NMR Spectrum Prediction

A number of [19]F databases and shift prediction software programs have been made available.[5,6,8,9,56] One of these, the ACD/FNMR database,[8] contains over 36 000 chemical shifts and 28 000 coupling constants associated with over 18 000 structures. All features described previously for the ACD/PNMR prediction software are contained within ACD/FNMR.

## 3.4.3  Nitrogen-15 NMR Spectrum Prediction

As with other NMR active nuclei the [15]N chemical shift is representative of the chemical environment. There has been considerable recent interest in the acquisition and utilization of long-range [1]H–[15]N heteronuclear shift correlation data at natural abundance, as shown by the publication of three major reviews on the subject in the last few years.[110–112] As described for the other nuclei access to a content database of structures and associated spectral parameters can accelerate the process of identification of an unknown. Electronic content databases of [15]N data are available from a number of sources[6,8,9] and the largest and most up to date source of data is associated with the ACD/NNMR Predictor program.

The ACD/NNMR Predictor database contains approx. 9500 chemical structures associated with over 22 000 [15]N chemical shifts and 4000 coupling constants. Data continue to be extracted from the literature on an ongoing basis and are checked according to a number of stringent criteria. An individual compound record includes the chemical structure, original literature reference, one or more [15]N chemical shifts, and, where available, associated heteronuclear coupling constants. The latter data are, unfortunately, only infrequently measured for [15]N. These data can be searched by different parameters including structure, substructure, and chemical shift. This database provides the foundation data for the derivation of HOSE code-based prediction algorithms for the $^{15}N$ nucleus.

As described earlier NMR prediction offers the possibility of structure verification based on chemical shifts. $^{15}N$ shift prediction in particular has been recommended to help optimize the parameters for the acquisition of spectra.[110] As usual, access is provided to the details of the prediction, as well as the possibility of creating a user database for both reference purposes and to facilitate and increase the accuracy of future predictions. To validate the performance of the $^{15}N$ chemical shift prediction algorithms a statistical analysis was performed.[110] Using a classical leave-one-out (LOO) approach, the $^{15}N$ shifts associated with approx. 8300 individual chemical structures contained within the ACD/NNMR v8.08 database were predicted. The resulting analysis gave a correlation coefficient of $R^2 = 0.97$ over 21 244 points. The standard error of approx. $\pm 15$ ppm is quite reasonable and affords a basis for setting the $F_1$ spectral widths in 2D NMR experiments to acquire long-range $^1H$–$^{15}N$ data.

To evaluate the capability of the ACD/NNMR spectrum prediction, Moser[113] selected a set of 56 nitrogen-containing structures with experimentally measured $^{15}N$ NMR chemical shifts. All compounds were absent from the ACD/NNMR database. A total of 100 chemical shifts were predicted. The calculated chemical shifts are plotted against the experimental shifts in Figure 3.24. The graph shows that the results obtained can be considered as quite satisfactory. The average deviation of the calculated chemical shifts from the experimental shifts is approx. 22 ppm. It should be acknowledged that this is based on a training set of only approx. 8500 compounds. It should be expected that the accuracy will continue to improve as the number of compounds contained in the training set increases in the future.

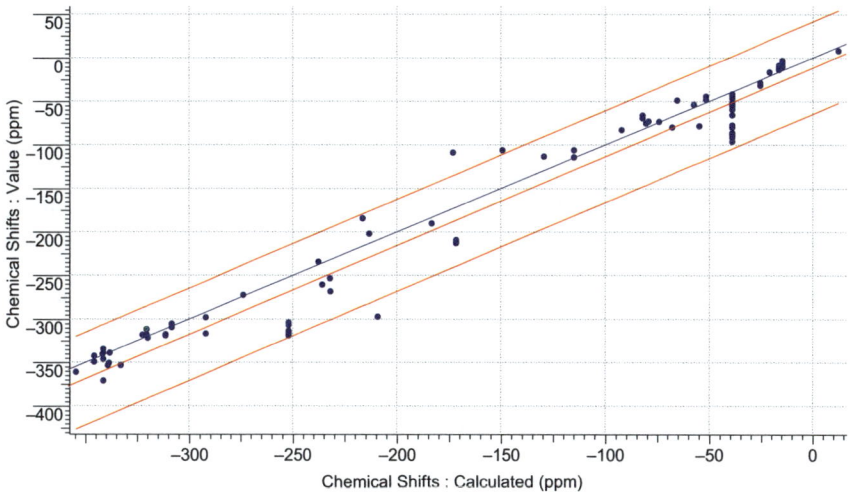

**Figure 3.24** The calculated and predicted $^{15}N$ NMR chemical shifts for 56 compounds (all are absent from ACD/NNMR database).

The $^{15}N$ NMR database is small at present relative to those available even for other heteronuclei. Despite the small size of the database, potential weaknesses in the $^{15}N$ shift prediction algorithms that are still being refined, and the variability of $^{15}N$ shifts as a function of solvent, temperature, and pH, $^{15}N$ shift calculations, nevertheless, provide a useful means of double checking intuitive estimates of $^{15}N$ chemical shift behavior for even experienced spectroscopists and can provide a vital starting point for experiment parameterization for less experienced investigators.

### 3.4.4   Other Heteronuclear NMR Predictions

A number of NMR databases and prediction algorithms are available for less commonly studied nuclei. These include a $^{29}Si$ system[56] and a $^{11}B$ system.[5,6] While reference databases for any NMR nuclei can be constructed using commercial software,[114] the derivation and validation of associated prediction algorithms has not been reported in the literature to the best of our knowledge.

## 3.5   Prediction of 2D NMR Spectra

2D NMR spectroscopy is now a routine tool used in virtually all modern NMR laboratories for the purpose of structure elucidation and verification. While skilled spectroscopists, in general, will find the interpretation of these data mundane in the majority of cases, the interpretation of 2D NMR data can still be a challenge for synthetic chemists. This is particularly true as 2D NMR data can be acquired robotically using an open access NMR spectrometers without any interaction with trained spectroscopists. To assist in the interpretation of these data, software has been developed that can predict various forms of both homonuclear and heteronuclear 2D NMR spectra for a combination of $^{1}H$, $^{13}C$, and $^{15}N$ nuclei.

When integrated with a 2D NMR processing package ACD/2D NMR Predictor allows direct comparison of experimental and predicted spectra of several types: homonuclear H,H-COSY and C,C-COSY (INADEQUATE); C,H HSQC (HMQC); N,H HSQC; and H,H and C,H J-resolved spectra (the latter two experiments are now only very rarely used). The user can compare predicted and experimental 2D NMR spectral pairs onscreen, synchronizing the axes as necessary, and overlaying them for direct comparison as necessary.

Since the 2D NMR prediction program uses the proton and carbon NMR prediction programs as a basis, it is possible to facilitate the 2D NMR predictions by utilizing the user-training capabilities available with the 1D NMR predictors. Fully automated verification of the correspondence between a proposed structure and one or more 2D NMR experimental spectra has been demonstrated.[115] Simpson *et al.*[116] applied the ACD/2D NMR Predictor program to produce 2D NMR databases containing spectra predicted from a series of chemical structures. These databases allow flexible searching *via* chemical structure, substructure, or similarity of structure, as well as by spectral features. The authors[116] used the biopolymer lignin as an example and demonstrated how a 2D NMR database of ~600 predicted 2D NMR spectra of lignin fragments could be easily constructed in ~2 days. Employing this database, many lignin fragments were identified in soil extracts through the use of various search tools and pattern recognition techniques. In this article[116] it was shown that such an approach could be used for the analysis of organic mixtures.

# References

1. M. E. Elyashberg, Y. Z. Karasev, E. R. Martirosian, H. Thiele and H. Somberg, *Anal. Chim. Acta*, 1997, **348**, 443.
2. L. A. Gribov and W. J. Orville-Thomas, *Theory and Methods of Calculation of Molecular Spectra*, Wiley, Chichester, 1988.
3. J. T. Clerc and H. A. Sommerauer, *Anal. Chim. Acta*, 1977, **95**, 33.
4. Scientists at Pfizer, Glaxo SmithKline and Novartis, (1996–2005), Private communications.
5. W. Robien, *Nachr. Chem. Tech. Lab.*, 1998, **46**, 74.
6. W. Robien, CSEARCH, http://felix.orc.univie.ac.at/~wr/csearch_server_info.html
7. L. D. Quin and A. J. Williams, *Practical Interpretation of P-31 NMR Spectra and Computer Assisted Structure Verification*, Advanced Chemistry Development, Inc., Toronto, 2004.
8. Advanced Chemistry Development. ACD/NMR Predictors. Prediction suite includes $^1$H, $^{13}$C, $^{15}$N, $^{19}$F, $^{31}$P NMR prediction, http://www.acdlabs.com, 2011.
9. Specinfo, Chemical Concepts GmbH, D-69442 Weinheim, Germany.
10. S. Berger and S. Braun, *200 and More NMR Experiments: A Practical Course*, Wiley, New York, 2004.
11. H. Kessler, P. Schmieder and M. Kurz, *J. Magn. Reson. Chem.*, 1989, **85**, 400.
12. A. Fürst and E. Pretsch, *Anal. Chim. Acta*, 1990, **229**, 17.
13. L. Chen and W. Pobien, *Anal. Chem.*, 1993, **65**, 12282.
14. K. L. Jensen, A. S. Barber and G. W. Small, *Anal. Chem*, 1991, **63**, 1082.
15. D. L. Clouser and P. C. Jurs, *Anal. Chim. Acta*, 1994, **295**, 221.
16. P. C. Jurs, J. W. Ball, L. S. Anker and T. L. Friedman, *J. Chem. Inf. Comput. Sci.*, 1992, **32**, 272.
17. W. Bremser, *Magn. Reson. Chem.*, 1985, **23**, 271.
18. L. Chen and W. Robien, *Chemom. Intell. Lab. Syst.*, 1993, **19**, 217.
19. H. N. Cheng and L. J. Kasehagen, *Anal. Chim. Acta*, 1994, **285**, 223.
20. H. Kalchhauser and W. Robien, *J. Chem. Inf. Comput. Sci.*, 1985, **25**, 103.
21. CS ChemDraw PRO, Cambridge Soft Corporation.
22. Upstream Solutions GMBH, NMR Prediction Products (SpecTool).
23. A. Fürst, E. Pretsch and W. Robien, *Anal. Chim. Acta*, 1990, **233**, 213.
24. R. C. Schweitzer and G. W. Small, *J. Chem. Inf. Comput. Sci.*, 1997, **37**, 249.
25. J. Zupan and J. Gasteiger, *Neural Networks for Chemists*, VCH, Weinheim, 1993.

26. J. Meiler, R. Meusinger and M. Will, *J. Chem. Inf. Model.*, 2000, **40**, 1169.
27. J. Meiler, W. Maier, M. Will and R. Meusinger, *J. Magn. Reson. Chem.*, 2002, **157**, 242.
28. V. Kvasnicka, *J. Math. Chem.*, 1991, **6**, 63.
29. J. P. Doucet, A. Panaye, E. Feuilleaubois and P. Ladd, *J. Chem. Inf. Comput. Sci.*, 1993, **33**, 320.
30. A. Panaye, J. P. Doucet, B. Fan, E. Feuilleaubois and S. R. El Azzouzi, *Chemom. Intell. Lab. Syst*, 1994, **24**, 129.
31. V. Kvasnicka, S. Sklenak and J. Pospichal, *J. Chem. Inf. Comput. Sci.*, 1992, **32**, 742.
32. L. S. Anker and P. C. Jurs, *Anal. Chem.*, 1992, **64**, 1157.
33. Y. Miyashita, H. Yoshida, O. Yaegashi, T. Kimura, H. Nishiyama and S. Sasaki, *J. Mol. (Struct. Theochem)*, 1994, **311**, 241.
34. B. E. Mitchell and P. C. Jurs, *J. Chem. Inf. Comput. Sci.*, 1996, **36**, 58.
35. D. L. Clouser and P. C. Jurs, *J. Chem. Inf. Comput. Sci.*, 1996, **36**, 168.
36. O. Ivanciuc, J.-P. Rabine, D. Cabrol-Bass, P. A. and J. P. Doucet, *J. Chem. Inf. Comput. Sci.*, 1996, **36**, 644.
37. Y. D. Smurnyy, K. A. Blinov, T. S. Churanova, M. E. Elyashberg and A. J. Williams, *J. Chem. Inf. Model.*, 2008, **48**, 128.
38. K. A. Blinov, Y. D. Smurnyy, T. S. Churanova, M. E. Elyashberg and A. J. Williams, *Chemom. Intell. Lab. Syst*, 2009, **97**, 91.
39. D. M. Grant and E. G. Paul, *J. Am. Chem. Soc.*, 1964, **86**, 2984.
40. E. Breitmaier and W. Voelter, *Carbon-13 NMR Spectroscopy*, VCH, Weinheim, 1987.
41. E. Pretsch, T. Clerc, J. Seibl and W. Simon, *Tables of Spectral Data for Structure Determination of Organic Compounds*, Springer-Verlag, Berlin, 1989.
42. K. A. Blinov, Y. D. Smurnyy, M. E. Elyashberg, T. S. Churanova, M. P. Kvasha, C. Steinbeck, B. A. Lefebvre and A. J. Williams, *J. Chem. Inf. Model.*, 2008, **48**, 550.
43. W. Bremser, *Anal. Chim. Act. Comp. Techn. Optimiz.*, 1978, **2**, 355.
44. A. B. Gray, J. G. Nourse, C. W. Crandell, D. H. Smith and C. Djerassi, *Org. Magn. Res.*, 1981, **15**, 375.
45. V. Schutz, V. Purtuc, S. Felsinger and W. Robien, *Fresenius' J. Anal. Chem.*, 1997, **359**, 33.
46. A. R. Leach and V. J. Gillet, *An Introduction to Chemoinformatics*, Kluwer Academic Publishers, Dordrecht, 2003.
47. J. Gasteiger and T. Engel, eds, *Chemoinformatics*, Wiley-VCH, Weinheim, 2003.
48. H. Satoh, H. Koshino, T. Uno, S. Koichi, S. Iwata and T. Nakata, *Tetrahedron*, 2005, **61**, 7431.
49. O. Ivanciuc, J.-P. Rabine, D. Cabrol-Bass, A. Panaye and J. P. Doucet, *J. Chem. Inf. Comput. Sci.*, 1996, 37, 587.
50. G. Hofle, N. Bedorf, H. Steinmetz, H. Reichenback and K. Gerth, *Angew. Chem. Int. Ed. Engl.*, 1996, **35**, 1567.
51. J. Meiler, *C_SHIFT*, http://www.krypton.org.uni-frankfurt.de/~mj, 1999.

52. J. Meiler, E. Sanli, J. Junker, R. Meusinger, T. Lindel, M. Will, W. Maier and M. Köck, *J. Chem. Inf. Comput. Sci.*, 2002, **42**, 241.
53. J. Meiler, B. Lefebvre, A. J. Williams and M. Hachey, *J. Magn. Reson. Chem.*, 2002, **157**, 242.
54. M. J. Frisch, G. W. Trucks, H. B. Schlegel, G. E. Scuseria, M. A. Robb, J. R. Cheeseman, V. G. Zakrzewski, J. A. Montgomery Jr., R. E. Stratmann, J. C. Burant, S. Dapprich, J. M. Millam, A. D. Daniels, K. N. Kudin, M. C. Strain, O. Farkas, J. Tomasi, V. Barone, M. Cossi, R. Cammi, B. Mennucci, C. Pomelli, C. Adamo, S. Clifford, J. Ochterski, G. A. Petersson, P. Y. Ayala, Q. Cui, K. Morokuma, P. Salvador, J. J. Dannenberg, D. K. Malick, A. D. Rabuck, K. Raghavachari, J. B. Foresman, J. Cioslowski, J. V. Ortiz, A. G. Baboul, B. B. Stefanov, G. Liu, A. Liashenko, P. Piskorz, I.Komaromi, R. Gomperts, R. L. Martin, D. J. Fox, T. Keith, M. A. Al-Laham, C. Y. Peng, A. Nanayakkara, M. Challacombe, P. M. W. Gill, B. Johnson, W. Chen, M. W. Wong, J. L. Andres, C. Gonzalez, M. Head-Gordon, E. S. Replogle and J. A. Pople, Gaussian, Pittsburgh, PA, 2001.
55. U. Stemberg, F. Koch and P. Losso, *Cosmos*, 2001, **4.52**.
56. Bio-Rad, Predict-It NMR, 2002.
57. S. Urban, J. W. Blunt and M. H. G. Munro, *J. Nat. Prod.*, 2002, **65**, 1371.
58. M. Anthony and P. Bartlett, *Neural Network Learning: Theoretical Foundations*, Cambridge University Press, Cambridge, 1999.
59. W. Press, S. Teukolsky, W. Vetterling and B. Flannery, *Numerical Recipes: The Art of Scientific Computing*, Cambridge University Press, Cambridge, 2007.
60. M. E. Elyashberg, A. J. Williams and G. E. Martin, *Prog. Nucl. Magn. Reson. Spectrosc.*, 2008, **53**, 1.
61. M. E. Elyashberg, A. K. Blinov, A. J. Williams, S. G. Molodtsov and G. E. Martin, *J. Chem. Inform. Model*, 2006, **46**, 1643.
62. M. E. Elyashberg, K. A. Blinov, S. G. Molodtsov, Y. D. Smurnyy, A. J. Williams and T. S. Churanova, *J. Cheminform.*, 2009, vol. 1:3, http://www.jcheminf.com/content/1/1/3.
63. C. Steinbeck, S. Krause and S. Kuhn, *J. Chem. Inf. Comput. Sci.*, 2003, **43**, 1733.
64. A Quality Check of the NMRShiftDB using the CSEARCH Algorithms http://nmrpredict.orc.univie.ac.at/csearchlite/enjoy_its_free.html
65. R. B. Schaller and E. Pretsch, *Anal. Chim. Acta*, 1981, **133**, 507.
66. R. B. Schaller, M. E. Munk and E. Pretsch, *J. Chem. Inf. Model.*, 1996, **36**, 239.
67. C. A. Shelley and M. E. Munk, *Anal. Chim. Acta*, 1994, **296**, 295.
68. J. Aires-de-Sousa, M. C. Hemmer and J. Gasteiger, *Anal. Chem.*, 2002, **74**, 80.
69. Y. Binev and J. Aires-de-Sousa, *J. Chem. Inf. Model.*, 2004, **44**, 940.
70. Y. Binev, M. Corvo and J. Aires-de-Sousa, *J. Chem. Inf. Model.*, 2004, **44**, 946.

71. T. G. Dietterich, in *The Handbook of Brain Theory and Neural Networks*, ed., M. A. Arbib, MIT Press, Cambridge, MA, 2002, pp. 405.

72. J. Sadowski, J. Gasteiger and G. Klebe, *J. Chem. Inf. Comput. Sci.*, 1994, **34**, 1000.

73. I. V. Tetko, *J. Chem. Inf. Comput. Sci.*, 2002, **42**, 717.

74. C. Lee, W. Yang and R. G. Parr, *Phys. Rev. B*, 1988, **37**, 785.

75. P. Cimino, L. Gomez-Paloma, D. Duca, R. Riccio and G. Bifulco, *Magn. Reson. Chem.*, 2004, **42**, S26.

76. A. Balandina, A. Kalinin, V. Mamedov, B. Figadere and S. Latypov, *Magn. Reson. Chem.*, 2005, **43**, 816.

77. N. J. R. Eikema Hommes and T. Clark, *J. Mol. Model.*, 2005, **11**, 175.

78. A. R. Katritzky, N. G. Akhmedov, J. Doskocz, C. D. Hall, R. G. Akhmedova and S. Majumder, *Magn. Reson. Chem.*, 2007, **45**, 5.

79. W. Migda and B. Rys, *Magn. Reson. Chem.*, 2004, **42**, 459.

80. K. W. Wiitala, C. J. Cramer and T. R. Hoye, *Magn. Reson. Chem.*, 2007, **45**, 819.

81. R. Infante-Castillo and S. P. Hernandez-Rivera, *J. Mol. Struct.*, 2009, **917**, 158.

82. M. Karabacak, A. Coruh and M. Kurt, *J. Mol. Struct.*, 2008, **892**, 125.

83. M. Karabacak, M. Cınar, A. Coruh and M. Kurt, *J. Mol. Struct.*, 2009, **919**, 26.

84. G. Barone, L. Gomez-Paloma, D. Duca, A. Silvestri, R. Riccio and G. Bifulco, *Chemistry*, 2002, **8**, 3233.

85. A. Bagno, F. Rastrelli and G. Saielli, *Chemistry*, 2006, **12**, 5514.

86. A. Balandina, D. Saifina, V. Mamedov and S. Latypov, *J. Mol. Struc.*, 2006, **791**, 77.

87. A. A. Balandina, V. A. Mamedov, E. A. Khafizova and S. K. Latypov, *Russ. Chem. Bull.*, 2006, **55**, 2256.

88. P. Wipf and A. D. Kerekes, *J. Nat. Prod.*, 2003, **66**, 716.

89. K. N. White, T. Amagata, A. G. Oliver, K. Tenney, P. J. Wenzel and P. Crews, *J. Org. Chem.*, 2008, **73**, 8719.

90. T. A. Johnson, T. Amagata, A. G. Oliver, K. Tenney, F. A. Valeriote and P. Crews, *J. Org. Chem.*, 2008, **73**, 7255.

91. C. Fattorusso, E. Stendardo, G. Appendino, E. Fattorusso, P. Luciano, A. Romano and O. Taglialatela-Scafati, *Org. Lett.*, 2007, **9**, 2377.

92. E. Fattorusso, P. Luciano, A. Romano, O. Taglialatela-Scafati, G. Appendino, M. Borriello and C. Fattorusso, *J. Nat. Prod.*, 2008, **71**, 1988.

93. S. D. Rychnovsky, *Org. Lett.*, 2006, **8**, 2895.

94. A. M. Sarotti and S. C. Pellegrinet, *J. Org. Chem.*, 2009, **74**, 7254.

95. C. A. Franca, R. P. Diez and A. H. Jubert, *J. Mol. Struct. (THEOCHEM)*, 2008, **856**, 1.

96. V. Barone, P. Cimino, O. Crescenzi and M. Pavone, *J. Mol. Struct.*, 2007, **811**, 323.

97. M. E. Elyashberg, K. A. Blinov and A. J. Williams, *Magn. Reson. Chem.*, 2009, **47**, 371.

98. M. E. Elyashberg, K. A. Blinov and A. J. Williams, *Magn. Reson. Chem.*, 2009, **47**, 333.
99. I. Stappen, G. Buchbauer, W. Robien and P. Wolschann, *Magn. Reson. Chem.*, 2009, **47**, 720.
100. P. A. M. Dirac, *History of Twenties Century Physics: Proceedings of The International School of Physics "Enrico Fermi". Course LVII.*, Academic Press, London, 1977.
101. D. B. Chesnut, *Chem. Phys. Lett.*, 2003, **380**, 251.
102. A. E. Aliev, D. Courtier-Murias and S. Zhou, *Mol. Struct. (THEO-CHEM)*, 2009, **893**, 1.
103. M. E. Elyashberg, K. A. Blinov, Y. D. Smurnyy, T. S. Churanova and A. J. Williams, *Magn. Reson. Chem.*, 2010, **48**, 219.
104. R. Infante-Castillo, L. A. Rivera-Montalvo and S. P. Hernandez-Rivera, *J. Mol. Struct.*, 2008, **887**, 10.
105. C. J. Pouchert and J. Behnke, *Aldrich Library of 13C and 1H FT-NMR Spectra*, 1993.
106. M. E. Elyashberg, K. A. Blinov, S. G. Molodtsov, A. J. Williams and G. E. Martin, *J. Chem. Inf. Comput. Sci.*, 2004, **44**, 771.
107. K. Dybiec and A. Gryff-Keller, *Magn. Reson. Chem.*, 2009, **47**, 63.
108. A. Bagno, F. Rastrelli and G. Saielli, *Magn. Reson. Chem.*, 2008, **46**, 518.
109. D. Sanz, R. M. Claramunt, A. Saini, V. Kumar, R. Aggarwal, S. P. Singh, I. Alkorta and J. Elguero, *Magn. Reson. Chem.*, 2007, **45**, 513.
110. G. E. Martin and A. J. Williams, *Ann. Rep. NMR Spectrosc.*, 2005, **55**, 1.
111. G. E. Martin and C. E. Hadden, *J. Nat. Prod.*, 2000, **65**, 543.
112. R. Marek and A. Lycka, *Curr. Org. Chem.*, 2002, **6**, 35.
113. A. Moser, ACD/Labs, private communication, 2005.
114. ACD/ChemFolder, Advanced Chemistry Development, Inc.
115. B. Lefebvre, *NMR Discussion Group*, Stevenage, UK, March 3, 2005.
116. A. J. Simpson, B. Lefebvre, A. Moser, A. J. Williams, N. Larin, M. Kvasha, W. L. Kingery and B. Kelleher, *Magn. Reson. Chem.*, 2004, **42**, 14.

CHAPTER 4

# Methods of Relative Stereochemistry Determination in CASE Systems

The biological activity of natural products and drug molecules is generally dependent on the relative stereochemistry of a molecule. Indeed, there are examples known where one stereoisomer can exhibit vastly different pharmacologic activity from the other stereoisomer. As an example, D-propoxyphene has analgesic activity, whereas the other optical isomer, L-propoxyphene, has antihistaminic activity. Another example is thalidomide where one enantiomer is effective against morning sickness, whereas the other is teratogenic. Stereoisomer considerations can influence reaction pathways and certainly reaction kinetics, with one form being favored over another. Generally, the final step in contemporary structure characterization efforts is to define the relative, and, if possible, absolute stereochemistry. NMR methods are generally well suited to the former, whereas the latter is generally obtained using chemical structure modification combined with NMR studies[1] or by X-ray crystallographic methods. NMR-based determination of relative stereochemistry is based on the nuclear Overhauser effect (NOE), which is dependent on the distance separating the cross-relaxing nuclides.[2] Typically, NOESY or ROESY 2D NMR experiments or their selective 1D analogs are used to provide the data for this analysis in rigid molecules. In the case of flexible molecules, considerable effort has been devoted to the development of J-based NMR methods that are used to measure long-range heteronuclear coupling constants that can then be used to assign the relative stereochemistry.[3] It was therefore attractive to develop a NOESY/ROESY-based algorithm that would be capable of determining the relative stereochemistry of an unknown molecular structure which was identified by a CASE system.

New Developments in NMR No. 1
Contemporary Computer-Assisted Approaches to Molecular Structure Elucidation
By Mikhail Elyashberg, Antony Williams and Kirill Blinov
© Royal Society of Chemistry 2012
Published by the Royal Society of Chemistry, www.rsc.org

At the same time, as discussed in the previous Chapter 3, QM approaches and particularly the GIAO option of the DFT method[4] are widely utilized as tools for NMR chemical shift calculations. It was shown that DFT-based methods of [13]C chemical shift calculation could be applied for the selection of a preferable structural hypothesis by means of comparing the predicted chemical shifts with those determined experimentally. The QM approach is also an efficient tool for evaluating the different conformers of flexible molecules, as well as the elucidation of the most probable stereoisomers.[4–8]

As was shown above (see Section 3.3), the accuracy of empirical methods of NMR chemical shift prediction is at least as good as, and sometimes better than, QM approaches. These methods are successfully used at the stage of structural hypothesis selection for which many examples will be given in later parts of this book. In this regard, a hypothesis arose[9] that empirical methods could also help in the preliminary selection of a set of the most probable stereoisomers for subsequent verification by additional experimental techniques as well as by QM chemical shift prediction. This may be possible since, if stereocenter configurations of structures are included into the ACD/CNMR database, then their known stereochemistry is taken into account by the NMR chemical shift prediction algorithms. The incremental (Inc) and neural network (NN)-based algorithms of chemical shift prediction also use the stereochemistry information related to the atoms included into 3–6-membered cycles.[10] A separate work by Elyashberg *et al.*[9] posed the question as to whether this information could be useful for the determination of relative stereochemistry. Examination of a structural test set containing natural products has shown[9] that empirical methods of [13]C chemical shift prediction, indeed, allow selection of a group of the most probable stereoisomers including the genuine stereoconfiguration. This fact provides grounds for suggesting that these methods can be used in CASE systems as a preliminary filter of the most probable stereoisomers.

In this Chapter we will consider two approaches to determine the relative stereochemistry implemented into the ACD/Structure Elucidator CASE system. The first is based on [13]C chemical shift prediction by empirical methods, whereas the second utilizes an algorithm to apply NOESY/ROESY-based energy minimization of the set of stereoisomers produced from a given structural formula.

# 4.1 Selection of the Most Probable Stereoisomer Set

The possibility of selecting the most probable stereoisomer was investigated using a series of examples. The examples were taken from recent literature (2007–2008) where novel structures and their relative stereochemistry were reported. Consequently, these structures were deliberately absent from the ACD/CNMR database which serves as a source of reference structures for NMR spectrum prediction.

The first example was taken from the article published by Fattorusso *et al.*[6] The authors[6] utilized GIAO DFT chemical shift computation to confirm the most probable stereoisomer of artarborol, **4.1**, a rare *nor*-caryophyllane derivative, isolated and structurally characterized by both 1D and 2D NMR spectroscopic methods.

**4.1**

To select the most probable stereoisomer the authors[6] carried out a series of investigations. Structure **4.1** contains five stereogenic carbons (numbered 1–5 on the structure) with four of them at junctions between the 9-membered ring and the small ring cycles, whereas both *cis*- and *trans*-junctions of rings adjacent to the nine-membered core are possible in natural caryophyllanes.

A combination of 2D ROESY experiments with Mosher's modified method[11] was used to assess the absolute configuration of C-2 (**R**) and allowed the authors[6] to reduce the total number of possible stereoisomers to four (Figure 4.1).

Further selection was made by analyzing the scalar coupling constants and additional spatial couplings across the entire molecule for which all candidate structures were subjected to a conformational search. As a result, structures **B** and **D** were rejected at the first step, structure **C** was then excluded, and, finally, stereoconfiguration **A** was assigned to artarborol. To support this stereochemical assignment each conformation of the stereoisomers **A** and **C** were fully optimized by the authors,[6] and the NMR chemical shifts were calculated using the GIAO option of the MPW1PW91/6-31G(d,p) DFT method.[12] A Boltzmann-weighted average of the [13]C NMR chemical shifts for all carbon atoms in the low-energy conformers was calculated for each configuration,

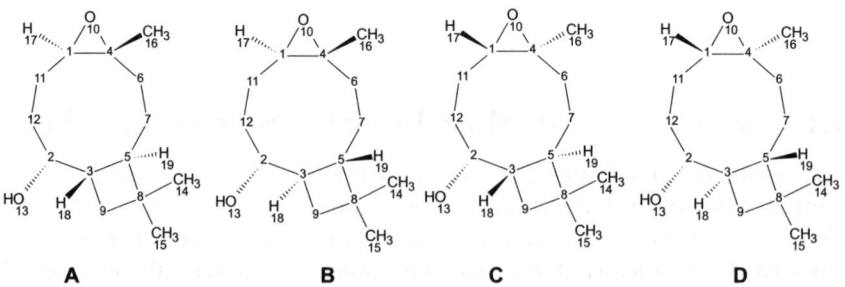

**Figure 4.1**   The four candidate stereoisomer structures of artarborol.

using the *ab initio* standard free energies as weighting factors.[13] The total processing time for each molecule was approx. 60 h (PC Pentium IV). A comparison of calculated chemical shifts with those determined experimentally for structures **A** and **C** showed that deviations were smaller for structure **A**, thereby confirming the validity of the solution.

Selection of the most probable stereoisomer was attained as a result of a comprehensive experimental and theoretical investigation of the compound and its conceivable 3D models. Elyashberg *et al.*[9] investigated what results would be obtained if the problem is solved using 1D and 2D (NOESY/ROESY) NMR spectra and the empirical chemical shift prediction methods described in Chapter 3.

To perform this analysis structure **4.1** was input into the system and all carbon and hydrogen atoms were supplied with experimental chemical shifts in accordance with the assignment of the authors.[6] Then, all $2^5 = 32$ stereoisomers were generated by the program and depicted using conventional designations for stereobonds. $^1$H and $^{13}$C chemical shifts were calculated for the complete stereoisomer set using the fragment-based approach within the Structure Elucidator program. In addition, $^{13}$C NMR chemical shifts were calculated using both NN and Inc approaches.

The average deviations of the predicted chemical shifts relative to the experimental shifts ($d_A$ = fragmental approach, $d_N$ = NN approach and $d_I$ = Inc approach) were calculated for each of 32 stereoisomers and all stereoisomers were ranked in ascending order of the $^{13}$C deviation values. Since the chemical shifts are insensitive to the absolute configuration of a stereoisomer and its inverse partner, the reduced ranked stereoisomer set was finally represented as a sequence of 16 stereoisomer pairs, each pair having equal deviations. Figure 4.2 shows the first 8 out of 16 "unique" stereoisomers ranked in ascending order of the average deviations calculated for $^{13}$C NMR spectrum. The remaining stereoisomers are characterized by $^{13}$C average deviations $d_A(^{13}\text{C})$ falling in the range between 2.49 and 2.90 ppm.

Figure 4.2 shows that the correct stereoisomer was distinguished both by its $^{13}$C and $^1$H average deviations. Experiences in the field of computer-aided structure elucidation have shown[14] that the $d_A(^1\text{H})$ deviation is a less reliable criterion compared with $d_A(^{13}\text{C})$ and it is usually only used for additional confirmation of the most probable structural isomer.[14–16] The difference between the deviations $d_A(^{13}\text{C})$ found for the second- and first-ranked structures is not large (0.2 ppm), but this value is frequently observed in the structure elucidation process when the "best structure" is selected.[14] It is worthy to note that in the stereoisomers 3, 4, 6 and 9, atoms H-17 and H-19 are situated on opposite sides of the macrocycle and are unlikely to be close enough in space to show a ROESY coupling. As Fattorusso *et al.*[6] made the final choice between structures **A** and **C** on the basis of comparison of differences between experimental and DFT calculated $^{13}$C chemical shifts of all carbon atoms, Elyashberg *et al.*[9] also compared these values (see Figure 4.3).

Figure 4.3 shows that the main difference between the chemical shifts calculated for structures **A** and **C** is observed for atoms 6 and 7. For structure **A**

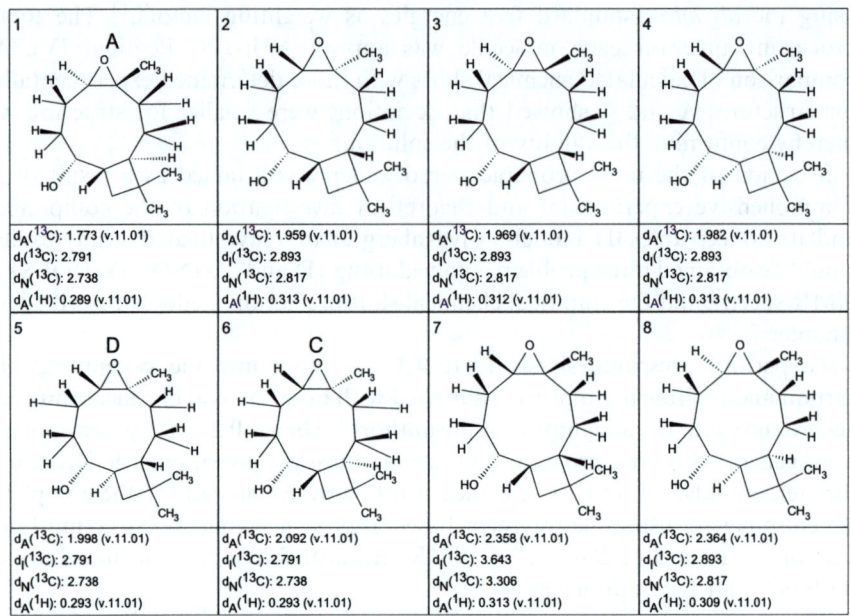

**Figure 4.2**  The first 8 out of 16 stereoisomers ranked in ascending order of the average deviation $d_A({}^{13}C)$.

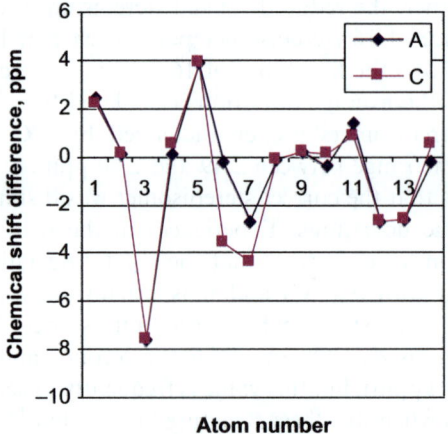

**Figure 4.3**  A comparison of the ${}^{13}C$ chemical shift deviations calculated for the carbon atoms contained in stereoisomers **A** and **C**.

the calculated values are markedly closer to the experimental values. The maximum prediction errors are shown for atoms 3 and 5 at the junction between the macrocycle and the four-membered ring. Stereoisomer ranking with $d_N({}^{13}C)$ and $d_I({}^{13}C)$ values, in general, supported the priority of stereoisomers **A–D**; these fell into the first four stereoisomers for which all $d_N(^{13}C)$ values and all $d_I(^{13}C)$ values proved to be equal.

The approach described here looks attractive due to its simplicity and high speed; the $^{13}C$ and $^1H$ chemical shift calculations by four methods for all 32 isomers took approx. 2 min on a Pentium IV, 2.8 MHz processor compared to 60 h per prediction as reported by the authors of the original paper.[6] It could be useful for the preliminary assessment of a full stereoisomer set and rejection of deliberately improbable structures when the analyzed molecule is relatively rigid. The reliability of such conclusions can be heuristically evaluated by visual comparison of the reference structures used for chemical shift prediction with the target structure. For instance, a series of structures containing the ring framework of artarborol were shown by the program when examining the chemical shift prediction protocol (see Section 3.1.3.3). It should be emphasized that the artarborol molecule was a newly identified compound and was absent from the library of structures included with the ACD/NMR prediction program. The reference structure **4.2** is the most similar structure to the artarborol structure under investigation:

**4.2**

It was demonstrated that removing structure **4.2** from the database did not influence the results: the deviation characteristic for the best stereoisomer was only slightly increased from 1.77 to 1.80 ppm.

The described approach was also applied to two new ketopelenolides **4.3** and **4.4** which were separated and scrutinized by the same research group.[17] The stereochemistry shown in structures **4.3** and **4.4** was determined by Fattorusso *et al.*[17] as a result of conformational analysis and the QM-based $^{13}C$ chemical shift calculation of the most probable stereoisomers. The calculations were performed in groups of four stereoisomers for each structure (**C1–C4** for structure **4.3** and **D1–D4** for structure **4.4**, see Figure 4.4). It has been shown that **C1** corresponds to stereoisomer **4.3** and **D1** to stereoisomer **4.4**.

**4.3**          **4.4**

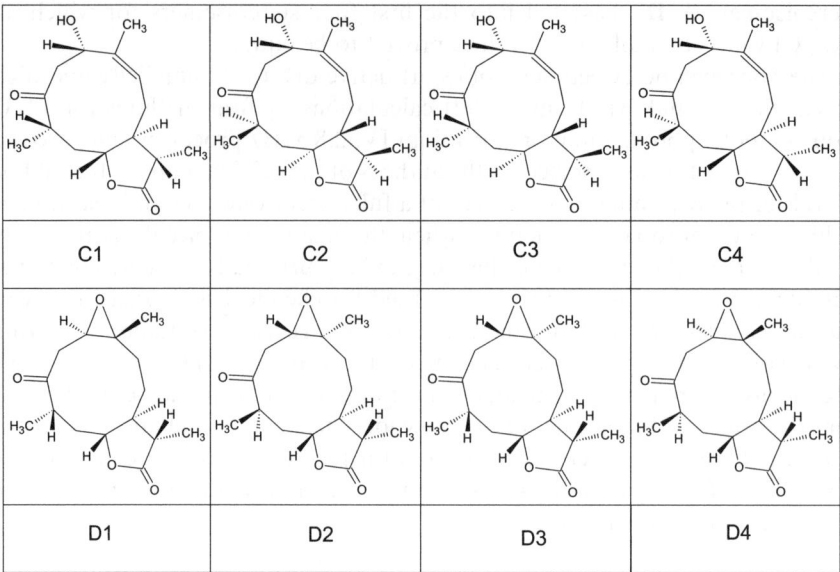

**Figure 4.4**  The most probable stereoisomers of structures **4.3** and **4.4** selected for detailed theoretical analysis.[17]

The Structure Elucidator was used to generate all possible stereoisomers for structures **4.3** and **4.4** (in both cases $N = 64$) and to perform NMR chemical shift calculations for all stereoisomers. [13]C chemical shift prediction using the fragmental method placed stereoisomer **C2** in first position in the ranked file and the genuine stereoisomer, **C1**, at the second position with a difference between the deviations of only 0.01 ppm. At the same time ranking the stereoisomers using the $d_N(^{13}C)$ values put stereoisomers **C1–C4** into the 1–4 positions with equal $d_N(^{13}C)$ and $d_I(^{13}C)$ values for all of the stereoisomers. For structure **4.4**, the stereoisomers were ranked by $d_A(^{13}C)$ values in the order they are displayed in Figure 4.4: first, **D1**; second, **D2**; third, **D3**; fifth, **D4**; *i.e.* the correct stereoisomer was placed in the first position and the other most probable stereoisomers selected in reference[17] were distinguished by the program as also deserving attention.

For preliminary evaluation of the generality of the described approach Elyashberg *et al.*[9] repeated the work using the structures of natural products belonging to a number of different classes, *i.e.* steroids, alkaloids, terpenes, cembranoids, *etc.* A set of such structures whose relative stereochemistry was recently described in a series of publications was chosen (see Table 4.1).

All selected structures were supplied with assigned experimental [1]H and [13]C NMR chemical shifts. For each molecule a full set of $N$ possible stereoisomers was generated, and the [13]C NMR chemical shifts of $N_{ds}$ differing stereoisomers ($N_{ds} = N/2$, $N = 2^n$, $n$ is the number of stereocenters) were calculated by all three of the mentioned algorithms. A stereoisomer file was ranked in the same way as in the artarborol case – in descending order of $d_A(^{13}C)$ values, and the position of the correct stereoisomer, as determined in a corresponding article, was detected in the ranked file. The result of each computational experiment was

**Table 4.1** Examples of structures for which sets of preferable stereoisomers were selected using empirical methods of $^{13}C$ NMR chemical shift prediction. The R and S designations shown in the structures correspond to the stereochemistry at the particular stereocenters.

| Example no. | Structure | $N_{ds}$, Number of stereo-isomers | $S_r$, Position of correct stereoisomer | Ref. |
|---|---|---|---|---|
| 1 | | 1024 | 1 | 18 |
| 2 | | 256 | 1 | 19 |
| 3 | | 32 | 1 | 20 |
| 4 | | 32 | 1 | 21 |
| 5 | | 64 | 1 | 22 |

**Table 4.1** (*Continued*).

| Example no. | Structure | $N_{ds}$, Number of stereo-isomers | $S_r$, Position of correct stereoisomer | Ref. |
|---|---|---|---|---|
| 6 | | 32 | 3 | 23 |
| 7 | | 128 | 3 | 24 |
| 8 | | 2048 | 3 | 25 |
| 9 | | 1024 | 3 | 26 |
| 10 | | 512 | 3 | 22 |

**Table 4.1** (*Continued*).

| Example no. | Structure | $N_{ds}$, Number of stereo-isomers | $S_r$, Position of correct stereoisomer | Ref. |
|---|---|---|---|---|
| 11 | | 64 | 3 | 27 |
| 12 | | 32 | 4 | 21 |
| 13 | | 256 | 8 | 28 |
| 14 | | 256 | 12 | 29 |

characterized by an $S_r$ value where $S_r$ is the number of stereoisomers for which the deviations $d_A(^{13}C)$ are less than or equal to the deviation calculated for the right stereoisomer. For instance, $S_r = 1$ means that the right stereoisomer was ranked first in the file with deviation $d_{A1}(^{13}C)$, and $d_{A1}(^{13}C) < d_{A2}(^{13}C)$, where $d_{A2}(^{13}C)$ is the deviation calculated for the stereoisomer ranked in second position. The notation $S_r = 4$ means that the correct stereoisomer is among the first four stereoisomers in the ranked file.

Table 4.1 shows that the suggested approach can indeed be used for selecting a set of the most probable stereoisomers from all possible members of the family. Even for rather complex structures, the preferable stereoisomer was ranked early in the set. Stereoisomer ranking using $d_N(^{13}C)$ is not as effective as $d_A(^{13}C)$, but nevertheless in this case the right stereoisomer most frequently fell into the set of the first eight ranked stereoisomers. Consequently, the NN approach can also be used for preliminary ranking of the stereoisomer file for subsequent spectrum prediction based on the fragmental method as is common in the Structure Elucidator system.[16] When NOESY/ROESY data were available from the corresponding articles, the common considerations using these data for the evaluation of structures presented in the top sets of structures ($S_r = 3–12$) showed that the right stereoisomer is the preferred one algorithmically also.

## 4.2    Computer-aided Determination of Relative Stereochemistry and 3D Models of Complex Organic Molecules from 2D NMR Spectra

In this Section we will describe an algorithm[30] implemented into the Structure Elucidator CASE system that allows for the determination of the relative stereochemistry of a molecular structure based on the NOE constraints. The program extracts NOE information from either NOESY and/or ROESY spectra and determines the molecular stereochemistry accordingly. The results of selective NOE or ROE experiments can also be used as inputs to the program. This process can be carried out for several of the most likely chemical structures produced during a structure elucidation by the expert system or performed on a structure proposed by the chemist.

The utility of NOESY/ROESY spectra for relative stereochemistry determination is based on a direct correlation between both the cross-peak volume integration and the internuclear distance. The peak intensity in NOE/ROE measurements has an inverse sixth power relationship. Consequently,[31] what can be referred to as a "strong" NOE is generally observed between pairs of hydrogens which are 1.8–2.5 Å apart. Responses of "medium" intensity usually correspond to an internuclear distance of 2.5–4.0 Å, whereas "weak" NOEs will generally be observed for larger distances if they are observed at all. NOE responses are not commonly observed for nuclei farther than 6.0 Å apart. These data can be used as constraints imposed on stereoisomer structures during the process of energy minimization.

Minimization algorithms deal with *numerical values* and, in this case, these numerical values are extracted from a set of NOEs overlaid on a 3D structure and examined for goodness of fit. The function describing this goodness of fit is called a *penalty function*. The better the solution, then the lower the value of the function. The function must exhibit the lowest value for the best-matching stereoisomer. Smurnyy *et al.*[30] suggested an appropriate function that can be minimized by calculation for *all* stereoisomeric structures or by using a genetic algorithm[32] to limit the number of stereoisomers that need to be investigated. To improve genetic algorithm convergence efficient methods of parameter optimization are suggested and compared. The algorithm is described below.

## 4.2.1 3D Structure Optimization with Molecular Mechanics

Optimizing the three-dimensional geometry of an organic molecule from a molecular graph is a well-known problem that, in the case of small molecules with less than 10–30 non-hydrogen atoms, can be performed using non-empirical QM methods. These methods are computationally cumbersome and are rarely used in analytical and drug discovery applications. Fortunately, for most small organic molecules the task can be treated with a classical mechanics approach ("molecular mechanics").[33]

Falk *et al.*[34] reported a molecular modeling procedure to determine the relative configuration of a chiral molecule from NMR data. The procedure used constrained molecular mechanics with the constraints being interproton distances derived from the experimental NOE data. High-temperature molecular dynamics allowed inversions at most stereocenters and the distance constraints guided the molecule into configurations consistent with the NOE data. For molecules with complex ring systems this approach can fail to invert certain centers with sufficient frequency and this was countered by allowing additional inversions of selected stereocenters. The procedure was proven on organic molecules of known stereochemistry, with 5–17 chiral centers, provided that the number of available constraints was at least twice the number of stereocenters. The procedure was shown to be tolerant of large errors in the estimated interproton distances and is reasonably rapid.

Smurnyy *et al.*[30] used a molecular mechanics method with the popular CharMM parameterization[34] as a starting point that is subsequently minimized to obtain the optimal molecular geometry.

Commonly, the full molecular energy ($E_{Total}$) is estimated by the following equation:

$$E_{Total} = E_{bonds} + E_{angles} + E_{dihedrals} + E_{improper} + E_{VdW} \qquad (4.1)$$

where $E_{bonds}$, $E_{angles}$, and $E_{dihedrals}$ stand for the energies of covalent bonds, valence, and dihedral angles respectively, $E_{VdW}$ is the van der Waals energy, and are calculated as follows:

$$E_{\alpha} = \sum k_{\alpha}(\alpha_{real} - \alpha_{ideal})^2 \qquad (4.2)$$

where $\alpha_{real}$ represents the current value of a structural parameter (*i.e.* bond length) and $\alpha_{ideal}$ is the "ideal" value of the same parameter as defined and tabulated in the original article.[34] In this expression the term $k_{\alpha}$ is a proportionality coefficient and $E_{improper}$ corresponds to the energy of "improper torsions", which are normal torsion angles with larger associated proportionality constants. The larger proportionality constants in this case serve to provide constraints that serve, for example, to keep benzene rings planar, *etc.* Improper torsions were also used to maintain stereocenter configurations while performing the molecular geometry optimizations.

For each chiral center, improper torsions, related to the energy value $E_{improper}$, are introduced as indicated on the drawing below.

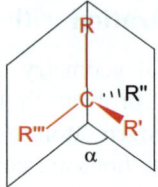

When a molecule is mirrored, the value of such a torsion change is energetically high so that freezing these angles prevents the stereocenters from being inverted. Finally, $E_{VdW}$ is calculated as follows[34] (the sum is taken over all possible pairs of atoms):

$$E_{VdW} = \sum_{i \neq j} k_{VdW}(a_{ij}R_{ij}^{-12} - bR_{ij}^{-6}) \tag{4.3}$$

and is an approximation of the van der Waals forces. Here the $a_{ij}$ and $b$ coefficients depend on atom types, $R$ is an internuclear distance and $k_{VdW}$ is a proportionality constant.

If distance restrictions derived from the NOE correlations are to be included into the calculation then $E_{GrandTotal}$ is used instead of $E_{Total}$:

$$E_{GrandTotal} = E_{Total} + E_{NOE} \tag{4.4}$$

where $E_{NOE}$ is an NOE penalty function characterizing how consistent the current geometry is with the set of NOE restraints. The next Section describes this penalty function.

## 4.2.2   NOE Penalty Function Calculation

To make use of NOESY correlations, a penalty function for a given pair of atoms has been chosen as follows.[35] The total *penalty function* is a sum of terms:

$$E_{ij}^{NOE} = k_{NOE} \begin{cases} \sqrt{1 + \dfrac{(D^2 - L^2)^2}{L^4}} - 1, & \text{if} \quad D < L \\ \qquad\qquad 0 \quad \text{if} \quad L \leq D \leq U \\ \sqrt{1 + \dfrac{(U^2 - D^2)^2}{U^4}} - 1 & \text{if} \quad U < D \end{cases} \tag{4.5}$$

where $D$ is the current value of an internuclear distance and $U$ and $L$ are upper and lower estimates, respectively, for the internuclear distance. These estimates are derived from the NOESY cross-peak intensity, as will be shown below.

The term $k_{NOE}$ is a proportionality constant that is consistent with other proportionality constants used elsewhere in the force field. For the CharMM parameterization values of 5–20 kcal/mol are suitable.

To obtain estimates for the values of $U$ and $L$, two schemes have been implemented. The final choice, however, depends on the quality of the spectra and is ultimately left to the user.

- The first scheme was suggested by Wüthrich.[31] When this approach is used NOESY peaks are separated into three groups according to their volumes: "strong" (15–100% of maximum the peak volume of the largest off-diagonal peak), "medium" (1–15% of maximal peak volume) and "weak" (everything else above noise level) according to the peak volumes. Corresponding internuclear distance constraints are assigned for the each group: 1.8–2.5 Å for strong, 2.5–4.0 Å for medium, and 4.0–6.0 Å for weak responses. Experiments have shown that using peak heights instead of volumes decreases the calculation accuracy because of how substantially peak widths can differ.
- Through-space interactions between nuclei can be regarded as magnetic dipole–dipole interactions. The equations that govern dipole–dipole relaxation define the intensity of such an interaction to be proportional to $r^{-6}$, where $r$ is the distance between the two protons engaged in the dipole–dipole relaxation process.[36] Consequently, the upper limit of the distance between two protons can be calculated as follows:

$$U = \left( \frac{r_{ref}^{-6}}{V_{ref}} V \right)^{-1/6} \tag{4.6}$$

where $V_{ref}$ is the volume of the reference peak and $r_{ref}$ is the corresponding internuclear distance between the reference peak and the peak of interest. In most cases, the cross-peaks between anisochronous geminal protons in a $CH_2$ group are used as reference peaks, since such peaks are almost always present, and the $r_{ref}$ value is well-established ($\sim 1.8$ Å). Because the intensity of a cross-peak is sensitive to many experimental parameters, eqn (4.6) can only be used to calculate the *upper* limit of an internuclear distance. The *lower* distance limit is usually set as the sum of the van der Waals' radii of two hydrogen atoms, *i.e.* 1.8 Å.

In proteins and larger peptides the concept of *ambiguous assignment* is widely used since it is difficult in these crowded spectra to unambiguously assign cross-peaks to a pair of atoms where there are chemical shift uncertainties due to overlap. Any ambiguous assignment is characterized by two sets of chemical shifts and any pair of items from either of these two sets may contribute to the observed cross-peak. For an ambiguously assigned peak, the $U$ and $L$ values can be calculated in the same way as an unambiguous one. Some difficulties arise while calculating the $D$ value (internuclear distance). It was suggested to use the $r^{-6}$-summed distance $\overline{D}$:[37]

$$\overline{D} = \left( \sum_{a=1}^{Nb} r_a^{-6} \right)^{-1/6} \tag{4.7}$$

instead of $D$. In eqn (4.7) summation is performed over all possible pairs of atoms. It can be shown that the final value is always smaller than the shortest internuclear distance. The contribution of any pair can be calculated as (obviously, the sum of $\chi_{mk}$ terms is equal to unity):[37]

$$\chi_{mk} = \frac{r_{mk}^{-6}}{\sum_{i,j} r_{ij}^{-6}} \tag{4.8}$$

The bigger the contribution from a particular pair (*i.e.* $\chi_{mk}$ value), then the more probable it is that the assignment of an NOE peak will be made to it. In the work by Smurnyy *et al.*[30] an ambiguous bond is treated as an unambiguous bond if one of the $\chi_{mk}$ values exceeds 95%.

### 4.2.3  Search by Running Over All Stereoisomer Structures

Natural product molecular structures run the gamut from conformationally rigid to highly flexible molecules. The most common, however, is the intermediate combination of both rigid and flexible substructures contained within a single molecule. For the molecules used in this study, Taxol and brevetoxin B (see below), there are a large number of unique conformers related to a flexible part of Taxol molecule, whereas the brevetoxin B molecules is very rigid. Since many natural products include conformationally rigid fused ring units, such molecules are very amenable to study using NOE or ROE experiments. For the study of flexible components of the molecules J-based analysis NMR methods have been developed in order to assign the configuration of the stereochemical centers.[3] For the stated purposes, Smurnyy *et al.*[30] assumed that there was a single "best" conformation assigned to each stereoisomer. Thus, in a case of a molecule with $N$ stereocenters there are $2^N$ stereoisomers to be inspected.

Experience shows that direct evaluation of the target function for each possible stereoisomer is possible for small molecules with a molecular mass of 100–200 Da and 2–7 chiral centers.

In such simple cases we can follow three steps in order to obtain a result:

1. For a given covalent structure with $N$ chiral centers, generate the full set of $2^N$ isomers.
2. Optimize each 2D structure into a 3D model where the stereocenter configurations are kept fixed with the aid of improper torsions as detailed earlier.

3. All models are ranked according to their penalty function values. The top-ranked stereoisomer is considered as the most probable.

Later in this Chapter the application of this approach to Taxol will be demonstrated.

## 4.2.4 The Genetic Algorithm (GA) Approach

Genetic algorithms represent an attempt to model an evolutionary process using purely stochastic means. The central concept of the approach is encoding of a series of control variables. The control variables are assigned to a particular solution referred to as a "chromosome". Usually a set of chromosomes is generated and this results in a pool of structures as an output.[32,38] In the most general scenario, the application of a genetic algorithm for solving the problem posed above can be described by the following five steps:

1. Encode the stereochemical information associated with a solution, the stereoconfiguration of the analyzed molecule, in the form of a chromosome. Specifically, a molecule within a chromosome is encoded by assigning a value of 0 to $R$-stereocenters and a value of 1 to $S$-stereocenters. The length of a chromosome is equal to $N$, the total number of stereocenters in the structure, and it is denoted as *CLength*.
2. Create a pool of chromosomes by random placement of ones and zeros in the $N$ positions of the chromosome vectors. The number of chromosomes included in the pool is determined during the process of genetic algorithm parameter optimization.
3. Perform *crossovers*. Two offspring are created from a pair of ancestors by exchanging all the bits following after a selected locus. For example:

| *Ancestors:* | *Locus selection:* | *Offspring pair:* |
|---|---|---|
| **0101101100011** | **0101101 100011** | **0101101** 100110 |
| 0111011100110 | 0111011 100110 | 0111011 **100011** |

More precisely, the operation should be referred to as a *one-point crossover* to differentiate it from a more sophisticated modification. A more realistic example can be found later in this Chapter where benchmark calculations are discussed.

4. Perform *mutations*. Mutation is simply a random change in a chromosome vector with a given probability, $P_M$, of one of the adjustable parameters. The purpose of the mutation stage is to provide insurance against an irrevocable loss of genetic information and hence to maintain diversity within the population. An example where mutation is performed for two bits of the chromosome is shown below.

| Initial chromosome: | Selecting bits for mutation: | Result: |
|---|---|---|
| 0111011100110 | 0111011100110 | 0101011101110 |

5. *Natural selection.* After a new solution (a set of chromosomes) is generated, the offspring with the worst penalty function values are eliminated to maintain the pool size. This is denoted by the variable *PoolSize*.

The algorithm, as described here, is the simplest implementation of a genetic algorithm. The intrinsic challenge of this method is the possible *degeneration* of the pool. This is defined as the situation in which the whole pool evolves from only a few ancestors. As a consequence, the diversity of the pool drops and the algorithm may become trapped in a local minimum. This is characteristic for complex structures with more than 10 chiral centers. However, even for smaller molecules pool degeneration may lead to the requirement of a bigger number of steps needed and therefore a longer run time. To increase the efficiency of this algorithm, enhancements have been implemented that are crucial when dealing with very complex ( > 10 chiral centers) molecules, and which may be optionally used for simpler cases.

1. *Diversity guided crossover.* In the standard implementation, crossover is performed with a given probability, $P_C$, over randomly selected pairs of ancestors. In the work[30] this step has been replaced with a *diversity-guided crossover* to maximize the genetic diversity of the offspring and to thereby minimize the number of generations needed to reach the solution. A similar approach was taken in an earlier work.[39] A diversity function, $F_{diff}$, the so-called Hammond distance function (see the original report[39] for further details) was introduced for a given pair of ancestors. This function is equal to unity, $F_{diff} = 1$, if half of the bits contained in the two chromosomes are different. The larger $F_{diff}$, the more diverse are paired chromosomes. The function is calculated for all randomly taken chromosome pairs, then *PoolSize* · $P_C$ pairs with the highest possible diversity function values are selected and crossover is performed on this set.

2. *One-point crossover versus uniform crossover.* In contrast to the *one point* crossover (see above) a *uniform* crossover implies an exchange at each bit position with a probability of 0.5. Some authors strongly believe in the superiority of this approach,[40] but in the work[30] the application of the GA approach to the solution of stereocenter determination has shown that *one-point crossover* exhibits much better results. The bottleneck in the implementation is *chemically reasonable encoding* of the chromosome which plays a decisive role in the algorithm convergence. Particularly, it is necessary to encode the stereocenters in a chromosome in accordance with their *order* and *proximity* in the real structure. The intention in this approach is to ensure that the local vicinity of most centers is not overly perturbed during the crossover process.

3. *Determining the local minima.* In general, crossovers can be thought of as tools to explore wide areas of conformational space and thereby assist in

the determination of a global solution. Mutations, in turn, help to step down to a local minimum, which may be isolated after a serendipitous crossover event. The authors[30] revealed that random mutations were computationally costly and not very efficient. The computational cost and lack of efficiency of random mutations is the primary reason that they have been replaced in this study with a simple function that tries to find a local minimum. This process works as follows. Assume the existence of a set of chromosomes as a starting point. For a given chromosome, *CLength* offspring are produced by subsequent mutation of different bits. The penalty function is then evaluated for each of the modified offspring chromosomes. If one of them is better than the ancestor, then the ancestor is replaced by the better chromosome and the procedure is repeated on this "best fit" offspring.

4. *Tournament selection.* According to the general GA scheme, the next step is the *natural selection.* For reference purposes the step responsible for the transition from crossover to natural selection is referred to as a "generation" and several generations form a "genetic run". The most powerful modification implemented in this work is a *tournament selection* scheme.[41] When two genetic runs are completed, a new pool of arbitrary size is produced by merging the two final pools and a new genetic run is performed. This process is referred to as one *tournament stage.* In the current version of software both one-stage and two-stage tournaments have been implemented. In the two-stage tournaments, four runs are performed in the first step, two in the second step, and one final run in the last step.

The algorithm developed to represent one genetic run is summarized below.

Randomly fill in population items;
**For** i = 1 **to** Generations Count **do**

> Calculate diversity function for each pair;
> Perform $P_C \cdot$ PoolSize crossovers;
> Randomly mutate $P_M \cdot$ PoolSize items (optional);
> Calculate the penalty function for each item;
> Select the 2 or 3 best chromosomes and perform a local minimum search (optional);
> Calculate the penalty function for new items;
> Sort pool;
> Delete items according to the penalty function value;

**End;**

## 4.2.5 An Example of a Genetic Run and Parameter Optimization for Taxol

Taxol (**4.5**) is a complex polyoxygenated diterpene natural compound. The first X-ray structure characterizing the stereochemistry of this compound was obtained in 1971.[42] Taxol has a 4:6:8:6 skeleton as well as pendant moieties.

Complex stereochemistry is abundant throughout the molecule and there are 11 stereocenters. In this study,[42] the authors dealt only with 9 of 11 stereocenters, since the other two are located on a flexible part of the molecule and this prevents NOE data from being observed for protons in this part of the structure. In such cases, J-based analysis techniques[3] would be applied for the determination of stereochemical information but the approach detailed here does not yet account for such input data.

**4.5**

Initially all stereoisomers were produced ($2^9 = 512$) for the structure and the penalty function value (PFV) calculated for every one in order to validate the accuracy of the penalty function calculation. These calculations were performed in 550 s using the PC Pentium IV computer operating at 2.8 GHz with 1 GB of RAM. The two lowest values from the penalty function were from the stereoisomer pictured here (**4.5**) and its mirrored twin which would be expected to show similar interatomic distances after a 3D optimization. The difference in the PFV values of this pair of stereoisomers is only 10% relative to their nearest neighbor in the rank order.

Following the protocol described earlier, the application of *crossovers* proceeds as represented in Table 4.2. Each stereocenter was associated with a chromosome locus and encoded with $0 = R$ and $1 = S$ as mentioned earlier. Initial chromosomes were generated randomly. This example illustrates that, even after one generation, the average PFV in the pool drops. Quantitatively the decrease in the average value of the PFV is a decrease of approx. 20% from 45.8 in the initial pool to 37.5 in the next. The best chromosome in the new pool has a PFV of 19. For Taxol and other species of similar complexity PFVs of 5–10 are typically achieved for the correct stereoisomers. Because mirrored stereoisomers are indistinguishable, all stereoisomers of the ordered file form a succession of corresponding pairs. For the present case PFVs for **4.5** and mirrored **4.5** are 5.7 and 5.9, respectively. Correct and mirrored stereoisomers have been ranked to first and second positions. The structure ranked third in the ordered file differs from **4.5** by one chiral center configuration and 3.1 PFV units. However, the PFV never drops to zero because of the uncertainty in determination of exact conformation for a given stereoisomer.

**Table 4.2** A crossover example. The initial pool and the first set of offspring are indicated, as well as which parents contribute to each offspring. For each chromosome, the PFV is given.

| Initial pool | | | New pool | | |
|---|---|---|---|---|---|
| No. | Chromosome | PFV | Parents | Chromosome | PFV |
| (1) | 111101101 | 35 | (5) + (6) | 110100 100 | 19 |
| (2) | 011101101 | 40 | (1) | 111101101 | 35 |
| (3) | 101111000 | 44 | (8) + (1) | 1011 01101 | 39 |
| (4) | 000000111 | 45 | (2) | 011101101 | 40 |
| (5) | 110100111 | 47 | (6) + (7) | 1111 01001 | 41 |
| (6) | 111101100 | 50 | (3) + (7) | 1011 01001 | 42 |
| (7) | 001001001 | 53 | (3) + (5) | 1011 00111 | 43 |
| (8) | 101111001 | 53 | (8) + (3) | 1011 11000 | 44 |

To investigate how the number of generations and pool size influence the overall performance, as well as algorithm convergence, 100 tests were run using the Taxol data as input. Ten generations were run with varying pool size. The results of testing three different modifications of the genetic algorithm and comparisons of their accuracy and speed are given below.

### 4.2.5.1   Plain Algorithm

The "plain algorithm" is the implementation of GA with *no mutation* or *tournament selection* and with the new pool formed only by the crossover. The results are summarized in Figure 4.5. With approx. 8 generations and a pool size of 12–15 items, the probability of success is >95%. It was implied that a run was successful if the final pool contains the right stereoisomer.

### 4.2.5.2   Complex Algorithm

In this implementation, the pool is mutated after performing a crossover. The purely random mutations (as described above) do not prove effective for this application. Mutations are performed with unity probability on degenerate positions in the chromosome. This modification can reduce the number of generations needed to 5 or 6. As seen in the case of the plain algorithm described above, small pool sizes (4–6) continue to show poor results and are not recommended (see Figure 4.6), whereas a pool size in the range of 8–10 again gave a >95% probability of success.

### 4.2.5.3   Tournament Selection

Due to the different workflow organization in the implementation of GA with tournament selection, it was difficult to directly compare the performance of this implementation to the previous two. Here the results of the one-stage tournament selection are given for comparison (*i.e.* two runs on the first stage

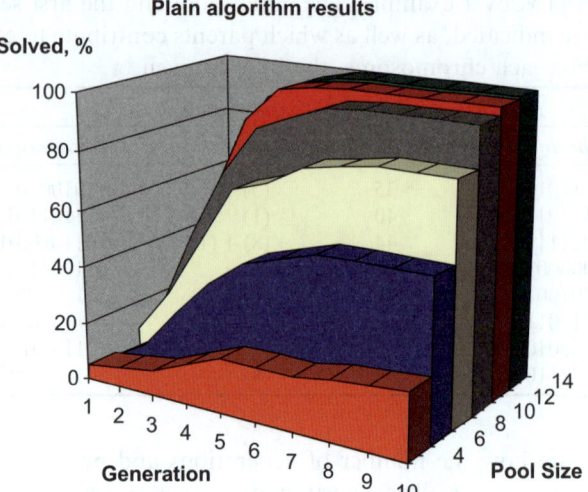

**Figure 4.5**    Results for the implementation of GA without mutation or tournament selection. The percentage of genetic runs where the structure is solved correctly is given as a function of the pool size and generation.

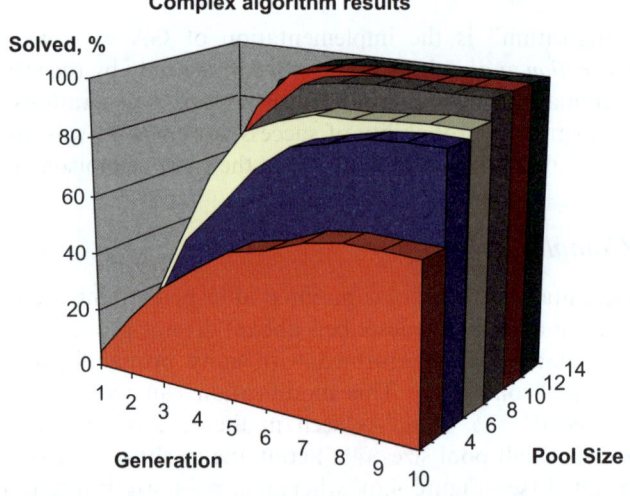

**Figure 4.6**    Results for the implementation of GA with mutation but without tournament selection. The percent of genetic runs where the structure is solved correctly is given as a function of the pool size and generation.

and the final genetic run). Each of the genetic runs included three generations, resulting in nine generations from the entire workflow. In Figure 4.7, the performance of the tournament selection algorithm is compared with the first two algorithms described above.

From the results presented here it is concluded that adding mutations and/or arranging a tournament selection is the best implementation of the genetic algorithm for this work. However, there is one more metric to be considered in this evaluation, namely the calculation speed. Table 4.3 summarizes the data related to the *speed* of the overall process. In the current version of software, up to 99% of computing time is spent on 3D optimization of the structures. The number of 3D optimizations may be used as an approximate measure of the total computing time. Since the quality of the results is the highest priority of the study, only those parameter combinations that lead to success in 95% or more of the runs will be discussed. From Table 4.3 it may be concluded that the genetic algorithm is superior to a deterministic approach, since the run with optimal parameter values is more than twice as fast as enumerating all possible isomers for this example – with optimal parameters around 200 3D

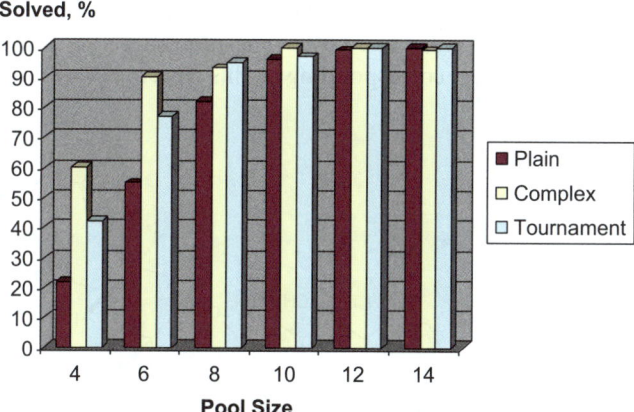

**Figure 4.7** Performance of the two-stage tournament selection algorithm compared with the plain algorithm (no mutations, no crossover) and complex (mutations after crossover).

**Table 4.3** Comparison of the speed of different algorithm modifications and parameter sets (for runs where >95% are successful). The fastest options are marked with an asterisk.

| Algorithm modification | Pool size | Number of optimizations |
| --- | --- | --- |
| Plain | 10 | 201* |
|  | 12 | 327 |
|  | 14 | 463 |
| Mutations | 10 | 275 |
|  | 12 | 368 |
|  | 14 | 471 |
| Tournament | 8 | 177* |
|  | 10 | 276 |
|  | 12 | 397 |
|  | 14 | 525 |

optimizations are performed *versus* as many as $2^9 = 512$ needed for enumerating all possible stereoisomers. It is concluded that the plain algorithm (with 10 chromosomes in the pool) and the tournament selection modification (with a pool size equal to 8) are the best options.

## 4.2.6  A Challenging Example: Brevetoxin B

To challenge the system the structure of brevetoxin B (**4.6**) has been examined. Brevetoxin B is the first and most prominent member of the brevetoxin family, produced by the dinoflaggelate *Ptychodiscus brevis* Davis (*Gymnodinium breve* Davis). This compound was isolated and characterized by spectroscopic and X-ray crystallographic means in 1981 by the groups of Lin and Nakanishi, and Clardy.[43,44] The highly complex molecular architecture is characterized by a novel array of ether oxygen atoms, regularly placed on a single carbon chain. This remarkable structure includes 11 rings, 23 stereogenic centers, and three carbon–carbon double bonds. The CPU time necessary for running over all ~8.4 million stereoisomers corresponding to this structure was estimated to be approx. 1 month.

**4.6**

The main problem with the genetic algorithm was the tendency of the algorithm to deteriorate toward pool degeneration. The search becomes trapped in a local minimum resulting in reasonably good, but not optimal, solutions. In the case of simpler structures such as Taxol, random mutations of the pool vectors can be used help to overcome this problem. In more complex cases improvements in the algorithm are necessary.

The standard GA implementation utilizing crossover and mutation did not provide positive results. Introduction of tournament selection resulted in only 30–35% of the runs being successful (20 trials were performed unless stated otherwise) – *i.e.* the right stereoisomer was present in the finally formed pool. As performing random mutations exhibited no improvement, the random mutation step was enhanced by including an alternative local minima operation (see Section 4.2.4). To save run-time this stage is performed for only 2–4 of the best structures in the pool. All local minima searches take about 10 min – approx. 6% of the total run time for the task (approx. 2 h

50 min). The rest of the time is consumed by 3D optimizations related to the crossover process.

Chemically reasonable encoding (*i.e.* stereocenters in physically close proximity in the structure are placed as close as possible in the chromosome) proved to be the most efficient improvement in the algorithm. The implementation is run with tournament selection, local minima detection, and chemically reasonable encoding 100 times and the right stereoisomer is determined in each final pool. An example run is summarized in Table 4.4. A pool size of 20 with four generations for each genetic run gave 28 generations in total. The Table indicates that the tournament selection scheme is necessary in the case of brevetoxin B, as the NOE penalty function steadily drops after each subsequent round of the tournament.

In Figure 4.8 the structure of brevetoxin B from X-ray studies (yellow) and the structure obtained from the best chromosome in the final pool for our system for stereochemistry determination (blue) are shown as superimposed structures. In this case, the configuration of all of the stereocenters in both structures are the same. Even the conformations of the most "flexible" 7- and 8-membered rings are similar. This demonstrates the power of the approach we have described to facilitate the identification of relative stereochemistry in a complex molecule containing multiple stereocenters.

**Table 4.4**   An example run workflow. Pool size = 20. Four generations are treated during each genetic run. Here, the algorithms of tournament selection and local minima finding are used. For each genetic run, a penalty function value and the number of incorrect stereocenters (best and average in a pool) are given.

|  |  | *Tournament Round 1* | | | | *Round 2* | | *Round 3* |
|---|---|---|---|---|---|---|---|---|
|  |  | *Run 1-1* | *Run 1-2* | *Run 1-3* | *Run 1-4* | *Run 2-1* | *Run 2-2* | *Run 3-1* |
| Penalty function | Best | 32 | 36 | 15 | 38 | 13 | 12 | 10 |
|  | Average | 43.2 | 46.9 | 37.9 | 54.7 | 20.3 | 22.3 | 15.5 |
| Number of incorrect stereocenters | Best | 3 | 2 | 5 | 3 | 1 | 1 | 0 |
|  | Average | 7.8 | 8.0 | 8.4 | 6.3 | 5.8 | 6.9 | 5.7 |

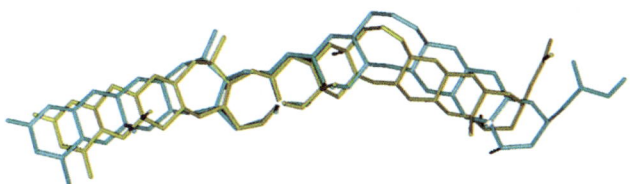

**Figure 4.8**   The X-ray crystal structure of brevetoxin B (yellow) and the 3D model of the best stereoisomer from the final pool (blue) of the stereochemistry determination system are superimposed. Small differences in the bond angles of some of the more flexible rings are present, but all stereocenters have been properly oriented.

# References

1. J. M. Seco, E. Quinoa and R. Riguera, *Chem. Rev.*, 2004, **104**, 17.
2. D. Neuhaus and M. Williamson, *The Nuclear Overhauser Effect in Structural and Conformational Analysis*, Wiley, New York, 2000.
3. R. T. Williamson, B. L. Marquez, W. H. Gerwick and E. K. Kover, *Magn. Reson. Chem.*, 2000, **38**, 265.
4. R. Ditchfield, *Mol. Phys.*, 1974, **27**, 789.
5. G. Bifulco, P. Dambruoso, L. Gomez-Paloma and R. Riccio, *Chem. Rev.*, 2007, **107**, 3744.
6. C. Fattorusso, E. Stendardo, G. Appendino, E. Fattorusso, P. Luciano, A. Romano and O. Taglialatela-Scafati, *Org. Lett.*, 2007, **9**, 2377.
7. A. B. Sebag, D. A. Forsyth and M. A. Plante, *J. Org. Chem.*, 2001, **66**, 7967.
8. A. B. Sebag, R. N. Hanson, D. A. Forsyth and C. Y. Lee, *Magn. Reson. Chem.*, 2003, **41**, 246.
9. M. E. Elyashberg, K. A. Blinov and A. W. Williams, *Magn. Reson. Chem.*, 2009, **47**, 333.
10. Y. D. Smurnyy, K. A. Blinov, T. S. Churanova, M. E. Elyashberg and A. J. Williams, *J. Chem. Inf. Model.*, 2008, **48**, 128.
11. I. Ohtani, T. Kusumi, Y. Kashman and H. Kakisawa, *J. Am. Chem. Soc.*, 1991, **113**, 4092.
12. C. Adamo and V. Barone, *J. Chem. Phys.*, 1998, **108**, 664.
13. G. Barone, D. Duca, A. Silvestri, L. Gomez-Paloma, R. Riccio and G. Bifulco, *Chemistry*, 2002, **8**, 3240.
14. M. E. Elyashberg, K. A. Blinov, A. J. Williams, S. G. Molodtsov and G. E. Martin, *J. Chem. Inf. Model.*, 2006, **46**, 1643.
15. K. A. Blinov, D. Carlson, M. E. Elyashberg, G. E. Martin, E. R. Martirosian, S. G. Molodtsov and A. J. Williams, *Magn. Reson. Chem.*, 2003, **41**, 359.
16. M. E. Elyashberg, K. A. Blinov, S. G. Molodtsov, A. J. Williams and G. E. Martin, *J. Chem. Inf. Comput. Sci.*, 2004, **44**, 771.
17. E. Fattorusso, P. Luciano, A. Romano, O. Taglialatela-Scafati, G. Appendino, M. Borriello and C. Fattorusso, *J. Nat. Prod.*, 2008, **71**, 1988.
18. P. T. Thuong, C. H. Lee, T. T. Dao, P. H. Nguyen, W. G. Kim, S. J. Lee and W. K. Oh, *J. Nat. Prod.*, 2008, **71**, 1775.
19. F. Lv, M. Xu, Z. Deng, N. J. de Voogd, R. W. M. van Soest, P. Proksch and W. Lin, *J. Nat. Prod.*, 2008, **71**, 1738.
20. G. E. Martin, B. D. Hilton, K. A. Blinov and A. J. Williams, *Magn. Reson. Chem.*, 2008, **46**, 997.
21. Y. Lu, C. Y. Huang, Y.-F. Lin, Z.-H. Wen, J.-H. Su, Y.-H. Kuo, M. Y. Chiang and J.-H. Sheu, *J. Nat. Prod.*, 2008, **71**, 1754.
22. Q.-W. Shi, F. Sauriol, O. Mamer and L. O. Zamir, *J. Nat. Prod.*, 2003, **66**, 1480.
23. H. M. Ge, B. Huang, S. H. Tan, D. H. Shi, Y. C. Song and R. X. Tan, *J. Nat. Prod.*, 2006, **69**, 1800.
24. C.-R. Zhang, S.-P. Yang and J.-M. Yue, *J. Nat. Prod.*, 2008, **71**, 1663.

25. A. Castro, J. Coll, Y. A. Tandro'n, A. K. Pant and C. S. Mathela, *J. Nat. Prod.*, 2008, **71**, 1294.
26. K. H. Jang, J.-E. Jeon, S. Ryu, H.-S. Lee, K.-B. Oh and J. Shin, *J. Nat. Prod.*, 2008, **71**, 1701.
27. K. P. Devkota, B. N. Lenta, J. D. Wansi, M. I. Choudhary and D. P. Kisangau, *J. Nat. Prod.*, 2008, **71**, 1481.
28. C.-C. Liaw, Y.-C. Shen, Y.-S. Lin, T.-L. Hwang, Y.-H. Kuo and A. T. Khalil, *J. Nat. Prod.*, 2008, **71**, 1551.
29. A. Hunyadi, G. Tóth, A. Simon, M. Mák, Z. Kele, I. Máthé and M. Báthori, *J. Nat. Prod.*, 2007, **70**, 412.
30. Y. D. Smurnyy, M. E. Elyashberg, K. A. Blinov, B. Lefebvre, G. E. Martin and A. J. Williams, *Tetrahedron*, 2005, **61**, 9980.
31. K. Wüthrich, *Angew. Chem. Intl. Ed.*, 2003, **42**, 3340.
32. M. Mitchell, *An Introduction to Genetic Algorithms*, MIT Press, Cambridge, MA, 1999.
33. U. Burkert and N. Allinger, *Molecular Mechanics*, ACS, Washington, 1982.
34. M. Falk, P. F. Spierenburg and J. A. Walter, *J. Comput. Chem.*, 1996, **17**, 409.
35. J. C. Smith and M. Karplus, *J. Am. Chem. Soc.*, 1992, **114**, 801.
36. R. J. Abraham, J. Fisher and P. Loftus, *Introduction to NMR Spectroscopy*, Wiley, London, 1988.
37. M. Nilges, M. Masias, S. I. O'Donohuge and H. Oschkinat, *J. Mol. Biol.*, 1997, **269**, 408.
38. J. R. Koza, *Genetic Programming on the Programming of Computers by Means of Natural Selection*, The MIT Press, Cambridge, MA, 1998.
39. H. Shimodaira, in *Proceedings of the Genetic and Evolutionary Computation Conference*, ed. W. Banzhaf, J. Daida, A. E. Eiben, M. H. Garzon, V. Honavar, M. Jakiela and R. E. Smith, 1999, p. 603.
40. W. M. Spears and K. A. De Jong, in *Proceedings of the Fourth International Conference on Genetic Algorithms*, ed. R. K. Belew and L. B. Booker, Morgan Kaufman, San Mateo, CA, 1991, pp. 230–236.
41. G. Harik, in *Proceedings of the Sixth International Conference on Genetic Algorithms*, ed. L. J. Eshelman, Morgan Kaufmann, San Mateo, CA, 1995, pp. 24–31.
42. M. C. Wang, H. L. Taylor, E. Monroe, P. Coggon and A. T. McPhail, *J. Am. Chem. Soc.*, 1971, **93**, 2325.
43. Y.-Y. Lin, M. Risk, S. M. Ray, D. Van Engen, J. Clardy, J. Golik, J. C. James and K. Nakanishi, *J. Am. Chem. Soc.*, 1981, **103**, 6773.
44. M. S. Lee, D. J. Repeta, K. Nakanishi and M. G. Zagorksi, *J. Am. Chem. Soc.*, 1986, **108**, 7855.

# Part II
# EXAMPLES OF CASE EXPERT SYSTEMS

CHAPTER 5

# CASE Expert Systems Based on 1D NMR Spectra

## 5.1  Introduction

Contemporary 2D NMR-based CASE systems were created not as a result of serendipitous invention, but as a result of the creative utilization of enormous experience and knowledge accumulated by many highly skilled researchers during the development of the first generation of expert systems (ES) based on 1D NMR, IR and MS spectra, the analytical technologies that were available at that time. 2D NMR-based expert systems have now replaced 1D NMR systems, but nevertheless learning about the fundamental operational modes of these systems provides insight into the approaches used to overcome many of the difficulties encountered in the development of efficient expert systems.

Among the expert systems created in the two decades spanning the 1970s to 1980s, probably the best known is the DENDRAL system.[1] This system is referred to as a "classical" example of an expert system in numerous publications on chemistry and books on artificial intelligence (AI). The popularity of the DENDRAL system may be related to the fact that this application was developed under the direction of the Nobel Prize winner J. Lederberg and noted scientists in the field of AI. That said, the authors of the project have published about 40 articles collectively addressing "The application of artificial intelligence in chemistry." Gray[2] critically analyzed the DENDRAL system and concluded that this AI system is far from rigorous in its approach. This judgment was hotly debated[3,4] with the leaders of the project. In contrast to the widespread opinion that the CONGEN program which was used as a structure generator in the DENDRAL[5] has found extensive application, Gray comments that it is in fact rarely used. The limited acceptance of CONGEN may be due to usability issues, a common issue with CASE applications.

New Developments in NMR No. 1
Contemporary Computer-Assisted Approaches to Molecular Structure Elucidation
By Mikhail Elyashberg, Antony Williams and Kirill Blinov
© Royal Society of Chemistry 2012
Published by the Royal Society of Chemistry, www.rsc.org

In this Chapter we will focus on 1D NMR-based expert systems that have appeared in the period from the end of the 1990s to the present. The efforts of researchers have been focused on improving methods for the extraction of structural information from spectra, to identify optimal strategies for computer-assisted structure elucidation and to increase the user-friendliness of the systems. Experience indicates that non-user-friendly expert systems with steep learning curves are unlikely to be widely utilized. Rather, the general expectations are that the system should:

1. Offer convenient methods for information interchange with the user, thereby allowing the chemist control over the various stages associated with solving a problem.
2. Offer an explanation to the user for each decision made by the program.
3. Inform the user of an "emergency" at any solution step and display the necessary messages to describe the nature of the challenge.
4. Offer advanced graphical tools for representing spectral and structural information to allow the user to quickly assess the current status of the program operation and facilitate the user's ability to make the appropriate decision.

A description of the STREC (RASTR) expert system developed to elucidate molecular structures using both NMR and IR spectra is available (see, for example, references[6,7]). This system was capable not only of revealing possible structures, but also of generating, constructing, and depicting the spatial models for all stereoisomers corresponding to a given structural formula. The STREC system was the basis of the user-friendly Bruker X-PERT system.[8,9] The most important elements of this expert system will be considered below, since many of these have been incorporated into several other expert systems.

## 5.2 The X-PERT System

The X-PERT system was created to elucidate the structure of organic molecules containing up to 30 skeletal atoms (C, N, O, P, S, B, Si, and the halogens) using a combination of IR, $^1$H, and $^{13}$C NMR spectral data.

### 5.2.1 Knowledge Base of X-PERT

The knowledge base of the system is comprised of a set of libraries containing molecular fragments and the chemical shift ranges of their characteristic spectral features. The libraries are generated by taking into account the fact that the fragments belong to different classes of chemical compounds and offer different responses to various spectroscopic techniques. Libraries belonging to different categories play different roles. The first category comprises universal libraries used for both fragment selection and structure verification by filtration. Libraries belonging to the second category are adjusted to apply to different types of spectroscopic techniques and are mainly used as filters for structure verification. Each fragment in every library is associated with its characteristic features in the IR and NMR spectra and these are stored in the knowledge base.

There are four hierarchically organized universal libraries.

1. A library of principal functional groups, *e.g.* C=O, CN, *etc.* exhibit the widest ranges of changes in the characteristic spectral features and are used for these groups irrespective of the nearest environment of a fragment. These libraries are mainly used for preliminary evaluation of the data to assign it to a particular class of chemical compounds and for recognition of the structure of non-additive compounds.
2. A library used for structure–spectrum interpretation [structural group analysis (SGA)] contains functional groups and information on the admissible nearest environment, *e.g.* C–C(=O)–C, C–O–C(=O), –CH$_2$–OH, *etc.* The library is used for structural group analysis and structure filtration.
3. A library of aromatic fragments containing 12 types of benzene ring substitution.
4. A library of phosphorus-containing fragments.

There are also a number of specialized libraries: a library of $^{13}$C NMR data; three libraries of $^1$H NMR data to reveal CH$_3$, CH$_2$, and CH groups with consideration given to the nearest environment; a library containing fragments (mostly cyclic) characterized by strong specific absorption bands in the IR spectrum (*e.g.* four- and five-membered cyclic ketones, lactones, lactams, *etc.*); and a library of fragments containing an OH group.

For each fragment, the system generates a "visit card" containing the following information: the fragment structure; the range of changes in the characteristic features in the IR spectrum (frequencies, half-height linewidths, and the intensities of the absorption bands) and NMR spectra (chemical shifts, coupling constants, signal multiplicities, and the corresponding numbers of C or H atoms); structural descriptors used to set the requirements for the fragment environment (additivity conditions) and to provide retention of its spectral features from one molecule to another. The descriptors characterize the terminal fragment atoms with free valence bonds. A descriptor also indicates the possibility of adding a hydrogen atom, heteroatoms (they can be enumerated), aromatic rings, and multiple bonds to a given atom. The possibility or impossibility of the formation of a multiple bond and the incorporation of the fragment into cyclic structures of a specified size is also indicated.

A descriptor can have three values: possible (P), forbidden (F), and obligatory (O). The program finds classes of topologically equivalent atoms with free bonds in the fragment structure and colors them using individual colors. Each colored atom is associated with a column of corresponding color in the table of descriptors. This makes it possible to easily form, interpret, and correct the values of descriptors characterizing equivalent atoms. The total number of fragments in the libraries is 450.

## 5.2.2   Structural Group Analysis

Experimental data assembled from the empirical formula or molecular weight and available experimental spectra can be input manually or imported using the Bruker OPUS and WIN-NMR programs. The system checks the data for consistency and prompts the chemist to double-check the reliability of the data in questionable situations. If the empirical formula is unknown or questionable, then the program generates all possible molecular formulae corresponding to the specified molecular weight while taking into account a series of spectral constraints imposed on the qualitative and quantitative composition of the substance.[10]

Once a molecular formula has been established, the system operates in an interactive mode allowing the experience and intuition of the chemist to be applied in addition to exploiting the capability of the computer to perform fast searches for necessary information and to complex logic combinatorial calculations. In this case, the basic requirement for an expert system is also met: at any step during the solution of a problem the program can display the reasons for a particular decision and the user can be prompted to select a particular option to progress further.

Prior to starting the operation, the program assesses the complexity of the problem and displays the appropriate messages concerning the method that will be applied to attempt a solution. If the problem can be solved automatically, then the program generates all structural isomers and produces a subset of possible structures by passing them through a system of filters, *i.e.* the system just operates in a straightforward manner corresponding to the general CASE strategy described in Section 1.4.

The first stage of the common method for solution is the generation of fragment sets. First, possible fragments are selected using the criteria of empirical formulae, spectral intervals, and the degree of unsaturation. The necessary libraries of fragments to be utilized at this stage are selected by the user. Fragments are selected by the program, assuming that the reliability of finding a fragment is 100%, if the experimental band falls at the center of the characteristic interval. Reliability decreases in a linear fashion as the interval bounds are approached and falls to zero outside of the bounds. A trapezoidal membership function[11,12] is chosen for the interval so that the decrease in reliability near the bounds relative to the increasing width of the interval will be gradual.

The selected fragments and their associated reliability indices are displayed on the screen. The *Show Spectra* procedure allows the experimental spectra and characteristic intervals for selected fragments to be represented in clear graphical form, allowing the user to assess the reliability of the fragments and to edit the list by adding, if necessary, user fragments. Next, a table with the interpretation data of all available experimental spectra is displayed. The user can analyze relationships between the fragments and correct the "reasoning" of the program using experience and *a priori* information. The final set of consistent logical equations is then used as input information for the inference engine. The result of solving the logical equations [function $f(A)$, see Section 1.7.1.3] is displayed as a set or several sets of fragments consistent with the spectra, empirical formula, and the degree of unsaturation for the data being analyzed. The program automatically adds sets of microfragments, such as $CH_3$, $CH_2$ *etc.*, which were not included as constituents of the structures of selected fragments. The user can also correct the selected fragment sets or compose other sets as necessary. The special purpose *Multiple Fragments* procedure allows for the inclusion of a fragment that is repeated in the structure, thereby reducing the dimensionality and complexity of the problem.

### 5.2.3   Structure Generation and Verification

The next step of the program is structure generation from the fragment sets. At this stage the possibility of imposing one or more of the structural constraints listed in Section 1.8.3 (GOODLIST/BADLIST constraints associated with the ring size, bond multiplicities *etc.*) is provided. An additional option enables or cancels "descriptor control" in the course of structure generation and is useful to preclude the generation of a large number of incorrect structures. Cancellation of the descriptor control is useful in those cases where repeated solution of the problem is necessary if the structures found are questionable. Structure generation is performed on the basis of the algorithms.[13,14] Structure counters allow the user to monitor the process and assess the efficiency of the constraints imposed on the program.

As mentioned above, structure generation using fragment sets and free atoms can lead to substructures whose presence contradicts either the spectra and/or concepts of organic chemistry. As a result, all of the generated structures are verified using the following five filters:

1. Check each isomer for consistency with experimental spectra.
2. Check for the number of signals in the $^{13}C$ NMR spectrum (should be used with caution).
3. Check for consistency with the principles of structural chemistry and stereochemistry, such as the presence of unlikely fragments and consistency with Bredt's rule.
4. Search for and exclude identical structures.
5. Predict the chemical shifts of the $^{13}C$ NMR spectra to reveal the most probable structures.

The system chooses the necessary libraries and then initiates a two-stage check. In the first stage, the list of fragments that *contradict the spectra* is generated. In the second stage, only contradictory fragments are sifted out using structural formulae and those structures in which at least one of these fragments is found are removed. The options used in the filtration stage allow the user to specify the parameters determining the severity of the imposed constraints (*i.e.* the degree of fuzziness of the interval bounds).

Structure verification using filters can be performed in both standard and supervised modes. The standard mode provides automated filtration (so-called "blind" filtration) and, in this case, only messages regarding the current status of the program operation and the number of "good" and "bad" structures at any particular time are displayed.

In the supervised mode, the filtration process is visualized and the user can both control the process based on visual cues, as well as obtain all of the information necessary to explain the decisions made by the system. Each structural formula to be checked is displayed. If no contradictory fragments are found in the structure (it should be kept in mind that only these are sifted), the structure is indicated as good. Otherwise, contradictory fragments found in the structural formula are colored and information regarding characteristic spectral features, including IR frequencies, chemical shifts, parameters associated with specific bands or signals not confirmed by the spectra, is displayed in the form of a "visit card" for the fragment. Depending on the number and type of contradictory fragments found in the structure, the user can reject the structure as unlikely or accept it for further checking. Structure filtration using this approach is fully controllable and understandable.

Checking the number of signals with different multiplicities in the $^{13}$C NMR spectra while in supervised mode is accompanied by displaying the data on the differences between the numbers of predicted and observed signals.

After passing the data through all filters the system automatically uses the answer file as input data for the Bruker WinSpecEdit program,[15] which then predicts the chemical shifts in $^{13}$C NMR spectra for the input structures. Then, the standard deviations, *s*, of calculated spectra from experimental ones are calculated and the structures are ranked in ascending order of *s*. Examples of structures elucidated using X-PERT system are shown in Figure 5.1.

An advantage of the X-PERT system is its relatively high level of "user-friendliness" that was included into the system to make it easier for the user to operate. A series of approaches realized in this system were implemented into modern CASE programs. Experience has shown (see examples in Figure 5.1) that the size of molecules that could be elucidated with the use of the X-PERT system did not usually exceed 20 skeletal atoms, a common situation for an expert system limited to the use of 1D NMR spectra. Authors of the Spec-Solv[16] system attempted to overcome this particular shortcoming.

## 5.3 The SpecSolv Program

The SpecSolv program was developed as an extension of the SpecInfo system,[17] a system intended as a solution for the storage and structural interpretation of NMR, IR, and mass spectra. Over 200 000 assigned $^{13}$C NMR spectra

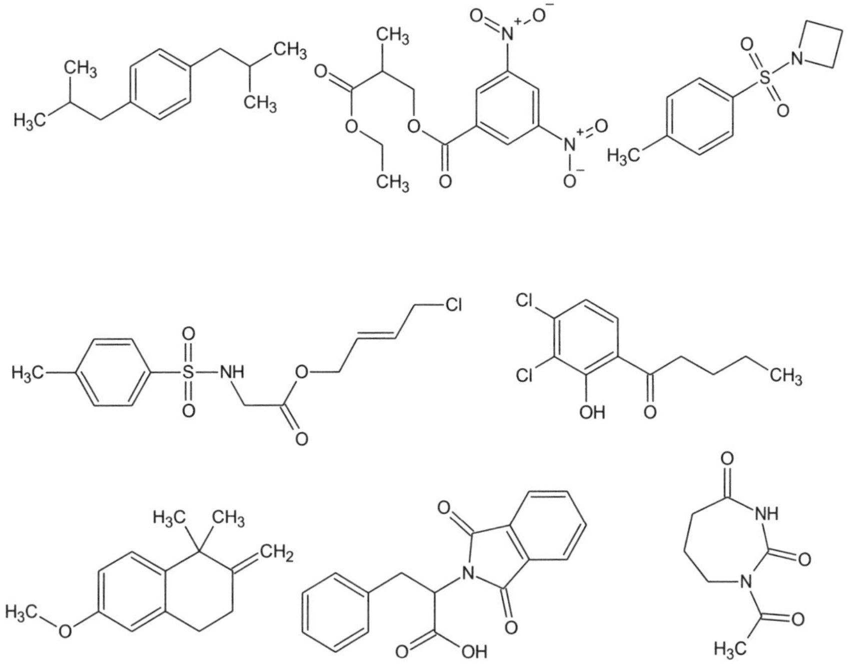

**Figure 5.1** Examples of structures elucidated using X-PERT system.

contained in the SpecInfo knowledge base served as the basis of the program. The SpecSolv system was designed to elucidate molecular structures using only a $^{13}$C NMR spectrum, and the knowledge of the molecular weight or molecular formula is not mandatory. The system uses the various parameters extracted from $^{13}$C NMR spectra – the chemical shifts, signal multiplicities, and intensities (the latter as the number of carbon atoms) – as input data. The knowledge base of the system contains a library of fragments and their associated subspectra. The library was created by automatically excising fragments from the structures stored in the SpecInfo database; this process produced more than 500 000 fragments encoded into HOSE codes[18] with respect to the central carbon atom. The library created was carefully reviewed by experts.

Information regarding a fragment includes the connectivity matrix and the HOSE codes of all heavy atoms and associated subspectra. The parameters of the $^{13}$C NMR spectra [chemical shifts, multiplicities, intensities, and root-mean-square deviation (RMSD) of the experimental chemical shifts relative to the predicted chemical shifts] are stored for each carbon atom. The RMSD values were calculated for the sets of equivalent substructures in the SpecInfo database.

The first stage of structure elucidation with the program is the usual fragment search by $^{13}$C NMR spectra. All subspectra in the library of fragments are compared with the experimental spectrum and fragments whose subspectra match corresponding regions in the experimental spectrum are selected. A typical case is exemplified by the identification of $\sim 500$ fragments for a molecule containing $\sim 25$ heavy atoms. The selected fragments are then ranked in descending order of the match factor, which is defined as the mean deviation of the chemical shift of each atom from the signals in the experimental spectrum.

Despite the fact that the program, in principle, can operate without knowledge of the molecular formula, it is necessary to specify the possible types of atoms, especially heteroatoms, since they cannot be immediately revealed by $^{13}$C NMR spectroscopy.

It is absolutely essential that a user should not be required to analyze all selected fragments simply due to the time-consuming nature of that requirement. Bad fragments should be specifically excluded during the structure generation stage. However, if *a priori* information is available, as frequently happens, it can be expected that analyzing the list will make it possible to exclude bad fragments, thereby reducing the time for solving the problem. The main feature of the structure generation algorithm used in the SpecSolv system is the *linkage of fragments by overlapping common atoms*. The largest fragment whose subspectrum is a best match for a particular region of the experimental spectrum is taken as the primary fragment and the program attempts to successively add all other fragments to this structural core. If a new fragment passes the plausibility test, then it is accepted and the newly enlarged fragment is then considered as the starting structure. Before adding the next fragment, the program generates the $^{13}$C NMR spectrum of the enlarged fragment. This is achieved by simple combination of the subspectra and provides good performance in terms of prediction time. Intermediate structures for which the deviations of the chemical shifts exceed a prescribed value are excluded. In this fashion, the final structure is being assembled at the same time as cross-validation of the intermediate structures is being performed. The absence of such a verification tool in other structure generators that admit an assembly of structures from overlapping fragments is certainly a drawback.

The assembly of the overall structure continues until all chemical shifts in the experimental spectrum are assigned and no free bonds remain in the calculated structure. The authors[16] report that usually ten generation cycles based on the first ten best fragments are sufficient to elucidate the structure. The ten best fragments are the largest substructures with the minimum deviation of the chemical shifts. This elucidation is possible, since the selected fragments are ranked by the match factor of their subspectra with the experimental spectrum. Most often, only one structure is found after an exhaustive search across the entire fragment space represented in the knowledge base. However, several isomers of similar structures can also be obtained, and, in general, if no empirical formula is given the generation of non-isomeric structures is also possible. In those cases when SpecSolv cannot find a solution to the problem, an appropriate message is generated and the program is terminated.

Will *et al.*[16] described the elucidation of two structures – **5.1** and **5.2**; both of them are representative of a size and structural complexity that is common for natural product structures.

| **5.1** | **5.2** |

Successful structure generation with the SpecSolv approach is possible only if the fragment library contains large overlapping fragments of molecules related to the resultant structure. Successful identification therefore requires the presence of a rather large number of large fragments related to the molecule under examination within the system database. However, even if the database contains appropriate fragments, the absence of common atoms will make generation of the proper structure impossible and thus there is a strong dependence of the result on both the diversity and the quality of the database. The program is also sensitive to the solvent in which the experimental spectrum is recorded and, in addition, does not recognize stereoisomeric forms. These factors can have a decisive effect on the results of comparison of predicted subspectra of intermediate substructures and the overall model spectrum with the experimental one. It was reported in an article entitled "Fully automated structure elucidation – a spectroscopist's dream comes true"[16] that the system elucidated the structure of 80% of all compounds analyzed in the authors' laboratory. With a failure rate of 20% the spectroscopist's dream remains unfulfilled, but this example does illustrate forward progress. To the best of our knowledge, no new publications describing the application of the SpecSolv system have appeared since 1996.

New approaches were required to facilitate the successful analysis of those cases where the SpecSolv program failed. Procedures for overcoming some of these failures have been implemented in the ACD/Structure Elucidator 1D system.[19]

## 5.4 The 1D ACD/Structure Elucidator System

The ACD/Structure Elucidator 1D system[19] integrates the recognition strategy developed by the authors of the SpecSolv system[16] and the conventional methodology used in the STREC (RASTR),[6,7] X-PERT,[8,9] CHEMICS[20–23] and other systems.

Two structure elucidation modes, an "overlapping common atoms" (OCA) mode and the "classical" expert system mode are supported. The OCA mode, to a great extent, is based on the general strategy suggested in the SpecSolv system. The knowledge base of the system is a library containing more than 1.5 million fragments and their $^{13}C$ NMR subspectra derived from >215 000 structures with assigned $^{13}C$ NMR spectra.

## 5.4.1   Operation in OCA Mode

The operation of the program, as illustrated in Figure 5.2, begins with the search for a $^{13}C$ NMR spectrum in the knowledge base. If a matching spectrum is found then the corresponding structure is displayed and the program is complete. If the search fails, then the structure is reconstructed from fragments.

First, the program selects fragments whose subspectra do not contradict the experimental spectrum. The found fragments are sorted in descending order of the number of skeletal atoms and the mean deviations of their subspectra from the corresponding chemical shifts in the experimental spectrum. At this stage, it is not feasible to accept or reject the presence of, for example, OH, NH, and CN groups or to differentiate between the $\delta_C$ of the O=C–N and O=C–O groups, using the $^{13}C$ NMR spectrum. Therefore, if an IR spectrum is available then filtration of selected substructures is performed using a library of functional groups (similar to that used in the X-PERT system) containing the most characteristic features of the IR spectrum. Analogous filtration can also be performed using fragments having clearly defined characteristic features in the $^1H$ NMR spectra. After filtration the number of selected fragments is generally reduced (in some instances, it is halved), ensuring the placement of representative fragments at the beginning of the list. Next, starting with the first fragment, they are checked for the possibility of overlap by common atoms to assemble the structure. As with the SpecSolv system, linking the fragments is accompanied by prediction of the $^{13}C$ NMR subspectrum of the enlarged fragment to verify the correctness of the assembly.

The problem is considered solved if, after checking all possible combinations of the fragments, the program generates one or more final structures. As noted above, the most promising fragments (*i.e.* the largest ones whose subspectra are good matches with the experimental spectrum) are placed at the beginning of the list. Consequently, the resultant structures are often generated within minutes of initiating the program. In those cases where there is good correspondence between the $^{13}C$ NMR spectrum predicted for the resultant structure(s) and the experimental spectrum the process can be terminated before completion of the exhaustive search for all combinations of the fragments. If an attempt to assemble a structure fails (which can happen when the knowledge base does not contain appropriate fragments with common atoms or alternatively when good fragments were not selected), the program automatically switches to the classical expert system mode to continue operations with *non-overlapping* fragments.

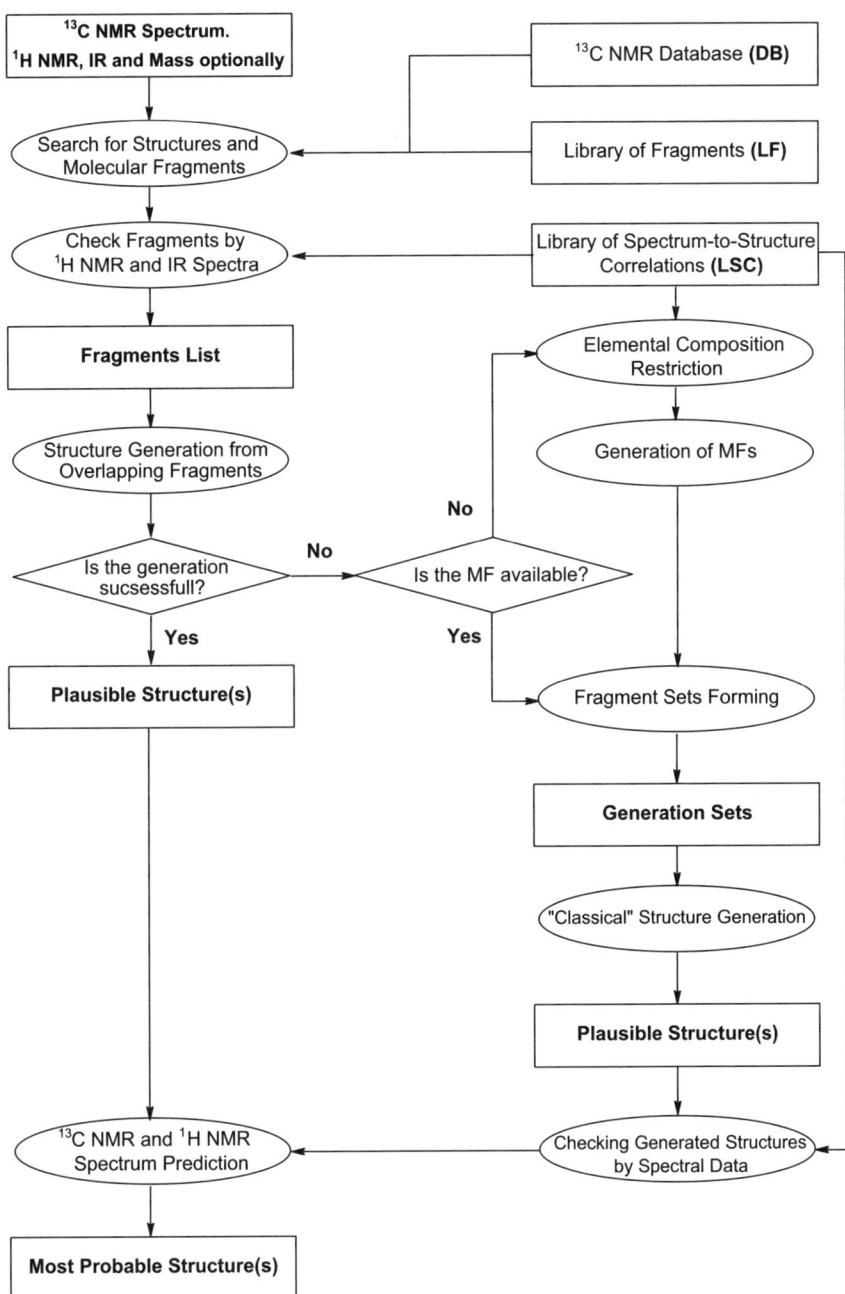

**Figure 5.2** Flowchart for the ACD/Structure Elucidator 1D expert system.

## 5.4.2 "Classic" Structure Generation

If the molecular formula is unknown, then the program determines the value of the formula or a set of possible formulae from the molecular mass and spectral data. One or several possible empirical formulae are then passed to the structure generator. The structure generator operates with atoms and fragments that have no common atoms and is based on a series of reported algorithms.[13,14,24] The fragment sets are composed as follows.

The first $N$ fragments in the primary list of selected and filtered fragments are chosen. Usually the chemist varies the $N$ value from 10 to 30 as a program option. Obviously, these are the largest fragments and this set almost always includes at least one or more fragments that are constituents of the chemical structure of the compound under study. Generally, small fragments, most often functional groups with their neighborhood environment, are placed at the end of the primary list following definite rules.

A user can add their own fragments to the sets and use all of the constraints described in Section 1.8.3. To facilitate this process the program displays a "generalized portrait" of the molecule under study, *i.e.* the frequency distribution of typical functional groups in the extracted fragments is displayed (a library of typical functional groups is included in the knowledge base of the system). A high frequency occurrence for a fragment obviously suggests that the group is likely to be a constituent of the molecule. Conversely, absent functional groups can be used as a basis to generate the BADLIST.

Fragment sets are formed and "classical" structure generation is initialized. The resultant structures are the input of the $^{13}C$ NMR prediction program (the ACD/CNMR Predictor program[25]). Post-prediction, the structures are ranked in ascending order of the mean deviations of the predicted spectra *versus* the experimental spectra. The most probable structure is selected as described above in Section 1.8.4. If the difference between the chemical shift deviations of the preferable structure and the subsequent structure is small (lies within the accuracy of the calculation of the $^{13}C$ NMR spectrum), then further verification of the solution is required. To this end, automated prediction of $^1H$ NMR spectra is performed using the ACD/HNMR Predictor program[25] (see Section 3.2.4). The experimental and calculated $^1H$ NMR spectra are compared by the user who can then make the final choice regarding the preferable structure or can make a decision regarding the necessity or advisability of acquiring additional experimental data. The calculated $^1H$ NMR spectra, integral curve, and $\delta_H$ values for all lines are displayed. If the predicted spectra of several structures appear to be equally close to the experimental spectrum, then the proper choice may be frequently made by using the $\Delta_{CH3}$ values for the $\delta_{(CH3)}$ singlets, where $\Delta_{CH3} = |\delta_{(CH3)exp} - \delta_{(CH3)calc}|$, and comparing the integrals for all multiplets. In most cases, at $\Delta_{CH3} \geq 0.2$–0.3 ppm the structure can be rejected. The resolution of questionable situations requires additional experiments.

Let us consider an example. The molecular mass ($M = 202$) and the $^{13}C$ and $^1H$ spectra for an "unknown" were input into a computer, and the fuzzy molecular formula $C_{11}H_{10}O_{(0-20)}N_{(0-10)}$ was suggested by the program as additional information. At the first stage, the molecular formula (MF) generator found only one MF, namely $C_{11}H_{10}O_2N_2$, which matched the experimental spectra. The spectral search was then carried out. As a result, 280 fragments were selected. When no structure was built in the OCA mode, the program switched to the classic generator. 1060 structures were generated from the first ten fragments shown in Figure 5.3. The first six structures ranked

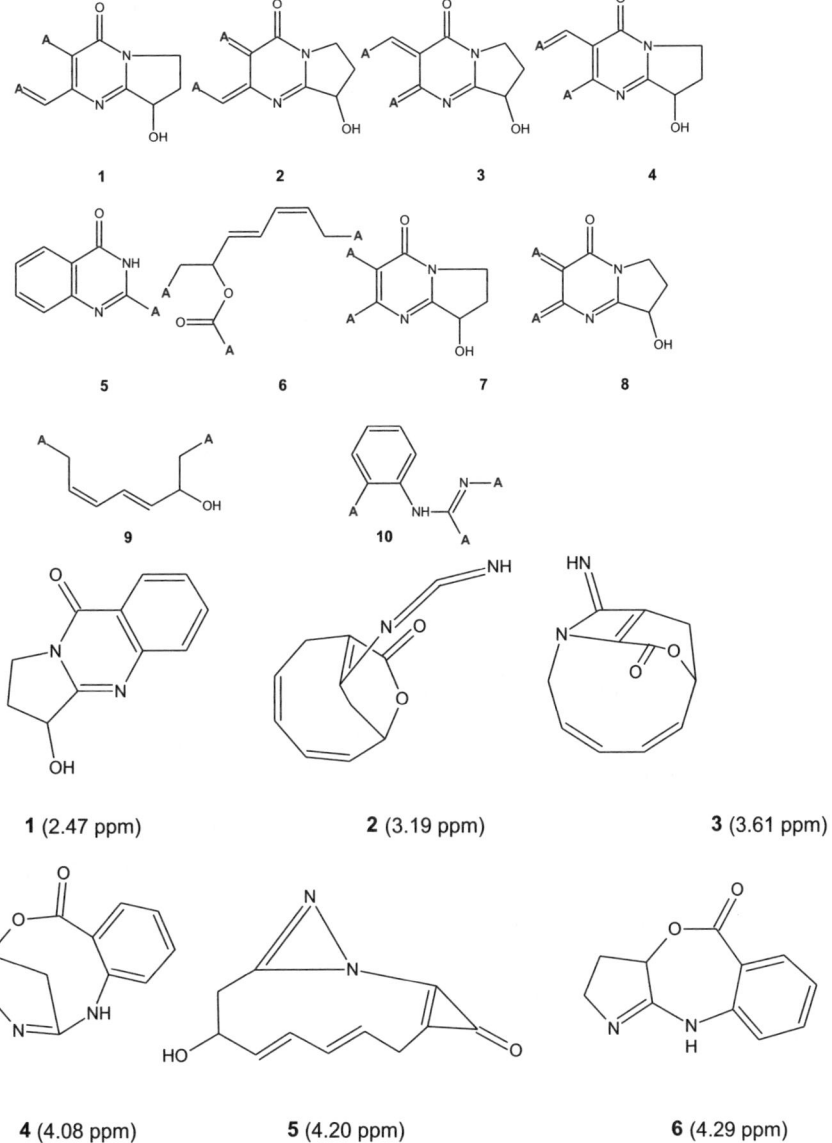

**Figure 5.3** The first ten fragments found in the database and the top structures from the output file ranked by $d(^{13}C)$ values.

by $^{13}$C NMR spectrum prediction are presented at the bottom of Figure 5.3. Comparison of the predicted and experimental $^1$H NMR spectra confirmed that structure #1 was the correct one.

The ACD/Structure Elucidator 1D system differs from other 1D NMR systems by allowing two interacting strategies to solve the problem, each based on the corresponding structure generator. If the program cannot find an adequate solution in either mode, then the process of solving the problem in a step-by-step mode under the supervision of a chemist can be followed. This offers an opportunity for the user to more directly influence the program. Testing of the system on a test set of 220 problems showed that a correct solution could be found in 90% of the cases, but the majority of structures were either of the same or slightly greater complexity than the molecules shown in Figure 5.1. Many unsuccessful attempts to elucidate structures of new natural products brought the authors[19] to conclude that the probability of finding a *complete* set of "good" fragments in the database to provide structure generation in OCA mode was slim, whereas switching to the branch of "classical" structure generation frequently led to enormous calculation times. The capabilities of 1D NMR-based expert systems are practically exhausted due to the lack of structural information carried by 1D NMR spectroscopy.

It should be noted that all 1D NMR expert systems developed by the end of the 20th Century operated on the basis of deterministic algorithms. However, some researchers[26] have suggested that the structures of large molecules can be recognized using stochastic algorithms, particularly genetic algorithms[27,28] that were assumed to be capable of removing restrictions intrinsic to the deterministic systems. This suggestion was experimentally verified by Meiler and Will,[29,30] and the results are described in the next Section.

## 5.5    Expert System GENIUS

In several reported studies[29,30] an attempt was made to examine the abilities of genetic algorithms (GAs) serving as an alternative approach to structure generation in an expert system designed for the elucidation of the structure of the products of organic synthesis based on their $^{13}$C NMR spectra. The attempt is of particular scientific interest, since a genetic algorithm had not been applied previously for this purpose.

For the application of the GENIUS system it is necessary to have both the molecular formula and the $^{13}$C NMR spectrum. Consider the main stages of structure elucidation, as the system's authors would suggest, applied to isoleucine (**5.3**), $C_6H_{13}NO_2$, as an example.

**5.3**

First, the adjacency matrix is *randomly* created from the molecular formula $C_6H_{13}NO_2$. For this example, the matrix is presented as:

|       | $C_1$ | $C_2$ | $C_3$ | $C_4$ | $C_5$ | $C_6$ | N | $O_1$ | $O_2$ |
|-------|-------|-------|-------|-------|-------|-------|---|-------|-------|
| $C_1$ | *     | 1     | 0     | 0     | 0     | 0     | 2 | 0     | 0     |
| $C_2$ | 1     | *     | 1     | 0     | 0     | 0     | 0 | 0     | 1     |
| $C_3$ | 0     | 1     | *     | 1     | 0     | 1     | 0 | 0     | 0     |
| $C_4$ | 0     | 0     | 1     | *     | 1     | 0     | 0 | 0     | 0     |
| $C_5$ | 0     | 0     | 0     | 1     | *     | 0     | 0 | 0     | 0     |
| $C_6$ | 0     | 0     | 1     | 0     | 0     | *     | 0 | 0     | 0     |
| N     | 2     | 0     | 0     | 0     | 0     | 0     | * | 1     | 0     |
| $O_1$ | 0     | 0     | 0     | 0     | 0     | 0     | 1 | *     | 0     |
| $O_2$ | 0     | 1     | 0     | 0     | 0     | 0     | 0 | 0     | *     |

It is easy to see that the matrix above corresponds to the structure **5.4** shown below:

**5.4**

The genetic code, or "chromosome", is constructed from this symmetric matrix by rearranging the upper triangular half matrix, marked in bold, into a vector:

**1000020010000011010001000000000000100**

This vector contains the bond state between all atom pairs of the molecule and is considered as the "genetic" code of the target structure. The molecular formula is used to generate a *random* set of *m* constitutions that fulfill that molecular formula. This set is evaluated by calculating the $^{13}C$ NMR spectrum and comparing it with the experimental data. The spectrum calculation is performed using the program C_SHIFT[31,32] on the basis of the artificial neural networks algorithm, and the $\Delta(^{13}C)$ value, the difference between the experimental and calculated spectrum, plays the role of a fitness function. The "fitness" of every single carbon atom *i* is the absolute deviation between its experimental and the corresponding calculated chemical shift value:

$$\Delta_I = |\delta^i_{calc}(^{13}C) - \delta^i_{exp}(^{13}C)|$$

The fitness of the whole molecular constitution formula is given by the RMSD over all *N* carbon atoms:

$$\Delta(^{13}C) = \sqrt{\frac{1}{N}\sum_{i=1}^{N}(\delta^i_{calc}(^{13}C) - \delta^i_{exp}(^{13}C))^2}$$

The multiplicity of a signal is also incorporated, if identified experimentally, and a generalized fitness function is used. Therefore this function is used in the *selection* process (see Section 4.2.4). The lower the $\Delta(^{13}C)$ value of a structure then the higher is its fitness and the higher the probability of participation in the *recombination* step of the genetic algorithm.

A new population of molecules is formed by recombining two parent molecules $m$ times. Two molecules from the parent generation are selected to form the *child* (offspring) molecule. The smaller the $\Delta^{13}C$ value of a molecule the higher is its probability to be considered as parent. This probability for a single molecule $j$ out of a population of $m$ constitutions is given by the expression:

$$p_j = [\Delta_j(^{13}C)]^{-1} / \sum_{i=1}^{m} [\Delta_j(^{13}C)]^{-1}.$$

After this selection, all possible (atom–atom) pairs in both parents are taken, and the bond type between them is analyzed (no bond, single, double, or triple bond). Randomly, one out of the two possibilities for every (atom–atom) pair is taken as the newly generated child structure. The vector representing the newly formed child constitution contains exactly one of the two possible values obtained at the corresponding positions in the parent molecule vectors. For example, from the parent molecules A and B, the child molecule C can be produced as the result of recombinations:

A

B

C

The corresponding vectors are shown below:

A: 100000210000100111100000000000000000
B: 100002001000001101000100000000000100
C: 100000211000100101000100000000000000

Since hydrogen atoms are not explicitly considered but are added to the free valences retrospectively, the molecular formula needs to be checked after a new child constitution is generated. If the number of potential hydrogen atoms in the generated constitution is not the same as that defined in the target molecular formula then bonds must be added or deleted until this deviation is corrected to zero. For this purpose the same function is used for *mutation* (see below). The structure must also be checked for connectedness. After both boundary conditions are fulfilled, the new molecule is accepted as a member of the newly formed population.

A *mutation* is implemented by simply modifying bonds. Two pairs of atoms are randomly selected and a bond is deleted (or the bond type is decreased by one) for the first pair, whereas, for the second pair, a bond is inserted. A deletion is always combined with an insertion, so that the total number of hydrogen atoms remains constant. Also, this process has to be controlled so that only one molecule and not a set of fragments are created.

By repeating the procedure of subsequent recombination and mutation, $m$ molecules for the child generation are created out of the $m$ parent molecules. Optionally, the $l$ fittest molecules of the parent generation replace the $l$ worst molecules of the child generation to ensure that the fittest constitutions are not lost. The recombination probability (RP) and the mutation probability (MP) are parameters systematically varied during the iterative calculation process. It is sufficient to predefine the RP and MP for certain evolutionary steps and to change the values linearly between these points to approach the defined values. This cycle of selection, recombination, and mutation is repeated until $\Delta(^{13}C)$ is minimized.

To develop a methodology for applying the algorithm and to evaluate its capabilities, the authors[29,30] performed a series of computational experiments with the isoleucine molecule (**5.3**). These experiments allowed optimization of the parameters $m$, $l$, $n$ (the number of parallel calculated populations), RP, MP, *etc.*

In a second experiment the previously optimized parameters were used to perform a fully automated structure elucidation for a set of 160 small molecules with 9–16 non-hydrogen atoms. The limitations of this method were examined by investigating larger molecules with up to 20 non-hydrogen atoms. For the second experiment, the substances were randomly selected from the SpecInfo database[17] and a uniform setup $n = 8$, $m = 32$, and $l = 8$ was chosen for all molecules. The following results were obtained: the correct solution was found for 69% of the structures, whereas for 31% the calculation is stopped either after 500 generated populations (time limit) or a molecule with a lower $\Delta(^{13}C)$ value than that for the target structure was created (accuracy limit). In both cases the algorithm is considered to have failed. In this effort, 85% of all molecules under 15 heavy atoms were predicted correctly, but the algorithm fails for 77% of the molecules with 15 and 16 heavy atoms. For 15% of the molecules containing 15 or 16 heavy atoms, GENIUS predicts false structures that have a smaller $\Delta^{13}C$ value than the correct solution. It was found that the structure elucidation of molecules having up to 20 heavy atoms was possible only under the condition that GOODLIST/BADLIST and user-defined fragments were used. Examples of problems solved by GENIUS[29,30] are presented in Table 5.1.

**Table 5.1** Molecular structures, parameters and results obtained for some example molecules solved by the genetic algorithm approach.

| Structure, molecular formula | No. heavy atoms | $\Delta(^{13}C)$ (ppm) | Possible no. of isomers[a] | $n$[b] | $m$[c] | $I$[d] | Excluded/included substructures[e] | Calculation time (min)[f] |
|---|---|---|---|---|---|---|---|---|
| 1 C$_6$H$_{13}$NO$_2$ | 9 | 1.12 | 23 946 | 1 | 32 | 8 | No constraints | <1 |
| 2 C$_6$H$_9$NO$_2$ | 11 | 1.63 | 89 502 542 | 5 | 50 | 25 | No constraints | 4 |
| 3 C$_{11}$H$_{12}$N$_2$O$_2$ | 15 | 1.44 | $\sim 10^{12}$ ($36 \times 10^9$) | 1 | 60 | 30 | No constraints | 20 |
| 4 C$_{16}$H$_{29}$NO$_3$ | 20 | 0.81 | $\sim 10^{12}$–$10^{14}$ ($66 \times 10^9$) | 64 | 64 | 16 | No constraints | 453 |

| | Structure | | | | | | | |
|---|---|---|---|---|---|---|---|---|
| 5 | $C_{15}H_{14}O_5$ | 20 | 1.77 | 64 | 64 | 16 | No constraints | 6704 |
| 5′ | $C_{15}H_{14}O_5$ | 20 | 1.77 | 1 | 64 | 16 | Included 2 fragments: (phenyl) | 3 |
| 6 | $C_{15}H_{12}N_2O_3$ | 20 | 1.89 | 16 | 128 | 32 | Included 2 fragments: (phenyl) | 18 |
| 7 | $C_{14}H_{23}NO_5$ | 20 | 1.36 | 8 | 32 | 8 | Included: (piperidine amide) | 3 |

[a]Total number of possible constitutions calculated or estimated in,[29,30] values in brackets, and by authors of the book. All approximations shown in brackets are significantly underestimated.

[b]Number of parallel calculated populations (n).

[c]Number of individuals in the populations (m).

[d]Number of best-ranked [small $\Delta(^{13}C)$ value] individuals in the parent generation that are conserved for the new child generation (l).

[e]Included structural fragments that have to be used in all generated structures and excluded structural fragments that are forbidden in all generated structures. Fragments C=C and X–X, where X is a heteroatom, were included in BADLIST.

[f]Total calculation time in minutes on a Pentium II 450 Mhz processor with 512 MB RAM.

The given examples show that the genetic algorithm basically copes with small molecules (up to 15 heavy atoms). Only two molecules containing 20 heavy atoms out of a total of 11 presented in the two articles[29,30] were elucidated without the application of user-defined fragments. Even in these cases, the probability of obtaining the correct structure is not high enough for general application of the program. This is due to the fact that the amount of structural information contained in $^{13}C$ NMR spectra does not allow the correct structure to be identified in a whole set of isomers whose number, by our estimates, is up to $10^{12}$–$10^{14}$ even for small molecules of 15 or 16 skeletal atoms (see Section 1.1). In addition, the calculation time is large, and molecules having 15–20 skeletal atoms can be identified in a reasonable time only when a significant part of the heavy skeletal atoms are absorbed by user-defined fragments (see, for example, structures #5 and #5′ in Table 5.1). Consequently, it can be stated that the genetic algorithm did not remove the restrictions inherent to deterministic expert systems based on 1D $^{13}C$ NMR spectra. The application of a genetic algorithm can be considered as, perhaps, a last attempt to retain 1D NMR spectroscopy as the only source of NMR data for structure elucidation with the aid of an expert system.

Thus experience has shown that the computer-assisted structure elucidation of real-world large molecules, particularly natural products, is possible only when 1D and 2D NMR spectral data are used in conjunction with expert systems and such systems will be reviewed in the next Chapter.

# References

1. R. K. Lindsay, B. G. Buchanan, E. A. Feigenbaum and J. Lederberg, *Applications of Artificial Intelligence for Organic Chemistry: The DENDRAL Project*, McGraw-Hill, New York, 1980.
2. N. A. B. Gray, *Chemometr. Intell. Lab. Syst.*, 1988, **5**, 11.
3. B. G. Buchanan, E. A. Feigenbaum and J. Lederberg, *Chemometr. Intell. Lab. Syst.*, 1988, **5**, 33.
4. N. A. B. Gray, *Chemometr. Intell. Lab. Syst.*, 1988, **5**, 37.
5. L. M. Masinter, N. S. Sridharan, J. Lederberg and D. H. Smith, *J. Am. Chem. Soc.*, 1974, **96**, 7702.
6. M. E. Elyashberg, V. V. Serov, E. R. Martirosyan, L. A. Zlatina, Y. Z. Karasev, V. N. Koldashov and Y. Y. Yampolskiy, *J. Mol. Struct.*, 1991, **76**, 191.
7. L. A. Gribov, M. E. Elyashberg and V. V. Serov, *Anal. Chim. Acta, Comp. Techn. Optimiz.*, 1977, **95**, 75.
8. M. E. Elyashberg, Y. Z. Karasev, E. R. Martirosian, H. Thiele and H. Somberg, *Anal. Chim. Acta*, 1997, **348**, 443.
9. M. E. Elyashberg, E. R. Martirosian, Y. Z. Karasev, H. Thiele and H. Somberg, *Anal. Chim. Acta*, 1997, **337**, 265.
10. M. E. Elyashberg, Y. Z. Karasev and E. R. Martirosian, *Anal. Chim. Acta*, 1999, **388**, 353.

11. V. V. Serov, L. A. Gribov and M. E. Elyashberg, *J. Mol. Struct.*, 1985, **129**, 183.
12. D. H. Rouvray, *Fuzzy Logics in Chemistry*, Academic Press, San Diego, 1997.
13. S. G. Molodtsov, *Commun. Math. Chem. (MATCH)*, 1994, **30**, 213.
14. S. G. Molodtsov, *Commun. Math. Chem. (MATCH)*, 1994, **30**, 203.
15. W. Maier, in *Computer-Enhanced Analytical Spectroscopy*, ed. C. L. Wilkens, Plenum Press, New York, 1993, vol. 4, pp. 37.
16. M. Will, W. Fachinger and J. R. Richert, *J. Chem. Inf. Comput. Sci.*, 1996, **36**, 221.
17. *Specinfo*, Chemical Concepts GmbH, D-69,442 Weinheim, Germany.
18. W. Bremser, *Anal. Chim. Acta, Comp. Techn. Optimiz.*, 1978, **2**, 355.
19. M. E. Elyashberg, K. A. Blinov and E. R. Martirosian, *Lab. Autom. Inf. Manag.*, 1999, **34**, 15.
20. S. Sasaki, I. Fujiwara, H. Abe and T. Yamaski, *Anal. Chim. Acta*, 1980, **122**, 87.
21. T. Oshima, Y. Ishida, K. Saito and S. Sasaki, *Anal. Chim. Acta*, 1980, **122**, 95.
22. K. Funatsu and S. Sasaki, *J. Chem. Inf. Comput. Sci.*, 1996, **36**, 190.
23. K. Funatsu, M. Nishizaki and S. Sasaki, *J. Chem. Inf. Comput. Sci.*, 1994, **34**, 745.
24. S. G. Molodtsov, *Commun. Math. Chem. (MATCH)*, 1998, **37**, 157.
25. Advanced Chemistry Development, ACD/NMR Predictors. Prediction suite includes $^1$H, $^{13}$C, $^{15}$N, $^{19}$F, $^{31}$P NMR prediction, http://www.acdlabs.com
26. J.-L. Faulon, *J. Chem. Inf. Comput. Sci*, 1996, **36**, 731.
27. M. Mitchell, *An Introduction to Genetic Algorithms*, MIT Press, Cambridge, MA, 1996.
28. D. E. Goldberg, *Genetic Algorithms in Search, Optimization, and Machine Learning*, Addison-Wesley, Reading, MA, 1989.
29. J. Meiler and M. Will, *J. Chem. Inf. Model.*, 2001, **41**, 1535.
30. J. Meiler and M. Will, *J. Am. Chem. Soc.*, 2002, **124**, 1868.
31. J. Meiler, R. Meusinger and M. Will, *J. Chem. Inf. Model.*, 2000, **40**, 1169.
32. J. Meiler, W. Maier, M. Will and R. Meusinger, *J. Magn. Reson.*, 2002, **157**, 242.

CHAPTER 6

# CASE 2D NMR-based Expert Systems

## 6.1  Introduction

A number of CASE systems have been developed that can use the molecular connectivity information contained within 2D NMR correlations. Over the past two decades these systems have been successfully challenged with a series of complex chemical structures. A number of general assumptions are made during the application of these systems and these should be considered carefully as the constraints can dramatically affect the successful application of the algorithms.

In accordance with the axiomatic approach described in Chapter 2, by default, expert systems utilizing 2D NMR spectral data assume that a $^1$H-$^1$H COSY connectivity is associated with couplings crossing one C–C bond (accepting that there are geminal correlations between protons attached to the same carbon), whereas HMBC connectivities describe heteronuclear correlations across one to two C–C bonds. Of course, 2D NMR spectra can contain signals corresponding to the interaction of spins over a greater number of bonds than those specified as the default in CASE programs.

Long-range correlations across three or more bonds ($^nJ_{HH,CH}$, n > 3) in homo- or heteronuclear 2D NMR spectra can be identified for many small molecules and are especially common in the spectra of molecules with rigid or strained carbon skeletons, as well as in the spectra of polyconjugated compounds.[1] It is commonly assumed that peaks with either high or medium intensities correspond to connectivities whose lengths are within the default limits expected for their corresponding 2D NMR spectra. Peaks of low intensity are generally assumed to be associated with longer-range interactions.

It is a common misconception that the cross-peak intensity directly represents the number of bonds separating coupling nuclei. This is *not* exclusively

New Developments in NMR No. 1
Contemporary Computer-Assisted Approaches to Molecular Structure Elucidation
By Mikhail Elyashberg, Antony Williams and Kirill Blinov
© Royal Society of Chemistry 2012
Published by the Royal Society of Chemistry, www.rsc.org

true, however, and examples abound of four-bond heteronuclear long-range correlations being stronger than three- and even two-bond correlations. The coupling depends to a large extent on the mutual spatial orientation of the intervening nuclei, as well as a series of other complicating factors.[1–3] The observation of long-range correlations is dependent not only on the nature of the chemical structure under study, but also on the nature of the observing experiment and the associated parameters used to probe these correlations.

Consider as an example a COSY spectrum displaying a cross-peak resulting from the interaction of two protons H(1), H(2) detected in a fragment H(1)-C1-C2-C3-H(2). By default, a CASE program will consider that such a cross-peak indicates that atoms C1 and C3 are separated from each other by only one bond. This will of course conflict with other spectroscopic data. As a result, this particular task will either not be solved or the solution will be wrong based on the default assumptions. To resolve this potential conflict such connectivities should be extended by one additional C–C bond. The main problem is the discrimination of how many of these connectivities should be lengthened.

Another typical reason for the occurrence of contradictory correlations is the presence of similar chemical shifts relating to structurally non-equivalent $^{13}$C and $^{1}$H nuclei, a situation known as accidental degeneracy and resulting in the superposition of peaks. For example, two identical chemical shifts $\delta_i$ and $\delta_k$ [$\delta_i = \delta_k = \delta(i,k)$] assigned to the non-equivalent carbon atoms $C_i$ and $C_k$ in a real structure can, for example, be separated by ten C–C bonds within the structure. Since a CASE program constructs all gradient-selected HMBC (gHMBC) connectivities containing the degenerate chemical shift pair $\delta(i,k)$, this will result in the generation of erroneous correlations reflecting the inter-actions of atom $C_i$ with protons isolated from atom $C_k$ by two or three bonds and *vice versa*. The identification and error handling of such correlations involves a series of specific challenges.

Since experimental 2D NMR data contain correlations of various lengths, even in those cases where the system default options accurately reflect the specific spin–spin couplings in a given structure, the purpose of spectral interpretation is to perform an elucidation based on the fuzzy data available. It has been demonstrated (*vide infra*) that structure generators that utilize 2D NMR data can cope with this challenge quite successfully if the data meet the "axioms" formulated in Chapter 2.

The issue is much more complicated if there is at least one correlation contained within the input data that corresponds to distant coupling interactions that are not supported by the default program option settings. As previously noted, the data set as a whole is contradictory under this situation. When the presence of contradictory correlations is unknown the input data are not only fuzzy, they are also potentially contradictory and vague. It can be expected that due to experimental parameter settings, the sample concentration or the threshold setting for visualization that occasionally certain expected resonances may not be observed in the experimental 2D NMR spectra. In a number of cases only data from a restricted number of 2D NMR experiments may be available (for example, a $^{1}$H-$^{15}$N 2D NMR correlation spectrum in a data set may be absent) and as a result the input data set can be *incomplete*. The incompleteness of data may result in a dramatic increase in the time required for structure generation. At times the generation process may be prohibitively long and the task will be abandoned and considered as unsolvable using an expert system.

Considering the complexity inherent to 2D NMR data, it is clear that usually an expert system will not be able to elucidate a molecular structure while operating in the *automatic* mode only. However, experience has also shown that many tasks can indeed be solved without the involvement of an expert spectroscopist. The challenges with the data are the primary driver behind the development of interactive methods for the automated elucidation of complex molecules from contradictory and fuzzy spectrum–structure data. These methods were developed on the basis of generalized experience accumulated while solving a large number of complex structures. It should be noted that the contradiction and incompleteness of 2D NMR data arise for different reasons. It is the presence of superfluous ("illegal") cross-peaks that generates contradictions, whereas it is the absence of crosspeaks that renders a data set incomplete.

Numerous experiments have enabled workers[4–10] to outline general strategies for the solution of problems based on 2D NMR data. These strategies include the detection and elimination of contradictions in the input data and the usage of additional parameters that reduce uncertainty around the input data. In order to successfully implement solutions to support these strategies a user-friendly interface was developed to allow a user to interact and participate with the process of structure elucidation at any stage. These strategies will be explained in Part III.

In this Chapter a series of 2D NMR-based expert systems described in the literature over the last two decades will be discussed. The systems chosen for discussion had previously demonstrated the possibility of elucidating structures common to organic chemistry and particularly for the analysis of natural products. These systems, with the exclusion of NMR-SAMS, have not been commercialized, but they paved the way to the development of the most effective strategies for the application of expert systems, which now permits complex real structural problems to be solved in many laboratories. We will first review the SESAMI system that was elaborated under the guidance of Munk, one of the first pioneers of CASE systems in the 1960s.

## 6.2   SESAMI-C

SESAMI (systematic elucidation of structure applying machine intelligence) includes a 2D NMR interpretation module, INTERPRET2D, which can be linked to the structure generators ASSEMBLE,[11] COCOA,[12] and the latest generator HOUDINI.[13] The general workflow of the system does not markedly depend on the type of structure generator used. The SESAMI-C system utilizing structure generation via the COCOA program is chosen here as the combination for discussion.

The SESAMI-C system is intended to determine molecular structures from a combination of $^1$H, $^{13}$C, and 2D NMR data, specifically $^1$H-$^1$H COSY, $^1$H-$^{13}$C HSQC (HMQC) and $^{13}$C-$^{13}$C INADEQUATE.

The flow diagram of the SESAMI-C system is shown in Figure 6.1. The system has been described in a number of publications.[14–19]

*SESAMI-C input.* The molecular formula and the 1D $^1$H/$^{13}$C and 2D NMR data are fed into the SESAMI-C structure generator prior to program execution. The $^{13}$C NMR data must include the chemical shift and multiplicity associated with each signal. SESAMI-C cannot handle accidental degeneracy resulting in accidental overlap of signals. Fewer carbon signals than atoms contained within the molecular formula suggests molecular symmetry in the

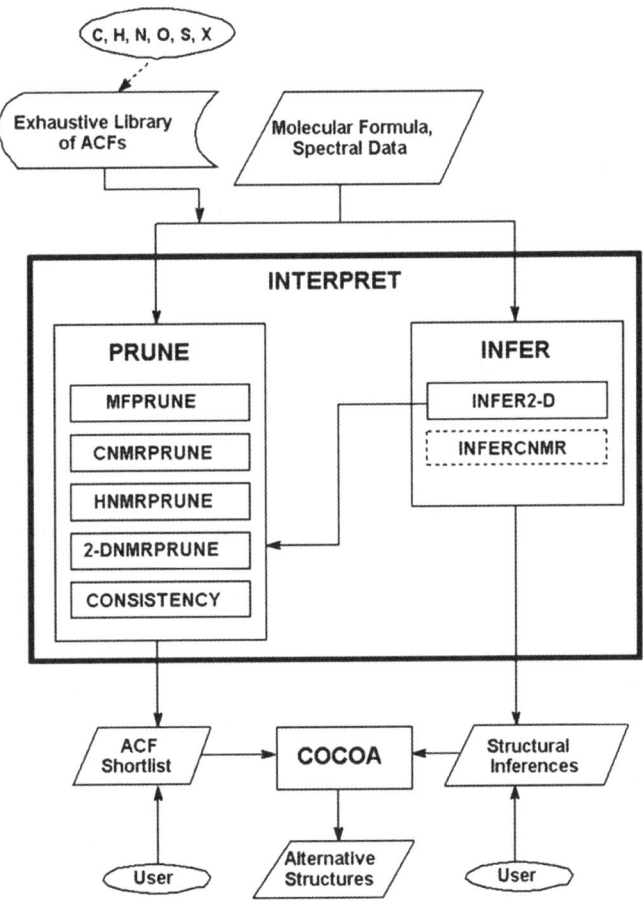

**Figure 6.1** The flow diagram describing the SESAMI-C system.

unknown. If two or more signals *are* known to overlap, then each can be given a slightly different chemical shift to remove the forced symmetry constraint, *e.g.* 33.45 and 33.46 ppm. The program will then treat these two signals as *separate and distinct*. The chemical shifts, integrals, and information regarding the presence of exchangeable hydrogens are required for the input of $^1$H NMR data.

The results of all types of 2D NMR experiments can be conveniently entered as pairs of correlated signals and the number of intervening bonds between the atoms, either as an exact number or a range. The range describing COSY and HMBC correlations is 2 to 3. If four or more intervening bonds are possible then the user must decide on the range to be set and should guess which correlation should be allowed to be of non-standard length.

INTERPRET is a two-track spectrum interpretation procedure. One track applies the PRUNE algorithm. The molecular formula and a series of spectroscopic data are reduced to a shortlist of uniformly sized, precisely defined structural fragments expected to be present in the unknown under study. These fragments serve as the structural building units for the structure generator, COCOA. They are referred to as ACFs: singly-concentrically-layered, atom-centered fragments (e.g. =CH-**CH₂**-O-) built of groups defining an element, its attached hydrogens, if any, and each of the partial bonds by which one element group is connected to another (*e.g.* -CH2-, =O). The initial exhaustive list contains all possible ACFs that can be derived from the following elements: C, H, O, N(III), S(II), and all halogens. The exhaustive ACF list excludes ACFs that would clearly confer chemical instability on compounds containing them (*e.g.* -CH$_2$-C(OH)$_3$) and includes approx. 5100 ACFs. PRUNE produces a set of possible structural building blocks for the unknown by deleting a series of ACFs from the set initially produced.

The ACFs removed are those that are incompatible with the molecular formula of the unknown, and incompatible with the $^1$H, $^{13}$C, and 2D NMR data. The ACF shortlist can be conveniently examined and pruned further in accord with the users' wishes. In practice, the ACF shortlist will usually contain more invalid ACFs that are not actually present in the unknown than valid fragments. However, most structures containing invalid ACFs are eliminated directly during the process of structure generation.

MFPRUNE initially removes ACFs that are incompatible with the molecular formula of the unknown. Furthermore, CNMRPRUNE and HNMRPRUNE use *databases* containing allowed $^{13}$C and $^1$H NMR chemical shift ranges, respectively, and signal multiplicities for the central carbon atom of each carbon-centered ACF, compare the ACFs surviving the MFPRUNE procedure with the observed $^1$H and $^{13}$C NMR spectra and delete those that are incompatible.

Surviving ACFs are then organized into groups based on the observed $^{13}$C NMR spectrum. For each chemical shift there is an associated list of ACFs [see eqn (1.6) in Chapter 1]. For each ACF the assigned central carbon chemical shift range and signal multiplicity matches those of the observed signal. Each group contains not only the carbon-centered ACFs, but also separate lists of compatible heteroatom-centered ACFs for each heteroatom contained in the unknown.

The INFER algorithm makes up the second track of the INTERPRET program and produces substructural inferences that serve as constraints for the structure generation process. Substructures are identified which are to be present or absent in the final structure of the unknown. Alternative interpretations of the data may also be produced. The output of the INFER algorithm may be viewed, edited, and supplemented by any structural information available to the user. The output generated by the two tracks of INTERPRET, the structural building units (ACFs), and the substructural constraints are then passed on to the structure generator to continue the process.

*INFER2D/2DNMRPRUNE.* The 2D NMR data are used as input to the 2DNMRPRUNE algorithm enhancing the ability of PRUNE to discriminate between valid and invalid ACF structural building units. INFER2D interprets all through-bond correlations and produces pair-wise connectivities in a tabular format. From this table the constraints are fed as inputs to the generator.

It is assumed that *unambiguous* connectivities between $^{13}C$ NMR signals are obtained from 2D NMR data. If the unknown does not possess symmetry (as determined by the number of carbon resonances), then INFER2D will attempt to build substructures from the carbon connectivities by looking for atom overlaps in the substructures. For example, several pairs of COSY connectivities can produce an atom chain which makes up a substructure, but INFER2D provides no information on the bond type or the nature of attached heteroatoms. If any symmetry is present, then more than one interpretation of the data is possible.

Alternative constraints are also generated by INFER2D. For example, C28.4 ~ C14.4/C28.4 ~ A ~ C14.4, where " ~ " represents any bond type and "/" is "or", and A represents any non-hydrogen atom. Since there is likely to be high ambiguity in the case of a set of such alternative constraints, no attempt to build discrete substructures is made.

The output of INFER2D also serves as the knowledge base for 2DNMRPRUNE and this is used later in the generation of the ACF shortlist. The 2DNMRPRUNE routine examines each surviving carbon-centered ACF for compatibility with inferences made by the INFER2D algorithm and produces a reduced ACF shortlist. The reduced ACF shortlist dramatically reduces the amount of computation necessary during the process of structure generation.

During the pruning process, a separate routine, CONSISTENCY, examines the list of surviving ACFs for conflicting structure information. The routine attempts to address any inconsistencies by deleting the ACF(s) causing the problem. For example, if *all* methyl-centered ACFs for a given unknown have either methylene or methine carbon atoms as first-layer neighbors, then all quaternary carbon-centered ACFs bearing methyl groups as first-layer neighbors should be deleted. The surviving ACFs are output as an ACF shortlist and

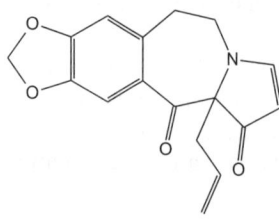

C$_{18}$H$_{20}$O$_5$, ref.[17]
$^1$H and $^{13}$C NMR, HMQC, HMBC, COSY.
Generator output: 6 structures

C$_{15}$H$_{22}$N$_2$O$_7$, ref.[15]
$^1$H and $^{13}$C NMR, HMQC, HMBC, COSY
Generator output: 4 structures
CPU time 5 min (unknown computer)

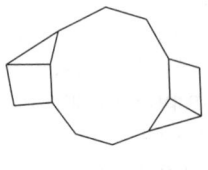

C$_{25}$H$_{38}$O$_3$, ref.[19]
$^1$H and $^{13}$C NMR, HMQC, HMBC, COSY
Generator output: 4 structures

C$_{15}$H$_{24}$O$_6$, ref.[19]
$^1$H and $^{13}$C NMR, HMQC, HMBC, COSY
User defined fragment: γ-lactone
Generator output: 2 structures

C$_{17}$H$_{15}$NO$_4$, ref. [17, 18]
$^1$H and $^{13}$C NMR, HMQC, HMBC, COSY
Generator output: 15 structures

C14H20, ref.[17]
$^1$H and $^{13}$C NMR, 2D INADEQUATE
Generator output: 7 structures

**Figure 6.2**   Examples of "unknown" structures elucidated by SESAMI-C. The molecular formulae and NMR spectra used as inputs, and the sizes of the output files are listed. A genuine structure was included into output files along with other competing structures.

may be reviewed and pruned further if desired. This shortlist is used as the input for the structure generator.

Example structures that were elucidated[15,17–19] using SESAMI-C are shown in Figure 6.2. In all of the cases, the 2D NMR data contained only correlations of default length. It can be concluded that the COCOA generator is capable of allowing the SESAMI-C system to elucidate structures of modest size and complexity.

# 6.3 SESAMI-H

In SESAMI-H, the HOUDINI structure generator[13] replaced the less efficient COCOA generator. Schulz *et al.*[16] described the SESAMI-H system and compared its efficiency with that of SESAMI-C. The following advantages of the HOUDINI generator over the COCOA system were reported:[13,16]

1. HOUDINI, in contrast to COCOA, requires only a single representation of the initial problem state, even in the presence of hybridization uncertainties.
2. HOUDINI accepts a set of structural inferences from INTERPRET and initially pre-processes them to produce a single, integrated representation of the collective information.
3. Relative to COCOA, information processing is superior and eliminates large families of invalid structures earlier in the structure generation process.
4. HOUDINI processes sets of alternative inferences in a concerted, rather than sequential, treatment. This is especially important, since much of the input to the structure generator is in the form of alternative inferences.

To compare the results obtained with the two variants of the system, Schulz *et al.*[16] selected seven natural products (Table 6.1), whose identification from 2D NMR spectra had been described in the literature.[19–24]

The results of testing the program using the compounds listed in Table 6.1 are presented in Table 6.2, in which the unknown structures with $k$, the number of structures in the output file, and $t_g$, the generation time values

**Table 6.1** Summary of input to SESAMI-H.[16]

| | | | 1D NMR | | 2D NMR | | |
| --- | --- | --- | --- | --- | --- | --- | --- |
| *Compound* | *Ref.* | *Mol. formula* | $^1HNMR$ | $^{13}C$ NMR | HMQC | HMBC | COSY |
| 1. DDP (18-dimethyl-14-diacetylpubescenine) | Ref. 20 | $C_{23}H_{37}NO_7$ | 31 | 23 | 26 | 78 | 26 |
| 2. Stenocarpine | Ref. 21 | $C_{21}H_{31}O_3N$ | 27 | 21 | 23 (24)[a] | 87 (89)[a] | 22 |
| 3. 1-Dimethylwinkleridine | Ref. 20 | $C_{22}H_{35}NO_6$ | 31 | 22 | 26 | 58 | 26 |
| 4. Guttiferone A | Ref. 22 | $C_{38}H_{50}O_6$ | 31 | 38 | 28 | 88 | 0 |
| 5. Lacinan-8-ol | Ref. 23 | $C_{15}H_{26}O$ | 18 | 15 | 17 | 71 | 22 |
| 6. α-Botryoxanthin | Ref. 24 | $C_{74}H_{112}O_2$ | 72 | 74 | 72 | 227 | 4 |
| 7. Syringolide | Ref. 19 | $C_{15}H_{24}O_6$ | 17 | 15 | 15 | 21 | 6 |

[a]Number of correlations shown in this Table in the original article.[21]

**Table 6.2**  The results of testing SESAMI-H.

| # Structure | k | $t_g$ (s) | Comments |
|---|---|---|---|
| 1 Stenocarpine[21] $C_{21}H_{31}O_3N$ | 1 | 0.6 | Problem solved in automatic mode. The presence of two $^4J_{HH}$ long-range correlations in the COSY spectrum is not mentioned in ref. 16. The corresponding connectivities probably were either corrected or deleted before the problem solution, because their existence is noted in the original article[21] ($^4J_{HH} = 1.6$ and $^4J_{HH} = 1.7$ Hz). |
| 2 1-Dimethylwinkleridine[20] $C_{22}H_{35}NO_6$ | 2 | 1.1 | Problem solved in automatic mode. The presence of one $^4J_{HH}$ long-range correlation in the COSY spectrum is not mentioned in ref. 16. The corresponding connectivity probably was either corrected or deleted before the problem solution, because its existence is noted in the original article.[20] A W-coupling gave $^4J_{HH} = 8$ Hz. |
| 3 Guttiferone A[22] $C_{38}H_{50}O_6$ | 4 | 12.9 | COSY data were not used. User-entered substructures are below: |

**Table 6.2**  (*Continued*).

| # Structure | $k$ | $t_g$ $(s)$ | Comments |
|---|---|---|---|
| 4 | | | Two W-couplings with $^4J_{HH}=2$ and $^4J_{HH}=3$ Hz were detected in the COSY spectrum by the authors of the original article.[23] |
| | 1 | <0.1 | 1. Two mentioned long-range COSY correlations were entered by hand. |
| | 1 | <0.1 | 2. All 22 COSY correlations were set as three or four intervening bonds. |
| Lacinan-8-ol[23] $C_{15}H_{26}O$ | | | |
| 5 | 1 | 3.9 | SESAMI-C was terminated after several days of calculations. |
| | | | The presence of one $^4J_{CH}$ HMBC correlation (2.08–15.3) is not mentioned in ref. 16. No information about the presence of this correlation exists in the original article.[24] Correction of this connectivity is possible only by comparison of the 2D NMR data with the assigned carbon atoms of the true structure. |
| α-Botryoxanthin[24] $C_{74}H_{112}O_2$ | | | |
| 6 | 2 | 10 | 1. To solve this problem, user-entered substructures (see Figure 6.3) were used. |
| | 1 | 42.8 | 2. No user-entered substructures were used. |
| | 3 | | |
| Syringolide[19] $C_{15}H_{24}O_6$ | | | |

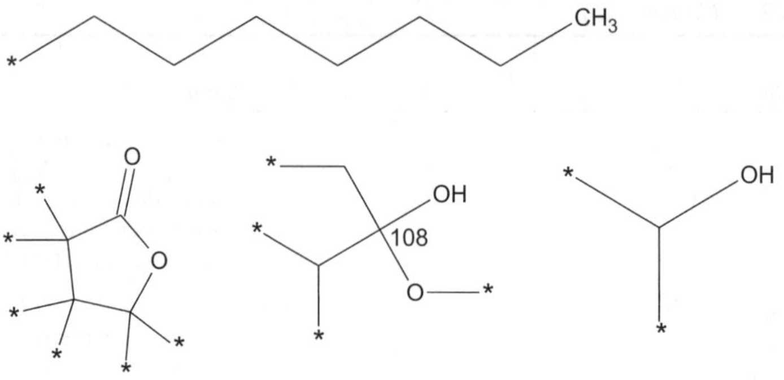

**Figure 6.3** User-defined substructures introduced during the structure elucidation of syringolide.[19]

obtained with SESAMI-H under standard conditions, are also shown in the Table.

Analysis of the results shows that if there are no contradictions in the 2D NMR data, then the SESAMI-H system can successfully determine the structures of quite large molecules (up to 75 skeletal atoms). The solution time can range from several fractions of a second to 10 s. This indicates that adopting a theory of constraints satisfaction[25] approach in HOUDINI accelerates structure generation relative to the COCOA generator.

We have repeated the analysis for all seven problems listed in Table 6.1 using the StrucEluc system;[5,6] the experimental data were reported in the publications[19–24] as inputs. Four of the seven problems were shown to contain correlations of a non-standard length. This situation was highlighted only in the publication studying lacinan-8-ol.[23] Problems 1, 2, and 5 (Table 6.2) can be solved only if these correlations are deleted or corrected. In the case of lacinan-8-ol, Schulz *et al.*[16] extended all connectivities in the COSY spectrum to $^{3-4}J_{HH}$ to solve the problem with non-standard correlations. According to our research (see Chapter 10), this approach is applicable only to very modest-sized molecules, such as lacinan-8-ol.

The work described here indicates that HOUDINI is a high-speed generator that can form the basis of an efficient expert system. The system would benefit from association with a program allowing the circumvention of problems originating from overlapping signals and allowing correct results, even in the presence of non-standard connectivities.

## 6.4 The Expert System CISOC-SES

The 2D NMR expert system CISOC-SES is described in a series of articles.[26–29] Its structure generator was optimized to process ambiguous information that can be extracted from 2D NMR connectivities. A consideration specific to this program is that there is no need for initial definition of the hybridization status of the skeletal atoms of the molecule under analysis. The definition of the hybridization states of the carbon atoms obviously reduces the number of probable structures and speeds up the generation process. This is vital for elucidation of complex natural products. In other words, the authors of the system tried to release users from their responsibility for setting the axioms regarding atom hybridization which are necessary to assemble structures (see Chapter 2). Since these axioms are crucial for structure generation, they can be only replaced by other suggestions (other axioms). The authors introduced this capability into their program. In order to enable the generation of large structures the researchers had to apply a wide range of heuristics aimed at mitigating the scope of searching.

As mentioned above, spectral imperfections and molecular symmetry frequently make it impossible to unambiguously assign two nodes (1D signals or atoms) to 2D-spectrum-derived distance constraints. In these cases the *ambiguous-node distance constraints* (ANDCs) are introduced. Peng *et al.*[26–29] focused their efforts on the utilization of these constraints in the framework of the suggested algorithm for structure generation. It should be noted that in the SESAMI system a very high emphasis is placed on the structure generation process from 2D NMR data which had already been processed by primary peak peaking and manual refinement.

The logical structure of CISOC-SES is shown in Figure 6.4. On the basis of the molecular formula, $^{13}C$ and DEPT spectral data (if available), one or more initial free-bond connection matrices (FBMX) are constructed to represent the bonding possibilities between the unsatisfied valences (called *free bonds*) of the skeletal atoms. The FBMX actually forms the search space for subsequent

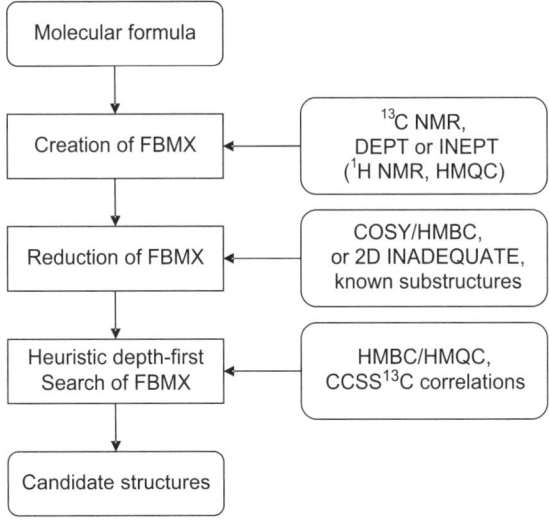

**Figure 6.4** Flowchart of the CISOC-SES structure generator.

structure generation. It should be noted that the hybridization states of the heavy atoms need not be explicitly specified in this system. The FBMX is then reduced by imposing constraints extracted from both the 1D and 2D NMR data. Structure generation is carried out during a heuristic depth-first search[30] of the FBMX matrices. This procedure is controlled by a series of empirical parameters set by the user. As a result of the introduction of a whole range of heuristic approaches the structure generation process can be shortened. However, such applications of both heuristic and empirical parameters require substantial control over the structure generation process on the part of the user. In other words, the chemist assumes responsibility over the feasibility and correctness of calculations within reasonable time limits. In our experience, the involvement of an expert spectroscopist in the structure determination process is almost unavoidable. The involvement of the user in the structure generation process and in the selection of parameters associated with the structure generator results from not postulating the appropriate "axioms" related to atom hybridization. We do believe, however, that the involvement should not extend beyond providing structural hypotheses nor should the user be required to decide on the choice of the structure generation algorithm.

The algorithms utilized in the CISOC-SES system have been illustrated using several examples to evaluate the capacity of the system for solving real-world problems.[27–29] A short description of some of them will be given.

*Example 1*[27]

$^1$H, $^{13}$C, HETCOR, COLOC, and COSY data were obtained for a molecule with molecular formula $C_{21}H_{29}NO$ (**6.1**). The intensities of both the 1D and 2D NMR peaks were manually classified as weak (1), medium (2), and strong (3). The existence of a C=O group, an isopropyl group, double bonds, and a C≡N group were indicated by the IR spectrum.

**6.1**

The NMR data and the molecular formula suggested that the "unknown" molecule was composed of the following skeletal atoms and groups 6(>C<), 7(>CH-), 2(-CH$_2$-), 6(CH$_3$), >N-, >O. Consequently, the total number of free bonds in the atom set was 60. Using this information a 60 × 60 FBMX could be constructed. In the second stage, the 2D COSY, HETCOR, and COLOC cross-peaks were interpreted and unified as a set of 41 $^{13}$C-$^{13}$C signal–signal connectivities (SSCNs), which were directly transformed into 41 C-C atom–atom connectivities (AACNs).

The data were analyzed by the program and an error message appears if connectivities with very small ($<0.02$ ppm) $^1$H chemical shift differences between two $^1$H peak pairs are detected. Since this leads to conflicting C–C connectivities, the latter were edited by hand and the corresponding distances associated with the $^1$H-$^1$H SSCNs were enlarged to 3–5 bonds. At this point, the user can influence the program by inputting known structural information by entering AACNs, especially those concerning heteroatoms derived from other spectral information (*e.g.* IR, UV, *etc.*). In this case the C=O and C≡N groups were added as two user-supplied AACNs in the same format as that of a spectra-derived AACN. A carbon atom was thus represented by its assigned $^{13}$C peak number. However, to enter the C≡N group as an input the user must choose one of two quaternary carbons with close chemical shifts, 121.57 or 123.40 ppm, both of which are equally probable. In order to obtain a correct solution both possibilities have to be verified.

In addition to the two user-supplied AACNs, other spectra-derived AACNs were distinguished, on the basis of which 15 fixed bonds were extracted and a reduced FBMX of $30 \times 30$ was established.

Prior to structure generation the user can modify a number of parameters to control the generation process: (1) the direction of structure generation (parameter GEN_FLAG = 0, 1, 2...); (2) the demanded level of satisfaction of long-range distance constraints (LRDCs) ($K_A$); (3) the maximum allowed number of violated LRDCs (MAX-ERR-LRDC); and (4) the maximum tolerable deviation of $^{13}$C chemical shifts ADD-C13-RNG, *etc.*

We believe that the number of variable parameters is too high. The lack of clear criteria for selecting parameters with respect to the available experimental data makes it difficult to choose values and the responsibility for any decision is consequently imposed directly on the user. The $K_A$ parameter determining the *speed of generation* is selected using an iterative procedure that requires quite a lot of computational time, even for a small molecule. This procedure cannot be avoided in practice, since the risk of missing a correct solution is too high.

In this experiment, the parameters MAX-ERR-LRDC and ADD-C13-RNG were all set to zero (default values), but some parameters were varied to identify their influences on the efficiency of structure generation. Under these conditions, a single correct structure **6.1** was always obtained, but the time for structure generation varied from 15 to 1400 s. The efficiency of structure generation was significantly improved when the search tree was properly weighted and rearranged on the basis of LRDCs. In this example, the MAX-ERR-LRDC value was set to zero. This did not hinder the identification of a correct result, since the two non-standard COSY correlations (10–21) and (4–9), indicated by arrows in the structure **6.1**, had been extended to 3–5 bonds.

The structure generator discussed here is inefficient when it operates with all parameters disabled. For example, we have shown that the StrucEluc system[5,6] generated one structure $10^4$ times faster under the same conditions and from the same initial 2D NMR data. It should be noted that StrucEluc was running on a Pentium III PC (500 MHz), whereas the original work was performed on a MicroVAX 3300 computer.

*Example 2*[28]

A standard set of 1D and 2D NMR data was used for the structure elucidation of compound **6.2** with molecular formula $C_{28}H_{34}O_7$.

**6.2**

To reduce the search space the following fragments were entered as user-supplied AACNs: C=O ($\delta = 204$ ppm), 2 O-C=O ($\delta = 169.90$ and 167.40 ppm), as well as the presence of a *furan ring* as suggested by the UV and IR data. One LRDC with a distance of exactly two C–C bonds was derived from a *weak* COSY peak. As previously mentioned, the intensities in 2D NMR spectra are not a reliable criterion for determining the true length of a correlation. An incorrect setting of the *exact* correlation length value therefore guarantees an incorrect solution or, at best, an incorrect signal assignment within the correct structure.

Structure generation gave ten candidate structures in 445 s with the most favorable combination of parameters. These candidates include the reported correct structure **6.2** and other possible alternatives that were excluded on the basis of NOE evidence in the literature. For comparison, structure generation took approx. 1300 and 3500 s to give 12 candidates when two other sets of parameters were applied. Hence, the appropriate choice of the $K_A$ parameter may exert a considerable influence both upon the time and results of the generation (if the parameter value is too big, the resulting file may be empty).

*Example 3*[27]

A new natural product with molecular formula $C_{22}H_{30}O_5$, **6.3**, was studied independently by hand and using the software.

**6.3**

As with Example 2, very few AACNs were available from the COSY data due to the presence of many quaternary carbon atoms in the molecule. Thirty C-C LRDCs were, however, extracted from the HMBC data. Carbonyl (C=O) and ester (O-C=O) groups with resonances at $\delta = 200.14$ and 169.82 ppm, respectively, were entered as user-supplied AACNs. Notably, if both C=C and OH groups are present in the molecule then the signal at 170 ppm may also be observed from the functional group C=C-OH. The first structure generator run gave no result, and it was suggested that erroneous peak-picking may be the issue since some HMBC peaks were very difficult to discern. By defining MAX-ERR-LRDC = 1 so that violation of one LRDC was permitted for each generated structure, the structure generation then provided three candidate structures, with one coinciding with the manually deduced structure **6.3**. The time for solution generation in this case was not listed in the article.

When the violated crosspeak was identified and corrected, then structure generation gave the same three candidate structures in approx. 1400 s. For comparison, structure generation took almost 160 h to give the four structures, including the additional candidate, when unfavorable parameters were set by the authors.[27]

Peng *et al.*[27] noted that as the size of the FBMX grows bigger, the constraining power of the chemical shifts decreases rapidly, since there are more carbon atoms with similar structural environments in a larger molecule. In contrast, the number of LRDCs increases in parallel to the size of a molecule. LRDCs therefore become the dominant constraints for structure generation when the FBMX is large. If conventional structure generation is used, with all user-defined parameters disabled, then the calculation time would be excessive.

Peng *et al.*[27] recognized that the CISOC-SES generator is unable to cope with large molecules without using the parameters suggested in the article. The efficiency of structure generators to produce a result without applying additional parameters is highly dependent on manual intervention. Another problem is whether or not the correct structure can be produced in those cases when $K_A > 0$ is used or whether or not the correct structure satisfies all of the LRDC constraints. The authors recommended that "a conservative user" should use a small or even zero $K_A$ to allow exhaustive structure generation. As mentioned previously, even for a modest molecule with a molecular formula of $C_{22}H_{30}O_5$, the generation time was approx. 160 h with a small value set for the $K_A$ parameter, indicating that this approach is severely limited at best. On the other hand, to convince "a conservative user" to apply the parameter settings suggested it would be desirable to present statistical data regarding the performance of the system by testing on a representative number of problems. Unfortunately, no such information was provided about the total number of problems solved. The main requirement for any CASE system is that the solution to a problem is valid. The possible loss of a correct structure is balanced by the time for calculation and it would be expected that many users would prefer a valid solution over a short calculation time.

Peng *et al.*[27] concluded that if the data are free of errors then a correct structure can *always* be generated and if conventional 2D NMR spectral data are employed then a manageable number (*e.g.* less than 50) of candidate structures can be produced in an acceptable time. These statements were not validated on a representative dataset.

The latest application of CISOC-SES[28] details the elucidation of the structure of a natural product betulinic acid **6.4** (MF $= C_{30}H_{48}O_3$) from its 2D NMR spectra.

**6.4**

The system has been updated and adapted for a SGI Indigo UNIX workstation. On this system a single correct structure was generated in approx.

1 min. For comparison, the same structure was obtained from the same data in automated mode with the StrucEluc system[5,6] in less than 1 ms when running on a PC Pentium III operating at 500 MHz.

The program CISOC-SES was originally commercialized and distributed under the name NMR-SAMS by Spectrum Research.[31]

## 6.5 The LSD Expert System

The LSD program[32–36] also exploits connectivity data available from 2D NMR correlations. The $^1$H-$^{13}$C correlations are complemented by substructural information derived from the analysis of $^{13}$C spectra and with data derived from HMBC spectra.

2D NMR data consisting of chemical shifts, J-coupling data, and homo- or heteronuclear correlations are converted into encoded data sets consisting of groups of resonances, specifically $^{13}$C peaks numbered in the order of decreasing chemical shifts in the 1D NMR spectra. The encoding of data for use by the LSD program is a manual process requiring careful inspection of the plotted spectra. In building the data sets, the first task is to correctly number the signals appearing in the 1D carbon and proton spectra. For the proton spectra, severely overlapping peaks are simply neglected. All carbon signals, however, must be identified and numbered. Degenerate signals are numbered collectively with HMBC correlations between any proton resonance and carbon nuclei handled by the program under the assumption that at least one member in the group should be responsible for the observed connectivity.

The HMQC, HMBC, and COSY spectra are each translated into sets of peak coordinates. The correlations observed are each assigned a pair of numbers designating a particular $^1$H-$^{13}$C or $^1$H -$^1$H combination that shares a common connection in the molecule. The interpretation of the HMBC spectra is somewhat biased toward the inclusion of artifacts and the following general rule is therefore applied: *all suspect data is discarded to avoid the risk of introducing uncertainties into the final analysis.* Analysis of the resulting sets of numbered correlation data allows the determination of the carbon connectivities in the molecule under examination. An accurate and complete structural solution will depend on the correct evaluation of the bonds linking correlated carbon nuclei. Other spectral data, such as $^1$H NMR, IR, UV, and MS fragmentation, provide an additional degree of refinement for building the data sets. Specifically these techniques can provide functional groups or substructures that impose constraints on the number and nature of the results set. Similarly, chemical information, such as starting materials or genus for a particular unknown natural product, can also provide valuable substructural fragments to bias the results.

The nature of information needed to derive the structure includes the molecular composition and the "atom status" or atom property for *every* non-hydrogen atom in the structure. The atom status is defined by an assigned number and the hybridization and valence states of the atom. For carbon nuclei this information is commonly obtained from either 1D or 2D multiplicity-edited spectra. Similar information also has to be provided for all heteroatoms contained in the structure (*e.g.* oxygen, nitrogen *etc.*), which are numbered arbitrarily. In those situations where the hybridization state is not easily determined, for example, where carbon nuclei resonating approx. 100 ppm can be either $sp^2$ or $sp^3$ in nature, then separate data sets must be constructed to take into account the various possibilities that can exist with regard to the atom status. Properties deduced from the analysis of spectral data, *i.e.*, chemical shifts, spin–spin couplings *etc.*, are also used as inputs to specify the *status of neighboring* atoms and thereby restrict the sets of possible atom–atom pairings that will result from the bond generation process.

The LSD program can build data sets according to both atom status and the input special properties data. The program initially treats correlated nuclei assigned with either one or two bond connectivity in the bond-forming process, which is then followed by the generation of bonds between pairs of atoms that do not appear in the correlation data. Similarly, if a particular bond or substructural fragment is known to be present, it can be introduced into the database as a collection of *subbonds* between *subatoms having their own numbering system* and used in the structure generation process. Only *one* substructure is allowed.

The data processing is based on a recursive generation and test algorithm, with a simple topological criterion used by the program to search out anti-Bredt structures[35] and discard them. In *partial mode*, the program generates *partial structures* that take account of all of the correlation data, allowing the possibility of completing these structures, *i.e.* adding in bonds at free positions to yield valid solutions, which are constantly verified. The output comprises lists of atom status and of the bonds generated in all structural solutions delivered by the program that satisfy the given constraints. These listings can then be directed to the input of other programs for either the 2D representation of structures or for 3D molecular modeling. Other available program options include single-step-driven operation and report generation, intended to provide a logbook of the resolution process.

Defining the status of all atoms is a strong constraint for the user. Any easily deduced bond must be introduced to LSD through a special command. Inconsistent or incorrect data leads LSD to fail, producing a message indicating the deepest analysis level which was reached. Introducing $^nJ$ COSY or HMBC correlations with $n > 3$ also leads LSD to fail. In order to avoid this situation, the weakest COSY and HMBC correlations are first entered as comments.

An invalid correlation can be eliminated to check if it is responsible for the program failure; consequently, the detection of invalid correlations is a problem shouldered by the user. A special command provides the maximum number of correlations that it is possible to eliminate so that at least one solution is produced. The features of the LSD program are demonstrated in more detail in a publication[36] by determining the molecular structure of aza-dirachtin (structure **3.2**, Chapter 3), a complex natural product with molecular formula $C_{35}H_{44}O_{16}$.

The application of the LSD program to the identification of a series of new natural products has been described in a number of works.[37–41] The structure elucidations of two natural products from 1D and 2D NMR data have been reported:[37]

6.5                     6.6

For compound **6.5**, only one structure was generated in 0.1 s. For compound **6.6**, the output file contained 32 structures. A biogenetic hypothesis was introduced, which eliminated 30 structures. Structure **6.6** was preferred on the basis of chemical considerations.

Almanza *et al.*[38] reported the elucidation of structures for the following two diterpenoids:

6.7                     6.8

The data set used for the LSD computer analysis of compound **6.7**, $C_{22}H_{22}O_8$, did not include any COSY data, but did contain the multiplicity and hybridization state of all the carbon and oxygen atoms, the HMQC and HMBC data, and bonds deduced from the presence of a ketone carbonyl and three ester/lactone groups. Carbons C-6, C-8, and C-20, based on the chemical shifts, were attached to oxygen atoms, whereas C-12 and C-28 were connected to quaternary carbons. The analysis gave nine structures. Four of these structures contained a cyclobutadiene ring, but no furan ring, and they were thus discarded. The correct solution, structure **6.7**, was present in the output file twice, each resultant structure having different $^{13}C$ assignments. Other possible structures were rejected due to an unrealistic value for C-3 (140.8 ppm) in the fragment shown below.

The data set for compound **6.8**, $C_{20}H_{20}O_6$, consisted of the atom status and the HMQC, HMBC, and COSY data. A long-range COSY correlation between H-2 and H-10 was eliminated, since the coupling constant was small, equal to 1.1 Hz. Two bonds were incorporated from two COSY correlations H-5/H-6 and H-14/H-15. The four C-O bonds of the two ester/lactone groups were also incorporated and the details regarding the neighbors of all carbon atoms were set. Compound **6.8** was identified by 2D NMR data as a unique solution.

The application of the LSD system to gibberellic acid (**6.9**), $C_{19}H_{22}O_6$, has been described.[39]

R$_1$ = Ac, R$_2$ = Palmityl

**6.9**                                                        **6.10**

A standard set of 2D NMR spectra was used and the program yielded six structures from which Nuzillard *et al.*[39] selected the correct structure manually. The generation time on an SGI Indigo computer was 0.1 s.

The structure of voamatim C (**6.10**), $C_{41}H_{58}O_{11}$, was derived[40] using the LSD program, and the structure of delevoin C (**6.11**), $C_{38}H_{48}O_{13}$, has also been identified[41] using LSD:

**6.11**

The examples reviewed here demonstrate that the system is quite capable of analyzing and solving complex chemical structures. The structure generation algorithm solves problems generally within a short time. The system is available for testing *via* the Internet.[42]

## 6.6 The COCON Expert System

A series of articles[43–46] describes the COCON system and examples of applying the system to the structure elucidation of organic molecules utilizing both 1D and 2D NMR data as inputs. This program is based on general principles of deterministic expert systems. The structure generation algorithm builds a molecule from substructures inferred by analyzing the 2D NMR spectra and/or from the substructures provided by a user. The application of ADEQUATE[47] spectra to allow the user to distinguish between $^2J_{CH}$ and $^3J_{CH}$ correlations is also desirable. The approaches used in this system will be briefly reviewed here.

The pre-processing stage includes the determination of general parameters such as the total number of bonds that will be formed. The information derived from the 2D-NMR data is reduced to its geometrical meaning. Specific inclusion or exclusion of bonds can be performed as an option. Specific bonds can be excluded on a case-by-case basis. Fixed bonds may result from the definition of substructures as well as from NMR-derived connectivity data. The pre-evaluation of COSY and 1,1-ADEQUATE data may define the absence of specific bonds, since bonds between atoms not showing COSY or 1,1-ADE-QUATE correlations are forbidden. For COSY correlations this rule is extended from proton-bearing carbon atoms to all proton-bearing atoms. The hydrogen atoms are not treated as separate atoms. The system is designed in a way that spectral information should not be considered if there is uncertainty about its interpretation. Since the generation speed and the number of structures in the output file are dependent on atom properties such as hybridization state and heteroatom neighbors, a set of rules was adopted for the determination of these parameters.

COCON considers $^{13}$C-NMR chemical shifts by excluding certain bonds between heteroatoms and carbon atoms in the following cases (a set of "axioms"):

$\delta_C < 150$ ppm: The carbon atom must not be connected to sp$^2$-hybridized oxygen or sulfur atoms.

$\delta_C < 130$ ppm: If the carbon atom is sp$^2$-hybridized, then it must not be connected to an oxygen atom.

$\delta_C < 105$ ppm: If the carbon atom is sp$^2$-hybridized, then it must not be connected to a nitrogen atom.

$\delta_C < 45$ ppm: The carbon atom must not be connected to an oxygen atom.

$\delta_C > 35$ ppm: If the carbon atom belongs to a methyl group, then it must not be connected to another carbon atom.

$\delta_C < 180$ ppm: The carbon atom must not be an allene carbon atom.

$\delta_C > 125$ ppm: The carbon atom must not be sp$^3$-hybridized.

$\delta_C > 115$ ppm: The carbon atom must not be sp-hybridized.

$\delta_C < 75$ ppm: The carbon atom must not be sp$^2$-hybridized.

These rules were stated on the basis of reference data in which the characteristic spectral intervals of different functional groups are generalized.[48]

To avoid the potential problem of misinterpreting $^4J_{CH}$ HMBC correlations as two- or three-bond interactions, Köck *et al.*[49] recommend running COCON in an iterative way, initially omitting low intensity HMBC cross-peaks in the first calculation. If COCON does not generate any solution in the next calculation, then the new correlations are not two- or three- bond distances. As noted previously, using 2D NMR peak intensities to evaluate the length of correlations is not reliable and the search for non-standard correlations at random by repetitive iteration is not an efficient approach to solving a structure.

Structure generation using the COCON computer program is controlled by both "permanent examinations" and by "appended examinations". The "permanent examination" process implies that intermediate structures are checked against the set of HMBC correlations. Other permanent examination processes include checking for the presence of geminal diols, which may occur in organic compounds, but may nevertheless be optionally excluded. The same is possible for cyclopropenoid structures. If the consistency between a generated substructure and the required connectivities or other criteria is detected by the "permanent examination" process, then the structure generator jumps to the next substructure of the same size. If the substructure agrees with all the required data, then this substructure is enlarged by usually one bond and checked again.

When all appropriate atoms are included, then one or more "appended examinations" are performed. COCON calculates every possible multiple-bond distribution and therefore provides every mesomeric form possible for a calculated constitution. Among the optional functions of COCON is a check for cyclobutadienoid substructures that are generally deemed to be unstable and that relies on exhaustive consideration of all multiple-bond systems possible for a certain structure. It should be noted that Bredt's rule is not automatically checked by the COCON program. Eventually, one of the structures may be accepted as output. The program terminates if every possible combination of direct neighbors with at least one overlapping atom has been fully examined.

The data sets of three well-known natural products, aflatoxin $B_1$, $C_{17}H_{12}O_6$ (**6.12**), 11-hydroxyrotenone $C_{23}H_{22}O_7$ (**6.13**) and haemoventosin $C_{15}H_{12}O_7$ (**6.14**) have been examined by COCON to demonstrate the difficulties of analyzing complex systems.[49]

**6.12**

**6.13**

**6.14**

For aflatoxin $B_1$, *theoretical* COSY and HMBC data sets from the proposed structure **6.12** were generated, whereas $^{13}C$ chemical shifts were extracted from the literature,[50,51] COCON generated 1004 molecular frameworks in $\sim 18$ min if the hybridization states present in aflatoxin $B_1$ were used as precise inputs. All calculations were performed on a SGI R10000 with a 195 MHz processor and with the source code of COCON compiled to run on a 64-bit computer. The number of possible structures and concomitant generation time were dramatically reduced by more than 98% ($k = 17$ and $t_g = 0.4$ s; $k$, number of structures in the output file; $t_g$, time consumed for the structure generation) when the 1,1-ADEQUATE correlations were added to the input data. Köck *et al.*[49] showed how the prediction of $^{13}C$ NMR spectra in combination with the system rules and taking account of specific chemical considerations allowed the elucidation of the correct structure. The example would probably be more demonstrative if real, rather than, synthetic 2D NMR experimental data were used.

The COSY- and HMBC-correlation data for 11-hydroxyrotenone were obtained experimentally. COCON generated 24 994 possible structures in 2 h 10 min when using the COSY and HMBC data and the hybridization states illustrated in structure **6.13**. To reduce the number of potential structures, fixed bonds for **6.13** were introduced. All fixed bonds are marked in bold on the structure. First, the C=O bond was fixed according to the available IR data and [13]C chemical shifts. With these restrictions, the number of possible structures, $k$, was 5148 and $t_g = 1$ h 8 min. Five more fixed bonds were introduced by the user and with these input data COCON generated 492 possible structures in 29 s. For all of the generated structures, the [13]C chemical shifts were calculated using the SpecEdit software.[52] Twenty-four of the 492 solutions showed an averaged deviation for the [13]C chemical shifts of from 1.48 to 2.91 ppm. The structure of 11-hydroxyrotenone showed the smallest averaged deviation overall, 1.48 ppm. By applying theoretical 1,1-ADEQUATE correlation data, the 492 structural proposals would be reduced to four and thereby demonstrates the usefulness of this procedure.

For haemoventosin (**6.14**), using the reported COSY and HMBC correlations,[53] as well as the five fixed bonds, shown in bold, COCON generated 938 possible structures. After introducing 1,1-ADEQUATE correlations, 206 possible structures were generated. As a result of the use of [13]C spectral prediction, a set of 49 structures was selected. The structure shown as **6.14** was not the best match in the SpecEdit calculations (it was ranked in fourth position), whereas three structures violating Bredt's rule were ranked first to third.

In order to evaluate the correctness of the proposed structure of oxepinamide A, $C_{17}H_{21}N_3O_5$ (**6.15**), COCON calculations[53] were performed based on a series of experimental 2D-NMR data including: COSY, [1]H-[13]C HMBC, [1]H-[15]N gHMBC, 1,1-ADEQUATE, and INADEQUATE. COCON quantifies the value of the connectivity information and generates a series of unbiased alternative structures.

**6.15**                                                        **6.16**

The $^4J_{NH}$ correlation between N-20 and H-6 was not used as input for the COCON calculations. The COCON analysis generated 42 possible structures, many of which were eliminated as they violated Bredt's rule or contained N-O bonds. There was no experimental evidence of N–O bonds. In addition, theoretical $^{13}C$ NMR chemical shift calculations for all structures were performed with SpecEdit.[52] Those structures for which the predicted $^{13}C$ NMR data deviated significantly from the experimental values were eliminated. The six best structures generated by the COCON system were analyzed manually and five of them were eliminated for different reasons. Finally, the structure proposed for oxepinamide A was recognized as the preferred structure following the COCON analysis.

The natural product ascomycin (**6.16**, $C_{43}H_{69}NO_{12}$) was used to demonstrate the ability of COCON to cope with the elucidation of large molecules.[46] The experimental NMR data included 52 $^1H$-$^1H$ COSY correlations (90% of those correlations that could theoretically be expected for the constitution of **6.16**) and 86 $^1H$-$^{13}C$ HMBC correlations (53%).

Despite an abundance of correlations to help define the structure, COCON generated 350 possible structures using predefined hybridization states for all atoms ($t_g = 48$ min 6 s). A SpecEdit calculation was carried out on all of the resulting COCON structures, and from the 350 structures ascomycin was ranked at position 10 with an average $\delta_C$ deviation of 4.35 ppm. The top 20 proposals remaining after the SpecEdit analysis are reviewed in the article.[46] Several of the 20 structures were excluded because of their anticipated chemical instability under laboratory conditions, *e.g.* carbamic acids. Finally, four structures were chosen as the most probable and the target structure was finally ranked by $^{13}C$ NMR deviation as the fourth in the list.

Junker *et al.*[46] found that the inclusion of 1,1-ADEQUATE correlations had a dramatic influence on reducing the number of generated structures, as well as on the reduction of the calculation time. If the filtering of structures cannot be performed based on experience, then the $^{13}C$ NMR shift predictions and assignments should be challenged. SpecEdit was shown to help in most cases to reduce even several thousand generated structures to only a manageable few candidate structures. The limit of this filter is, of course, the quality of both the chemical shift assignment procedures and the similarity of the reference substructures contained within the $^{13}C$ chemical shift library.

Urban *et al.*[54] used the COCON program for structure confirmation of a new alkaloid, coproverdine (**6.17**), $C_{15}H_{11}NO_6$:

**6.17**          **6.18**          **6.19**

The initial 510 structures that were calculated were reduced to eight that fit the experimental data best. Of these eight structures, that proposed for coproverdine was the best fit for the experimental data in terms of HMBC correlation matches, as well as providing the closest match of observed and calculated $^1$H and $^{13}$C NMR shifts. We have also carried out the coproverdine structure elucidation with the StrucEluc system using the spectral data presented in the article.[54] It was found that the HMBC data contained five non-standard correlations (all of them $^4J_{CH}$), so the problem was solved using the Fuzzy Structure Generation approach.[4,55] Unfortunately, no information concerning the method used for overcoming the problem of non-standard connectivities was presented by Urban *et al.*[54]

In another work,[56] the structure of N-methyldibromoisophakellin (**6.18**), $C_{12}H_{14}Br_2N_5O$, a newly isolated natural product, was also confirmed using the COCON application.

Lysek *et al.*[57] employed COCON for the structure elucidation of diformyl-flustrabromine (**6.19**), $C_{16}H_{21}BrN_2$. On the basis of connectivity information derived from HSQC, HMBC, and COSY NMR experiments, the program calculated two structures fulfilling all of the required constraints. Due to the absence of experimental HMBC correlations of the methyl protons, an alternative structure was generated in which the bromo- and methylamino- substituents have exchanged places. A calculation of the $^{13}$C chemical shift spectrum and biosynthetic considerations favored structure **6.19**.

The cited examples show that COCON can be used for the structure confirmation and structure elucidation of natural products of fairly large size, for example ascomycin (56 heavy atoms).

## 6.7 The LUCY Expert System

As with other systems discussed earlier, the LUCY system can use 2D NMR data for computer-aided structure elucidation.[58] Input data can include an empirical formula, 1D $^{13}$C NMR, DEPT-90 and -135, HMQC, and HMBC data. The use of $^1$H-$^1$H COSY data is considered optional but useful, as the presence of a COSY spectrum is sometimes crucial for solving the problem. The user can input details regarding whether or not a carbon atom is attached to a heteroatom as shown by its chemical shift or its coupling constants. The CH$_x$ ($x = 0$–3) fragments and the heteroatoms are combined to produce all possible interpretations of the long-range $^1$H-$^{13}$C couplings in question. The algorithm is based on the hypothesis ("axiom") that the HMBC correlations represent only $^2J_{CH}$ or $^3J_{CH}$ couplings. $^4J_{CH}$ couplings are not considered. If a valid interpretation of all HMBC signals is determined, then the structure generator is used to complete the connectivity matrix. The resulting structures are then checked using user-defined selection criteria (for example, exclusion of rings of certain size). The results are then displayed as a structure formula with only connectivities shown, not bond orders.

The performance of the system was demonstrated on structures **6.20** and **6.21**. Monochaetin, **6.20**, with molecular formula $C_{18}H_{20}O_5$ was elucidated using data previously utilized to test the SESAMI system.[17]

**6.20**                                        **6.21**

When HMBC and COSY data were used to analyze monochaetin, the SESAMI system produced six structures in the output file. The same experimental data were used with the LUCY system, except that, additionally, an O-C=O group and two ketone groups were entered during pre-processing. LUCY identified the correct structure as the only result within ~1 min using a Pentium PC operating at 100 MHz.

For polycarpol, **6.21**, with the molecular formula $C_{30}H_{48}O_2$, six structures were identified by the LUCY system from the HMBC and COSY data. The program took 2 h to complete the calculation. For comparison purposes, the StrucEluc system[5,6] identified the structure of polycarpol in 0.16 s using a 500 MHz Pentium III PC.

The LUCY system was applied to the structure elucidation of two new natural products[59] – **6.22**, $C_{13}H_{16}O_3$, and **6.23**, $C_{13}H_{14}O_3$:

**6.22**                                        **6.23**

The structures were elucidated in approx. 30s each using a PC with a 100 MHz Pentium processor. Comparing LUCY with other programs in terms of speed and complexity of molecules that have been elucidated suggests that LUCY is not one of the more powerful CASE programs. For example, structure generation in the StrucEluc system[5,6] for all four tasks (structures **6.20–6.23**) took only a fraction of a second using a 500 MHz Pentium III.

## 6.8 Stochastic Algorithms for Structure Generation in 2D NMR-Based Expert Systems

As shown above, expert systems allowing computer-assisted structure elucidation based on spectral data, including multi-dimensional NMR data, have proven valuable in the elucidation of complex organic molecules. These systems are capable of elucidating the chemical structures of even complex natural products. Most publications presented examples that showed the capabilities of the expert systems in terms of solving tasks that would commonly be dealt with by qualified spectroscopists.

The systems described in this chapter are *deterministic*, because they use discrete mathematical methods, such as logic, combinatorial analysis, and graph theory, combined with heuristic approaches for exhaustive generation of all isomeric structures satisfying a set of structural constraints. Experience accumulated as a result of the application of expert systems to many examples has demonstrated that such systems derive direct value out of the huge amount of chemical structure information contained within spectral data.

A huge number of isomers corresponding to molecules containing 30–100 skeletal atoms resulted in some researchers[60–62] declaring that, in principle, deterministic systems would not be able to analyze the entire space of potential isomers and therefore fail in the elucidation of large structures. As an alternative, they suggested stochastic algorithms for structure generation, specifically simulated annealing[60,61] and genetic algorithms.[62,63] The number of skeletal atoms in a molecule that could be identified by deterministic systems was limited by these authors[61,62,64] to 30 or less. Examples of large molecules that were successfully elucidated and reported, but above the limit of 30 atoms, were considered to be exceptions.[61] One could expect that, in order to support their position, proponents of stochastic algorithms would cite successful examples of stochastic algorithms applied to molecules with more than 30 skeleton atoms. This has not happened to the best of our knowledge.

During the past 40 years of the development of deterministic systems a large number of methods have been elaborated to overcome the "problem of dimensionality". Molecules with 40–80 skeletal atoms are fairly typical examples for structure elucidation using deterministic systems. With the effective application of the structural constraints provided by 2D NMR spectra, as well as the use of sophisticated databases and spectra-structural correlations, it is possible to significantly reduce the resulting output set of a deterministic system. The sum total of all of these innovations introduces elements of artificial intelligence into expert systems, thereby supplying many abilities that are absent from a human expert. The program starts with a huge, but restricted, space for all possible isomers containing, of course, structures complying with the constraints imposed by spectral data and/or introduced as additional information. The challenge is to reveal these selected structures and select the most probable.

Stochastic methods of structure generation are of interest, since any such algorithms should be comprehensively studied in terms of their potential advantages in expert systems. It is not improbable that the use of techniques peculiar to deterministic systems in combination with stochastic algorithms could lead to the creation of *hybrid* systems that, due to synergistic effects, could successfully compete with purely deterministic systems.

To check the potential applications of stochastic algorithms two versions of the system SENECA[61,62] were developed, one of which[62] used genetic algorithm[65,66] and the other[61] which used simulated annealing.[67] The systems were tested using three molecules with the molecular formulae $C_{15}H_{28}O_2$, $C_{18}H_{20}O_5$, and $C_{30}H_{48}O_2$. From the examples discussed, it cannot be concluded that the SENECA system will be able to tackle problems which are too large for deterministic algorithms.

Despite these results, the potential use of stochastic algorithms remains of interest. The successful selection of a scoring function allows the algorithm to effectively direct the search to an isomer subspace where the structures comply with the constraints imposed by the 2D NMR data. An advantage of the approach is that very few assumptions are made prior to the structure elucidation run. Large fragments do not need to be deduced from the spectral data in the pre-processing run and no particular hybridization states are assumed. This does not indicate that additional data may not be necessary in certain situations, for example, when a large molecule has a structure with a deficit of hydrogen atoms and a significant number of heteroatoms. Deterministic systems can successfully solve tasks such as this, as shown by many examples (see Part III). There is, however, a stage of molecular structure elucidation that has been shown to demand the application of a genetic algorithm. The application of genetic algorithms to determine the relative stereochemistry of large rigid molecules elucidated using CASE systems was discussed in Chapter 4.

# References

1. H. Gunther, *NMR Spectroscopy*, Wiley, Chichester, 2001.
2. A. E. Derome, *Modern NMR techniques for Chemistry Research*, Pergamon, Oxford, 1987.
3. A. P. Marchand, *Stereochemical Applications of NMR Studies in Rigid Bicyclic Systems*, Verlag Chemie International, Deerfield Beach, FL, 1982.
4. S. G. Molodtsov, M. E. Elyashberg, K. A. Blinov, A. J. Williams, E. R. Martirosian, G. E. Martin and B. Lefebvre, *J. Chem. Inf. Comput. Sci.*, 2004, **44**, 1737.

5. K. A. Blinov, D. Carlson, M. E. Elyashberg, G. E. Martin, E. R. Martirosian, S. G. Molodtsov and A. J. Williams, *Magn. Reson. Chem.*, 2003, **41**, 359.
6. M. E. Elyashberg, K. A. Blinov, S. G. Molodtsov, A. J. Williams and G. E. Martin, *J. Chem. Inf. Comput. Sci.*, 2004, **44**, 771.
7. K. A. Blinov, M. E. Elyashberg, S. G. Molodtsov, A. J. Williams and E. R. Martirosian, *Fresenius' J. Anal. Chem.*, 2001, **369**, 709.
8. M. E. Elyashberg, K. A. Blinov, A. J. Williams, S. G. Molodtsov and E. R. Martirosian, *J. Nat. Prod.*, 2002, **65**, 693.
9. G. E. Martin, B. D. Hadden, C. E. Russell, D. J. Kaluzny, J. E. Guido, W. K. Duholke, B. A. Stiemsma, T. J. Thamann, R. C. Crouch, K. A. Blinov, M. E. Elyashberg, E. R. Martirosian, S. G. Molodtsov, A. J. Williams and P. L. J. Schiff, *J. Heterocycl. Chem.*, 2002, **39**, 1241.
10. M. E. Elyashberg, K. A. Blinov, E. R. Martirosian, S. G. Molodtsov, A. J. Williams and G. E. Martin, *J. Heterocycl. Chem.*, 2003, **40**, 1017.
11. M. Badertscher, A. Korytko, K.-P. Schulz, M. Madison, M. E. Munk, P. Portmann, M. Junghans, P. Fontana and E. Pretsch, *Chemom. Intell. Lab. Syst.*, 2000, **51**.
12. B. D. Christie and M. E. Munk, *J. Chem. Inf. Comput. Sci.*, 1988, **28**, 87.
13. A. Korytko, K.-P. Schulz, M. S. Madison and M. E. Munk, *J. Chem. Inf. Comput. Sci.*, 2003, **43**, 1434.
14. B. D. Christie and M. E. Munk, *Anal. Chim. Acta*, 1987, **200**, 347.
15. M. E. Munk, *J. Chem. Inf. Comput. Sci.*, 1998, **38**, 997.
16. K. P. Schulz, A. Korytko and M. E. Munk, *J. Chem. Inf. Comput. Sci.*, 2003, **43**, 1447.
17. B. D. Christie and M. E. Munk, *J. Am. Chem. Soc.*, 1991, **113**, 3750.
18. M. E. Munk, V. K. Velu, M. S. Madison, E. W. Robb, M. Baderstscher, B. D. Christie and M. Razinger, in *Recent Advances in Chemical Information II*, ed. H. Collier, Royal Society of Chemistry UK, Cambridge, 1993, pp. 247.
19. M. E. Munk, M. S. Madison, K.-P. Schulz and A. Korytko, in *CIC, Thirteenth Workshop, Nov. 13–15,* Bad Durkheim, Germany, 1998.
20. G. Almanza, J. Bastida, C. Codina and G. de la Fuente, *Phytochemistry*, 1997, **45**, 1079.
21. G. E. Martin and L. R. Mesia, *Phytochemistry*, 1997, **46**, 1087.
22. R. Gustafson, J. W. Blunt, M. H. G. Munro, R. W. Fuller, T. C. McKee, J. H. Cardellina II, J. B. McMahon, G. M. Cragg and M. R. Boyd, *Tetrahedron*, 1992, **48**, 10093.
23. Y. Fukushi, C. Yajima, J. Mizutani and S. Tahara, *Phytochemistry*, 1998, **49**, 593.
24. S. Okada, I. Tonegawa, H. Matsuda, M. Murakami and K. Yamaguchi, *Phytochem.*, 1998, **47**, 1111.
25. E. Tsang, *Foundations of Constraints Satisfaction*, Academic Press, London, 1993.
26. C. Peng, S. Yuan, C. Zheng and Y. Hui, *J. Chem. Inf. Comput. Sci.*, 1994, **34**, 805.

27. C. Peng, S. Yuan, C. Zheng, Y. Hui, H. Wu, K. Ma and X. Han, *J. Chem. Inf. Comput. Sci.*, 1994, **34**, 814.
28. C. Peng, G. Bodenhausen, S. Qiu, H. H. S. Fong, N. R. Farnsworth, S. Yuan and C. Zheng, *Magn. Reson. Chem.*, 1998, **36**, 267.
29. C. Peng, S. Yuan, C. Zheng, Z. Shi and H. Wu, *J. Chem. Inf. Comput. Sci.*, 1995, **35**, 539.
30. J. Gasteiger and T. Engel, ed., *Chemoinformatics*, Wiley-VCH, Weinheim, 2003.
31. http://www.specres.com.
32. J.-M. Nuzillard and G. Massiot, *Tetrahedron*, 1991, **47**, 3655.
33. J.-M. Nuzillard, W. Naanaa and S. J. Pimont, *Chem. Inf. Comput. Sci.*, 1995, **35**, 1068.
34. J.-M. Nuzillard and G. Massiot, *Anal. Chim. Acta*, 1991, **242**, 37.
35. J.-M. Nuzillard, *J. Chem. Inf. Comput. Sci.*, 1994, **34**, 723.
36. S. V. Ley, K. Doherty, G. Massiot and J.-M. Nuzillard, *Tetrahedron*, 1994, **50**, 12267.
37. J.-M. Nuzillard, J. D. Connolly, C. Delaude, B. Richard, M. Zeches-Hanrot and L. Le Men-Olivier, *Tetrahedron*, 1999, **55**, 11511.
38. G. Almanza, L. Balderama, C. Labbe, C. Lavaud, G. Massiot, J.-M. Nuzillard, J. D. Connolly, L. J. Farrugia and D. S. Rycroft, *Tetrahedron*, 1997, **53**, 14719.
39. J.-M. Nuzillard, *J. Chim. Phys.*, 1998, **95**, 169.
40. D. A. Mulholland, M. Randrianarivelojosia, C. Lavaud, J. M. Nuzillard and S. L. Schwikkard, *Phytochemistry*, 2000, **53**, 115.
41. D. A. Mulholland, S. L. Schwikkard, P. Sandor and J. M. Nuzillard, *Phytochemistry*, 2000, **53**, 465.
42. http://www.univ-reims.fr/Labos/UPRESA6013/GNOSIE/LSD.
43. T. Lindel, J. Junker and M. Koeck, *Eur. J. Org. Chem.*, 1999, 573.
44. T. Lindel, J. Junker and M. Köck, *J. Mol. Model.*, 1997, **3**, 364.
45. M. Köck, J. Junker, W. Maier and M. Will, *Eur. J. Org. Chem.*, 1999, 579.
46. J. Junker, W. Maier, T. Lindel and M. Köck, *Org. Lett.*, 1999, **1**, 737.
47. B. Reif, M. Köck, R. Kerssebaum, H. Kang, W. Fenical and C. Griesinger, *J. Magn. Reson.*, 1996, **118A**, 282.
48. E. Breitmaier and W. Voelter, *Carbon-13 NMR Spectroscopy*, VCH, Weinheim, 1987.
49. M. Köck, J. Junker, W. Maier and M. Will, *Eur. J. Org. Chem*, 1999, 579.
50. R. S. Iyer, M. W. Voehler and T. M. Harris, *J. Am. Chem. Soc.*, 1994, **116**, 8863.
51. K. G. R. Pachler, P. S. Steyn, R. Vleggaar, P. L. Wessels and D. B. Scott, *J. Chem. Soc., Perkin Trans.*, 1976, **1**, 1182.
52. W. Maier, in *Computer-Enhanced Analytical Spectroscopy*, ed. C. L. Wilkens, Plenum Press, New York, 1993, vol. 4, pp. 37.
53. G. N. Belofsky, M. Anguera, P. R. Jensen, W. Fenical and M. Kock, *Chemistry*, 2000, **6**, 1355.
54. S. Urban, J. W. Blunt and M. H. G. Munro, *J. Nat. Prod.*, 2002, **65**, 1371.

55. M. E. Elyashberg, K. A. Blinov, A. J. Williams, S. G. Molodtsov and G. E. Martin, *J. Chem. Inf. Model.*, 2007, **47**, 1053.
56. M. Assmann, R. W. van Soest and M. Kock, *J. Nat. Prod.*, 2001, **64**, 1345.
57. N. Lysek, E. Rachor and T. Lindel, *Z. Naturforsch.*, 2002, **57C**, 1056.
58. C. Steinbeck, *Angew. Chem., Int. Ed. Engl*, 1996, **35**, 1984.
59. C. Steinbeck, V. Spitzer, M. Starosta and G. von Poser, *J. Nat. Prod.*, 1997, **60**, 627.
60. J.-L. Faulon, *J. Chem. Inf. Comput. Sci.*, 1996, **36**, 731.
61. C. Steinbeck, *J. Chem. Inf. Comput. Sci.*, 2001, **41**, 1500.
62. Y. Han and C. Steinbeck, *J. Chem. Inf. Comput. Sci.*, 2004, **44**, 489.
63. J. Meiler and M. Will, *J. Am. Chem. Soc.*, 2002, **124**, 1868.
64. C. Steinbeck, *Nat. Prod. Rep.*, 2004, **21**, 512.
65. D. E. Goldberg, *Genetic Algorithms in Search, Optimization, and Machine Learning*, Addison-Wesley Reading, MA, 1989.
66. M. Mitchell, *An Introduction to Genetic Algorithms*, MIT Press, Cambridge, MA, 1996.
67. S. Kirkpatrick, C. D. J. Gerlatt and M. P. Vecchi, *Science*, 1983, 671.

# Part III
# EXPERT SYSTEM: STRUCTURE ELUCIDATOR

# The Knowledge Base of the Structure Elucidator CASE System

A series of databases containing hundreds of thousands of chemical structures supplied with assigned NMR spectra of different nuclei are available (see Chapter 3). A great number of structures, along with their MS and IR spectra, are also collected in commercial databases[1–3] (reviewed in reference[4]). The structural diversity of these databases can provide great value for researchers, especially for those who are interested in characterizing newly isolated or synthesized organic compounds. If the unknown is not new and its structure and spectra are already present in a database then identification can be successfully performed by searching the spectra against the database. Even in those cases when the unknown is absent from the library of reference compounds, the spectral search can yield a set of structures similar to that under analysis according to the common spectroscopic principle "similar structures produce similar spectra". The spectral search is also helpful for dereplication which answers the first question that arises when a sample is isolated: is the compound new?

The databases described in the literature are usually utilized for the goals outlined and are used as standalone resources. We will describe the ACD/NMR database in detail because it is an integral part of the CASE system Structure Elucidator (StrucEluc)[5] discussed throughout this book. At the same time a description of this database allows the reader to gain understanding regarding general methods for manipulation and processing of spectrum–structural information, which are common for all stores of spectral reference data.

In Chapter 6, we showed that structure elucidation by 2D NMR data is, in principle, possible with an *ab initio* approach. In practice this method fails

New Developments in NMR No. 1
Contemporary Computer-Assisted Approaches to Molecular Structure Elucidation
By Mikhail Elyashberg, Antony Williams and Kirill Blinov
© Royal Society of Chemistry 2012
Published by the Royal Society of Chemistry, www.rsc.org

without the application of axiomatic knowledge including different spectrum–structure correlations and chemical rules (see Chapter 2). Molecular structure elucidation is a complex logical combinatorial process, including treatment of experimental data combined with both *factual* and *axiomatic* knowledge. Therefore, a high-performance CASE expert system must rely on a knowledge base (KB) that contains both *factual* and *axiomatic* knowledge.

In this Chapter we will consider the main features of the knowledge base implemented into ACD/Structure Elucidator and explain the algorithms developed to create such a knowledge base and how it is applied.

# 7.1 Structure and Content of Factual Knowledge ACD/NMR Database

The factual knowledge of the StrucEluc system contains the following set of databases:

- A database containing 426 000 chemical structures with their $^{13}C$ and $^{1}H$ NMR chemical shifts assigned to the corresponding carbon and hydrogen atoms.
- A fragment library containing 2 375 000 fragments supplied with their $^{13}C$ and $^{1}H$ NMR subspectra with chemical shifts assigned to the corresponding carbon and hydrogen atoms.
- A database containing 207 000 structures supplied with $^{13}C$ and $^{1}H$ NMR chemical shifts assigned to their corresponding carbon and hydrogen atoms. The database is adjusted to support $^{13}C$ and $^{1}H$ chemical shift prediction using a HOSE code-based algorithm.
- Databases containing structures supplied with $^{15}N$ (22 000), $^{19}F$ (18 000) and $^{31}P$ NMR ($>28$ 000) chemical shifts assigned to nitrogen, fluorine, and phosphorus atoms respectively. The databases include data for $^{15}N$, $^{19}F$ and $^{31}P$ NMR chemical shift predictions calculated using HOSE code-based, incremental (PLS) and neural nets algorithms.

A database containing more than $\sim$426 000 chemical compounds associated with their $^{13}C$ and $^{1}H$ NMR spectra, with the chemical shifts assigned to their corresponding carbon and hydrogen nuclei is a repository of factual information and can be used independently for either structural or spectral ($^{13}C$ and $^{1}H$ NMR spectra) searching. The database includes both natural products and compounds produced by organic synthesis. All of the data were taken from scientific articles and they encompass practically all compounds whose NMR spectra were published in leading chemical journals (*e.g. Journal of the American Chemical Society, Journal of Organic Chemistry, Journal of Natural Products, Organic Letters etc.*) since 1990. This work continues iteratively and the number of reference structures constantly increases. To provide high reliability of the data, all of the information related to a structure–spectrum pair is carefully checked using both NMR spectrum prediction and human expert inspection. If any contradictions between a structure and its spectrum are revealed the structure is declared as questionable and is not included in the database. The number of structures stored in the database is small in comparison with just the total number of registered chemical substances, approx. 60 million based on the recent announcement by the Chemical Abstracts Service (*i.e.* less than 1% of this total number and this is only *registered* substances). However, practically all basic classes of natural products, as well as many artificially synthesized organic molecule classes, are presented in the database.

The following information is given for each compound:

1. Structural formula, molecular mass, solvent, references to the articles from which the data were obtained, ID of the structure.
2. $^{13}$C chemical shift ranges specified for each group of equivalent carbon atoms existing in a molecule. For a given compound, a definite chemical shift associated with the group of equivalent carbon atoms is shown. Spectra of the same compound registered under different experimental conditions (kind of spectrometer, solvent used, *etc.*) are usually not fully coincident, and this is the reason why spectral intervals are used. The chemical shift is also characterized by the intensity of the corresponding signal which is quantified by the number of equivalent carbon atoms.
3. $^{1}$H chemical shift ranges specified for each group of equivalent hydrogen atoms and the number of hydrogen atoms assigned to each chemical shift.
4. Indexes generated regarding chemical composition and chemical shifts values in the $^{13}$C and $^{1}$H spectra of a compound. These are used to accelerate searching of structures using spectral data (see Section 7.3).

An example of a database structure with the assigned $^{13}$C chemical shifts and accompanying information is shown in Figure 7.1. Note that the stereochemistry of the molecule is represented by the accepted up/down convention of wedge and hashed-wedge bonds.

The distribution of the number of molecules based on the number of skeletal atoms across the full structure database is shown in Figure 7.2.

It might be expected to see a smooth structure distribution, but four weak maxima are distinctively observed on the curve around the maximum. We hypothesized that the reason may be in the different origins of the structures contained in the database. To investigate this question, we extracted two sets of structures from the database. One of them was composed of approx. 17 000 natural products and another from approx. 21 500 synthesized compounds, the latter being selected on the grounds that they were published in journals such as *Synthesis* and *Synthesis Communications*. The distribution of compounds included into these two sets as a function of the number of carbon atoms is shown in Figures 7.3 and 7.4.

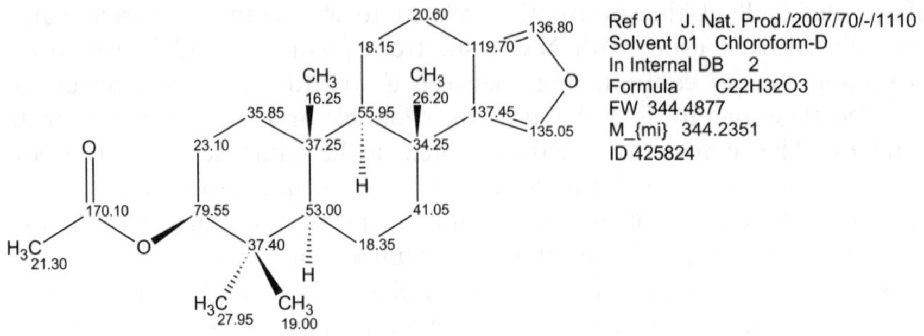

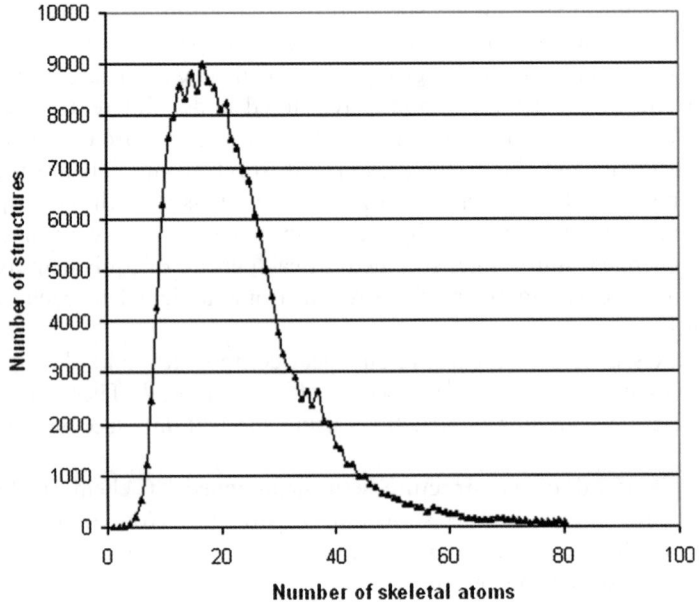

**Figure 7.1** An example of a structure presented in the database with accompanying information.

**Figure 7.2** The distribution of molecules contained in the full structure database as a function of the number of skeletal atoms. The maximum of the distribution is around 17 skeletal atoms.

The Figures demonstrate that the distributions are very different. Synthetic compounds are distributed smoothly and the bell-shaped curve has a distinct maximum at approx. 12 carbon atoms, whereas the distribution of natural products resembles a spectrum containing sharp narrow lines at the following carbon atom numbers: 10, 15, 20, and 30. As the distance between the "lines" equates to exactly five carbon atoms, it is reasonable to suggest that these clusters originate from the large number of terpene compounds for which

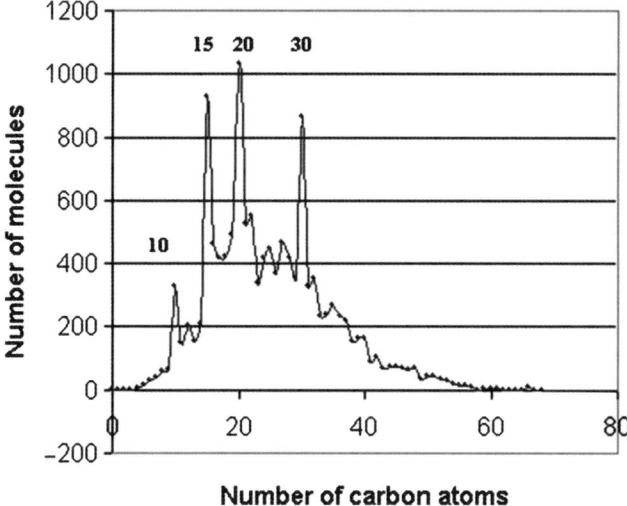

**Figure 7.3**   The distribution of natural products as a function of the number of carbon atoms.

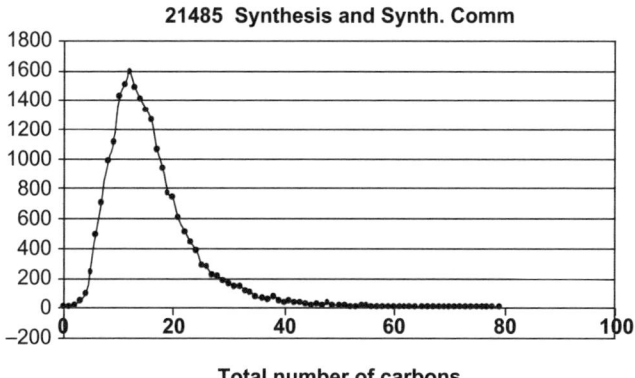

**Figure 7.4**   The distribution of synthesized compounds as a function of the number of carbon atoms.

molecular compositions differ by five isoprene carbons. The inspection of the data validated this hypothesis.

# 7.2   Structural Searching Using a $^{13}$C NMR Spectrum

The structure search algorithm to allow searching of the database was developed to take into account the properties of an experimental $^{13}$C NMR spectrum as much as possible. The signals in $^{13}$C NMR spectra are generally narrow and well-resolved peaks. An assumption can be made that each signal in the spectrum corresponds to one group of equivalent carbon atoms. On forming a search request, proper weighting is given to the following peculiarities of carbon NMR spectra:

1. The values of chemical shifts can change dependent on the conditions under which the spectra were acquired (temperature, solvent, *etc.*). A chemical shift tolerance is therefore postulated.
2. Relative integral peak intensities and peak heights do not always correspond to the number of related carbon atoms. This causes difficulties in determining the number of carbon atoms assigned to a given signal. To circumvent this problem, the possibility to indicate the number of potential nuclei was provided.
3. Spectra can contain signals produced by impurities and different artifacts. At the same time signals from some carbon atoms may not appear in the spectrum, a condition that is rather common for quaternary carbons in low-sensitivity spectra. To account for these issues, a lack of or excess number of signals compared with the number of carbon atoms in the molecular formula can be defined in the search results.
4. If the information about the number of hydrogen atoms attached to a given carbon atom is available from the experimental data (DEPT, INEPT, or HSQC), then it is also defined in the search results.

The algorithm comparing the experimental and reference spectra present in the database includes the following main steps:

1. For each signal in the experimental spectrum, a corresponding chemical shift is looked for in the reference spectra. The search begins from the experimental signal to which the maximum number of equivalent carbon atoms is assigned. The signal search in the reference spectrum is considered as successful if the following conditions are satisfied: (a) intensity of the reference signal is not less than the number of atoms in the group of equivalent atoms; (b) the number of hydrogen atoms attached to the corresponding carbon atom (multiplicity) must coincide in both spectra compared; and (c) the difference between numbers of chemical shifts in both spectra must not exceed a definite value set in the options.
2. If signal correspondence is established, then the signals are excluded from further analysis. However, it can happen that the signal intensity in the requested experimental spectrum is higher than that following from the number of equivalent carbon atoms in a reference structure. In this case the signal is not excluded from further searching, but its intensity is diminished by a corresponding value. This approach accounts for signal overlap and to obtain a correct solution in such cases.
3. The number of non-assigned chemical shifts in the reference or experimental spectra, $n_{na}$, that show an excess or lack of resonances is set in the options of the request. The structure is considered to be in agreement with the experimental spectrum if the number of non-assigned chemical shifts does not exceed the $n_{na}$ value.

A structural search can be carried out under the following additional constraints:

1. Possible elemental compositions of the unknown compound, for example, $C_{20-22}H_{42-46}O_{1-2}$.
2. A tolerance in the molecular mass value.
3. The maximum feasible sum of distances between the corresponding peaks in the experimental (retrieval request) and reference spectra. This value is averaged as a sum of the differences divided by the number of peaks in the experimental spectrum. The average value usually does not exceed 5 ppm.

# 7.3 Methods of Spectral and Structural Search Optimization

Since the number of structures in the database is large ($\sim$426 000) and continues to steadily grow, the comparison of a query spectrum with all spectra is a time-consuming procedure, even on modern fast computers. To reduce the retrieval time, different methods of search optimization can be used, the main being the primary indexing scheme. Indexing allows the program to quickly reject structures which do not correspond with the query spectrum. A unique identifier utilized for indexing is generated for each structure. In so doing, the two following types of indexing are used as follows.

*Indexing by elemental composition.* A matrix is created in which the number of rows is equal to 10 [in accordance with the number of chemical elements (H, C, O, N, S, P, F, Cl, Br, and I)], while the number of columns is equal to the total amount of compounds in the database. An index in the matrix is therefore a column corresponding to the structure number, the ID, and it is described by a vector containing the number of atoms corresponding to the quantity of each of the chemical elements in the molecular formula as illustrated in Table 7.1.

During the search the possible elemental composition indicated in the search query is compared with the molecular formulae listed in the indexes. This approach rejects structures whose elemental compositions conflict with the query.

*Indexing by chemical shifts.* The full $^{13}$C chemical shift range (0–250 ppm) is divided into intervals (usually of 3–5 ppm) and a matrix similar to that described earlier is created. The number of rows corresponds to the number of chemical shift intervals (80–100), whereas the number of columns is equal to the number of structures in the database. The matrix is filled with ones and zeros where 1 denotes the presence of chemical shifts in the given interval and 0 denotes their absence. The query spectrum is automatically coded in the same manner using the intervals adopted for the database. The comparison of values contained in the intervals in both kinds of spectra (the query spectrum and all reference spectra) is performed during the query process. This procedure

**Table 7.1**  An example illustrating structure indexing in the database.

|  | Indexes | | | |
|---|---|---|---|---|
| *Atoms* | *ID = 1* | *ID = 2* | *ID = 3* | ...... |
| H | 16 | 72 | 80 | |
| C | 15 | 49 | 52 | |
| O | 6 | 10 | 8 | |
| N | 0 | 6 | 8 | |
| S | 0 | 0 | 1 | |
| P | 0 | 0 | 0 | |
| F | 0 | 0 | 0 | |
| Cl | 0 | 0 | 0 | |
| Br | 0 | 0 | 0 | |
| I | 0 | 0 | 0 | |

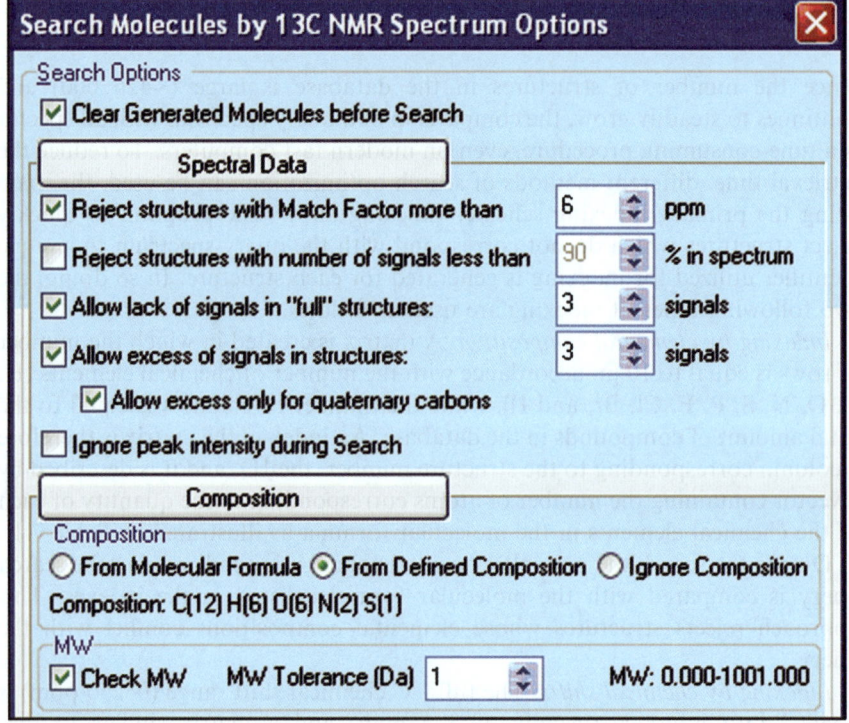

**Figure 7.5**  An example of the search request.

quickly rejects structures which either have chemical shifts absent from the query spectrum or *vice versa*.

An example of a search query is shown in Figure 7.5. It follows from the Figure that the $^{13}$C NMR spectrum of a compound with molecular formula $C_{12}H_6O_6N_2S_1$ is searched in the database with the following parameters for the query: use the defined composition CHONS; the tolerance of molecular mass is 1 Da; an allowed excess or lack of signals is equal to 3, whereas the excess is permitted only for quaternary atoms; the average deviation between two spectra does not exceed 6 ppm. As a result one structure, **7.1**, satisfying the query was found in the database. The $^{13}C$ and $^{1}H$ chemical shifts are shown as labels on the atoms.

**7.1**

# 7.4 Fragment Library

As well as the database, a fragment library (FL) incorporated into the system is a very important source of spectrum–structural information. The content and size of the fragment library depends both on the number of initial structures and on the algorithm used for library creation. The fragment library implemented into the StrucEluc system contains 2 375 000 fragments produced from 425 000 structures. Each fragment is accompanied by the following information:

1. Structural formula and the molecular mass.
2. The ranges of $^{13}C$ chemical shifts corresponding to each group of equivalent carbon atoms. Every range contains the minimal and maximal values of the chemical shifts occurring for the given group of atoms in the current fragment.
3. The ranges of $^{1}H$ chemical shifts corresponding to each group of equivalent hydrogen atoms.
4. Information on the topological distance (number of chemical bonds) from each carbon atom to the nearest label of an *abstract* atom, A, placed on the end of a free bond. For instance, in the fragment:

the distance between the $CH_3$ group and the abstract atom A is four bonds. The information is necessary to take into account the influence of possible environments on a chemical shift.

5. An index similar to that used in the database, which allows speeding up the retrieval process.

## 7.4.1 The Algorithm for Creating the Library of Fragments with $^{13}$C NMR Subspectra

In contrast to a molecule, a fragment contains atoms possessing one or more free valences. Free valences of these types of atoms can be described by two basic methods: (a) as a set of possible bonds ("beams") ongoing from an atom, and (b) as bonds of definite multiplicity directed to abstract atoms (for instance, C=A, C≡A). The first variant is more general, but in this case the state of atom hybridization is ambiguous. The second variant permits explicit definition of the atom hybridization and for this reason is more preferable. Examples of different ways to designate free valences on fragment atoms are shown in Table 7.2.

Chemical shifts are known to strongly depend on atom hybridization, so if the first variant was used chemical shift ranges of atoms having free valences would be unacceptably wide. That is the reason to employ the second approach of storing fragments in the library.

As tens (or even hundreds) of fragments can be created from a medium-sized structure, producing fragments from a large database, such as the ACD/ database, should be performed automatically. The number of ways to break bonds in a chemical structure increases exponentially with the size of a molecule and, consequently, an attempt to solve the problem simply using an exhaustive search leads to an extremely large number of fragments. Moreover, the majority of fragments produced in this manner contains many atoms having ambiguous environments and results in too broad ranges of chemical shifts related to these atoms. This problem can be overcome by employing a "spherical" scheme of fragment selection. For each atom, a "sphere of environments" is defined which contains atoms separated by $n$ chemical bonds from a "central" atom. An example illustrating this approach as applied to structure **7.2** is shown in Figure 7.6 where CH$_3$ is a "central" atom.

**7.2**

**Table 7.2**  Examples of different ways to designate free valences on terminal fragment atoms.

Utilizing a spherical approach for fragment selection provides a condition that at least a part of the fragment atoms, the "central" atoms, will have narrow chemical shift ranges.

The algorithm for fragment creation from a structure includes the following main steps:

1. Environment spheres are determined for each atom contained within the structure.
2. Fragments with a specified number of spheres $n_{sph}$ ($n_{sph}$ varies usually from 2 to 3) are selected.
3. Repeated identical fragments are removed.

In the process of fragment selection, atoms situated in the next $[(n_{sph} + 1)^{th}]$ sphere are replaced by non-defined (abstract) atoms. Chemical shifts are conserved for each atom in a produced fragment. Figure 7.7 shows fragments obtained on the basis of the "central" atom $CH_3$ when its environment is defined by the second and third spheres ($n_{sph} = 3$).

All fragments obtained as a result of structure **7.2** "fragmentization" by selecting the second and third spheres, and then consecutive utilization of all atoms as "central" atoms, are presented in Figure 7.8.

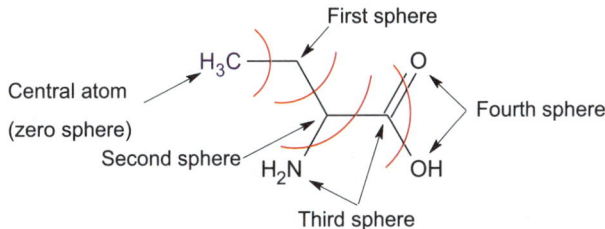

**Figure 7.6** The selection of an atom environment sphere for creating fragments.

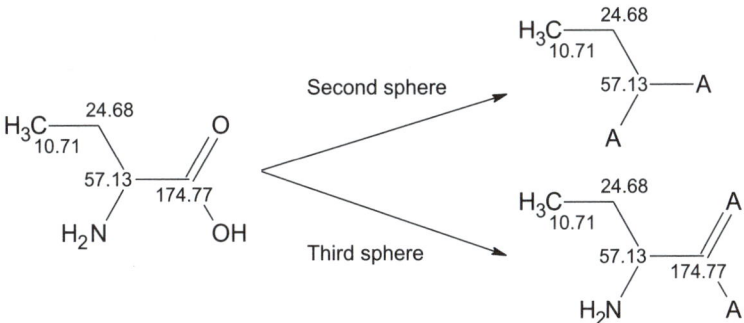

**Figure 7.7** Introducing non-defined (abstract) atoms into fragment structure.

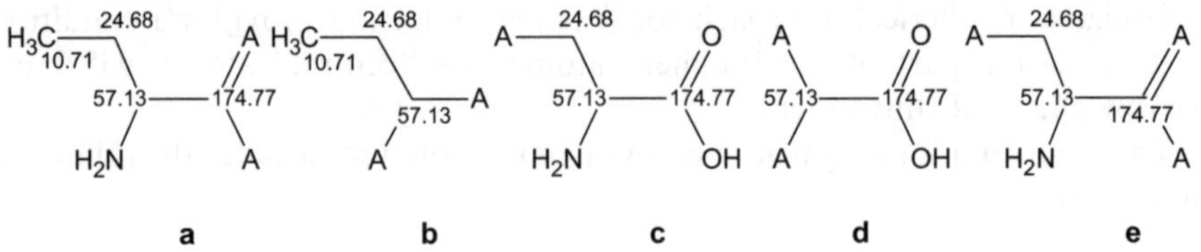

**Figure 7.8** All fragments obtained as a result of structure **7.2** "fragmentization" by selecting the second and third spheres and consecutive utilization of all atoms as "central" atoms.

Since the fragments included into the fragment library must be carriers of spectrum–structural information, they have to satisfy chemical common sense and provide relatively small changes of chemical shift ranges when a fragment is transferred from one molecule to others. Figure 7.8 shows that not all fragments meet these requirements. Thus, in fragments **a** and **e**, the cutting of the double bond C=O and ordinary bond C–OH destroys the carboxyl group. As a result, the chemical shift range of a carbon atom obtained from a COOH group will be too wide, because equivalent fragments will also be produced by cutting the C=C bond. Fragment **b** loses its chemical sense as the two abstract atoms labeled as A and connected to C(57.13) make the chemical shift range of this carbon atom extremely broad. To conserve the chemical shift ranges close to those characteristic for definitive atom environments,[6] some empirical rules are applied. The following kinds of chemical bond are not allowed to be cut:

1. Bonds existing between heteroatoms (groups: $NO_2$, $RSO_2R$, $SO_3R$, *etc.*).
2. Multiple bonds between carbon and heteroatoms (C=O, C≡C, C≡N, *etc.*).
3. Bonds between carbon and heteroatoms, if the number of heteroatoms connected with a given carbon atom is greater than one (CCl3, O–C–O, *etc.*).

These intuitively sensible rules can be generalized as follows: it is forbidden to cut bonds from carbon atoms to heteroatoms. The result of applying these rules to the fragment set presented in Figure 7.8 is displayed in Figure 7.9.

The obtained fragments look more chemically sensible and have acceptable chemical shift ranges assigned to the carbon atoms. Nevertheless, some kinds of fragments can not be selected when a spherical scheme for fragment selection is used. For instance, it is impossible to create the fragment:

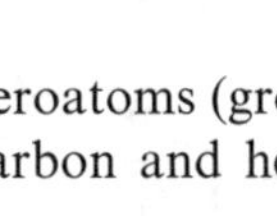

as the benzene ring will be lost. Moreover, information on important structural fragments included in natural products (as a rule those are cyclic systems) can

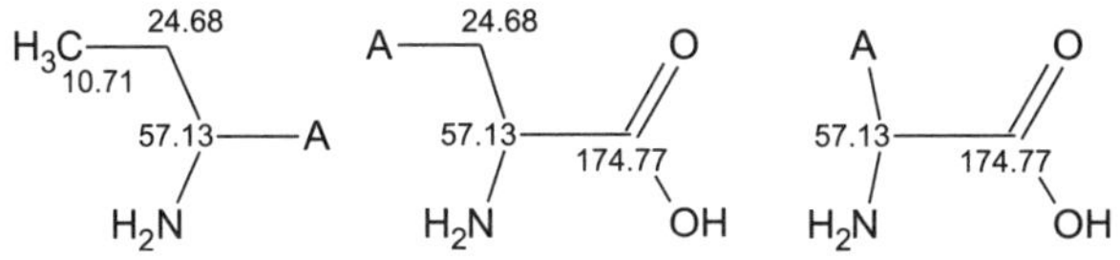

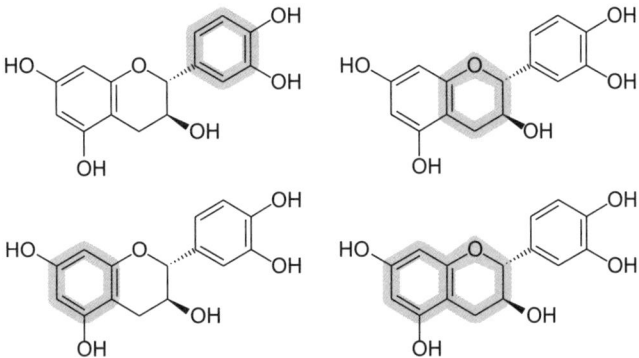

**Figure 7.9** A modified set of fragments obtained as a result of fragmenting structure **7.2**.

**Figure 7.10** Substructures of the catechin molecule, highlighted using a gray color, which plays the role of the "central" atoms.

be lost. To remove this problem, the algorithm was modified: bonds, cycles, and combinations of cycles were allowed to serve as conditional "central atoms". For example, Figure 7.10 shows the fragments (highlighted by a gray color) of the catechin molecule which, along with atoms and bonds, will be accepted by the algorithm as "central atoms". The described approach allows significant structural parts of natural products to remain intact.

Some identical fragments may be produced from different structures as a result of creating the fragment library. In this case the chemical shifts of the corresponding atoms which belong to identical fragments are used to provide for the subsequent determination of the chemical shift ranges. A range can be determined by simple indication of the minimum and maximum values, as well the result of statistical data processing. It is assumed that chemical shifts assigned to equivalent atoms belonging to identical fragments obey a normal distribution.

Fragment distributions in the fragment library as a function of the number of carbon atoms and skeletal atoms are presented in Figures 7.11 and 7.12.

The Figures show that fragment sizes vary over a wide range between several and eighty atoms. The overwhelming majority of fragments fall in the intervals 1–25 (carbon atoms) and 1–35 (skeletal atoms). The bell-shaped distributions have maxima at 10 and 20 atoms for the number of carbon and skeletal atoms respectively.

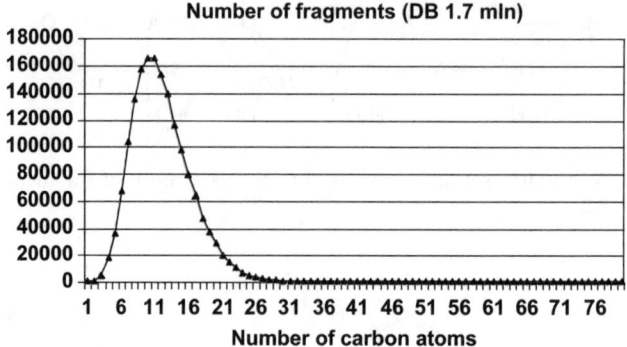

**Figure 7.11**   Distribution of 1.7 million fragments included into the library as a function of the number of carbon atoms.[7]

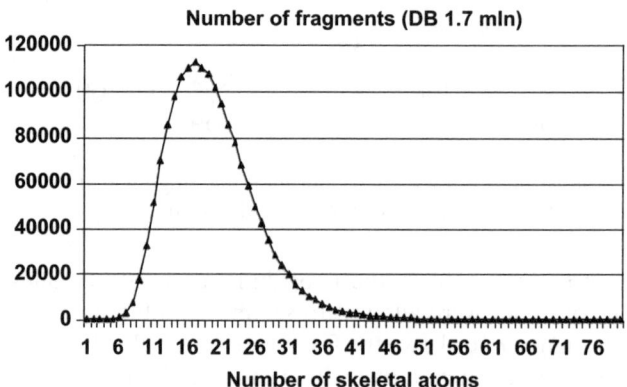

**Figure 7.12**   Distribution of 1.7 million fragments included into the library as a function of the number of skeletal atoms.[7]

## 7.4.2   Fragment Searching in the Fragment Library Using a $^{13}$C NMR Spectrum

If the structure of a *new* chemical compound must be established, the unknown will be absent from the database. However, structures stored in the database generally do contain some fragments of the unknown molecule and in most cases such fragments may be found in the database for the majority of new organic compounds. Some information therefore exists for parts of the novel structure in the database. However, this information is distributed across many molecules, each of which contains some fragments of the unknown structure. Figuratively speaking, the structural information regarding the molecule under analysis is spread across the database and the challenge is to extract it. The search for structural fragments corresponding to $^{13}$C NMR subspectra is a problem which is somewhat more complicated than the search for full structures. The primary reason is that chemical shifts are affected by the influence of neighboring atoms and spectral features can be significantly different, even in similar compounds.

Nevertheless, the algorithm for searching for fragments using a $^{13}$C NMR spectrum of an unknown compound as an input is very similar to the approach used to search for full structures. The main difference is that the distance from a given atom to an abstract atom is taken into account in the search process. The possible chemical shift range for an atom is calculated from the chemical shift range stored in the fragment library, with the mentioned distances to the abstract atoms being accounted for. The maximum possible range is set in the search options and its value is usually equal to 20 ppm. At the same time, the range cannot be narrower than a chemical shift interval defined in the search options for atoms of a given type. The following values are commonly used: 12 ppm for atoms, having free valence bonds (one bond distance to the abstract atom); 6 ppm for atoms separated from abstract atoms by two bonds; and 3 ppm for other atoms, as illustrated in Figure 7.13.

The chemical shift interval used to index fragments in the library differs from that employed for the structural database (10 ppm) and is 3 ppm. This leads to a decrease in the number of fragments found that belong to structures in the database. This drawback is compensated by significant improvements in the search speed. However, this does not influence the subsequent stages of structure elucidation as the set of found fragments is deliberately redundant and some of the fragments overlap.

During the process of searching for fragments $^{13}$C NMR spectra, the algorithm verifies whether the $^1$H NMR spectrum of the unknown is in agreement with the $^1$H chemical shifts assigned to the hydrogen atoms of the fragment. In this case, the chemical shift intervals postulated for hydrogens attached to the carbon having free valences is equal to 2 ppm, whereas for other atoms this value is equal to 1 ppm.

Options for the search are shown in Figure 7.14.

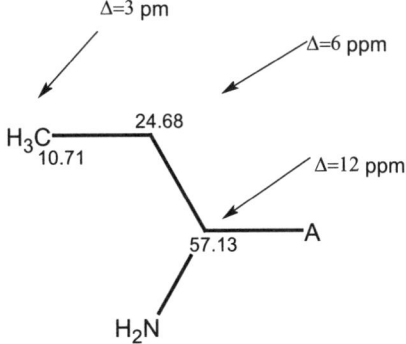

**Figure 7.13** Chemical shift ranges commonly used during the $^{13}$C NMR spectral search.

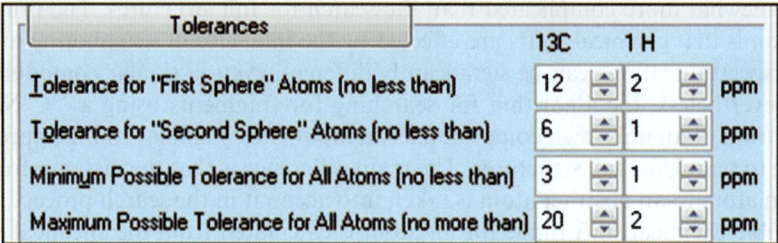

**Figure 7.14**   Options available to perform searches for fragments. The tolerances for the $^{13}$C and $^{1}$H chemical shifts are shown.

# 7.5   Composition of Axiomatic Knowledge

## 7.5.1   Fragment Libraries

The axiomatic knowledge of the system is formed from a series of fragment libraries, correlation tables, algorithms for the generation of spectral constraints based on different levels of spectrum prediction, as well definite chemical rules. This knowledge can by systematized in the following way:

- A fragment library containing substructures (mainly functional groups) accompanied with the appropriate characteristic spectral feature ranges in $^{13}$C, $^{1}$H and IR (optionally) spectra. The library is used mainly for filtering the output structural files of StrucEluc (we will call it a filter library).
- A correlation table containing substructures and their characteristic spectral ranges in $^{13}$C and $^{1}$H NMR spectra. The atom property correlation table (APCT) is used for assignment of carbon atom properties – atom hybridization state and the possibility of neighboring with a heteroatom.
- A library containing the most typical functional groups in organic chemistry. The library is used for preliminary structural group analysis and building a "generalized portrait" of the unknown compound.
- A library of fragments that are considered as unlikely in organic chemistry under most conditions.
- Spectral constraints imposed by $^{1}$H, $^{13}$C, $^{15}$N, $^{19}$F, and $^{31}$P NMR chemical shift prediction using HOSE code-, neural net-, and incremental-based approaches.
- Common sense chemical rules used for imposing constraints on the structures generated by the expert system (Bredt's rule and some geometrical constraints).

The filter library comprises a set of molecular fragment libraries ordered in a hierarchical manner. Every fragment is accompanied by intervals of characteristic feature variations in $^{13}$C NMR, $^{1}$H-NMR, and IR spectra. In the NMR spectrum the chemical shift intervals, multiplicities, coupling constants, and integrals are used as characteristic attributes of a fragment. The intervals of characteristic frequency, half-height bandwidth and intensity variations are ascribed to fragments possessing peculiar spectral features in the IR spectrum. The characteristic spectral features of the fragments were taken from various spectroscopic sources (for example, reference[6]) and then checked carefully using the spectral database of StrucEluc.

The libraries (there is no limitation on their number) can be classified into the two following categories: libraries containing well-known chemical functionalities and those which consist of substructures chosen depending on common approaches to the structural interpretation of spectra of different types.

## 7.5.2 Universal Libraries

### 7.5.2.1 Library of Principal Functionalities (LPF)

The library of principal functionalities (LPF) contains the most important functional groups such as $>C=C<$, $>C=O$, N–H, aromatic rings, *etc.*, with the intervals of the characteristic features in all three different kinds of spectrum. For example, the carbonyl group, contained in any environment, can have a $^{13}C$ NMR signal in the range 150–220 ppm and a strong IR absorption band within the interval of 1630–1880 $cm^{-1}$. This library is used at the first stage of filtering to reject structures containing the main functional groups in any environments if the structure is not confirmed by spectra.

### 7.5.2.2 Library of Functional Groups (LFG)

This library also contains common functional groups, but in this case a more precise description of the group and its surrounding environment is given (for instance, C–**CO**–C, C–**CO**–O–C, **CO–OH**, C–**CO**–C=C, C–**CO**–Ar, *etc.*), and combined IR and NMR spectral features are specified.

### 7.5.2.3 Library of Aromatic Fragments (LAF)

This library contains all 12 kinds of aromatic ring substitution with the corresponding combined spectral features in IR, 1H-NMR, and 13C-NMR.

## 7.5.3 Specialized Libraries

### 7.5.3.1 Library Specialized for $^{13}C$-NMR (CNMRL)

This library includes fragments consisting, as a rule, of a central carbon atom in different environments. The CNMRL has proven to be very effective in the filtering of generated structures.

### 7.5.3.2 Three Libraries Specialized for $^1H$-NMR (HNMRL)

These libraries (CH3-Lib, CH2-Lib, and CH-Lib) are adjusted for the 1H-NMR detection of CH3, CH2, and CH groups with regard to different environments and states of carbon atom hybridization. If the fragment environment includes some functional groups (for instance, C=C–**CH2**–C=O,

**CH3**–C=O, Ar–**CH**–C), then the $^{13}$C-NMR and IR features of these groups are considered as essential spectral properties of the corresponding fragments.

At present the total number of fragments included into the axiomatic knowledge base of StrucEluc is approx. 450.

## 7.5.4   Fragment "Visit Card"

To provide for the creation, refinement, and further development of the filter library, a system of routines has been worked out. The full information about any fragment is concentrated in a "visit card", as in the X-PERT system described in Chapter 5. The visit card displays the graphical presentation of a fragment structure, the structural descriptors of its environment, the degree of unsaturation and the tables of characteristic intervals in $^{13}$C and $^1$H NMR, and IR spectra. To introduce the structural fragment, it is drawn in the graphical edit field of the screen. If a fragment exists in several stereoisomeric forms (*cis–trans*, endo–exo, *etc.*), then all stereoisomeric substructures, together with their spectral features, are entered into the knowledge base as independent (and marked in a special way) fragments. An example of such a "visit card" for the **CH$_3$-CH$_2$-C=O** fragment is represented in Figure 7.15.

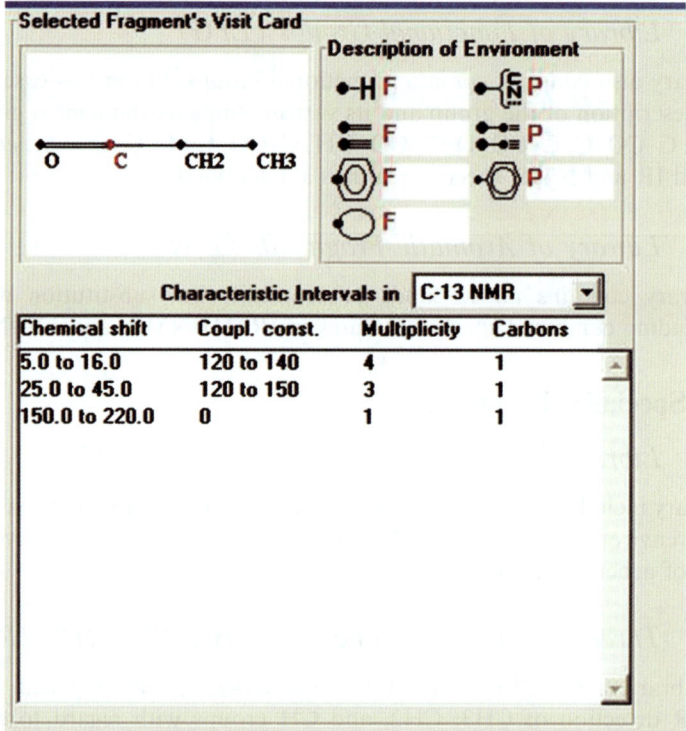

**Figure 7.15**   The "visit card" of the fragment **CH$_3$-CH$_2$-C=O**.

Note that characteristic ranges of $^{13}$C chemical shift inherent for an ethyl group connected to a carbonyl are defined here, whereas a carbonyl group is allowed to have any neighbor and show a signal in wide interval of chemical shift variation.

# 7.6 Structural Filter

The structural filter enables the chemist to select the structures which are both in accordance with constraints imposed by the user and some restrictions following from the main regulations of structural chemistry and stereochemistry. Three groups of constraints are provided for structural verification, as follows.

1. *Substructural constraints*. This group of constraints consists of three fragment libraries, two of them (GOODLIST and BADLIST) being empty before the problem starts running, and the third library is a list of unlikely fragments. The GOODLIST (BADLIST) libraries may contain obligatory (forbidden) fragments introduced by the chemist according to *a priori* information, experimental data or theoretical considerations during problem solving. The input of fragment structural formulas is performed using the integrated structure editor, ACD/ChemSketch.

The constraints imposed on the molecular skeleton can be introduced in the generalized form of "abstract fragments", for example, X=X, X=X=X, X=X-X=X. One should bear in mind that the GOODLIST/BADLIST fragments belong either to the knowledge base or to the user introduced fragments and are available to be overlapped.

2. The *unlikely fragment library* is a list of a limited number of fragments which are unlikely to exist in terms of general organic chemistry and stereochemistry (for example, triple bonds and the allene substructure in small cycles); different highly strained unsaturated polycyclic structures such as those shown below:

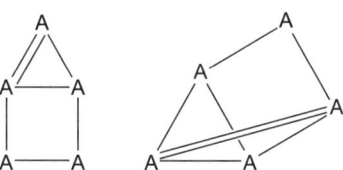

and some fragments containing heteroatoms O-O-O, OH-C-OH, *etc.*). If some unlikely fragments seem to be probable in the context of a given problem, the chemist may delete them from the library before the check is started. It should be noted that the skeletons of fragments included into a structural filter can be entered into the computer as "abstract fragments".

3. *Control upon the Bredt's rule*. If the molecular formula, degree of unsaturation and the generator fragment set allow for the generation of bridged structures with double bonds, then the appearance of structures defying Bredt's rule is quite possible. Bredt's rule states: bi-cycles containing a double bond at the nodal atom of the bridge can exist if the larger of the two rings containing the double bond includes not less than eight atoms in the cycle. According to this rule structure **7.3** is unrealistic, whereas structure **7.4** in principle may exist. Structural filtering by Bredt's rule eliminates all contradictory structures.

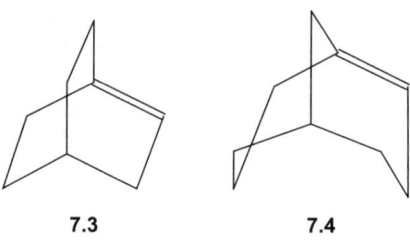

**7.3**                    **7.4**

Our experience has shown that the structural filter is very useful not only as a verification aid, but also as a tool for "step-by-step" selection of the most probable structures, when the answer file is large.

# 7.7   Atom Property Correlation Table (APCT)

An atom property correlation table (APCT) was generated from the system knowledge base. The table contains carbon atom-centered fragments with the corresponding intervals of the $^{13}$C NMR chemical shift variation for the central carbon atom and the ranges of $^1$H chemical shifts corresponding to hydrogens attached to the central atom. The program uses this table for the automatic assignment of the hybridization ($sp^3$, $sp^2$, $sp$, *not defined*) to all carbons and for assessing the possibility of their neighboring heteroatoms (*forbidden, obligatory, not defined*). The mark "*not defined*" is assigned to a parameter if several conceivable possibilities are equally probable.

To optimize the chemical shift intervals assigned to the fragments present in the axiomatic knowledge (filter library and APCT), 215 000 structures contained within the database of assigned $^1$H and $^{13}$C NMR spectra were used. The database structures were filtered with the help of the fragment libraries. If any contradictions were detected between a structure and the characteristic spectral intervals associated with a fragment, then the program provided a corresponding message. On the basis of these messages, the intervals were either modified to resolve the contradiction or were left unchanged. Changes were not made in those cases where the contradiction was caused by the presence of "anomalous" chemical shifts in a structure, a so-called "exotic" structure. An example of such a structure[8] is illustrated below:

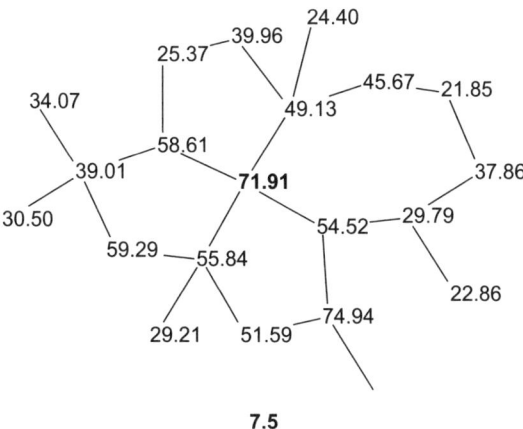

**7.5**

In structure **7.5** the chemical shift of the quaternary carbon atom that is common to all four cycles is 71.91 ppm, a value which is typical for a carbon neighboring an oxygen atom.

The assumption was made that such anomalous shifts were quite rare. The risk of overlooking the correct structure in rare cases was justified by the possibility of solving a great number of problems using chemical shifts corresponding to the common values similar to those known from the literature. Some fragments characterized by specific chemical shift values were placed into a *library of exceptions* that was applied as part of the structure filtration process using a specially derived algorithm. For example, in structure **7.6**:

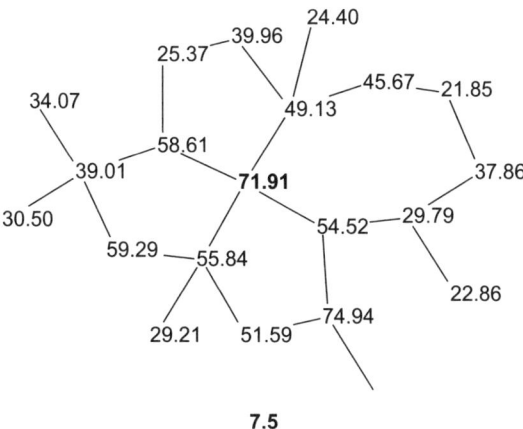

**7.6**

the anomalously small carbon chemical shift value of the $CH_3$ group (7.20 ppm) is typical of a methyl group attached to a benzene ring with two neighboring substituents of the type -O-R. The reason for this is that oxygen atoms influence carbon shifts at the β position moving the resonances upfield. These effects are cumulative in a manner similar to β branching. The fragment **7.7** shown below was therefore introduced into the library of exceptions:

**7.7**

Analysis of the fragment tables revealed that spectral filtering libraries and correlation tables (APCTs) allowed the solution of tasks with a negligible risk of overlooking the correct structure. Of the structures, 98% present in the system database withstood a verification challenge by both $^{1}$H and $^{13}$C NMR

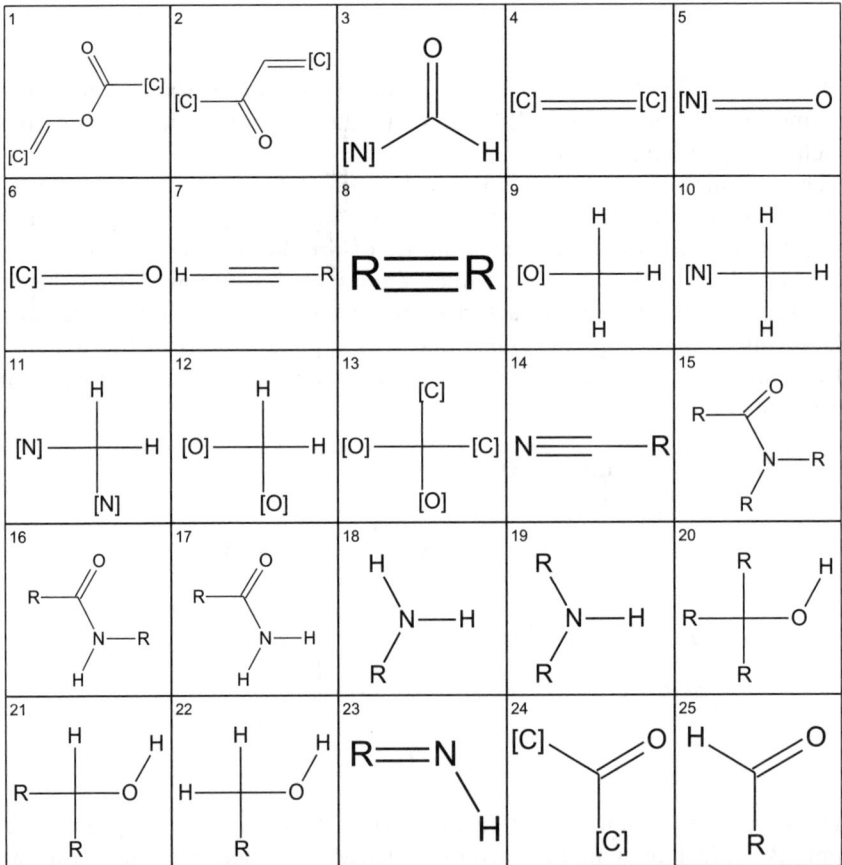

**Figure 7.16** Examples of typical functional groups included in the library.

spectra using spectral filters and ACPTs. Taking into account the high degree of diversity of the structure library, it can be expected that the spectral filtering of the output file is a procedure that offers only a small risk of losing the actual structure.

## 7.8 Library of Typical Functional Groups

To help chemists to initially relate the unknown to some chemical classes, the program displays a "generalized portrait" of the analyzed molecule, *i.e.* the distribution of typical functional groups with the frequency of their occurrence in fragments found as a result of database search by $^{13}$C NMR spectrum. A special library of typical functional groups is used to form this generalized portrait (see examples in Figure 7.16). The high frequency of occurrence of a group indicates that it is present in a molecule with higher probability. The groups that are completely absent can be used to form a BADLIST.

Experience has shown that filtering the fragments found in a database allowed us to use this procedure for getting hints as to the presence in a molecule of functional groups that cannot be detected directly from the $^{13}$C and $^1$H NMR spectra. For instance, if it becomes evident that a significant number of fragments contain an -NO2 group, then the possibility of the presence of this group in the molecule under analysis should be taken into account.

## References

1. *Specinfo*, Chemical Concepts GmbH, Weinheim, Germany.
2. *NIST/EPA/NIH. NIST Mass Spectral Library*, U.S. Government Printing Office, Washington, DC, 2005.
3. *Wiley Registry of Mass Spectral Data*, Wiley, New York, 2010.
4. L. Borland, M. Brickhouse, T. Thomas and A. W. Fountain III, *Anal. Bioanal. Chem.*, 2010, **397**, 1019.
5. M. E. Elyashberg, K. A. Blinov, S. G Molodtsov, Y. D. Smurnyy, A. J. Williams and T. S. Churanova, *J. Cheminform.*, 2009, http://www.jcheminf.com/content/1/1/3.
6. E. Pretsch, T. Clerc, J. Seibl and W. Simon, *Tables of Spectral Data for Structure Determination of Organic Compounds*, Springer-Verlag, Berlin, 1989.
7. M. E. Elyashberg, K. A. Blinov, A. J. Williams, S. G. Molodtsov and G. E. Martin, *J. Chem. Inf. Model.*, 2006, **46**, 1643.
8. R. T. Weavers, *J. Org. Chem.*, 2001, **66**, 6453.

CHAPTER 8

# Primary Data Processing: Preparation, Input and Checking

## 8.1 Data used for Structure Elucidation

There are various types of data that can be used in order to perform computer-assisted structure elucidation. In particular, as has been described in the previous Chapters, 2D NMR is the essential technique for the elucidation of complex chemical structures and, due to its inherent complexity in terms of processing and manipulation, it is this form of NMR that provides significant demands to the software. The 2D NMR structure generator discussed previously (see Chapter 6) requires as input a set of atoms and connectivities between them. The generator also takes into account the associated chemical shifts of the atoms as well as a series of different structural restrictions. The base set of atoms is usually obtained from the molecular formula while the connectivities between the atoms are revealed by combining the data encoded into the 1D and 2D NMR spectra.

The following spectra are commonly used in structure elucidation:

- Mass spectra mainly provide the molecular mass and, as a result, access to the molecular formula for the compound under examination. Under certain circumstances, a mass spectrum can definitively define the elemental composition, whereas in other cases they can provide a series of potential molecular formulae. The fragmentation of the compound represented by a mass spectrum provides access to further detail regarding the molecular composition in terms of key molecular fragments present in the compound under study.
- 1D NMR spectra contain information about atoms included in the structure in terms of their electronic environments, their proximity to each

New Developments in NMR No. 1
Contemporary Computer-Assisted Approaches to Molecular Structure Elucidation
By Mikhail Elyashberg, Antony Williams and Kirill Blinov
© Royal Society of Chemistry 2012
Published by the Royal Society of Chemistry, www.rsc.org

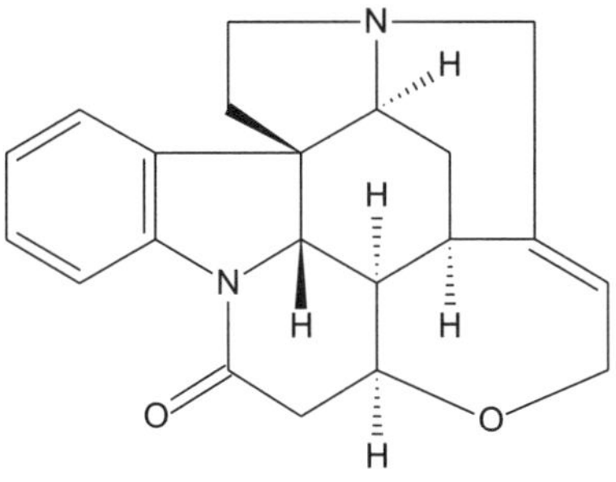

**Figure 8.1** The chemical structure of strychnine.

other both in terms of skeletal connections and through-space interactions, as well as details regarding internal barriers of rotation.

- 2D NMR spectra can provide information about both through-bond and through-space interactions between atoms and are the most informative, especially when multiple types of 2D NMR spectra are acquired and analyzed in parallel.

Data preparation is a key part of both manual and automated structure elucidation. Almost any error made during this procedure can lead to erroneous structures being derived as a result of the elucidation process. Data preparation should therefore be done as rigorously as possible. Data preparation consists of two main steps – the determination of an atom list and the determination of the connectivities between atoms. Existing algorithms for structure generation can automatically correct some errors in connectivities between atoms, but it is almost impossible to rectify mistakes in the list of atoms.

Most procedures described in this Chapter will be illustrated using the example of strychnine, a model compound that has been studied many times in multiple laboratories and is both challenging to elucidate without 2D NMR spectroscopy but, at this time, is well understood. Strychnine has the molecular formula $C_{21}H_{22}N_2O_2$ and the structure shown in Figure 8.1.

## 8.2 Molecular Formula

Almost all modern structure generation algorithms (see Chapters 5 and 6) require a list of atoms as an initial data input. The molecular formula (MF) is a compact representation of this list and is an absolute requirement in order to perform structure elucidation. Several algorithms for structure generation without knowledge of the molecular formula are described in the literature,[1,2] but they have a relatively strict area of applicability and cannot be used for the determination of an unknown compound.

Usually a monoisotopic (exact) mass is used to determine the molecular formula and is determined by analyzing the mass spectrum for the mass ($m/z$)

of the molecular ion. It should be noted that a molecular ion peak is not always present in the mass spectrum. This is more common in electron impact ionization spectra, but spectra obtained using other (more "mild") ionization methods, for example, electrospray, usually do contain the molecular ion peak. Mass spectra obtained using the positive ion electrospray ionization (ESI) method contain the adduct peak of the protonated molecular ion. In addition to the protonated ion, usually called $[M+H]^+$, other adducts include other ions such as $Na^+$, $K^+$ or $NH_4^+$, denoted as $[M+Na]^+$, $[M+K]^+$, and $[M+NH_4]^+$. This information should be taken into account when generating molecular formulae; *i.e.* the monoisotopic mass should be corrected as appropriate.

The number of molecular formulae corresponding to a given monoisotopic mass can vary depending on the accuracy of the mass determination and the possible element composition. For example, 302 molecular formulae correspond to the monoisotopic mass of strychnine (334.1681 Da) if the accuracy of the mass determination is to within 0.5 Da and only the elements C, H, N, and O are allowed. If the number of carbon atoms is restricted to 21, corresponding to the number of peaks in the 1D NMR carbon spectrum, then the number of molecular formulae is reduced to eight. Similar constraints in the number of potential formulae occur when the accuracy of mass determination increases. When the accuracy is 0.05 Da, then the number of molecular formulae is 91. However, when measured to an accuracy of 0.005 Da, the number of formulae decreases to ten and only one molecular formula can be found when the accuracy is 0.0005 Da. In practice for both methods, the restriction of elemental composition using other data and increasing mass accuracy, are used simultaneously to identify a single molecular formula in most cases. In some relatively rare cases, when unambiguous determination of a molecular formula is impossible, the structure elucidation process can be run several times using different molecular formulae.

# 8.3   Molecular Connectivity Diagram (MCD)

The list of atoms and the connectivities between them is the initial input data used for structure generation. A molecular connectivity diagram (MCD) is a convenient graphical representation of this information and is generally used to review and edit the initial data if necessary. In this Chapter the MCD format will be used to illustrate relationships between spectral data and structural information. At the starting point when only the molecular formula is known an MCD obviously contains only the atoms.

Figure 8.2 displays the MCD corresponding only to the molecular formula of strychnine – $C_{21}H_{22}N_2O_2$. The positions of the atoms on the MCD correspond to their positions in the structure of strychnine in order to simplify understanding. In real-world elucidations, it is almost impossible to assign atom coordinates to the MCD before structure elucidation has been performed.

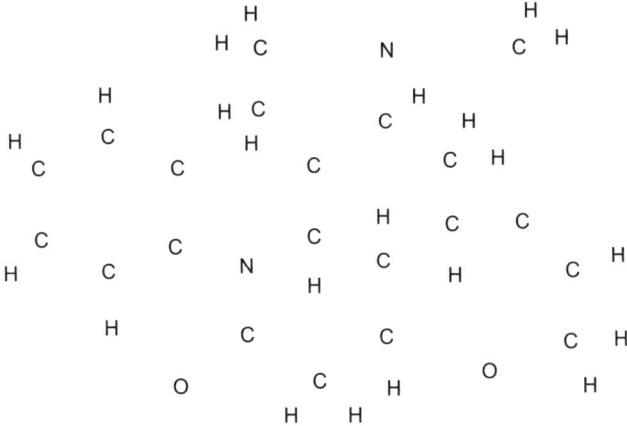

**Figure 8.2** A MCD from the molecular formula only. The MCD contains almost no structural information and the number of structures corresponding to this MCD is extremely large; atom coordinates correspond to the structure of strychnine shown in Figure 8.1.

## 8.4 NMR Spectra

While the molecular formula provides the complete set of atoms making up a structure all of the following stages for elucidation of a structure are dedicated to establishing the connectivities between these atoms. NMR spectra are used to establish these connectivities and the following NMR spectra are commonly used in computer-assisted structure elucidation:

- 1D NMR: standard $^{13}$C and $^{1}$H, DEPT135, DEPT90, and APT $^{13}$C.
- 2D NMR: HSQC, HMBC, and COSY are used to determine through bond connectivities, whereas ROESY or NOESY spectra are used to assist in the determination of stereochemistry. 1,1-ADEQUATE, 1,n-ADEQUATE, and INADEQUATE spectra can be used in some specific cases when the spectrum–structure information is especially scarce. In most cases, the following set of data is used: $^{13}$C, $^{1}$H, HSQC, HMBC, and COSY.

After the acquisition of a set of experimental spectra, they are processed to convert them from the time domain into the conventional "frequency domain" form. This processing can be performed with any processing software, as there are no specific requirements in preparing the spectra for the StrucEluc system. Conventional Fourier transform, zero filling, linear prediction, and the application of relevant weighting functions are used. In most cases, automated phase correction can be performed. For HSQC and HMBC spectra the following acquisition and processing scheme is recommended: (a) acquire spectra with 1024 (*t2*) and 256 (*t1*) data points, and (b) for *t1* points apply linear prediction to 512 points and zero fill to 1024 points; if the amount of substance is small and the signal-to-noise ratio is low, then the number of *t1* increments can be reduced as appropriate.

## 8.4.1    1D NMR Spectra

$^1$H NMR spectroscopy uses a combination of chemical shifts, coupling constants, and integral signals to determine the environments and number of hydrogens in particular groupings. In practice, signal multiplicity is used only in simple cases, because the same information can generally be extracted from multiplicity-edited 2D NMR spectra in a more easy fashion. Figure 8.3 shows the aromatic region of a strychnine $^1$H spectrum.

A $^{13}$C spectrum contains information about carbon nuclei. In an ideal case, each peak in the spectrum corresponds to a single carbon atom or, in the case of symmetry in a structure, to a group of symmetrical (equivalent) carbon atoms. Of course, in reality, degenerate signals with the same resonance frequency can occur. Certain types of carbon spectra, specifically DEPT or APT are acquired in a manner that encodes information regarding the number of hydrogen atoms connected to each carbon atom. This information can significantly simplify structure elucidation. The carbon spectrum of strychnine is displayed in Figure 8.4.

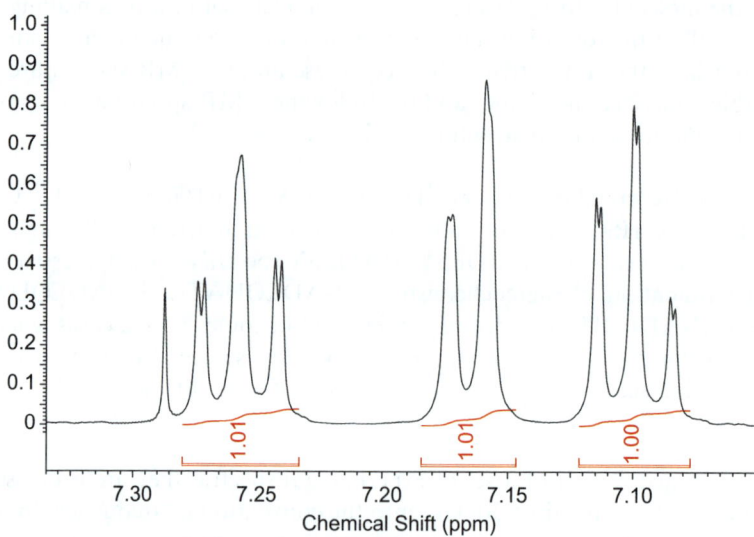

**Figure 8.3**    An example of a proton spectrum. The region of the spectrum displayed corresponds to the aromatic protons of strychnine. The integral value is used to determine the number of hydrogens corresponding to each signal. The multiplicity of the signals can be used to obtain information about neighboring coupling nuclei. In this particular case, the spectrum shows signals for three different hydrogen atoms. One hydrogen corresponding to signal 7.15–7.18 ppm has one vicinal proton neighbor, whereas two other protons have two vicinal proton neighbors each.

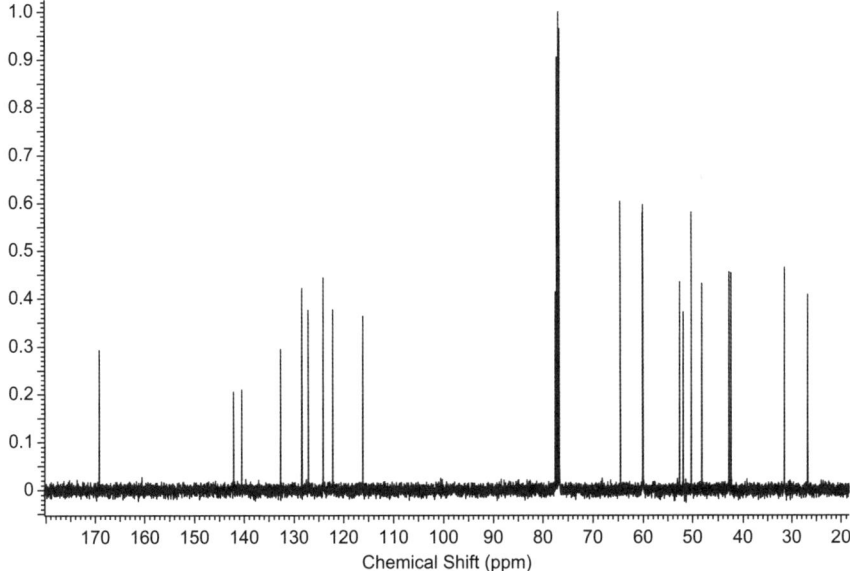

**Figure 8.4** The carbon spectrum of strychnine contains 21 signals (plus three solvent peaks). Each signal corresponds to a unique carbon atom.

In practice, two main cases can be distinguished: when the amount of sample is enough for the acquisition of a 1D carbon spectrum and, alternatively, when it is almost impossible to acquire a carbon spectrum. Data preparation and analysis are slightly different in each of these cases.

The simplest case is when a carbon spectrum is available. The digital resolution of a 1D carbon spectrum is usually enough to unambiguously resolve all carbon resonances and determine the chemical shifts of all carbon atoms in the structure under examination. Figure 8.5 displays an MCD after adding the information obtained from a carbon spectrum. This MCD differs from that shown in Figure 8.2 only by the presence of the chemical shifts associated with all carbon atoms. These additional data alone do not contain a lot of structural information, but they will be used in later stages to extract connectivity information from the 2D NMR spectra.

In this case the chemical shifts for all carbon atoms are known and this allows us to correctly assign a significant number of the carbon resonances and from there make use of the 2D NMR correlations.

Carbon spectra can provide information about the number of attached hydrogen atoms and this information can significantly simplify structure generation, and, in addition, is useful when interpreting an HSQC spectrum. Figure 8.6 contains different DEPT (distortion-less enhancement by polarization transfer) spectra. DEPT135 contains peaks associated with $CH_3$, CH (positive signals), and $CH_2$ (negative signals) groups. On the other hand, DEPT90 spectra contain peaks for CH groups only. DEPT135 spectra, relative

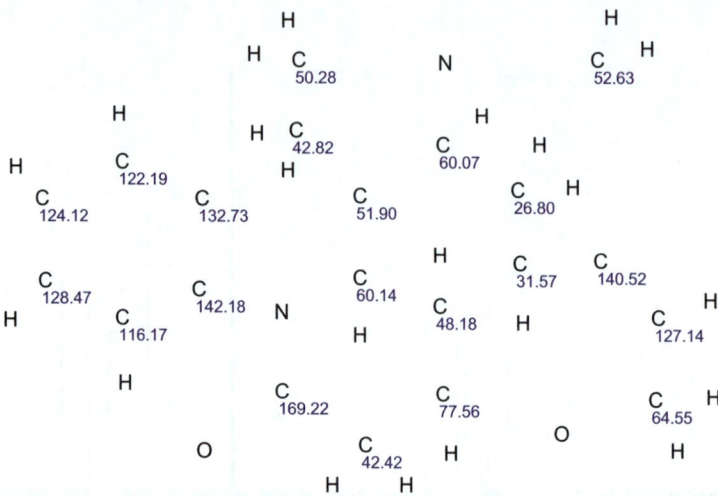

**Figure 8.5**   Chemical shift information obtained from a carbon spectrum. The carbon chemical shifts are displayed near the carbon atoms.

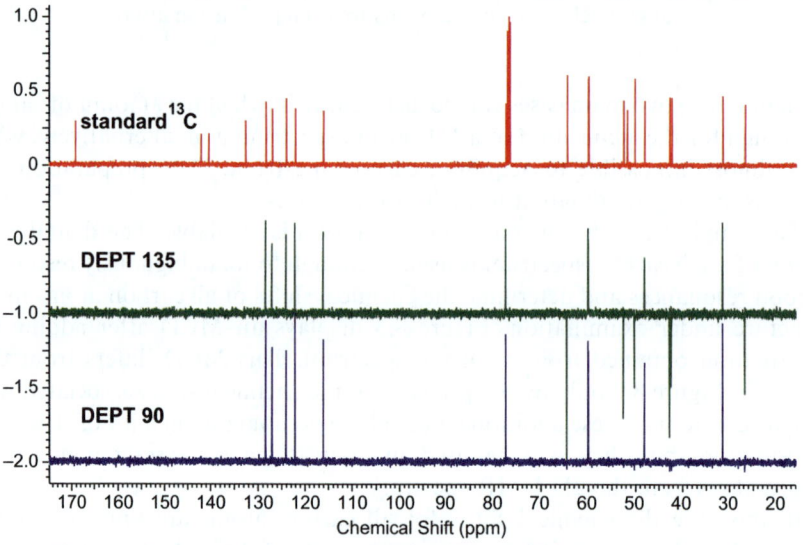

**Figure 8.6**   A stack plot of both conventional and DEPT spectra. From top to bottom, a standard carbon spectrum (red) contains all signals, the DEPT 135 spectrum (green) shows positive CH and $CH_3$ and negative $CH_2$ signals and the DEPT 90 spectrum (blue) contains CH signals only.

to normal carbon spectra, allow for the identification of quaternary carbons as their signals are absent in the DEPT135 spectrum, whereas the $CH_2$ carbons are negative signals in a DEPT135 spectrum. By comparing DEPT135 with DEPT90 spectra, CH signals, present in both spectra, and $CH_3$ signals, absent

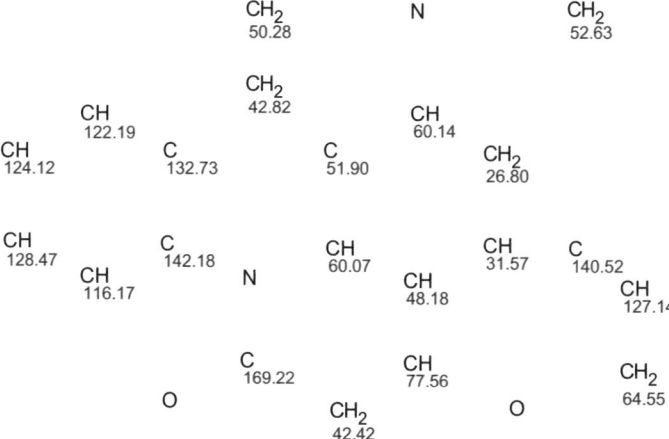

**Figure 8.7** The MCD after adding information extracted from DEPT spectra. The MCD is significantly simplified relative to that available from the carbon spectrum only and all hydrogen atoms are now associated with the corresponding carbon atoms.

in DEPT90, can be distinguished. The APT (attached proton test) experiment can also be used instead of a DEPT experiment. In this experiment the peaks corresponding to CH and $CH_3$ groups are shown as positive responses, whereas the C and $CH_2$ signals are negative. The multiplicity of carbon atoms can also be determined using other methods described briefly later.

Figure 8.7 shows the MCD after the inclusion of information from DEPT spectra. The number of hydrogen atoms connected to each carbon atom is now incorporated into the MCD.

## 8.4.2 2D NMR spectra

### 8.4.2.1 Peak-Picking HSQC Spectra

Peaks in HSQC spectra represent direct bonds between carbon atoms and directly connected hydrogen atoms. This is key information to enable transfer between H-H correlation spectra (COSY, *etc.*) and C-H correlation spectra (HSQC, HMBC, *etc.*) to produce C-C connectivities which can then be used for the purpose of facilitating structure generation. In practice, two types of experiment are used to derive similar information: HSQC and HMQC. Both types of spectra contain the same information and are invaluable for structure elucidation. The acronym HSQC will be used in all descriptive text below to describe both HSQC and HMQC simply because HSQC spectra are now employed more regularly.

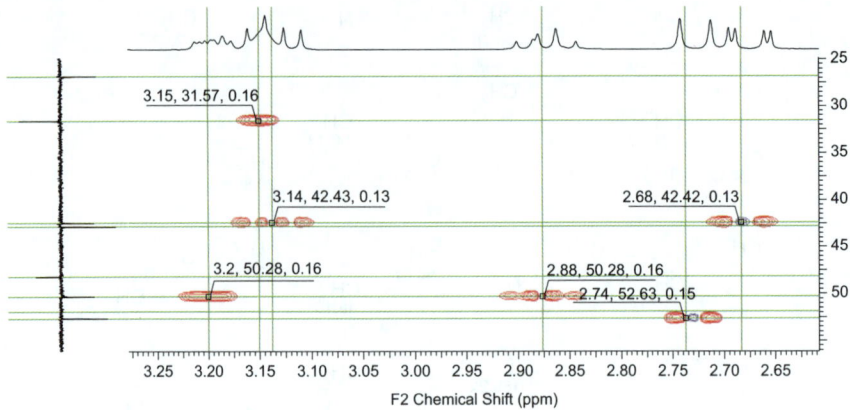

**Figure 8.8** A section of an HSQC spectrum with peaks corresponding to $CH_2$ groups.

Figure 8.8 shows a slice of an HSQC spectrum with associated labeled signals. Several important points should be kept in mind during the peak picking of signals. The most important point is that signals in 2D NMR spectra often appear as "multiplets", since a signal can have several maxima, even when the digital resolution is low. As a result, it is very important to peak pick the whole signal instead of a series of several local maxima, as it is the whole signal that corresponds to a single connectivity.

Another important point is that a single carbon atom cannot have more than two peaks in a HSQC spectrum, whereas CH and $CH_3$ groups can only have one peak in a HSQC spectrum. The hydrogen atoms in a $CH_2$ group can have different chemical shifts and therefore two different peaks can correspond to a $CH_2$ group in a HSQC spectrum. The presence of two different hydrogen chemical shifts for $CH_2$ groups is a consequence of the different magnetic environments for each hydrogen atom. This is common in cyclic structures where free rotation is impossible and the magnetic environment is not averaged. It is also observed for $CH_2$ groups with stereogenic hydrogen atoms. Figure 8.8 contains several peak pairs corresponding to $CH_2$ groups.

The position of peaks is also very important. In those cases when a carbon spectrum is available the chemical shifts of the carbon atoms are known. The green horizontal gridlines displayed on the spectrum in Figure 8.8 correspond to the carbon chemical shifts. To remove possible ambiguity and further simplify structure elucidation peak labels should be set as horizontal gridlines. In this case the chemical shift values for carbon atoms will be the same in all spectra and this significantly simplifies data analysis.

The relationship between the chemical shifts of hydrogen and carbon is the main result of peak picking an HSQC spectrum. This is used to convert H-H connectivities to C-C connectivities and to provide input to the structure generator algorithm. In addition, the chemical shifts of the hydrogen nuclei are established. This information is then used during the analysis of the other 2D NMR spectra.

In theory the chemical shifts for the hydrogen nuclei can be obtained directly from the $^1$H spectrum, but in many cases it is not so simple. The signals in proton spectra are relatively wide and overlap with other peaks occurs often. In this case the determination of accurate chemical shifts is challenging. Peak overlap in HSQC spectra is quite rare, however, and the determination of proton chemical shifts is a much easier task in this case. Figure 8.9 displays sections of a HSQC spectrum and corresponding 1D HNMR spectrum. The peaks are well resolved in the HSQC spectrum, while severe overlaps are observed in the proton spectrum. The proton chemical shifts can easily be determined from the position of the peak center in the HSQC spectrum. The MCD containing information extracted from the HSQC spectrum is presented in Figure 8.10.

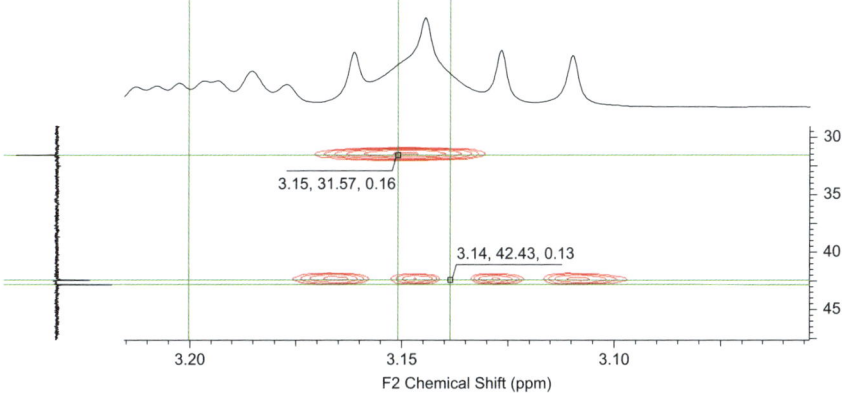

**Figure 8.9**    A section of a HSQC spectrum showing well-resolved peaks.

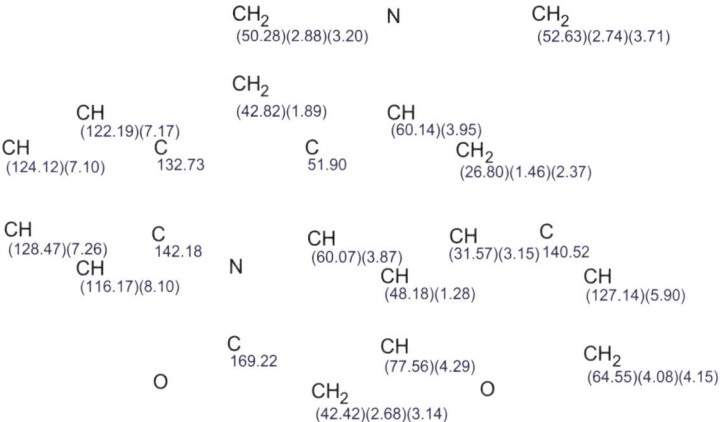

**Figure 8.10**    The MCD after adding information extracted from the HSQC spectrum. The numbers in the first parentheses are the chemical shifts of the carbon atoms and in the second and third parentheses are the chemical shift(s) of the hydrogen atom(s).

## 8.4.2.2   Peak Picking of COSY and HMBC Spectra

Both COSY and HMBC, in addition to HSQC, are the main 2D NMR experiments used in structure elucidation. A COSY spectrum contains information about the connectivity between hydrogen atoms. Figure 8.11 displays a section of a COSY spectrum.

At this stage the positions of almost all hydrogen peaks are already known as the result of peak picking of the HSQC spectrum, and peak picking of the COSY spectrum is relatively easy at this point. Peaks are set to the intersection of the gridlines where each gridline corresponds to the chemical shift of a hydrogen resonance. A COSY spectrum is diagonally symmetrical and contains peaks in the diagonal, but these do not contain any structural information and can be completely ignored.

Figure 8.12 shows the MCD with connectivity information obtained from the COSY spectrum. Generally, a COSY spectrum can contain peaks which are

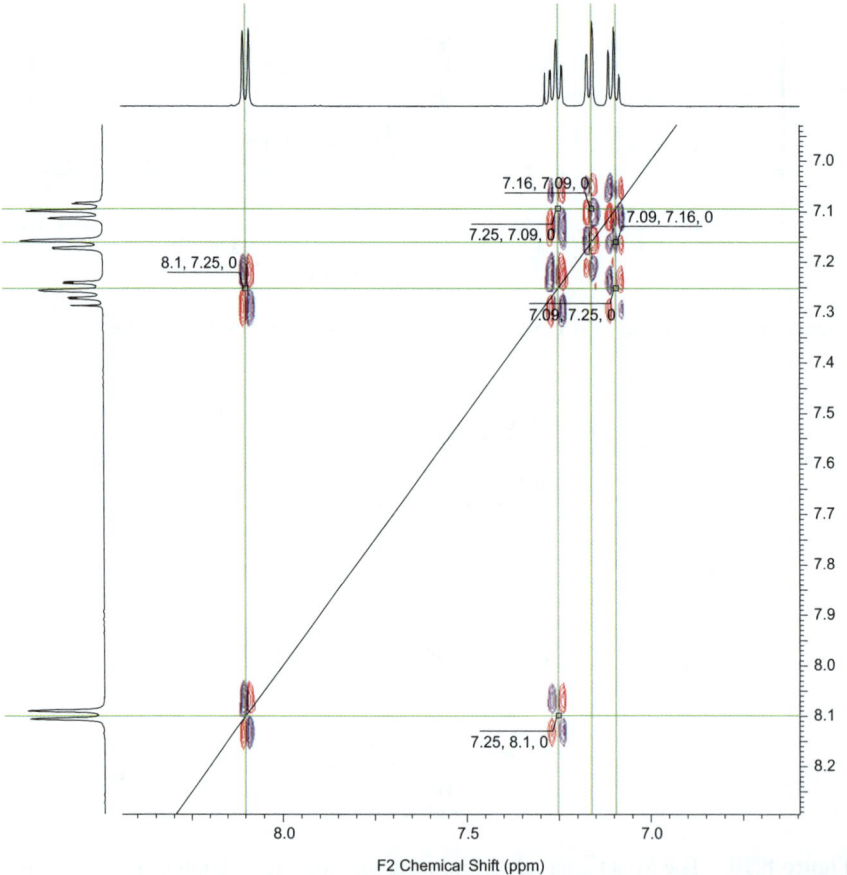

**Figure 8.11**   A section of the COSY spectrum of strychnine (aromatic region displayed). Some partially overlapped signals are resolved using gridlines.

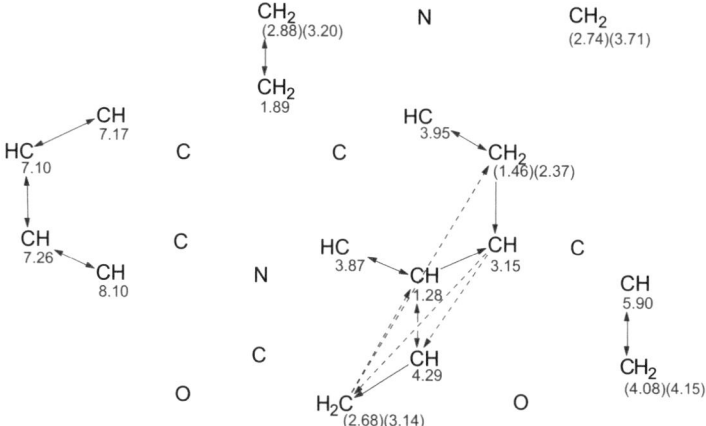

**Figure 8.12**   The MCD with information extracted from the COSY spectrum. Only the vicinal ($^3J$) connectivities are displayed. Some connectivities cannot be assigned unambiguously to one atom and are displayed as dashed lines.

produced by hydrogen atoms separated by two bonds ($^2J$) to five bonds ($^5J$). There is, unfortunately, no simple way to unambiguously separate connectivities by $J$ values. This produces problems in structure generation, as an unrecognized five-bond correlation treated, as a three-bond correlation during structure generation will likely produce an incorrect structure or make structure generation impossible (*vide infra*). Geminal ($^2J$) and vicinal ($^3J$) correlations are generally the most intense signals in COSY spectra, with long-range ($^4J$ and $^5J$) correlations usually much less intense. The simplest way to separate connectivities is therefore to use the peak intensities (volume). In practice, the best solution for COSY is to pick only the most intense peaks which *should* correspond to the $^2J$ and $^3J$ connectivities and to ignore less intensive peaks. In addition, long-range correlation peaks can be almost completely suppressed during processing by using appropriate weighting functions.[3]

Some of the correlations displayed in Figure 8.12 are illustrated as dashed lines and represent peaks that cannot be unambiguously assigned to one hydrogen atom, as two or more hydrogen atoms have almost degenerate chemical shifts. Generally, the structure generator can use ambiguous connectivity information, but in most cases this ambiguity can be removed by setting peaks to exact gridline intersections. Figure 8.13 displays a section of the spectrum with ambiguous peak position.

After removing ambiguous correlations, the MCD contains all theoretically possible $^3J$ COSY correlations (Figure 8.14) and all correlations between $CH_n$ carbons are now present in the MCD.

The MCD displayed in Figure 8.14 has 11 bonds extracted, but a residual set of 20 bonds remain unidentified. Additionally, six carbon atoms do not have any connectivity to other atoms. Five of these are quaternary carbons. The connectivities of this type of carbon atom cannot be determined from either

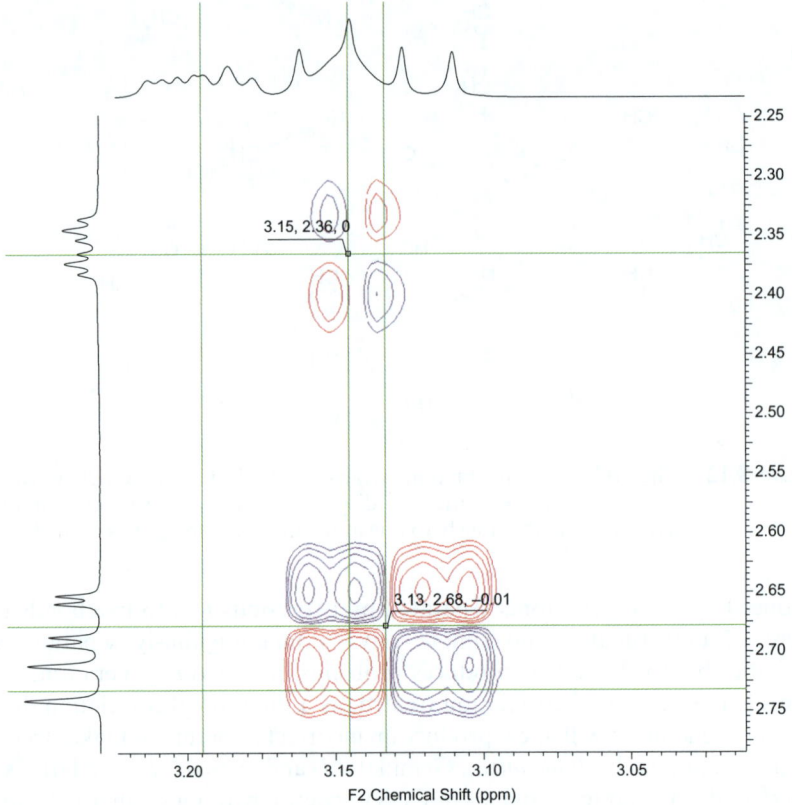

3.15, 2.36, 0

3.13, 2.68, -0.01

2.25
2.30
2.35
2.40
2.45
2.50
2.55
2.60
2.65
2.70
2.75

3.20    3.15    3.10    3.05

F2 Chemical Shift (ppm)

**Figure 8.13**   Resolving ambiguity in a COSY spectrum by placing peaks to gridline intersections.

COSY or any other $^1H$-$^1H$ 2D NMR spectrum and a long-range $^1H$-$^{13}C$ HMBC spectrum must be used to determine the connectivities of the quaternary carbons and connect parts of the structure partially defined within the COSY spectrum. Technically, HMBC is equivalent to an HMQC experiment, but optimized to examine small coupling constants (for example, 6–8 Hz instead of 120–150 Hz in HMQC) and peaks corresponding to $^2J$ and $^3J$ CH coupling constants are therefore observed. Practically, it is almost impossible to distinguish peaks corresponding to $^2J$ and $^3J$ connectivities, since the ranges of possible values of $^2J$ and $^3J$ couplings intersect significantly. Several experiments[4–7] have been demonstrated to distinguish $^2J$ and $^3J$ correlations. For example, the H2BC experiment[6,7] can distinguish between $^2J$ and $^3J$ correlations, but only in those cases where the correlations are observed between $CH_n$ carbons. This restriction makes the H2BC experiment useful but limited in its applicability, since the same information can be obtained from a COSY spectrum. Another significant problem with HMBC spectra is that in some cases, specifically for narrow proton signals such as those of Me and *t*-Bu groups, $^4J$ and even $^5J$ correlations

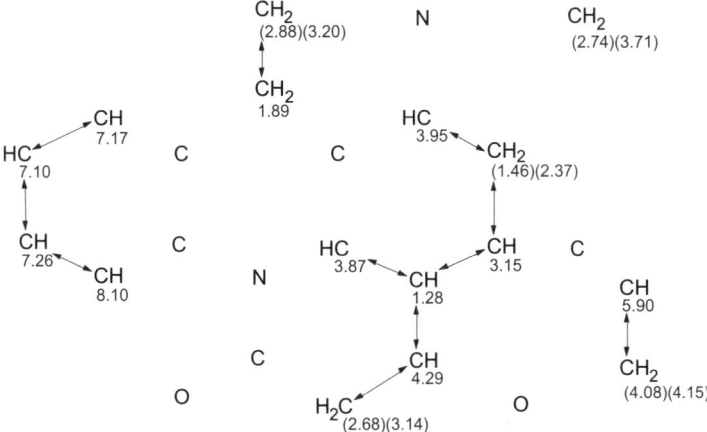

**Figure 8.14** The MCD with information extracted from the COSY spectrum. All ambiguity has been resolved and now only unambiguous connectivities are present in the MCD. The numbers in parentheses are the $^1$H chemical shifts of the corresponding hydrogen atoms.

can appear in the 2D NMR spectra. In practice, almost all HMBC spectra contain at least one $^4J$ correlation. In many cases, these $^4J$ correlations are significantly less intense than the $^2J$ and $^3J$ signals and can be identified by their peak intensity. However, for sharp resonances as discussed above $^4J$ correlations can be intense and it is almost impossible to determine the real $J$ value of the correlation.

Figure 8.15 shows a section of the HMBC spectrum (green) superimposed with the HSQC spectrum (red). The HMBC spectrum contains several $^4J$ connectivities, which are marked on the figure as blue rectangles.

Figure 8.16 displays the MCD containing connectivities from both COSY (blue) and HMBC (green and magenta) spectra. The HMBC spectrum of strychnine contains ten $^4J$ correlations (magenta). In this case, a 10% threshold level is used to separate the connectivities between the $^{2-3}J$ and $^4J$ correlations.

Data preparation for conventional elucidation is essentially complete at this point. If the structure is not very large and contains a large enough number of hydrogen atoms then these data ($^{13}$C, DEPT135, DEPT90, HSQC, COSY, and HMBC) are often enough to perform a successful structure elucidation. For the data presented in Figure 8.16, the structure generator[8,9] implemented into StrucEluc produces one correct structure in less than 1 s. For more complex problems, described below, some additional spectral data and data preparation may be required.

## 8.5 Complex Problems

In many cases a full set of data is not available or the structure is too complex and cannot be solved unambiguously from the conventional set of data that is

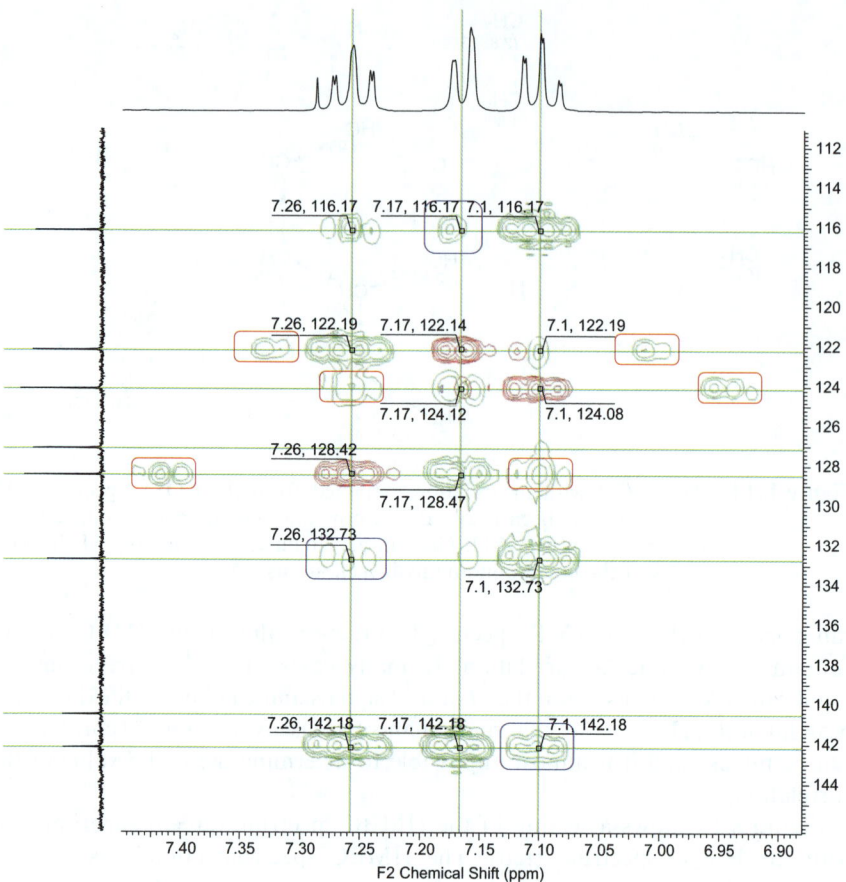

**Figure 8.15** A section of the HMBC spectrum overlaid with the HSQC spectrum. HSQC peaks are displayed in red and HMBC peaks are displayed in green. $^4J$ HMBC correlations are highlighted by blue rectangles. The residual HSQC responses which appear in the HMBC spectrum are highlighted by red rectangles. Some of these residual HSQC responses accidentally appear in the gridline intersections and can therefore produce false HMBC correlations.

available. In many cases where sample limitations lead to sensitivity issues $^{13}C$ and DEPT spectra are unavailable and the $^{13}C$ chemical shifts need to be extracted from the 2D NMR data. Alternatively, some additional experiments may need to be used to perform structure elucidation.

## 8.5.1 When Carbon Spectra are Unavailable

When the amount of substance available leads to sensitivity limitations and preclude the acquisition of a carbon and DEPT spectra in a reasonable time, then the number and type of carbon atoms need to be extracted from the HSQC

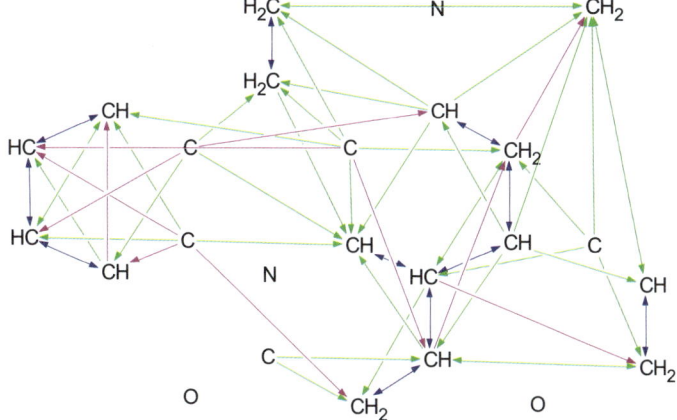

**Figure 8.16** The MCD with information extracted from both the COSY and HMBC spectra. The COSY connectivities are displayed in blue, $^{2-3}J$ HMBC in green and $^4J$ HMBC as magenta. HMBC correlations were separated into $^{2-3}J$ and $^4J$ by intensity using a 10% intensity threshold.

and HMBC 2D NMR spectra. This process cannot be completely automated in all cases and sometimes requires several iterative attempts. The ideal approach is to utilize multiplicity-edited HSQC spectra instead of DEPT spectra, as this also allows the determination of the number of hydrogen atoms connected to each carbon atom and to use the carbon projections from HSQC and HMBC to determine all carbon shifts. The CH and $CH_3$ groups are displayed as positive responses and the $CH_2$ groups are negative responses in a multiplicity edited spectrum. An example spectrum is shown in Figure 8.17 and the MCD information obtained from this spectrum is displayed in Figure 8.18.

It is not possible to distinguish the CH and $CH_3$ peaks from this spectrum (this spectrum contains the same information as DEPT135). This problem can be solved in two ways. An additional HSQC spectrum which encodes information about the CH peaks only could be acquired (this spectrum would then contain the same information as a DEPT90 spectrum). However, in practice another method is usually used. $CH_3$ and CH peaks can generally be distinguished by peak shape or by the intensity of the integrals corresponding to the signals in proton spectra. This last approach is applicable only in those cases when the corresponding signals do not overlap with other signals in the proton spectrum. In those cases when none of these methods help to determine the multiplicities, then the structure generator can generate structures for all of the possible variants of the number of hydrogen atoms attached to the carbon atoms in question. This approach requires some additional time for structure generation but it is, generally, possible.

To detect all carbons within a structure using 2D NMR spectra only is not a trivial task[10] for the following reasons. The digital resolution of 2D NMR spectra is relatively low and if the chemical shift difference between carbons is less than the digital resolution, then it is very difficult to determine the presence

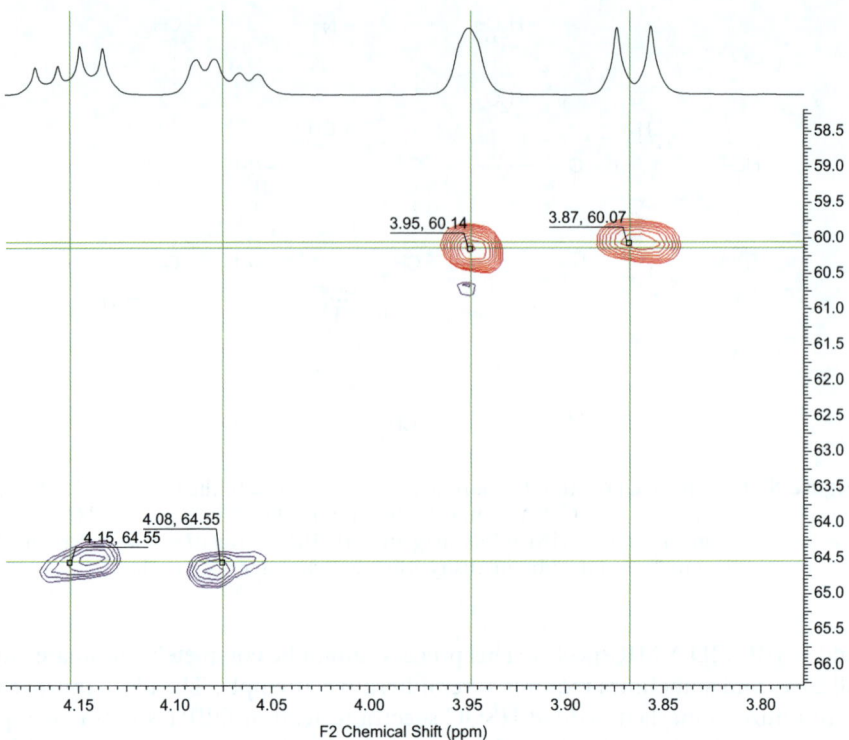

**Figure 8.17** A section of the multiplicity-edited HSQC spectrum for strychnine with the blue (negative) peaks corresponding to $CH_2$ groups and the red (positive) peaks corresponding to CH or $CH_3$ groups. The peaks in the top right-hand corner of the spectrum have very similar carbon chemical shifts but can be assigned to different carbon atoms, since only $CH_2$ groups can produce two peaks in a HSQC spectrum but the peaks displayed in the top right-hand corner are from either CH or $CH_3$ groups.

of several carbon atoms because they will appear in the spectrum as one peak in the carbon projection. Figure 8.19 displays a region of the HMBC spectrum which shows peaks from two almost undistinguishable carbons. In this particular case these carbons can be separated using information from multiplicity-edited HSQC (Figure 8.17). The HSQC spectrum has two positive peaks at approx. 60 ppm on the carbon axis. These peaks cannot be from a $CH_2$ group, because they are both positive, whereas $CH_2$ peaks should be negative. It can be concluded that two different carbon atoms with similar chemical shifts therefore exist in the structure.

Some additional problems may appear in such cases. If two carbon atoms have similar chemical shifts and are mistakenly processed as one carbon, then the correlations from both carbons will be assigned to a single carbon center. If these two carbon atoms are significantly distant from each other in the structure then the correlations become very long and will not correspond to the correlations in the real structure. As a result, the correct structure will not be

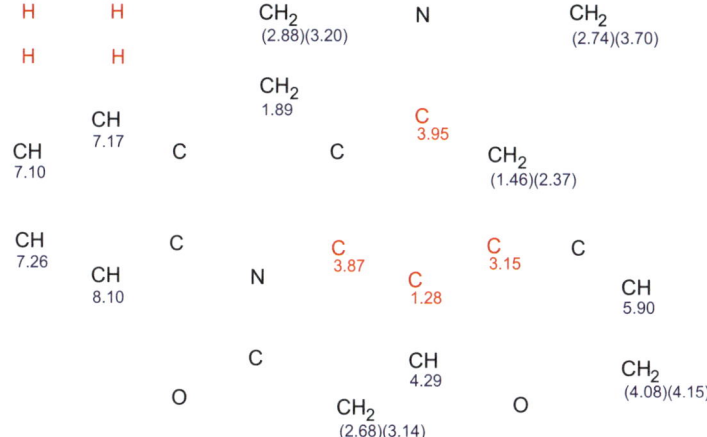

**Figure 8.18** The MCD including data obtained from a multiplicity-edited HSQC spectrum. In this case all of the $CH_2$ carbons are known, since they are unambiguously identified from the sign of the peak (positive or negative). Some of the CH carbon resonance can be determined based on empirical rules: the $^1H$ chemical shifts of $CH_3$ groups have an upper limit on the chemical shift scale of approx. 5 ppm. CH multiplicity can be assigned to all $CH_n$ groups with proton chemical shifts of more than 5 ppm. The remaining four carbon atoms, highlighted in red, have unknown multiplicity, but can all be assigned to CH groups based on the number of residual hydrogen atoms; only four hydrogen atoms are left over (placed in the upper left-hand corner of the MCD).

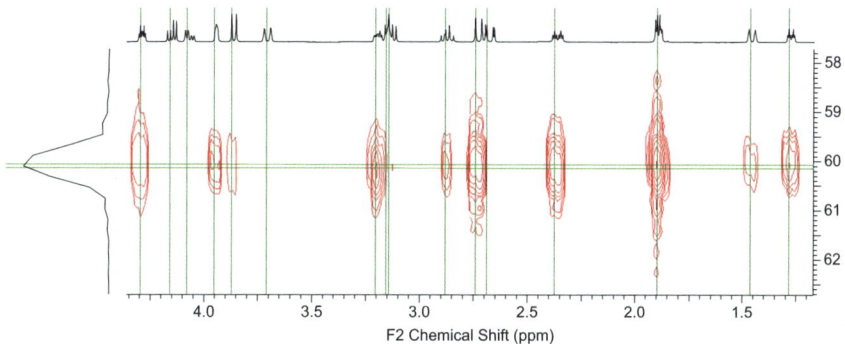

**Figure 8.19** The cross-peaks associated with two almost degenerate carbon atoms in the HMBC spectrum of strychnine. The horizontal gridlines correspond to the chemical shifts of the carbon atoms obtained from the carbon spectrum. If a carbon spectrum is not available, then these carbons would be indistinguishable as shown by the projection to the left of the figure where the peaks of both carbons merged into a single peak.

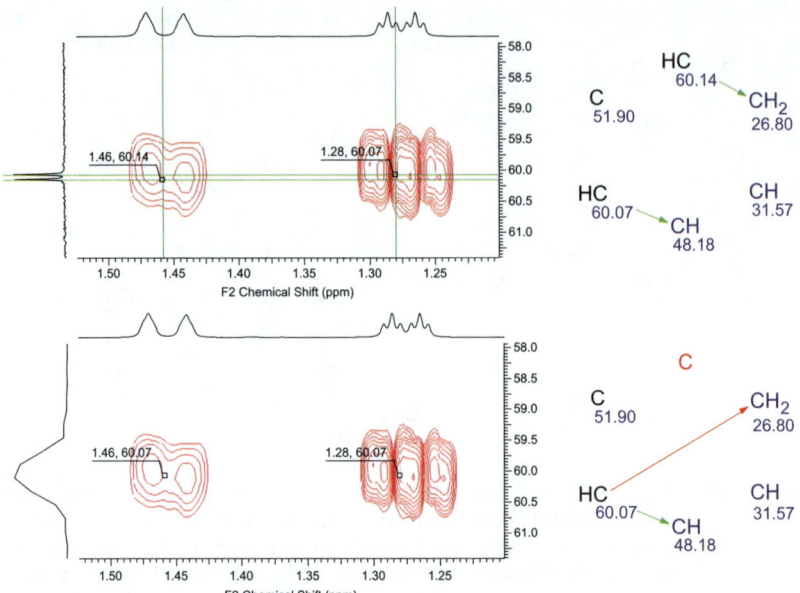

**Figure 8.20**   The correct and incorrect assignments of HMBC peaks to carbon atoms. The upper part of the Figure contains a section of the spectrum with correct peak picking and with peaks assigned to different carbons and the fragment of the MCD with connectivities corresponding to these peaks. The lower part of the Figure displays that case when only one carbon is identified. All correlations are assigned to a single carbon atom and one correlation, highlighted in red, becomes long and does not correspond to the actual structure.

generated from this data. Figure 8.20 shows two MCDs, one with the correct correlations and another with incorrect correlations mistakenly assigned to improper carbon atoms. In the second case some correlations become very long and clearly do not correspond to the actual structure.

In certain cases some quaternary carbons have no correlations at all in the HMBC spectrum, for example, in those cases when all of the neighboring carbon atoms do not have attached hydrogens. In this case, this carbon can be found by comparing the molecular formula and the number of carbon atoms already present in the spectrum. The structure generator is capable of producing structures, even when some of the carbon atoms are not present in the 2D NMR spectra. However, the number of generated structures can be significantly larger in this case.

## 8.5.2   When Spectral Data is not Sufficient to Elucidate Structures

The majority of modern 2D experiments commonly used to perform structure elucidation are "proton-detected" and show correlations either between hydrogens (COSY) or between hydrogens and carbons. If a structure contains a relatively large number of heteroatoms, or has a small number of hydrogen atoms, then a conventional dataset (HSQC, HMBC, and COSY) will not be enough to unambiguously determine the structure. In these cases some additional experiments can be used. Some examples are listed below.

### 8.5.2.1 $^{15}N$-$^{1}H$ 2D NMR Spectra

For compounds containing nitrogen atoms, long-range heteronuclear $^{15}$N-$^{1}$H experiments can be extremely valuable in helping to determine chemical structure. Examples of the application of $^{15}$N-$^{1}$H HMBC have been described by Hilton and Martin.[11] For example, consider the structure of posaconazole (molecular formula = $C_{37}H_{42}F_2N_8O_4$) displayed in Figure 8.21.

The structure contains a large number of nitrogen atoms, a common situation for pharmaceutical compounds, and it is very hard to elucidate the structure without using $^{15}$N data. In the work reported by Hilton and Martin,[11] more than 104 million structures were generated in more than 25 h using 26 HSQC, 29 COSY and 78 $^{1}$H-$^{13}$C HMBC correlations. Structure generation was interrupted because no reasonable structures were generated in that time. After adding information from a $^{1}$H-$^{15}$N HMBC spectrum (20 correlations), 87 structures were generated in 6 h. The correct structure of posaconazole was found in the first three positions (three equivalent structures with different chemical shift assignments) after ranking the structures according to the correspondence between the experimental and calculated chemical shifts. This clearly represents the dramatic impact of having such data available and the investment in time to acquire the data can mean the difference between a successful elucidation exercise and failure.

### 8.5.2.2 ADEQUATE Spectra

Generally, an INADEQUATE 2D spectrum is the best source of data to use for structure elucidation. INADEQUATE spectra contain the connectivities between carbon atoms and, in most cases, almost all bonds in a structure can be

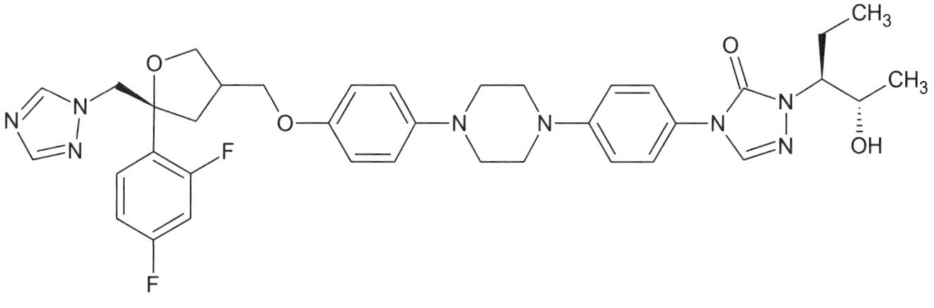

**Figure 8.21** The structure of posaconazole.

determined from this experiment. Unfortunately, due to the low natural abundance of the $^{13}$C isotope, the sensitivity of the INADEQUATE experiment is too low and it is used only very rarely in practical structure elucidation. Several more sensitive alternatives of the INADEQUATE experiment have been suggested. The ADEQUATE experiment is the most useful alternative. It is much more sensitive, as it is a $^1$H-detected experiment but the sensitivity of this spectrum is still low relative to either HSQC or HMBC (because it requires $^{13}$C-$^{13}$C magnetization transfer). The ADEQUATE experiment was not widely used until 2010 due to the lack of sensitivity. Only several examples of the application of ADEQUATE for structure elucidation have been described in the literature. The experiment has become more popular as cryoprobes have become widely available, and the enhanced sensitivity of cryoprobes allows the acquisition of ADEQUATE spectra in a reasonable time.

The ADEQUATE experiment is much more sensitive than INADEQUATE, but ADEQUATE spectra do not contain correlations between quaternary carbons. ADEQUATE spectra generally contain the same information as HMBC but, unlike HMBC, $^2J$ and $^3J$ can be distinguished, as several different kinds of ADEQUATE experiment exist. Most important is the 1,1-ADE-QUATE spectrum that contains only $^2J$ CH correlations which correspond to one-bond correlations between carbons. The ADEQUATE experiment is a good compromise of sensitivity and structural information.

Figure 8.22 displays the MCD containing connectivities from a 1,1-ADE-QUATE spectrum. Of the possible 31 bonds, 20 are unambiguously defined from the 1,1-ADEQUATE spectrum. This is significantly different from a COSY spectrum, in which only 11 bonds are distinguished. Only two bonds between quaternary carbons and nine bonds to heteroatoms remain undefined.

1,1-ADEQUATE correlations define almost all bonds between carbons and remove most ambiguity from the HMBC connectivities. This significantly

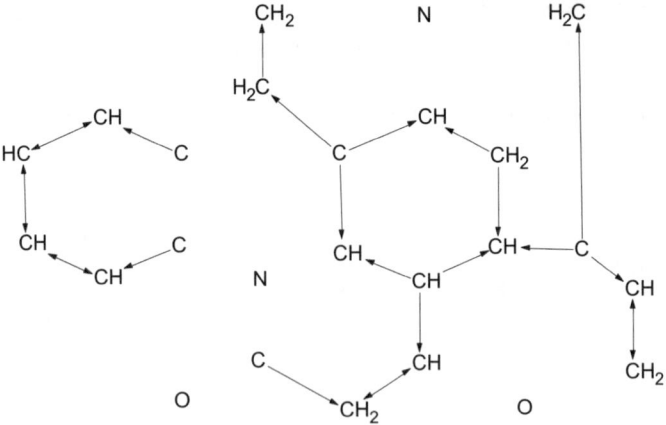

**Figure 8.22**   The MCD with information extracted from a 1,1-ADEQUATE spectrum.

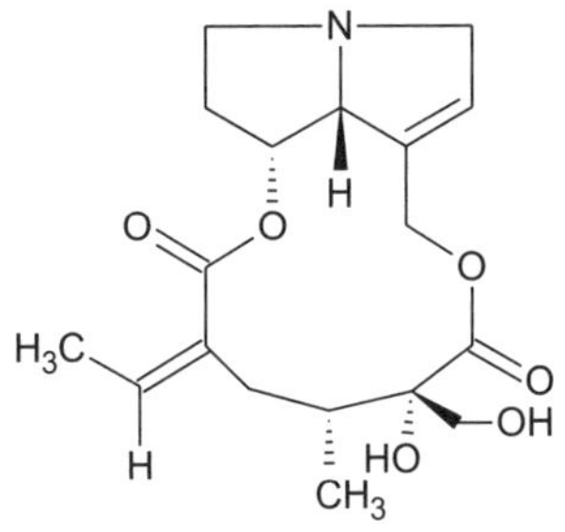

**Figure 8.23**   The structure of retrorsine.

simplifies structure generation, thereby reducing generation time and the number of generated structures. A comparison of structure elucidation with and without 1,1-ADEQUATE data has been published by Cheatham *et al.*[12] The authors investigated the influence of a 1,1-ADEQUATE spectrum on the structure elucidation of retrorsine (Figure 8.23) as model example. 10,921 structures were generated in 70 min using a combination of COSY and HMBC correlations, and only eight structures were generated in less than 1 s using a combination of 1,1-ADEQUATE and HMBC data. In this case the addition of 1,1-ADEQUATE data reduced the generation time by about four orders of magnitude.

Recently, the use of ADEQUATE spectra has become very popular in structure elucidation and a comprehensive review of ADEQUATE-type experiments and its application to structure elucidation has been published by Martin.[13]

# 8.6   Checking MCD for Consistency

## 8.6.1   Forming the MCD

To provide a complete and clear pattern of the properties of the skeletal atoms and the connectivities between them (*via* the MCD), the program places skeletal atoms together with hydrogen atoms attached to skeletal atoms ($CH_3$, $CH_2$, CH, and C groups, as well as OH and NH, if identified by the user from $^1$H NMR and 2D NMR spectra) in a display window as shown in Figure 8.16.

HMBC and COSY connectivities are shown in the MCD as "fuzzy" sub-graphs (fragments) connecting carbon atoms and/or carbon and nitrogen atoms when the corresponding $^1$H-$^1$H COSY and $^{15}$N HMBC data are available. Different possible distances, if known, between the connected atoms are marked on the MCDs with specific colors. In accordance with the main axioms (see Section 2.2) the lengths of the COSY and HMBC connectivities are taken as *default* (standard connectivities) to be one and one to two bonds correspondingly, indicating the number of bonds between the skeletal atoms. In StrucEluc the standard length HMBC connectivities are colored green and the COSY connectivities are colored blue (see Figure 8.16). This scheme permits 'at-a-glance' knowledge of whether two atoms are linked by exactly one bond, exactly two bonds, or a specific range of bonds. Information from different 2D NMR experiments can be viewed or suppressed by clicking the appropriate buttons on the toolbar.

The program analyzes the $^{13}$C and $^{1}$H chemical shifts of the $CH_n$ groups ($n = 0$–3) and automatically sets, if possible, the parameters to show the most probable hybridization and the possible heteroatom neighborhood for each carbon atom. To carry out this procedure special atom property correlation tables (APCTs; see Section 7.7) are used. The relationships to neighboring heteroatoms are marked as *"forbidden"* (*fb*), *"at least one"*, *"at least two"*, *"at least three"*, *"four"*, and *"not defined"* (*nd*). The possible states of atom hybridization are designated as $sp^3$, $sp^2$, $sp$, *not sp,* and *"not defined"*. To ease the visual recognition of the type of hybridization of a given atom each type is marked in its own specific color. It is essential to note that both $^{13}$C and $^{1}$H NMR chemical shifts are taken into account by the program when setting the atom parameters. Those descriptors for the carbon atoms allow the system to analyze 2D NMR data and to efficiently apply restrictions during the process of structure generation.

If a distinct multiplet is observed in the $^{1}$H NMR spectrum from a structural block (C-*i*)H$_n$ then the total number of hydrogen atoms attached to carbons adjacent to the (C-*i*) carbon is set. This property is determined by the chemist after visual analysis of the $^{1}$H spectrum pattern and after taking into account the coupling constants (if measured). The atom properties should be set and edited with great caution because an erroneous assumption (a wrong "axiom") leads to the exclusion of the correct structure from the output file. All structural constraints presented in the MCD are used during the structure generation.

Figure 8.24 shows a window where all properties of a particular CH group are presented as an example. Here a carbon atom with a chemical shift of 118.9 ppm is in a $sp^2$ hybridization state, and its connection with a heteroatom is forbidden. Since the signal at 6.22 ppm is a distinct singlet in the $^{1}$H NMR spectrum the number of hydrogens attached to the carbon atoms closest to the C(118.9) carbon is set by the user to be equal to zero. The latter constraints speed up the structure generation process significantly because the generation of structures where this constraint is violated will be suppressed. The chemist can then edit these parameters using other available information. For example, if the molecule belongs to the CHNO class and the carbon atom C(175.5) is marked as $sp^2$/*at least two* then the system will only generate O-C$=$O, N-C$=$O, O-C$=$N, and O-C$=$N fragments on the basis of that atom. The chemist is also offered the opportunity to draw bonds of any multiplicity between the atoms and to set some functional groups (for instance, C$=$O, O-C$=$O, O-H, *etc.*). If a molecule contains heteroatoms of only one type and there are free H atoms, O-H, N-H, NH$_2$, *etc.*, then these bonds may also be drawn in. This provides a quick and intuitive mechanism for entering structural information evident from the $^{1}$H NMR and/or IR spectra without the need for inputting exhaustive tables of numerical data. All structural constraints presented in the MCD are used during the structure generation process.

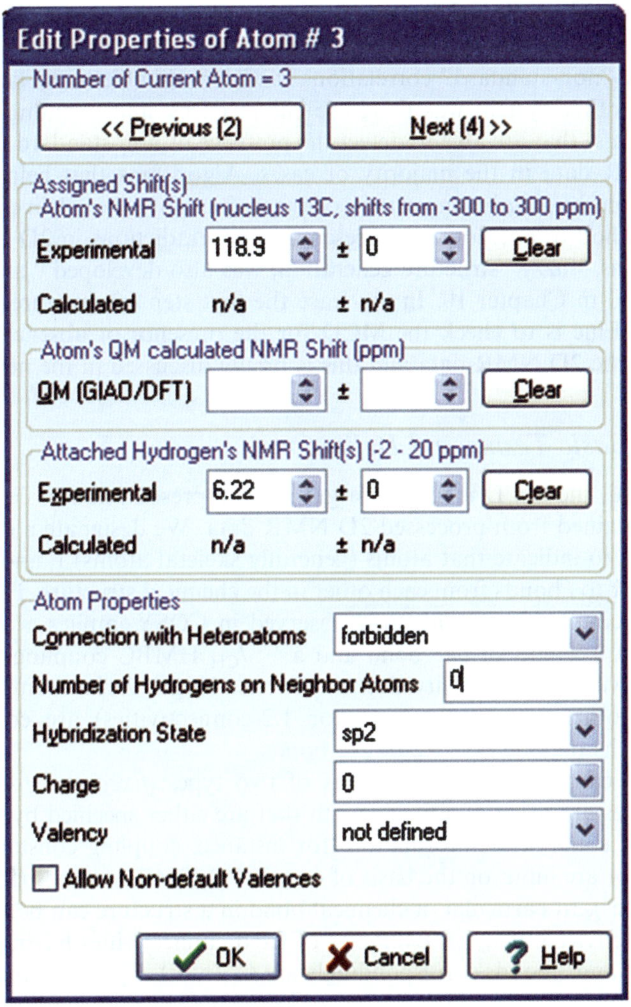

**Edit Properties of Atom # 3**

Number of Current Atom = 3

| << Previous (2) | Next (4) >> |

Assigned Shift(s)
Atom's NMR Shift (nucleus 13C, shifts from -300 to 300 ppm)

Experimental  118.9 ⇕ ± 0 ⇕  Clear

Calculated  n/a  ± n/a

Atom's QM calculated NMR Shift (ppm)

QM (GIAO/DFT)  ⇕ ±  ⇕  Clear

Attached Hydrogen's NMR Shift(s) (-2 - 20 ppm)

Experimental  6.22 ⇕ ± 0 ⇕  Clear

Calculated  n/a  ± n/a

Atom Properties

Connection with Heteroatoms  forbidden ▾

Number of Hydrogens on Neighbor Atoms  0

Hybridization State  sp2 ▾

Charge  0 ▾

Valency  not defined ▾

☐ Allow Non-default Valences

| ✓ OK | ✗ Cancel | ? Help |

**Figure 8.24**  A window showing an example of setting the properties for a CH group.

## 8.6.2 Checking MCD for Consistency

The conclusion from Chapter 6 is that 2D NMR-based expert systems are capable of solving very complicated problems if the spectral data are free of contradictions. Since all of the expert systems described in Chapter 6 are adjusted by default to account for the coupling constants $^{2-3}J_{HH}$ and $^{2-3}J_{CH}$ which are common for the corresponding COSY and HMBC correlations (referred to as "standard" correlations in Chapter 2), contradictions will appear when at least one correlation of > 3 bonds results in a response in the 2D NMR data. Despite recent developments to aid in the identification of the correlation lengths,[4-7] there is presently no routine NMR technique that is capable of distinguishing couplings of different lengths in a reliable fashion. Therefore, the development of theoretical methods for 2D NMR data analysis that identify the presence of "non-standard" correlations is of considerable importance.

The StrucEluc system was therefore enhanced by implementing algorithms and programs[9] that are able to detect the presence of non-standard correlations in 2D NMR data in the majority of cases. Algorithms that help to remove contradictions by lengthening certain connectivities have been delivered. A more general method to overcome the presence of contradictions in 2D NMR data, referred to as "fuzzy" structure generation, was also developed[14] and is described in detail in Chapter 10. In any case the first step of structure elucidation using StrucEluc is to check the MCD for the presence or absence of contradictions in the 2D NMR data and this is briefly discussed in the next Section.

### 8.6.2.1   Basic Terms and Definitions

As discussed, the MCD is used as a graphical representation of connectivities that are obtained from processed 2D NMR data. We designate a connectivity as $C_{kl}(v_i-v_j)$ to indicate that atoms (generally skeletal atoms) $v_i$ and $v_j$ are at a distance of $k$ to $l$ bonds from each other in the chemical structure. For instance, the vicinal coupling constant $^3J_{HH}$ observed in COSY implies a connectivity $C_{11}(v_i-v_j)$ of a length of *one* bond and a $^{2-3}J_{CH}$ HMBC coupling constant is associated with a connectivity $C_{12}(v_i-v_j)$ whose length varies from one to two bonds. In other words, $C_{12}(v_i-v_j)$ (or 1,2-connectivities) are connectivities whose lengths may be 1-1, 1-2 or 2-2 bonds.

A MCD can contain connectivities of two types: *fixed* and *formal*. Fixed connectivities are those of a fixed length that are either specified by the user on the basis of some prior information (for instance, coupling constant values if measured) or are input on the basis of certain 2D NMR methods (for example, see reference[4]). In particular, a chemical bond in a structure can be defined as a fixed connectivity $C_{11}(v_i-v_j)$. Examples of fixed connectivities having lengths of exactly 1, 2, and 3 bonds correspondingly are $C_{11}(v_i-v_j)$, $C_{22}(v_i-v_j)$, and $C_{33}(v_i-v_j)$.

Formal connectivities are those drawn in the MCD by the program itself on the basis of 2D NMR data analysis and with default correlation lengths assumed for each type of 2D spectra. In particular, default connectivities $C_{11}(v_i-v_j)$ are produced from COSY data and $C_{12}(v_i-v_j)$ from HMBC data. Formal connectivities can then be divided into two categories: standard and non-standard ones.

Formal connectivities that correspond to the structure of the analyzed molecule will be called standard connectivities. If all connectivities extracted from the 2D NMR data are standard correlations, then the program will be capable of generating a correct structure.

If a 2D NMR spectrum contains correlations of a *non-default* length (exceeding the default upper limits) and their lengths were not identified explicitly by the chemist, the corresponding formal connectivities are called *non-standard* (NSC). Note that in the case when there is no prior information about the presence of non-default correlations all formal connectivities are considered as standard. Obviously *formal non-standard* connectivities contradict the structure of the molecule under study.

Examples of non-standard COSY correlations (depicted as two C-C bonds with blue double-ended arrows) and HMBC (depicted as three C-C bonds with a green one-ended arrow) are shown below:

The non-standard COSY connectivity $C_{22}$(C3-C4) and HMBC connectivity $C_{33}$(C5-C2) will be accepted by the program as $C_{11}$(C3-C4) and $C_{12}$(C5-C2) connectivities correspondingly, if their true lengths were not specified by the chemist on the basis of the 2D NMR experiments. Note that if the $C_{12}$(C5-C2) connectivity is accepted as a standard correlation then it allows the C2 atom to occupy all positions on the benzene ring except its real position 2.

Since the structure generator is capable of utilizing connectivities of default (standard) lengths, then non-standard connectivities, if present in a MCD diagram, will either yield no result since no structure can be generated, or an invalid result where the results file does not contain the correct structure. This happens if the algorithm considers the topological distances between the corresponding skeletal atoms (specified by the inherent chemical shifts) of a non-standard connectivity equal to standard length. In reality, this distance is longer. Non-standard connectivities therefore appear as standard lengths to the software algorithms. Non-standard connectivities can also appear due to heavy overlap of the NMR resonances or accidental degeneracy and this can lead to permuted pairs of assignments.

We assume that the fixed connectivities, unlike the formal ones, are standard and correct in all cases. A set of atoms in the structure separated by a distance of $k$ bonds from the $v$ atom will be denoted as the $k^{th}$ *layer* of the $v$ atom environment. The set of atoms in the structure separated by less than or equal to $k$ bonds from the $v$ atom will be the $k^{th}$ sphere of the $v$ atom environment.

Assume that an MCD is created on the basis of a set of 2D NMR spectra for an unknown and the true lengths of some of the connectivities are determined. In such a general case there is a set of fixed and formal connectivities. The task is therefore to detect the non-standard connectivities or, at least, locate these by detecting small groups of connectivities that contain non-standard responses. A correct structure determination of the compound under study can be achieved only after the correction or elimination of *all* non-standard connectivities to make the 2D NMR data consistent.

Experience indicates that all non-standard connectivities can be partitioned into at least four different types which are dependent on the extent to which they contradict the rest of the connectivities:

1. Explicit non-standard connectivities that contradict the environment of a skeletal atom and some fixed connectivities that refer to this atom.

2. A group of connectivities containing at least one non-standard connectivity. Atoms and atom pairs with a set of connectivities that contain
   non-standard ones.
3. A group of connectivities referring to an atom and contradicting other
   connectivities which refer to other atoms.
4. Implicit non-standard connectivities that are not obviously contradictory
   but, nevertheless, prevent the elucidation of a correct structure.

The mathematical methods that allow for the detection and removal of
contradictions appearing from the presence of each type of non-standard connectivity have been described by Molodtsov and co-workers.[9] These methods
provide a logical analysis of a full set of connectivities produced from the collective 2D NMR data. The mathematical algorithm is rather sophisticated and
we refer the reader to the original work[9] for details. Here we will consider the
simplest case of detecting explicit non-standard connectivities and explain the
main approach on which the logical analysis of spectral data is based.

## 8.6.2.2   Explicitly Non-Standard Connectivities

An explicit non-standard connectivity can be detected if an atom exists whose
fixed connectivities fully define all atoms and all bonds between them in the $r^{th}$
sphere of this atom environment. If there are formal connectivities $C_{kl}(v_i-v_j)$,
where $l \leq r$, defined for this $v$ atom, then it is possible to determine the
appropriateness of each of the connectivities. For instance, the drawing below
shows that HMBC connectivities $C_{12}(C1-C2)$ and $C_{12}(C1-C3)$ are of a standard
length, whereas the HMBC connectivity $C_{12}(C1-C4)$ is obviously non-standard.

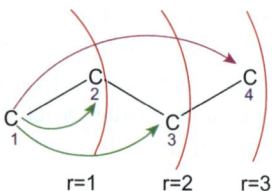

r=1          r=2          r=3

Furthermore, the validity of a formal connectivity $C_{kl}(v_i-v_j)$ may sometimes
be determined, even if not all the atoms and bonds in the first sphere of the $v_1$
atom are defined. It is only possible if a set of allowed environment spheres is
defined for each of the C-atoms. These spheres are defined using the APCT, as
described in Chapter 3, and is based on the chemical shifts of the appropriate
carbon atoms.

It should be noted that there is a significant difference between the ambiguity
associated with the HMBC or COSY correlations and that associated with the
APCT. Specifically, the HMBC correlations are ambiguous in nature and
correspond to either two or three bonds in length. The non-standard correlations are, of course, even more ambiguous. At present, StrucEluc uses the
structure generator to search for all possible variants for the different distance

lengths of the standard correlations within a reasonable time frame. The ambiguity involved in the elucidation process using APCT data has a different character and is a consequence of the fuzziness of the knowledge regarding the interrelation between units used to construct the structure and their associated spectral features. The diversity of the different kinds of atom environments means that it is impossible to enumerate all conceivable atom combinations and their related spectral features. It is therefore necessary to determine some typical atom combinations and associate assigned ranges of spectral features which characterize them. The boundaries of these spectral ranges are fuzzy since such a combination could have a spectral feature (for instance, a $^{13}C$ chemical shift) that is observed slightly out of the limits of a specific range. While using the APCT can be deemed to be a risky procedure, it is difficult to avoid since it helps overcome the possibility of a combinatorial explosion and possible enormous structure generation times. Such a situation is inevitable in the case of large molecules when there are a large number of atoms and no assumptions can be made about the nearest neighbors of a given carbon atom having specific chemical shifts. Consider an illustrative example, as described below.

Assume that the figure below represents a fragment with the fixed bonds indicated for the $v_1$ atom environment (in the figure the atoms are denoted by their numbers and the bond with a heteroatom provided for by the APCT, is represented by a dashed line):

Such a fragment, in principle, can be revealed using COSY and TOCSY experiments. Assume that there is a formal HMBC connectivity $C_{1-2}(v_1-v_6)$ and the APCT indicates that the $v_6$ atom has no bonds with heteroatoms. In this case the $v_6$ atom cannot obviously be located both in the first and second layers of the $v_1$ atom (the connectivity C6–Het–C1 is forbidden by the APCT) and the $C_{12}(v_1-v_6)$ connectivity is, therefore, non-standard.

Explicit non-standard connectivities can be removed either by simply deleting them or by increasing the maximum possible distance between the atoms. Replacing the $C_{12}(v_1-v_6)$ connectivity with $C_{13}(v_1-v_6)$ in the above example ensures that it is no longer non-standard. Explicit non-standard connectivities are the primary targets for detection and removal during the elucidation process. To reveal the presence of non-standard connectivities in more complicated cases, cases 2–4 mentioned above, the determination of the maximum admissible numbers of atoms situated in the first and second layers of the analyzed atom plays an important role.

Obviously, it is easy to determine the maximum possible number of skeletal atoms (refer to as "atoms" below) in each layer and therefore in each sphere of a given atom with a definite valence. The $CH_3$ group, for example, always has one atom in the first layer, and no more than three atoms in the second layer as seen in the drawing shown below:

$$H_3C-\overset{\overset{\displaystyle C}{|}}{\underset{\underset{\displaystyle C}{|}}{C}}-C$$

The second sphere of a $CH_3$ group therefore can contain no more than four atoms. Thus, if a $CH_3$ group has more than four different 1,2-connectivities, at least one of these must be non-standard. The maximum number of atoms in the second sphere of the $CH_2$ group illustrated is eight:

The data regarding the maximum possible number of atoms in the second sphere are used as a criterion for detecting non-standard connectivities in the atoms and atom pairs with a relatively large number of formal 1,2-connectivities. The algorithm analyzes all connectivities related to the current atom and estimates the number of atoms which would be situated in the second sphere if all connectivities are assumed to be of standard length. If the number of atoms exceeds the maximum admissible value, the algorithm perceives this as a hint to the presence of at least one non-standard connectivity associated with the atom and confirms this hypothesis by deeper analysis of the connectivity sets.

If non-standard connectivities are not revealed following deeper analysis, it does not necessarily indicate that they are really absent. It may happen that among the atoms for which the algorithm fails to detect the presence of non-standard connectivities that *groups of atoms* possess the following property: the union of connectivities included in the groups must contain a non-standard connectivity and, at the same time, the atoms included in the group may not have common connectivities.

In the work by Molodtsov et al.[9] consideration was given to groups containing only two atoms and having 1,2-connectivities. This is for the following reasons: (a) the number of possible groups formed from more atoms would be rather large, and (b) in those cases where a group containing many atoms with non-standard connectivities was selected, the union of connectivities belonging to these atoms would, in turn, contain a great number of connectivities. This reduces the significance of selecting a group of atoms, because the smaller the set of connectivities among which non-standard connectivities exist, the greater is the probability of removing any contradictions in the initial connectivity set.

If a MCD contains $N$ atoms then the number of different groups consisting of two atoms is equal to $N \cdot (N - 1)/2$. In reality, only those groups where both atoms contain 1,2-connectivities can be considered. As mentioned above, we consider such atoms for which the presence of non-standard connectivities is not revealed at the first stage of checking. As an example, assume that two atoms A and B are examined and at least one of them has an associated non-standard connectivity. The algorithm can fail to detect the presence of the non-standard connectivity during the first step. However, when these atoms are unified in a group of two atoms, the algorithm can detect the presence of a non-standard connectivity in the connectivity set formed from the connectivities belonging to *both* atoms. The corresponding computational procedures have been reported elsewhere.[9]

Assume that explicit non-standard connectivities, as well as atoms and groups of atoms having non-standard connectivities, are not found in the initial or renewed (corrected by the program) set of connectivities. Nevertheless, when the presence of contradictory data prevents the generation of chemical structures corresponding to all connectivities, it may be possible to detect the presence of non-standard connectivities in the initial connectivity set. Moreover, it is possible to identify an atom containing non-standard connectivities.

A bond distribution for an atom $v$ is termed *accurate* if the bond distributions of all of the atoms including the $v$ atom, are *admissible*. If *all* of the accurate bond distributions at the $v$ atom are analyzed, then it may be possible to determine the *obligatory* and *forbidden* bonds between the $v$ atom and other atoms. If *all* of the accurate bond *distributions* of the $v$ atom contain any bonds, then all of these bonds are defined as *obligatory* ones. Bonds that are absent in the entire accurate bond distributions are *forbidden*. Detecting obligatory and forbidden bonds in a chemical structure is based on the assumption that all the given connectivities are standard.

During the process of looking for non-standard connectivities it is necessary to sequentially conduct a bond analysis of all atoms. If, when analyzing bonds at the $v$ atom, it is found that there is no accurate bond distribution of the given atom, it means that the connectivities of the $v$ atom are at variance with the connectivities of the atoms analyzed earlier. It follows that the assumption regarding the validity of all connectivities is not true and as a result the MCD will therefore contain some non-standard connectivities. It is worthy to note that the bond refinement of any atom directly influences the bond analysis of all the other atoms. Hence, after the bond analysis of the last atom, it is necessary to start analyzing the bonds of the first atom. Bond analysis is only complete when new obligatory and forbidden bonds are no longer identified as a result of the successive bond analysis of *all* atoms.

In practice, the atom for which the contradiction was identified during bond analysis often contains non-standard connectivities. This follows from the fact that the number of atoms having non-standard connectivities is usually small. Most frequently, bond analysis is carried out initially for atoms with standard connectivities. As a result the correct obligatory and forbidden bonds are determined in the structure. When a contradiction is identified during the bond analysis of a recurrent atom, it is very probable that the non-standard connectivities of a given atom are in conflict with those belonging to atoms which have already been analyzed. In any case, the connectivities of the found atom are then lengthened to attempt removal of the contradictions, after which the check for the presence of non-standard connectivities is restarted.

## 8.7 Fuzzy Structure Generation

There are situations (see the examples in Chapters 10 and 12) when building structures corresponding to all of the defined connectivities, including the non-standard ones, is possible. In these cases it is impossible to identify the non-standard connectivities and this leads to *invalid* solutions. It is possible, even after seemingly successful checking of the data for contradictions that non-standard connectivities will nevertheless remain unnoticed by the algorithm. Their presence can only be identified in indirect ways, including: (a) large value(s) for the $^{13}$C NMR spectral deviations[14,15] calculated for the most probable structure(s); (b) inconsistencies between the most probable structure and additional experimental data (for instance, NOESY, ROESY, *etc.*); (c) comparison of the chemical shifts and multiplicities of the experimental and calculated data of $^{1}$H NMR spectra; and (d) by checking the structures using infrared correlation tables (if an IR spectrum is available) and/or interpretation of a mass spectrum, if available. The non-standard connectivities that do not prevent the structure from being built are termed *implicit* non-standard connectivities.

As the identification of implicit non-standard connectivities cannot be guaranteed, since any of the given connectivities may be non-standard, a method of removing non-standard connectivities was developed that, in many cases, allows the identification of a valid solution even in this situation.

To remove implicit non-standard connectivities, the following approach has been implemented. Some connectivities are declared, as a series, to be *suspicious* and are lengthened or eliminated. Each time the structure generation is initiated with a renewed connectivity set. This process is termed fuzzy structure generation (FSG).

Let $n$ be the total number of connectivities in the 2D NMR data and $m$ be the number of connectivities that are suggested to be non-standard. In this case it is necessary to consider $\binom{n}{m} = \frac{n!}{m!(n-m)!}$ different sets of $m$ connectivities that will be declared as suspicious. If all (without exception) combinations of connectivities are used for structure generation, then the calculation time increases dramatically as the number of tasks resulting in structure generation sharply increases with the rise of the $m$ value.

This "suspicious connectivity" concept has a number of applications. Declaring members of the connectivity sets to be suspicious is sometimes useful for searching for atoms and pairs of atoms with non-standard connectivities, as well as the direct determination of the presence of non-standard connectivities. In these cases, the program usually lengthens or deletes all connectivities belonging to the atoms selected during data analysis. In so doing both non-standard and standard connectivities are deliberately lengthened or deleted, which correspondingly leads to an increase in the number of structures generated. If only the definite connectivities [related to atoms at which the presence of non-standard connectivities are revealed, termed *found atoms* (FA)] are lengthened or deleted, then the number of generated structures will be considerably lower. In some cases the total time for structure generation will decrease in spite of the repeated initiation of the generation process. This is explained by the fact that most connectivity sets that do not contain connectivities emanating from the found atoms are excluded from consideration. Furthermore, during the atom-bond analysis process and the lengthening or deleting of definite connectivities, a greater number of obligatory and forbidden bonds can be established relative to the case when all the connectivities at the *found atoms* are lengthened or deleted. In this case the structure generation time with the new connectivity sets can sharply decrease.

It is worth noting that in the presence of atoms with non-standard connectivities, sets of connectivities declared as suspicious must contain connectivities belonging to all *found atoms*. This indicates that the considered number of suspicious connectivity sets will decrease. The methodology and strategy of fuzzy structure generation will be demonstrated with examples in Chapter 10.

# References

1. H. Masui and H. Hong, *J. Chem. Inf. Model.*, 2006, **46**, 775.
2. M. Will, W. Fachinger and J. R. Richert, *J. Chem. Inf. Comput. Sci.*, 1996, **36**, 221.
3. G. E. Martin and A. S. Zekter, *Two-Dimensional NMR Methods for Establishing Molecular Connectivity*, VCH Publishers, New York, 1988.
4. V. Krishnamurthy, D. Russell, C. Hadden and G. E. Martin, *J. Magn. Reson.*, 2000, **146**, 232.
5. T. Sprang and P. Bigler, *Magn. Reson. Chem.*, 2003, **41**, 177.
6. N. T. Nyberg, J. O. Duus and O. W. Sorensen, *J. Am. Chem. Soc.*, 2005, **127**, 6154.
7. A. J. Benie and O. W. Sorensen, *J. Magn. Reson.*, 2007, **184**, 315.
8. S. G. Molodtsov, *Commun. Math. Chem. (MATCH)*, 1998, **37**, 157.
9. S. G. Molodtsov, M. E. Elyashberg, K. A. Blinov, A. J. Williams, E. R. Martirosian, G. E. Martin and B. Lefebvre, *J. Chem. Inf. Comput. Sci.*, 2004, **44**, 1737.

10. P. Sandusky, *J. Nat. Prod.*, 2007, **70**, 1895.
11. B. D. Hilton, W. Feng and G. E. Martin, *J. Het. Chem.*, 2011, **48**, 948.
12. S. F. Cheatham, M. Kline, R. R. Sasaki, K. A. Blinov, M. E. Elyashberg and S. G. Molodtsov, *Magn. Reson. Chem.*, 2010, **48**, 571.
13. G. E. Martin, *Ann. Rep. in NMR Spec.*, 2011, **74**, 215.
14. M. E. Elyashberg, K. A. Blinov, S. G. Molodtsov, A. J. Williams and G. E. Martin, *J. Chem. Inf. Model.*, 2007, **47**, 1053.
15. M. E. Elyashberg, K. A. Blinov, S. G. Molodtsov, A. J. Williams and G. E. Martin, *J. Chem. Inf. Comput. Sci.*, 2004, **44**, 771.

# Approaches to Algorithmic Structure Elucidation

## 9.1 Introduction

Two generations of the StrucEluc expert system (ES) have been described previously. The first generation system, StrucEluc-1,[1] was developed for the structure elucidation of organic molecules from 1D $^{13}$C NMR spectra and was found to be successful for the elucidation of molecules in the mass range of 150–300 amu and containing up to 20–25 skeletal atoms (see Section 5.4). In this Chapter we will consider the second generation StrucEluc system that is capable of elucidating the chemical structure of much larger molecules, up to a mass of 1500 amu to date and containing more than 100 skeletal atoms. Typically, this task is mainly accomplished by the analysis of 2D NMR spectral data. In general, the system has been designed to elucidate structures containing up to 250 skeletal atoms. The capabilities of the StrucEluc-2 system in terms of general utility as a tool for the structure elucidation of complex molecules, especially natural products, has been demonstrated in many publications (see review[2]).

It should be emphasized that the large series of problems described in the first publication[3] regarding StrucEluc-2 were solved using only heteronuclear (HMQC, HMBC or COLOC) and homonuclear (H-H COSY) 2D NMR correlations and without using any additional structural information. In this mode of operation, referred to as the *common mode*, the system creates connectivities from the spectral data and generates all possible structures in accordance with the default settings for the number of intervening bonds between corresponding skeletal atoms and with atom properties, including the state of hybridization and the possibility of neighboring heteroatoms taken into account.

Further investigations challenged StrucEluc-2 with problems that could not be solved or proved to be very time consuming due to a lack of information in the 2D NMR data (see Section 9.5). In these cases it proved necessary to

New Developments in NMR No. 1
Contemporary Computer-Assisted Approaches to Molecular Structure Elucidation
By Mikhail Elyashberg, Antony Williams and Kirill Blinov
© Royal Society of Chemistry 2012
Published by the Royal Society of Chemistry, www.rsc.org

introduce additional structural information, if available, to facilitate the elucidation process. In the real world, it is common for a chemist or spectroscopist faced with elucidating a structure to have prior knowledge of reaction components in a synthesis, knowledge of the class of compounds that may have been isolated or even hypothetical structures for validation rather than full elucidation from no information.

The hypothesis that the utilization of molecular fragments found from the system knowledge base or potential substructures proposed by the chemist would be helpful to circumvent the difficulties was tested. Such a *fragment approach* has been used in a number of first generation expert systems (see Section 5.2). It is based on correlation tables containing substructures and their associated characteristic intervals for specific spectral features, *e.g.* chemical shift variation (for example, see references[4–6]). In contrast, systems including both SpecSolv[7] and StrucEluc-1[1] employ databases containing substructures and their associated $^{13}$C NMR subspectra. At present, the StrucEluc database contains over 405 000 chemical structures and more than two million substructures. The database continues to grow as further literature data are added. The value of including substructures directly into the elucidation process is that a fragment, considered as a macroatom, can absorb a significant number of the skeletal atoms and leads to a reduction in the complexity of the problem. This results in acceleration of the structure generation process, which is typically the most time-consuming stage of the structure elucidation process.

Nevertheless, even in those cases case when 2D NMR data are employed, the utilization of molecular fragments is hampered by the fact that *all* of the carbon atoms existing in a fragment utilized in solving the problem *must* be supplied with chemical shifts. Moreover, the values of these chemical shifts must be as close as possible to the observed values for the atoms of the corresponding fragments in the experimental $^{13}$C NMR spectrum of the unknown compound under study. In addition, the accommodation of one or more fragments within a set of connectivities derived from the 2D NMR data is a problem that requires the development of new algorithms. To our knowledge, no attempts have been made to investigate the possibility of using molecular fragments for structure generation from 2D NMR connectivities.

In this Chapter we will discuss different strategies for applying the StrucEluc system. Special attention will be focused on methods that allow for the utilization of fragments stored in both the $^{13}$C NMR database [found fragments (FFs)] and those introduced by the user [user fragments (UF)] in combination with 2D NMR data. In order to explain the appropriate strategy and the methodology for the utilization of fragments, a series of challenging tasks related to the chemistry of natural products will be considered.

## 9.2 Functional Scheme of the System

The flowchart represented in Scheme 9.1 provides an overview of the system. As shown, the system consists of three interrelated branches, each having access to the system database.

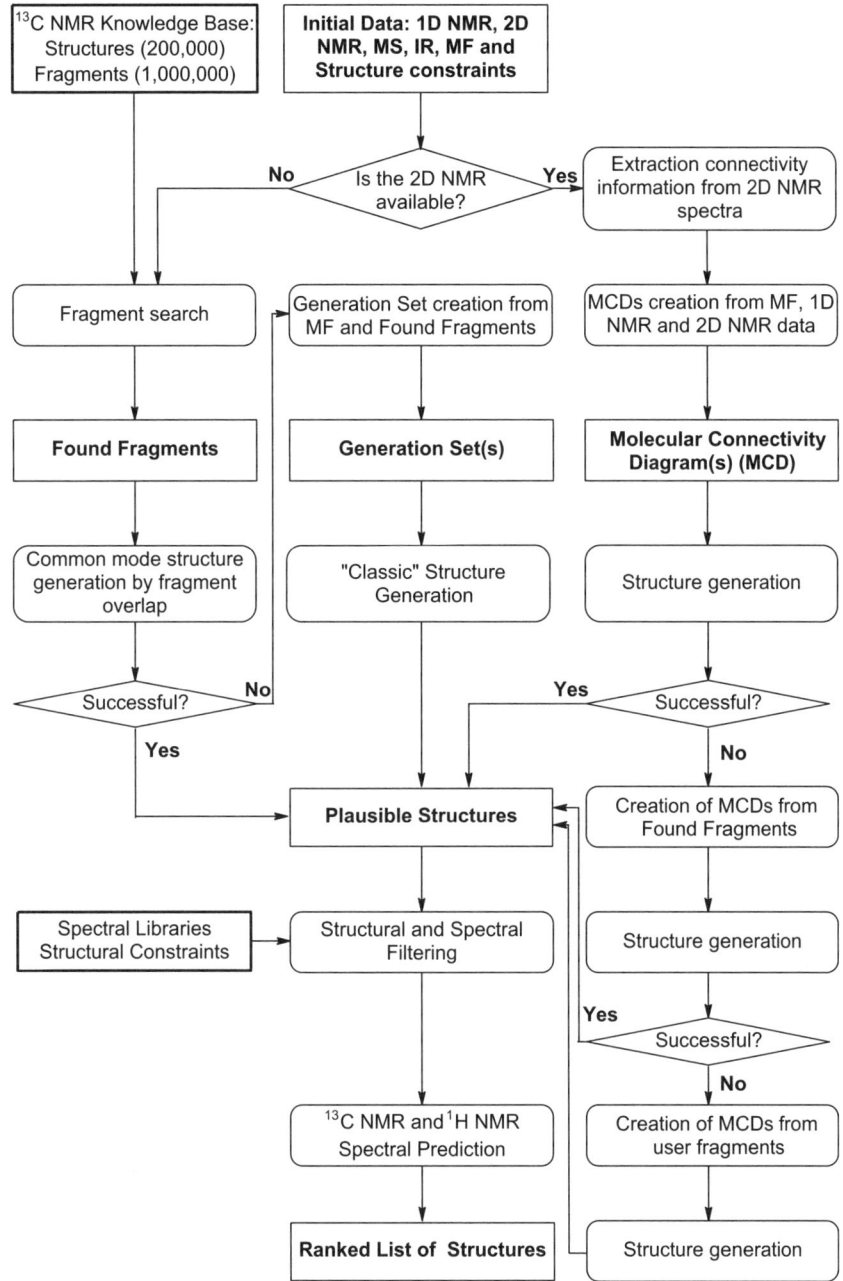

**Scheme 9.1**   General scheme of the StrucEluc system.

Depending on the initial data available, and the complexity of the molecule being analyzed, the system offers a wide range of methods for solving a problem. Although a 2D NMR-based method of structure elucidation is the main approach utilized, we will outline briefly the operation of all branches of the scheme to show the flexibility of the system.

## 9.2.1   System Operation with 1D NMR Spectra

### 9.2.1.1   A Fragment Superposition Mode

In this mode, first described in reference,[7] a $^{13}$C NMR spectrum is used as the main experimental data input for the elucidation process. The chemical shifts, multiplicity, and intensity (number of carbon atoms) of signals are specified. Quite frequently, an experimental $^{13}$C NMR spectrum is sufficient to elucidate the structure of a molecule of up to 20–25 skeletal atoms. However, IR and $^1$H NMR spectra, if available as supplementary data, can certainly assist with the solution of the problem as explained later. In principle, the program is capable of working *without* either a molecular mass (M) or molecular formula (MF).

The program starts with the search of the entire $^{13}$C NMR spectrum in the knowledge base (KB). If the spectrum is found in the database, then the corresponding structure is displayed; otherwise a search of the fragment library (FL) is carried out. As a result of the search in the fragment library, the program displays $L$ fragments that have corresponding subspectra that do not contradict the experimental spectrum. The retrieved fragments are then ranked in order of decreasing number of carbon atoms. If $^1$H NMR and IR spectra are available, the program can filter the selected substructures with the help of the spectral filter library, which can noticeably reduce the number of selected fragments and move the "good" fragments to the beginning of the list. The possibility of merging the fragments by overlapping common atoms is checked with the intent of assembling these fragments into a final structure. After the merging of fragments, the $^{13}$C NMR subspectra of the integrated fragment are predicted to help identify whether the fragments have been merged correctly.

If the result of checking all of the possible fragment combinations is the detection of a structure having no free bonds, then $^{13}$C NMR spectrum prediction and signal assignment are performed. Each structure is then verified in accordance with all available spectral data, general rules of structural chemistry, and any constraints imposed by the chemist. If several structures are generated then they are ranked in order of increasing $d$ value which is calculated as the sum of deviations found for each $^{13}$C NMR signal divided by the total number of signals in the spectrum. If an attempt to assemble a structure fails, then the program automatically switches to the "*classical*" mode of the expert system that requires the availability of the molecular formula. The X-PERT[4] system serves as an example of this "classical" approach.

### 9.2.1.2   The "Classical" Approach

The "classical" approach for the StrucEluc system starts with a molecular formula and the formation of sets of fragments. The methods have been described previously.[8] The first *l basis* fragments, ($l = 10$–$30$), $l \leq L$, are chosen from the list of selected fragments. They will be the largest fragments.

The subspectrum of the first basis fragment is compared with the experimental spectrum. The chemical shifts that were not originally identified in the experimental spectrum are associated. The first *basis* fragment is then combined with combinations of smaller fragments to provide complete spectral interpretation. The sets from the second, third, *etc.* basis fragments are similarly formed. Each set is then checked for correspondence with the suggested molecular formula. A chemist may add sets of fragments and apply appropriate restrictions common to most "classical" expert systems: GOODLIST, BADLIST, minimum and maximum ring cycle sizes, multiplicities of bonds, *etc.* (see Section 1.8.3). The program is capable of displaying a "generalized portrait" of the analyzed molecule. This overview is a distribution of functional groups included in the typical functional group library (TFGL) by the frequency of their occurrence in the fragments selected from the fragment library. Groups that are completely absent and those that occur most frequently are correspondingly used to automatically form the BADLIST and GOODLIST.

All of the generated structures are checked for their correspondence with the available spectra, as well as with the rules of organic chemistry and stereochemistry. Finally, to choose the most probable structure, $^{13}$C NMR spectra of candidate structures are predicted and the structures are ranked in order of increasing average deviations of the predicted spectra from the experimental data. The method that finds the most probable structure is described below in Section 9.2.2.4. It has been shown that the system operating in this 1D NMR mode can generally elucidate structures of medium size, up to 20–25 skeletal atoms.

## 9.2.2 Molecular Structure Elucidation from 2D NMR Spectra

### 9.2.2.1 The "Common" 2D NMR Mode

If an analyzed molecule is large and is related to newly isolated materials, then it is probable that both the fragment superposition and the "classical" approaches will fail to elucidate a final structure. Experience has shown that the reliable elucidation of the structures of large molecules (containing > 20–25 skeletal atoms) is generally impossible without employing 2D NMR data.

The next module of the StrucEluc system is based on a number of programs developed for elucidating a molecular structure from a combination of 2D NMR spectra. The most typical combination providing the basis for structure determination includes H-H COSY, HSQC/HMQC, and HMBC. The StrucEluc system presently operates with the following 2D NMR methods: H-H COSY, HSQC/HMQC, HMBC, ROESY, NOESY, ADEQUATE and INADEQUATE. Other methods can also be used by the system through a flexible procedure that allows input and processing of experimental 2D NMR data. In addition to spectral data the program also needs the molecular formula to proceed in this mode.

After checking the molecular connectivity diagram (MCD) for the presence of contradictions, the data are collected in connectivity tables and graphically presented as an MCD are used as the input information for the 2D structure generator. Structures are generated under constraints determined from the molecular formula, the MCD diagram and any additional constraints which may be introduced by the chemist. The structure generator is based on mathematical algorithms developed by Molodtsov.[9–11] Generated structures are verified using the approaches discussed previously: filtering with LSC, GOODLIST, BADLIST, *etc.* [13]C NMR spectrum prediction is performed for all structures included in the output file and the structures are ranked in order of increasing *d* value using the method described in detail in Section 9.2.2.4.

The system can also be used for verification of the proposed structure and the associated [13]C and [1]H NMR signal assignments by validation of all available 2D NMR spectra. If any contradictions are found, for instance, if the distance between a pair of carbon atoms in a proposed structure is greater than that postulated from a particular connectivity, the program displays a textual message detailing the probable cause(s) of the conflict(s) and showing the connectivities in graphical form.

## 9.2.2.2 Application of Fragments in Combination with 2D NMR Data

Computer-assisted structure elucidation using 2D NMR data is quite efficient for the elucidation of structures of complex natural compounds.[3] However, if the structural restrictions imposed by the MCD are not sufficient for the generation of a reasonable number of possible structures within an appropriate time, it is to be expected that the utilization of molecular fragments can help facilitate the solving of the problem. Commonly, appropriate fragments to aid in the solution of a problem can be found in the knowledge base. The main advantage of these fragments is that all fragment carbon atoms are supplied with the [13]C NMR assignments obtained from the full structures that were used for creation of the fragment database.

The first step of the process is a fragment search of the knowledge base. As a result, a set of *L* FFs is selected. The next step is the creation of the MCDs using the FFs. For this purpose, either all of the FFs, or any selected number of them, can be incorporated by the operator. The main idea of the algorithm that implements this procedure is as follows. Prior to creation of the MCD, the chemist defines the number of fragments, $l$ ($l \leq L$), that will be used in this procedure and sets an error, $E$, that defines the maximum difference allowed between the chemical shifts of the fragment carbons and the corresponding values observed in the experimental spectrum under study. It is important to note that both parameters, $l$ and $E$, are closely interrelated and choosing the most efficient values may be a matter of trial and error.

The [13]C NMR subspectrum of each fragment is compared with all experimental chemical shifts. The number of hydrogen atoms attached to a carbon atom is taken into account during this process. Consider a fragment contains $f$ carbon atoms and an arbitrary atom $C_i$ of the fragment has a chemical shift $\delta_i$ ($i = 1/f$) and multiplicity $m_i$. Suppose that the experimental chemical shifts $\delta_{i1}$, $\delta_{i2}, \ldots \delta_{iq}, \ldots \delta_{ip}$ meet the conditions $|\delta_i - \delta_{iq}| \leq E$ and $m_i = m_{iq}$. Then, all possibilities of substituting the $\delta_i$ values for the experimental values $\delta_{i1}$, $\delta_{i2}, \ldots \delta_{iq}, \ldots \delta_{ip}$ must be verified.

If the conditions $|\delta_i - \delta_{iq}| \leq E$ and $m_i = m_{iq}$ hold for all $f$ carbon atoms, then the given fragment is recognized as a candidate for inclusion in the process of creating the MCD. If this condition does not hold then the fragment is excluded from consideration. It is possible that one experimental chemical shift $\delta_{iq}$ can also substitute chemical shifts assigned to several carbon atoms within the fragment. All rearrangements of the experimental chemical shifts within the corresponding carbon atoms of the fragment should be considered. The chemical shift distribution of each carbon that produces a conceivable assignment of a given fragment carbon atoms has to be verified. During the verification process, the program checks whether the carbon atom assignments correspond to the experimental chemical shift correlations comprising the skeletal atoms making up the fragment. The fragments that survive the test are then included in the set of *prospective* fragments.

The more skeletal atoms that are "absorbed" by the fragments the shorter is the process of structure elucidation. With this in mind, an algorithm that combines the prospective fragments within one MCD was developed. To realize this procedure, all possible combinations of prospective fragments are searched and only combinations that are in agreement with the experimental 2D NMR correlations are chosen. The fragment combinations that pass this examination form a set of prospective fragment combinations. These fragments are then "projected" on to the MCDs together with any remaining free atoms. The user can then visually analyze these MCD diagrams.

The total number of MCDs, $n_{MCD}$, depends on the following parameters which are defined by the user: $l$, number of FFs which will be used for the creation of MCDs ($l \leq L$); $n_f$, the minimal number of fragments that must be present in each MCD; $q$, the minimum percentage of all skeletal atoms that must be absorbed by the fragments present in each MCD.

As noted above, in the general case, the more atoms from the fragments consumed by the MCD, the greater the likelihood that the process of structure generation from a given MCD will be more time efficient.

The speed of structure generation depends on the size of the molecular fragments. If the number of small fragments composing the MCD is large enough then this will speed up the generation. Structure generation is much faster when the MCD is comprised of a small number of big fragments. Depending on the size of the molecule being analyzed and the size of fragments placed at the beginning of the ranked list of FFs, the $n_f$ value is usually defined as a number from 1 to 4. The most efficient results are obtained if $q$ is significant, generally 40–60%.

The conclusion of all further verification procedures is a check of all produced MCDs for contradictions. The program offers an option that deletes all

MCDs that are recognized as contradictory. The diagrams remaining after checking can be used in the structure generation process. The user has the opportunity to omit the connectivity verification because contradictory MCDs will be detected anyway and rejected in the process of structure generation. Moreover, for the process of structure generation the user can select one or more MCDs that are attractive to the user who may have prior knowledge of a particular structure class or target structure. To alleviate having to choose a preferable MCD they are automatically ranked in order of the increasing number of free carbons. In this way it is possible to select a series of appropriate MCDs, starting from that ranked first.

The number of MCDs produced from a given set of fragments can be rather large (sometimes the $n_{MCD}$ value is greater than 1000). To provide the possibility to edit a big set of MCDs we elaborated a special procedure which allows one to transfer all changes made in one MCD to all MCDs in the set. In particular, it is possible to specify options which transfer atom coordinates, zoom factor, atom properties, manually drawn connectivities, and connectivities automatically modified during the MCD checking for the presence of contradictions.[12] This procedure essentially alleviates using *a priori* information in the fragment mode of structure elucidation.

In the process of analyzing a novel compound it is entirely possible that there will be no fragments in the database that will reduce the magnitude of the challenge. It is natural in such cases to expect that the introduction of user-defined fragments may help to form the MCDs. The main qualitative difference between a FF, and a UF, is that the FF already contains carbon atoms with assigned chemical shifts, whereas the carbon atoms of the UF have no carbon chemical shift assignments. Two ways have been suggested to introduce UFs into the program:

a. Calculate the carbon chemical shifts of the fragment using the "*accurate*" (HOSE code-based) method (see Section 9.2.2.4).
b. Search the knowledge base for fragments that *comprise* the UF.

It is likely that fragments from at least one of the two sources would be available for use by the program.

### 9.2.2.3 Choice of E Value

In the process of MCD creation from fragments, the $E$ value is of great importance since it markedly influences the result of applying the fragments. There are a number of principles governing the selection of the $E$ value. As a rule, the smaller the value of $E$ then the smaller the number of MCDs, $n_{MCD}$, created from FFs. The advantage of a small number of MCDs is, of course, that it can reduce the time for structure generation, $t_g$. At the same time, $t_g$ is also a function of the *fragment dimensions*. Larger fragments generally shorten the structure generation process. However, if a fragment is large and correspondingly contains many assigned carbon atoms, then as a consequence it is not as likely that *all* carbon atoms, especially the terminal ones, of a large fragment will fit the experimental shifts, thereby satisfying a narrow interval for $\pm E$. The program automatically sets the $E$ value for terminal atoms equal to 12 ppm to account for this issue.

Large fragments are the most useful, but to utilize them in the structure elucidation process a large $E$ value is necessary. A large $E$ value can correspondingly increase the $n_{MCD}$ value. The optimal approach would be to set a large enough $E$ value and select only MCDs containing large fragments for the structure generation process. This principle therefore justifies manual (user) or automatic rejection of MCDs containing small fragments. With testing it has been shown that the optimal program parameter controlling the minimum number of carbon atoms in the fragment used for creation of the MCD should be set to a value of 5. Unfortunately, it is impossible to determine an optimal $E$ that is valid for a diverse range of problems. The value of $E$ should be optimized for each task by gradually increasing the $E$ value starting from 0.5 ppm.

There is one more situation that requires relatively large $E$ values. Experience has shown (see Section 9.4.1.2) that the program sometimes accepts incorrect fragments as legitimate when the $E$ value is small. Obviously, if the magnitude of the error is sufficiently large, the program will select fragments of increased structural diversity.

## 9.2.2.4 Selection of Preferable Structure

The StrucEluc system provides a four-step procedure for identifying the most probable structure in the output file.

- First step. $^{13}$C NMR spectra are predicted for all generated structures using an incremental method, the so-called "fast" method, and $d_I$ values, the average deviation of an experimental $^{13}$C NMR spectrum (or the chemical shifts derived from corresponding projections of 2D NMR spectra) versus predicted chemical shifts, are calculated.
- Second step. Duplicate structures in the output file are deleted. Among the generated structures there are usually duplicates that differ from each other only in terms of the assignment of the chemical shifts to different carbon atoms. If this possibility is not appropriately considered when deleting isomorphic structures, then the structure with the correct assignment of the chemical shifts could conceivably be the deleted isomorphic structure. To avoid this eventuality, the system executes a special procedure for duplicate removal. For each duplicate family only the structure that has the minimum $d_I$ value is retained in the file as "the best representative" of the family. After duplicates are removed, the structures are then ranked by the $d_I$ value and sorted in ascending order. The smallest $d_I$ value indicates the best match between the experimental and calculated spectra and this structure will therefore be the first in the list. Experience shows that the fast calculation of $^{13}$C NMR spectra and

their subsequent ranking usually places the correct structure as the first or second in the list. Only in rare instances will the correct structure be listed below fifth place. Such a preliminary ranking of the big resulting files can help to reject hundreds and thousands of structures that are known to be unsuitable.

- Third step. $^{13}$C chemical shift prediction is carried out using a neural network (NN) algorithm and the structural file is ranked again with $d_N$ deviations. If the resulting file is extremely large, the calculations can be applied only to the first several thousand structures. As a result of this step the preferable structure is selected with greater reliability.

- Fourth step. During the fourth stage, $^{13}$C NMR spectra are calculated for the first 10 to 50 structures of the ranked file using a fragmental method based on the HOSE code approach.[13] The average deviation values between the experimental and calculated values ($d_A$) are found and the structures are again rank-ordered. Subsequent ranking increases the probability of moving the correct structure to the first position in the list. For additional control over the correct choice of the output structure, the HOSE code-based proton chemical shifts can be predicted and displayed together with the corresponding deviation value $d_H$. If the experimental $^1$H NMR spectrum is not recorded, the chemical shifts are automatically found from 2D NMR projections. For proton NMR prediction, the predicted proton–proton couplings are enhanced by 3D optimization of the structure. The position of the correct structure in the file determines its rank depending on the type of ranking parameter, *i.e.* $d_N$, $d_A$, $d_I$, or $d_H$ correspondingly. The "rates" of the correct structure in the ranked file are denoted as $r_N$, $r_A$, $r_I$, and $r_H$. If the correct structure is the first in the list ranked by $d_A$ values, then $r_A = 1$. As a rule, the final structural ranking is carried out according to increasing $d_A$ and $d_N$ values, whereas magnitudes of the $d_I$ and $d_H$ parameters serve as secondary aids for estimating reliability of the correct structure selection.

When the structures at the beginning of the ranked file possess different structural elements, prediction of the MS match factor ($m_i$, where $i$ is the position of a structure in the ranked file) may also be useful for the confirmation of the preferable structure. The system utilizes a routine that is capable of calculating the percentage of peaks in the experimental MS spectrum that can be interpreted on the basis of a given structure. The calculation of the MS match factor is relatively time consuming, so it is used only in cases when the difference $\Delta_{(2-1)} = d_A(2) - d_A(1)$ is small. Here $d_A(1)$ and $d_A(2)$ represent the deviations corresponding to the first and second structures in the ranked file.

In ambiguous cases, it may be useful to display the calculated $^1$H NMR spectra because of the complexity of some multiplet patterns. Also, to facilitate structure analysis in the output file, the StrucEluc system is supplied with a feature that calculates structural similarity coefficients.[14] In this way if the investigator has an idea of the class of structure under investigation he or she can use this structure as an input to allow rank ordering relative to the structural similarity of the results file. To demonstrate the main features of the approaches suggested above a number of examples will be cited.

## 9.3 How Does the System Work?

To demonstrate all of the various methods for structure elucidation currently available in StrucEluc we consider the elucidation of a chemical structure from the 1D and 2D NMR spectra of a rather simple molecule, 2,4,4'-trihydroxydihydrochalcone (**9.1**), isolated and characterized by Gonzales and co-authors.[15] A molecular formula of $C_{15}H_{14}O_4$ was deduced from the high-resolution MS spectrum. The NMR spectral data ($^1$H and $^{13}$C NMR, DEPT, HMBC) obtained by Gonzales *et al.*[15] and used here are represented in Table 9.1.

For the explanations of the analysis to be clear, the target structures of the examples described in this Chapter will be displayed, if necessary, with already assigned carbon chemical shifts.

**9.1**

**Table 9.1**  NMR data for compound **9.1** in pyridine-d$^5$.

| Atom number | $\delta_C$, multiplicity | $\delta_H$ | HMBC |
|---|---|---|---|
| 1 | 119.68, s | | 7.02, 6.79 |
| 2 | 157.82, s | | 7.30 |
| 3 | 104.22, d | 7.02 | 6.79 |
| 4 | 158.66, s | | 7.30 |
| 5 | 107.02, d | 6.79 | 7.02 |
| 6 | 131.42, d | 7.30 | |
| 7 | 129.59, s | | 7.16 |
| 8 | 131.27, d | 8.17 | |
| 9 | 116.14, d | 7.16 | |
| 10 | 163.54, s | | 8.17 |
| 11 | 116.14, d | 7.16 | |
| 12 | 131.27, d | 8.17 | |
| 13 | 36.68, t | 3.54 | 3.43 |
| 14 | 26.12, t | 3.43 | 7.30, 3.54 |
| 15 | 198.80, s | | 8.17, 3.43 |

## 9.3.1    Utilization of 1D NMR Spectra

### 9.3.1.1    Atom Overlapping Mod

When the "Search Fragments by CNMR Spectrum" command was performed, the program picked out $L = 180$ fragments and ranked them according to decreasing molecular formula. The first 20 fragments from the list are shown in Figure 9.1. Note that all carbon atoms of the fragments stored in the

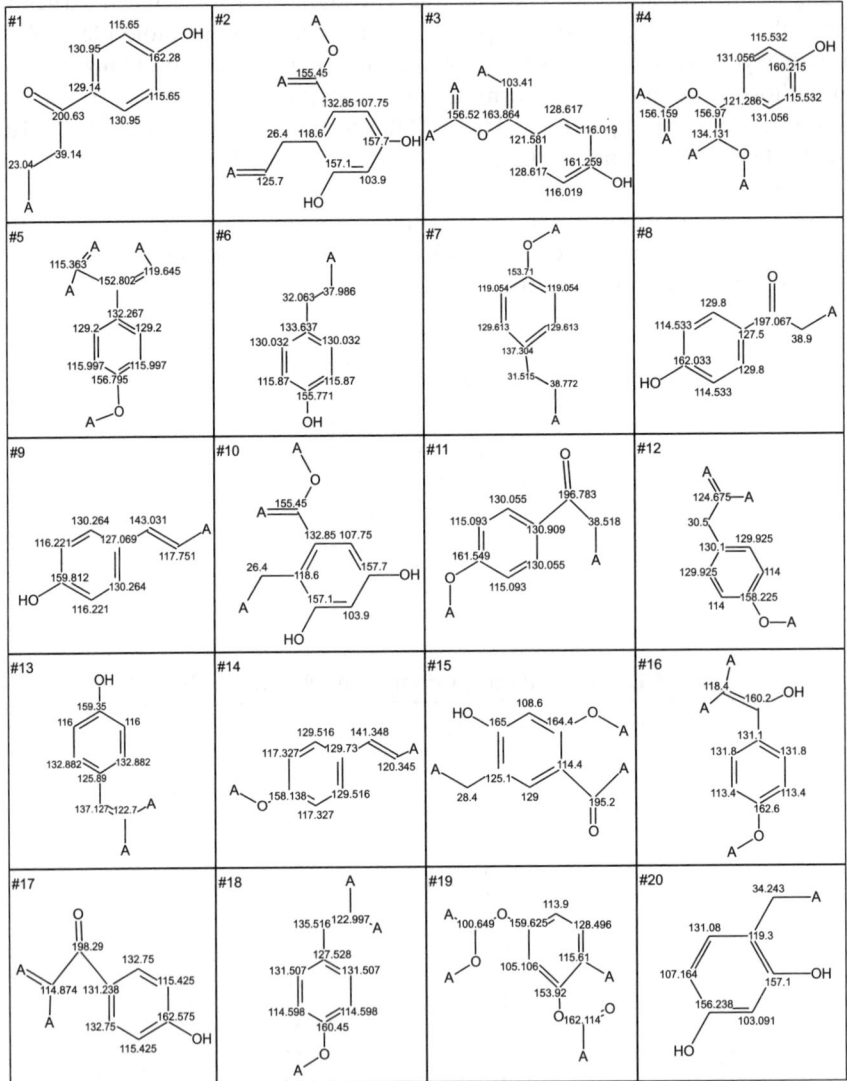

**Figure 9.1**    The first 20 fragments obtained in the FF list during the elucidation of compound **9.1**.

database are supplied with $^{13}$C NMR assignments taken from the full structures that were used to create the fragment database (see Chapter 7).

The comparison of fragments with structure **9.1** shows that fragments #1 and #20 exist in the molecule under investigation and have one overlapping group, which is a $CH_2$. Atom overlapping structure generation (AOSG) accompanied by structure filtering with spectral libraries was performed. This is the application of the first system branch. The program combined the FFs using at least one *overlapping* atom. As a result, the program produced only one structure coinciding with compound **9.1** with a structure generation time of $t_g = 45$ s (all calculations were performed on a PC Celeron operating at 500 MHz, Windows 98, RAM 128 Mb). A single structure was generated in 4 s, whereas the remaining time was consumed by attempts to assemble other conceivable structures from the first 30 fragments, this number being the default for the assembly process. A single correct structure was obtained from the 1D $^{13}$C NMR spectrum in fully automatic mode.

### 9.3.1.2 *"Classical" Mode*

An attempt was made to apply the second system branch to the data utilizing the algorithms in the so-called "classical" generation mode based on combining *non-overlapping* fragments and free atoms. According to the methodology described previously (see Chapter 5) the program automatically selected the first 20 FFs ($l = 20$ is the default setting in this case) from the list, and then created 223 fragment sets. Classical structure generation was performed and the spectral libraries and internal BADLIST were used for structure filtering during the generation process. As a result, the number of structures generated, $k$, gave 112 resultant molecules, while the generation time, $t_g$, was 1 min 30 s. After the removal of identical structures, the number of structures was reduced to 73. This is denoted by the representation of $k = 112 \rightarrow 73$ and will be used throughout this book. The $^{13}$C NMR prediction and structure ranking allowed the program to reliably distinguish the correct structure as the first structure. The difference between the first and second structure is given by the difference between $d_A(1) = 1.64$ ppm and $d_A(2) = 3.98$ ppm which gives $\Delta_{(2-1)} = d_A(2) - d_A(1) = 2.34$ ppm. Further details regarding the process of evaluating the structure reliability will be discussed in Section 9.9.

## 9.3.2 Application of 2D NMR Data Analysis

### 9.3.2.1 The "Common" 2D NMR Mode

The possibility of elucidating an unknown structure from the 2D NMR data in an automated or semi-automated manner is certainly attractive. In this case, the user would not need to make any assumptions that would serve as constraints for the system. However, to demonstrate the role of different structural constraints two common modes of solving a given problem will be

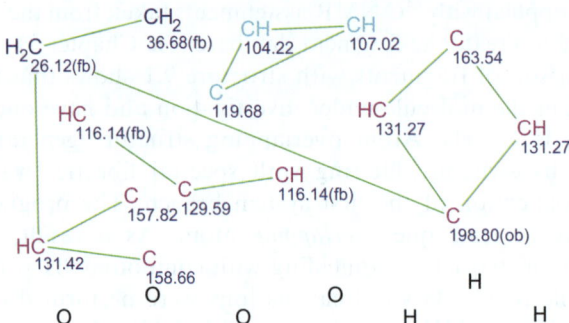

**Figure 9.2**   MCD created for compound **9.1**. The following colors are used for
                marking the atom hybridization: not defined, pale blue; sp$^2$, violet; *sp* $^3$,
                blue. Obligatory neighboring with a heteroatom is noted as (ob), and
                prohibition of neighboring as (fb).

considered: without user intervention and with the influence of the user by
including additional information.


**9.3.2.1.1   Solution Without any User Assumptions.** As described earlier, the
program automatically created and displayed a MCD. The MCD reflecting
the HMBC data from Table 9.1 is shown in Figure 9.2.

   The atom properties were automatically determined for all carbon atoms by
the program using the atom property correlation table (APCT) (see section
9.2.2.1). Note that the hybridization of carbon atoms C(104.22), C(107.02), and
C(119.68) colored in pale blue was not specified because these atoms can exist
either in *sp²* or *sp³* states (the possibility of O–C–O fragment was taken into
account). For other carbon atoms in the benzene rings the parameters of
hybridization were automatically set to *sp²*, whereas for the carbon with a
chemical shift of 198.8 ppm atom the properties were set to *sp²/at least one*. As
a result, the number of structures generated and filtered was $k = 64864 \rightarrow 8 \rightarrow$
$4 \rightarrow 1$, with $t_g = 2$ min 20 s. Four identical structures had differing chemical shift
assignments, however, [13]C chemical shift prediction allowed the program to
select the best (and correct) structure with a minimum deviation ($d_A = 1.14$
ppm). The chemical shift assignments suggested by the program coincided with
the assignments performed by the authors.[15] The relatively large $t_g$ value for
this simple molecule is explained by utilization of HMBC data only (a COSY
spectrum was not cited by the authors[15]), as well as by the presence of atoms for
which hybridization was not specified.


**9.3.2.1.2   Imposing Additional Constraints.** One manner by which to formulate additional structural restrictions is to search for functional groups in
the fragments selected from the database. This may enable the formation of

a "generalized portrait" of the molecule (see Chapter 5). When the command "Search Functional Groups" is applied, the program "sifts" the typical functional group library described earlier through the structures of the $L$ FFs. As a result, both the GOODLIST and the BADLIST can be created manually or in an automated fashion, and the functional groups are available for viewing on the screen along with numbers characterizing the probability of their presence in the analyzed molecule.

In this example the following oxygen-containing groups were created in the BADLIST:

$$\text{O—CH}_2\text{—O} \qquad \underset{\underset{\displaystyle O}{|}}{\overset{\overset{\displaystyle C}{|}}{\text{C—C—O}}} \qquad \underset{\underset{\displaystyle O}{|}}{\overset{\overset{\displaystyle O}{|}}{\text{C—C—O}}}$$

This indicates that it is highly unlikely that carbon atoms having the $sp^3/(at\ least\ two)$ property (*i.e.* O–C–O) are present in the molecule which is under consideration. In addition, the presence of three hydroxyl groups and one carbonyl group are evident from the MCD. With this in mind for the second run, the following additional constraints were imposed: the properties of carbon atoms having chemical shifts between 104.22 and 131.42 ppm were replaced with ($sp^2/forbidden$). For the three carbon atoms with chemical shifts observed between 157.82 and 163.54 ppm the properties were set to $sp^2/at\ least\ one$ and the following result was obtained: $k = 80 \rightarrow 32 \rightarrow 4 \rightarrow 1$, $t_g = 0.2$ s. When these additional assumptions were employed $t_g$ decreased from 2 min 20 s to 0.2 s, a change of a factor of 700.

As was mentioned earlier (see Chaper 2), the alleviation of the difficulty of solving the problem by increasing user intervention suffers from a shortcoming. If at least one of the user-defined constraints ("axiom") is erroneous, the solution obtained will be incorrect and the structural output file will not contain the right molecule. Even with this shortcoming, the investigator's knowledge, experience, and intuition should be used as much as possible to facilitate the elucidation process.

As the structure elucidation process belongs to the class of *inverse problems* (see Section 1.5), the chance of fully replacing human intellect is unlikely at best. Moreover, in accordance with the Bohr principle of complementarity, the methodology of computer-assisted structure elucidation includes two major elements that complement each other. They are the deterministic logic of the computer and the knowledge and intuition of the investigator. The interaction of these elements in the process of solving the problem is what gives rise to the synergistic effect that allows the elucidation of complex molecules. It is therefore necessary to find a rational way of combining connectivities deduced from the experimental 2D NMR data with additional information from a scientist to obtain a solution to the problem in a reasonable time. Certainly, employing the fragments from the database can be used to minimize user intervention and, consequently, the risk of obtaining an incorrect solution.

## 9.3.2.2 Utilization of Fragments Presented in the Knowledge Base

Consider again the process of structure elucidation of the 2,4,4'-trihydroxy-dihydrochalcone molecule using the HMBC spectrum and fragments stored in the database. As mentioned above, $L = 180$ and the first 20 fragments in the list are shown in Figure 9.3. In this example the following options for creation of MCDs were specified: $l = L = 182$; $E = 0.5$ ppm; $n_f = 2$; $q = 50\%$. The $n_{MCD}$ value was equal to 6. To illustrate the mechanism of fragment combination, Figure 9.3 displays one of the created MCDs and the target structure.

Two "good" fragments existing in the fragment list form a combination that absorbs all skeletal atoms of the molecule. This is a fortuitous situation. After creation of the MCDs, the "Check MCDs" command was executed and indicated that all MCDs are consistent. It should be noted that the right structure with the correct chemical shift assignment can be generated from a combination of these fragments only if the carbon chemical shift assignment of the fragments corresponds to the assignment inherent to the genuine structure, because connectivities are defined for experimental chemical shifts. The result is that the elucidated structure has carbon assignments that coincide with those intrinsic to the molecule under investigation. Structure generation was performed from all six MCDs. During this process no constraints were entered for the atom properties. The single and correct structure was generated in less than 1 s. As shown the application of fragments drastically shortened the generation time. Compared to the first run

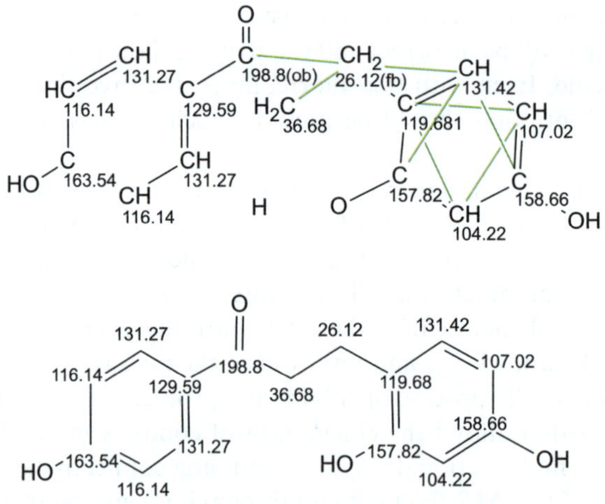

**Figure 9.3** One of the MCDs created from the FFs obtained for compound **9.1**. Structure **9.1** is shown below the MCD to help in understanding the MCD.

performed in the common mode, the reduction in time is about three orders of magnitude. Obviously, in this example, the application of FFs indeed obviated any user intervention in the solution process and provided a dramatic reduction in the processing time.


## 9.4   Examples of Structure Elucidation in the Common Mode

Ge *et al.*[16] isolated and determined the structure of a new unusual natural product named hopeanolin. To challenge StrucEluc we used the published 1D and 2D NMR data[16] to elucidate the "unknown" structure. Eighty HMBC and three COSY correlations were input into the program and the MCD was created. Atom hybridization was automatically set for all carbons except eight CH atoms and two quaternary C atoms with chemical shifts in the range 90–120 ppm. In so doing, the program took into account that chemical shifts observed in this region can be assigned either to C=C or to O–C–O carbons. A single obvious constraint was imposed: three $sp^2$ carbon atoms with chemical shifts between 170 and 180 ppm were marked as having at least one neighboring oxygen each. No non-standard correlations (NSCs) were detected in the 2D NMR data by checking the MCD. The results of the structure generation and filtering were: 68 structures were generated in 12 s, and nine structures were stored after filtering ($k = 68 \rightarrow 9$, $t_g = 12$ s). $^{13}$C NMR chemical shifts were predicted for all structures using all of the fragment-based, incremental and NN approaches. The four structures at the top of the ranked structural file are shown in Figure 9.4, in which the "best" structure is hopeanolin.

Shigemori and co-workers[17] isolated two new diterpenoids containing an eight-membered lactone ring. The structures were elucidated utilizing a combination of spectral data. In their work, the authors[17] present a table

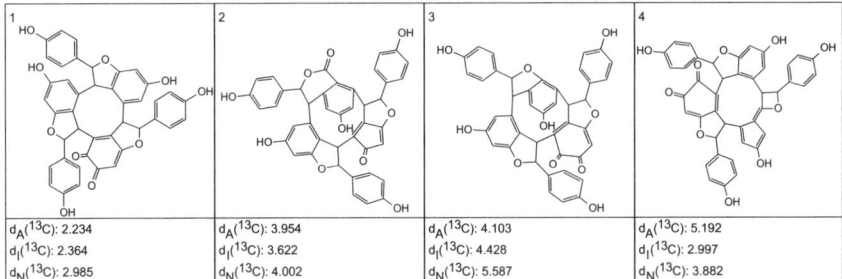

**Figure 9.4**   The four structures at the top of the ranked structural file. The first ranked structure is identical to the structure of hopeanolin as determined by the authors.[16]

containing $^1$H, $^{13}$C, HMQC and HMBC spectra of one of the diterpenoids, compound **9.2**, with a molecular formula of $C_{20}H_{30}O_3$:

**9.2**

These data were used to challenge the StrucEluc system during the first stage of its development. Having formed the MCD, the program checked connectivities between the skeletal atoms for the presence of contradictions. The data passed all tests, but when structure generation was initiated the process stopped immediately and no structures were generated. This may be caused by the presence of contradictions that cannot be recognized by the program (see Chapter 2 and the next Chapter) or the presence of chemical shifts which are out of range of the characteristic intervals of the spectral filter libraries (LSC) in the experimental $^{13}$C NMR spectrum. The latter can easily be checked simply by repeating the generation process with the spectral filter switched off. When the process was repeated with the spectral filter switched off, one structure identical to **9.2** was generated that was characterized by the following devia-tions: $d_A = 1.52$, $d_I = 2.32$, $d_N = 2.42$.

Since the StrucEluc system can be trained, it is valuable to determine which spectral features cause outliers to the corresponding ranges. This may be examined by using the filter options that provide three degrees of severity for the characteristic intervals: *tight*, *medium*, and *loose*. The *tight* filter corresponds to an interval width commonly used for the present fragment. The two other options automatically widen the ranges according to specific rules. The earlier versions of the system used the *tight* option by default, which ensured the highest selectivity of the spectral filter. When fast and accurate methods of chemical shift prediction were implemented into the program, it became insensitive to the dimension of the output file and a *loose* option is used by default now, which allows for the preservation of the genuine structure in the result file.

To understand the reason for rejection of the molecule in our example, the resulting structure was filtered initially with the *medium* degree of severity and then the *loose* one. The program reported that the C=C–CO–O fragment was found in the structure and that the conjugated carbonyl group is expected to have a characteristic interval of 160.0–170.0 ppm. In reality, the carbonyl in the experimental spectrum was observed at 171.3 ppm. In this case a slight shift of the C=O signal outside the interval range results in loss of the diterpenoid under study. To prevent the rejection of compounds of this class in future, the corresponding interval was increased to 172 ppm. When the *loose* option was used as a default, no rejection of the true structure by the StrucEluc spectral filter was observed.

As a comparison of structure generation results, the StrucEluc and COCON systems have been compared using materials published by Junker *et al.*[18] They described the structure elucidation of ascomicyne, **9.3**, ($C_{43}H_{69}NO_{12}$) using COSY (51) and HMBC (79) correlations:

**9.3**

In the *common mode*, with the default atom properties set, spectral filtering and application of APCT during structure generation, StrucEluc generates only one correct structure within less than 1 s. When atom hybridizations were set, but checks by heteroatom neighboring and spectral filter were switched off, the results gave $k = 3916 \rightarrow 1926$, $t_g = 11$ s, $r_A = r_I = r_N = 1$, $d_A(1) = 0.92$ ppm, $\Delta_{(2-1)} = 1.10$ ppm. Spectral filtering of the output file reduced the output file markedly: $k = 1926 \rightarrow 321$.

Under similar conditions, the COCON program produced the following results: $k = 350$, $t_g = 48$ min, $r = 2$, $d(1) = 3.51$ ppm, $d(2) = 3.56$ ppm. These results were obtained using an SGI R10000 running at a clock speed of 195 MHz. Under essentially the same conditions, the StrucEluc structure generator runs markedly faster and, in this example, the correct structure selection proved to be more successful ($r = 1$).

# 9.5 Examples of Structure Elucidation in Fragment Mode

## 9.5.1 Azadirachtin

Experimental data for the next example were obtained from a report devoted to the structure elucidation of the insect antifeed azadirachtin (**9.4**).[19] In this publication the elucidation was performed with the assistance of the LSD program.[20] The molecular formula $C_{35}H_{44}O_{16}$ and the 1D $^{13}$C, COSY (15), HMQC and HMBC (66) spectra were used as inputs for StrucEluc.

Initially, an attempt was made to elucidate the structure using the common mode and without any editing of the MCD. The generation process was very time consuming and the program was halted after several hours. Since the molecule is rather large and complicated, containing 51 skeletal atoms, Ley et al.[19] set the properties of all carbon atoms and assumed the presence of four ester and three hydroxyl groups to elucidate the structure with the assistance of the LSD program. When we made the same assumptions and drew the corresponding functionalities in the MCD we obtained the results: $k = 5424 \rightarrow 542 \rightarrow 542$, $t_g = 3.4$ s, and the correct structure was distinguished by its $d_A = 1.95$ ppm value ($r_A = 1$).

**9.4**

The main disadvantage of the solution to this problem using the approach just cited is that it required a great many assumptions to be made and numerous constraints were imposed by the user. This shortcoming can, however, be overcome by using a fragment database search to provide structural core components for StrucEluc to work from.

An initial check indicated that the azadirachtin molecule was absent from the full structure database. A fragment search resulted in a list of fragments with $L = 2885$. The larger substructures are of the most interest for this process and the first 500 fragments were chosen for creation of the MCDs. The minimum value of the chemical shifts error was set to $E = 0.5$ ppm. No further assumptions or constraints were imposed.

With these starting conditions the program created eight MCDs. Examination of the MCDs showed that only two contained the large fragments **9.5** and **9.6** which could influence the process of problem solving.

**9.5**                    **9.6**

All MCDs, except the two selected above, were deleted. These two MCDs passed the confirmatory connectivity check and structure generation was performed without any constraints. The results produced were: $k = 630 \rightarrow 105 \rightarrow 105$, $t_g = 1.2$ s, and the priority of the correct structure was obvious since $r_A = r_I = r_N = 1$. It is notable that when the generation was performed from the MCD containing the largest fragment, **9.5,** only three structures including the correct one were generated in 3 s.

With this approach a *large* and *complex* natural product molecule was identified both without any atom property constraints and without the input of either carbonyl or hydroxyl groups in contrast to the numerous constraints imposed in the initial report.[19] There were also no assumptions regarding the ring sizes contained within the molecule.

## 9.5.2 Cycloshermilamine D

The experimental data for the next example was taken from Koren-Goldshlager *et al.*[21] The authors performed the structure elucidation of cycloshermilamine D (**9.7**), a recently isolated alkaloid. The molecular formula is $C_{21}H_{16}N_4O_1S_1$ (MW $= 372$). In their article,[21] the authors presented the

COSY, HMQC, and HMBC correlations obtained from the 2D NMR experiments.

**9.7**

Cycloshermilamine D is a proton-deficient molecule (the ratio of heavy atoms to protons is approx. 2:1); the problem is further complicated by the presence of three different heteroatoms including S (an element with a valence which can vary as 2, 4, or 6). The number of correlations detected by the authors proved to be very small (eight correlations were revealed by the COSY experiments and 20 from the HMBC experiment). When structure generation was initiated, an enormous number of structures were produced within the first several minutes. The generation process was aborted, since it was evident that the introduction of additional structural information would be necessary to help solve the problem.

The fragment search by $^{13}$C NMR spectrum gave a list with $L = 1371$. Many attempts were made to create MCDs from the FFs using different $E$ values, but the number of MCDs turned out to increase very quickly (for instance, at $E = 4$ ppm, $n_{MCD} \approx 12\,000$). As none of these MCDs resulted in a structure, the time-consuming process of creating and checking the connectivities from the FFs was aborted. It became clear that only the introduction of a UF could help one to solve the problem.

The main difference between a FF and a UF is that the FF already contains assigned carbon atoms, whereas the carbon atoms of a UF have no associated assignments. In our example, the presence of the following fragment was assumed:

**9.8**

This hypothesis can be justified by the fact that substructure **9.8** is present in congenerous compounds – shermilamines D and E,[22] as well as in segolins,[23] which were isolated and identified earlier. Such assumptions can be made when families of structures are under analysis.

The HOSE code-based ("accurate") $^{13}$C NMR chemical shift prediction of substructure **9.8** was performed and five MCDs were created from this UF with $E = 7$ ppm. The result of checking the MCDs for contradictions gave $n_{MCD} = 5 \rightarrow 2$. At this stage structure generation was performed with only one constraint of $R_c = 5–6$ imposed ($R_c$, allowed cycle sizes), taking into account that the compound under investigation, as Koren-Goldshlager *et al.*[21] supposed, belongs to the pyridoacridines. The structure generation process delivered: $k = 1987 \rightarrow 373$, $t_g = 8$ min 25 s. After structure ranking by the $d_I$ values, "accurate" $^{13}$C NMR spectra were calculated for the first 20 ranked structures. As a result the correct structure delivered the best match (minimal $d_A$ value).

Another method of utilizing a user-defined fragment was also examined. Fragment **9.8** was searched through the list of FFs to select those items contained within the list which contain the UF. The selected FFs are shown in Figure 9.5. As indicated in the Figure, the fragments produced from different molecules during the process of creating the fragment library have slightly different chemical shift assignments. These fragments can be used to create an MCD which avoids chemical shift calculation for the user-defined fragment.

An attempt to create MCDs from this set of six FFs was pursued. At $E = 0.5$ ppm, $n_{MCD} = 18 \rightarrow 4$ resulted. Structure generation was started with only one constraint imposed on the ring sizes: $R_c = 5–6$. The result was: $k = 291 \rightarrow 52$; $t_g = 2$ min 7 s. Fragmental $^{13}$C NMR spectral prediction was performed for the first 20 structures ranked previously by the $d_I$ values. The true structure was again correctly distinguished as shown in Figure 9.6. It is worth noting that structures #2–#4 are very similar to the determined structure, but they are significantly distinguished from cycloshermilamine D by deviation differences.

The results described in this example are indeed significant, since the problem initially seemed to be hopeless due to the lack of correlations. Experience

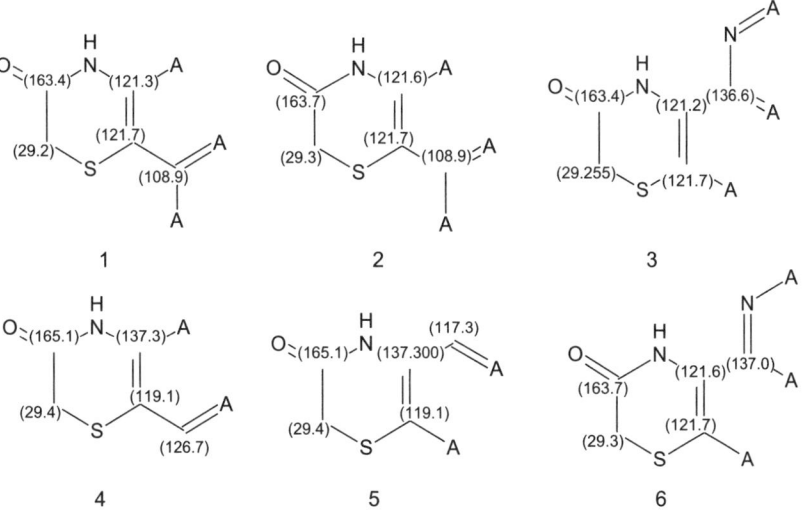

**Figure 9.5** Library fragments containing the user-defined fragment (**9.8**).

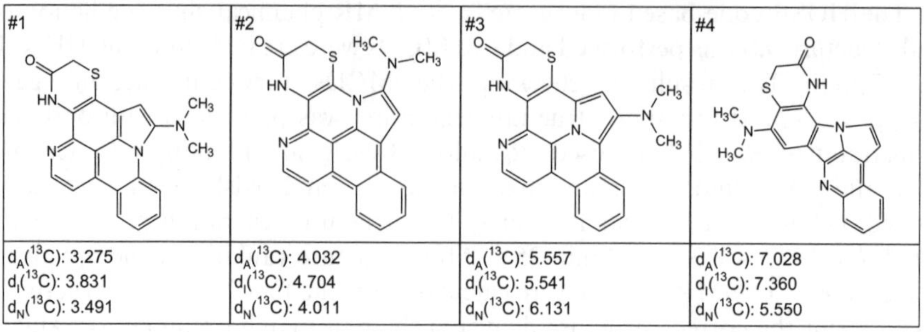

| #1 | #2 | #3 | #4 |
|---|---|---|---|
| | | | |
| $d_A(^{13}C)$: 3.275<br>$d_I(^{13}C)$: 3.831<br>$d_N(^{13}C)$: 3.491 | $d_A(^{13}C)$: 4.032<br>$d_I(^{13}C)$: 4.704<br>$d_N(^{13}C)$: 4.011 | $d_A(^{13}C)$: 5.557<br>$d_I(^{13}C)$: 5.541<br>$d_N(^{13}C)$: 6.131 | $d_A(^{13}C)$: 7.028<br>$d_I(^{13}C)$: 7.360<br>$d_N(^{13}C)$: 5.550 |

**Figure 9.6**  The initial structures in the ranked answer file generated from the HMBC and COSY connectivities and using the FFs shown in Figure 9.5.

indicates that the $^{13}C$ NMR calculation based on HOSE codes is capable of assigning carbon atoms for a UF with a degree of precision that allows the fragment to be accepted by the program as a useful starting point. It has also been shown that FFs can be used to supply a UF with *realistic* carbon chemical shifts. Therefore, despite the fact that the list of FFs was not used for direct MCD creation, the structural and spectral information stored in the system database, nevertheless, proved to be valuable for UF assignment and, eventually, for obtaining the solution to the problem.

## 9.5.3 Lyngbyabellin A

An attempt to determine the structure of lyngbyabellin A (**9.9**), a novel cytotoxic compound isolated and investigated by Luesch *et al.*,[24] was undertaken using StrucEluc. Compound **9.9** has a molecular formula of $C_{29}H_{40}Cl_2N_4O_7S_2$ and a molecular mass of M = 690. In this example, 2D NMR data obtained from COSY (22), HMQC, and HMBC (60) experiments, as well as a list of IR frequencies, were taken from the original work.[24]

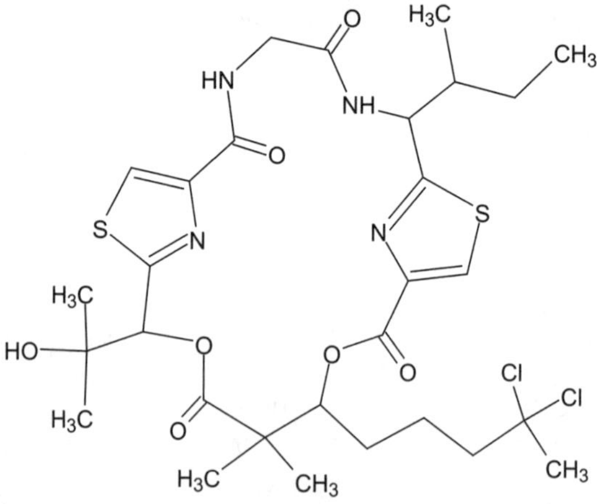

**9.9**

## 9.5.3.1 Common Mode

The molecule under investigation is fairly complex. It contains four types of non-carbon skeletal atoms, N, O, S, and Cl to give a total of 15 heteroatoms and two thiazole rings. Sulfur can exist in different valence states and as a result can introduce a large number of possible isomers during structure generation.

Several attempts were made to solve this problem. Many different forms of user intervention were given including specifying atom properties and drawing C=O bonds. The structure generation process was aborted after generating more than 150 000 structures within a 24 h period. With this in mind, it was concluded that only the application of the fragment approach could be helpful for solving the problem.

## 9.5.3.2 Utilization of Fragments from the Knowledge Base

For this example the number of FFs was fairly large: $L = 7427$. Initial viewing of the FFs indicated that the fragments listed at the start of the file had $^{13}$C chemical shifts (presented as a bar graph) that were very close to those observed experimentally. For the first fragment the general pattern of both the experimental and predicted spectra appeared similar when they were visually compared (see Figure 9.7). In addition, this fragment contained 19 carbon atoms (from a total of 28 skeletal atoms). It was decided that these coincidences were likely not serendipitous. It was natural to suggest that this and related fragments could be associated with the structure under investigation.

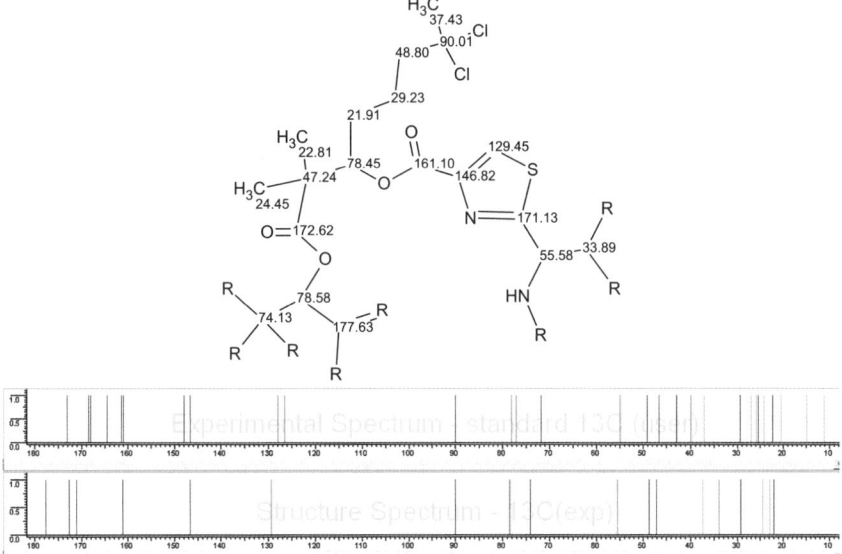

**Figure 9.7** Fragment #1 in the FFs list. The experimental $^{13}$C NMR spectrum of the compound **9.9** is displayed in the upper window and the fragment sub-spectrum in the lower window.

With these assumptions the process of creating connectivities from the first 25 fragments ($l = 25$) was initiated using $q = 60\%$, $n_f = 1$, and an initial $E$ value of 0.5 ppm. The program created no MCD for $E$ values lying in the interval 0.5–9 ppm, whereas 144 diagrams were created using $E = 10$ ppm. The absence of any vibrational frequency in the region near 2600 cm$^{-1}$, as well as the $^1$H NMR spectrum suggested that an SH group can be introduced into the user BADLIST. Structure generation was performed but no structures were produced from these MCDs.

Successively increasing the $E$ value to $E = 14$ ppm produced 272 MCDs when 128 new MCDs were produced. This indicates that 272 different distributions of the carbon assignments are possible at this given error value ($E$). Only 32 MCDs survived a check for contradictions. Eight of these contained the first fragment and the others were composed of two fragments. Structure generation was performed using a single constraint – four-membered rings, which are uncommon in natural products, were forbidden. The results gave $k = 13568 \rightarrow 11 \rightarrow 11$ and $t_g = 29$ s. The correct structure has minimum deviations for all spectral predictions: $d_A(1) = 1.20$ ppm, $d_I(1) = 0.94$ ppm and $d_N(1) = 0.77$ ppm, whereas $r_A = r_I = r_N = 1$ and the difference $\Delta_{(2-1)} = d_A(2) - d_A(1) = 0.56$ ppm. The $\Delta_{(2-1)} = 0.56$ ppm value indicates the reasonable reliability of the solution as described in detail in Section 9.7.

When structure generation was repeated without any limitation on ring cycle sizes, 25 structures were generated and the spectrum prediction again endorsed the priority of the correct structure. In order to determine the similarity of the generated structures to the real structure Tanimoto coefficients, $C_{sim}$, were calculated for structural similarity.[14] The structures were ranked in decreasing $C_{sim}$ values and the first four are shown in Figure 9.8. The structure most similar to the correct one gave $C_{sim} = 1.000$ and was the second in the ranked output file in accordance with the spectral data. The third and fourth structures have similarity coefficients equal to 0.9955. At the same time all four structures are characterized by different deviations, which demonstrates directly the high "structural" resolving power of the methods based on $^{13}$C NMR spectrum prediction.

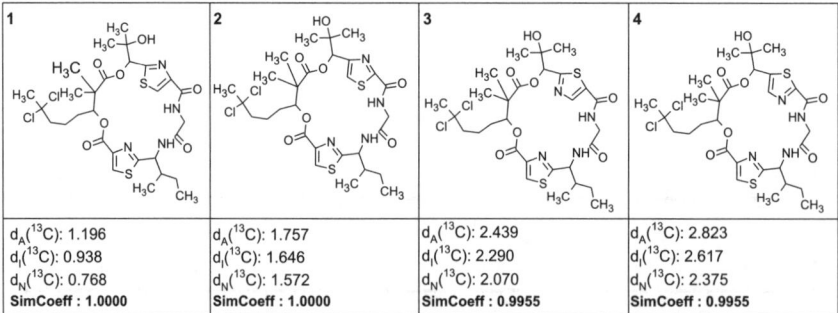

| 1 | 2 | 3 | 4 |
|---|---|---|---|
| $d_A(^{13}$C): 1.196 | $d_A(^{13}$C): 1.757 | $d_A(^{13}$C): 2.439 | $d_A(^{13}$C): 2.823 |
| $d_I(^{13}$C): 0.938 | $d_I(^{13}$C): 1.646 | $d_I(^{13}$C): 2.290 | $d_I(^{13}$C): 2.617 |
| $d_N(^{13}$C): 0.768 | $d_N(^{13}$C): 1.572 | $d_N(^{13}$C): 2.070 | $d_N(^{13}$C): 2.375 |
| SimCoeff : 1.0000 | SimCoeff : 1.0000 | SimCoeff : 0.9955 | SimCoeff : 0.9955 |

**Figure 9.8**   The elucidated structures most similar to the correct structure. Those displayed are selected from the beginning of the ranked answer file and ranked in decreasing $C_{sim}$ values.

The example shows that a visual comparison of the $^{13}$C NMR spectrum of large FFs ranked by the program at the top of the list of FFs with the experimental spectrum can aid in revealing fragments which can be suitable for creating the MCDs. This approach may be utilized to solve a difficult problem without making risky assumptions. If, however, the top ranked fragments have $^{13}$C subspectra significantly differing from the experimental spectrum, the chance of problem solving from the FFs decreases because in this case only small appropriate fragments may exist in the list (small fragments hardly can help to reduce the problem dimension).

## 9.5.4 Paradisin C

Ohta *et al.*[25] detail the isolation and identification of a new natural compound, paradisin C (**9.10**). The elucidation was based on high-resolution FAB-MS data which gave a molecular formula of $C_{42}H_{46}O_{11}$ and using 2D NMR spectra HMQC, COSY (10), and HMBC (63).

**9.10**

An attempt to solve the problem in the common mode failed. The structure generation process continued for over 24 h and was aborted at that time. Searching through the knowledge base using the $^{13}$C NMR spectrum as the input gave 2946 FFs. As in the previous example, it was determined based on visual comparison that the carbon chemical shifts of the first fragments in the list were close to the experimental values. This provided a basis to set the conditions for forming the MCDs which were created using $l = 500$. With $E = 5$ ppm, $q = 40\%$ and $n_f = 3$ the program created $n_{MCD} = 384$ MCDs. The process of structure generation without any additional restrictions resulted in $k = 2000 \rightarrow 16 \rightarrow 15$, $t_g = 28$ s, $d_A(1) = 1.42$ ppm, $d_I(1) = 1.58$ ppm, $d_N(1) = 1.64$, $\Delta_{(2-1)} = 1.00$ ppm and $r_{all} = 1$.

The examples considered in this section indicate that using fragments found in the database allows the user to solve problems that seemed to be impossible for the system to cope with in the common mode. Nevertheless, it may happen that even good FFs do not reduce the dimension of the problem significantly enough as to make it solvable.

# 9.6   Utilization of Both UFs and FFs

Using fragments from the knowledge base often leads to solution of the problem, but, unfortunately, the knowledge base is restricted to published data and may not include suitable fragments specific to proprietary chemistries. In this Section two examples related to the structure elucidation of peptides will be considered that could not be solved utilizing fragments found in the knowledge base. In this case it was demonstrated that these problems could be solved with the help of additional data, including UFs.

## 9.6.1   Apramide G

Luesch et al.[26] extracted and identified a new metabolite apramide G (**9.11**) whose molecular mass was determined as M = 976 and a molecular formula of $C_{52}H_{80}N_8O_8S$. In the publication,[26] tables of data are listed that contain $^1H$, $^{13}C$ NMR spectra and 2D NMR correlations from COSY (32), HMQC, and HMBC (71). Various methods of solving this problem were investigated. An initial attempt to solve the problem in the "Common Mode" indicated that the processing time would be excessive. The structure contains a thiazole ring and therefore a sulfur atom with all of its valence issues. The structure also contains a C≡C group. For the first attempt at elucidation the structural data obtained by the authors from the spectrum analysis were used.

**9.11**

## 9.6.1.1 Common Mode

Checking the MCD revealed that there were no non-standard connectivities. The authors of the article[26] had found that for the fragment $CH_2-C\equiv CH$, one COSY correlation corresponded to a $^4J_{HH}$ coupling. This was taken into account when creating the MCD. Attempts were made to solve the problem in the common mode using the following data obtained by the authors:[26] the molecule contained six $NCH_3$ groups (six singlets in the $^1H$ NMR spectrum) and one $CH_2-C\equiv CH$ group (a triplet in the $^1H$ NMR spectrum at 1.77 ppm with a coupling constant of $^4J_{HH} = 2.4$ Hz provided evidence for the presence of the terminal acetylene group). A correlation between $\delta_H$(1.77 ppm) and $\delta_C$(68.9 ppm) suggested *sp* hybridization. Instead of the *nd/nd* property this atom and the quaternary C(84.3 ppm) were manually attributed with the *sp/fb* parameter. The maximum bond multiplicity was set to 3 during the structure generation process.

After 16 h of operation in the common mode the program did not generate any structures that complied with the structural and spectral filters. This result was not unexpected. The formal number of double bonds in the structure under investigation is 17 and the molecule is proton deficient. In addition, a major difficulty in this case is the size of the molecule, 69 skeletal atoms, and the fact that the molecule contains a significant number of heteroatoms, 17 in total and including eight nitrogens and a sulfur atom. Competition between the nitrogen and oxygen atoms to be neighbors of the carbon atoms for which such vicinity is allowed leads to a significant increase in the number of candidate structures.

## 9.6.1.2 Utilization of UFs

The authors[26] identified from the experimental data the presence of a thiazole ring. This is supported by two doublets in the $^1H$ NMR spectrum at 6.56 and 7.45 ppm ($J = 3.2$ Hz), although this could just as easily have been a furan ring based on this data. With this assumption ("axiom"), a thiazole fragment, **9.12,** was entered as a UF. The chemical shifts shown in structure **9.12** were calculated using the HOSE code-based method and MCD creation was initiated. Note that the calculated shifts are very close to the experimental ones in structure **9.11**.

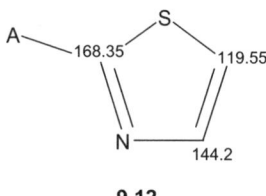

**9.12**

With $E = 2.5$ ppm, one MCD was created containing this fragment and all constraints used in the common mode were applied, however, these restrictions were still insufficient to solve the problem and the generation process was again aborted after 2 h. It was clear that the process would be too time consuming and an attempt was therefore made to use fragments from the system knowledge base.

## 9.6.1.3   Utilization of FFs

After a fragment search in the database, the number of FFs was $L = 13\,934$. In order to prune the fragments that are unlikely and to move prospective fragments to the top of the list, OH and NH groups were placed into the BADLIST as there were no absorptions in the IR spectrum above 3000 cm$^{-1}$. The fragments were then filtered with simultaneous exclusion of ionic structures, since these were improbable in the analyzed molecule. The result gave $L = 13\,934 \rightarrow$ 5121 and the number of fragments was reduced by a factor of almost three, which demonstrates the efficiency of using structural information easily obtained from the IR spectrum. Visual investigation of fragments at the start of the list revealed that there were some fragments that showed good correspondence with the experimental $^{13}$C NMR spectrum.

The first ten fragments were selected from the list to form the MCDs. To confirm whether the problem could be solved with only the FFs the UF (thiazole ring) was deleted from the fragment list before starting the MCD generation process. The following options were used for creation of the MCDs: $l = 10$, $E = 2.5$, $n_f = 1$, $q = 30\%$. The program created 12 MCDs and each contained fragment **9.13**, but with different distributions of the chemical shifts. In structure **9.11** this fragment is highlighted with blue bold lines. Compare the chemical shifts of the structure and the fragment and obvious similarities exist. At the same time we see that chemical shifts for the fragment **9.13** can be divided into sets containing similar values (for instance, 170.7, 169.6, 171.2), which leads to a great number of possible shift permutations.

**9.13**

Before structure generation, atom properties were adjusted in the first MCD and automatically transferred to other diagrams. After 3 h of structure generation, the process was stopped for the same reasons as described in the previous step. To solve the problem in a reasonable time, UFs, in combination with FFs, were considered.

## 9.6.1.4   Combining UFs and FFs

With settings of $E = 2.5$, $n_f = 2$, $q = 30\%$ the program created 12 MCDs, each containing the UF **9.12** and the FF **9.13**. Structures were then generated with one restriction only, a ring cycle size of $R_c = 5$–6. The results are $k = 283391 \rightarrow 1346 \rightarrow 738$, $t_g = 7$ min, and $r_{all} = 1$. The first two structures are shown in Figure 9.9.

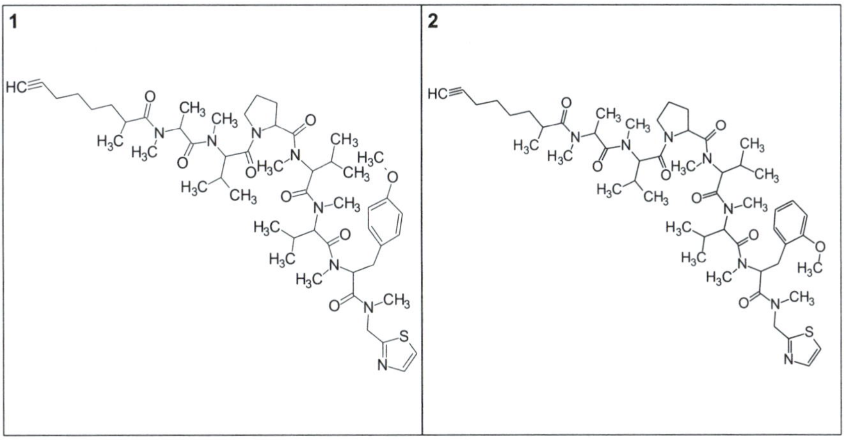

$d_A(^{13}C)$: 1.128                    $d_A(^{13}C)$: 1.538

**Figure 9.9** The two top structures from the answer file (apramide G). The structures differ only in the positions of the OCH$_3$ group on the benzene ring.

The second structure differs from the first only by the position of the substituent on the benzene ring. It can be concluded that only the combined usage of both UFs and FFs allowed solving this problem. The unresolved question is obviously how much information can or should be extracted by a scientist to feed into the program initially?

## 9.6.2 Hetcochlorin

A similar approach allowed the elucidation of hetcochlorin (**9.14**), a novel natural product of molecular formula $C_{27}H_{34}Cl_2N_2O_9S_2$ which was isolated and characterized in a report by Marquez and co-workers.[27] The spectral data included HMBC data with 53 connectivities and COSY with four connectivities.

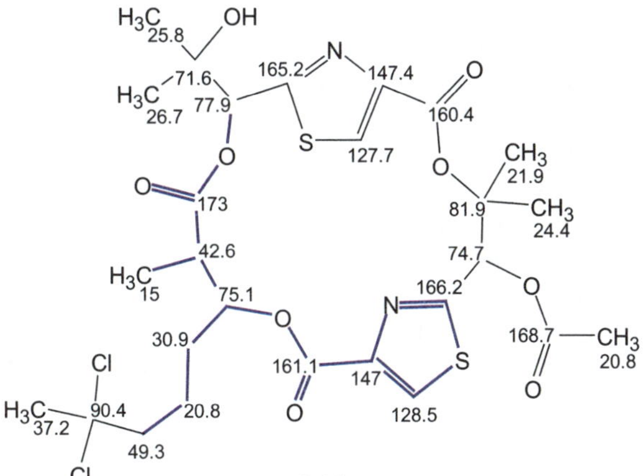

**9.14**

The molecule contains two thiazole rings and, based on earlier experience, it was evident that the structure could likely only be elucidated with the help of UFs and fragments found in the database. The authors[27] suggested the presence of two thiazole rings and four carbonyls from the 1D $^1$H and $^{13}$C NMR spectra.

A search through the knowledge base gave $L = 3401$. Visual comparison of the chemical shifts of the carbon atoms of the first fragment in the list, substructure **9.15,** with the experimental values, indicated they were similar. This is evident by comparing structure **9.14** and fragment **9.15** as highlighted by blue lines.

**9.15**

To complete the investigation, two MCDs were obtained from this fragment with $E = 12$ ppm. Attempts to generate structures from these MCDs failed and the structure generation process was aborted. It was evident that the usage of UFs would again be necessary. Since fragment **9.15** already contained one thiazole ring, an additional thiazole ring was introduced according to the authors' assumption prior to creation of the MCDs. The program created 12 MCDs with settings defined as $E = 12$ ppm, $q = 50\%$, $n_f = 2$. Structure generation without any restrictions resulted in: $k = 5848 \rightarrow 3893 \rightarrow 2967$, $t_g = 22$ s, $r_{all} = 1$. This approach, based on the usage of fragments from various sources, allowed for the elucidation of a complex molecule when the utilization of both the common mode and fragments from the fragment library only failed.

Several problems were identified where the best ranked structure did not coincide with that suggested by researchers. Particularly challenging were large cyclopeptide molecules, for example, phoriospongin A (Capon *et al.*[28]), **9.16,** as shown below.

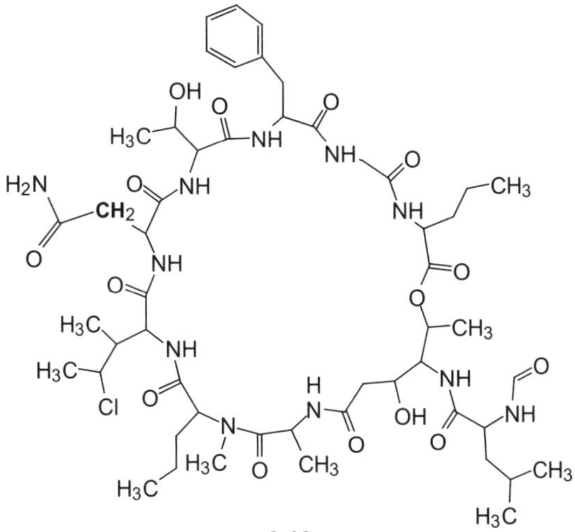

**9.16**

This molecule has the formula $C_{52}H_{82}N_{11}O_{15}Cl$ with 79 heavy atoms, including 27 heteroatoms of three different types. The spectral data[28] produced over 258 000 structures in 17 h. Of these structures, 624 were non-isomorphic. After ranking according to the $d_A$ value, the correct structure was number 2. The following structure (**9.16a**) was selected as *the best one*, while $d_A(1) = 1.54$ ppm, $d_A(2) = 1.59$ ppm and the difference $\Delta_{(2-1)} = 0.05$ ppm is very small:

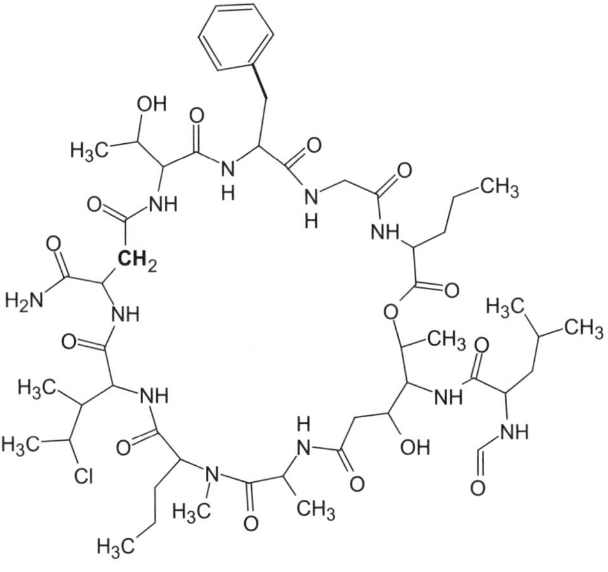

**9.16a**

Structure comparison allows one to see that the only difference between **9.16** and **9.16a** is in the position of the CH$_2$ group in the left portion of the molecule. This example suggests that the larger the molecule being studied, the less structural "resolving power" the deviations offer in terms of identifying the correct structure. The reason for this is probably the leveling of the deviation values due to the presence of a large number of carbon atoms having similar properties. In this case, the top ranked structures have very similar calculated spectra that are characterized by regions containing many close chemical shifts. To solve such problems and determine the actual structure, additional information is generally required. Nevertheless, when the structure of durhamycin A (C$_{62}$H$_{92}$O$_{28}$, 90 skeletal atoms) was elucidated[29] using StrucEluc, the correct structure was distinguished in the output file of 66 structures by all three methods of $^{13}$C NMR spectrum prediction.

# 9.7    Application of a User Database for the Structure Elucidation of Challenging Problems

During the course of the development of the StrucEluc software a number of complicated and challenging structure problems were studied. Typically, in these problems when attempting the structure elucidation in the common mode, structure generation was not accomplished after more than 48 h. Attempts to solve these problems using fragments found in the system knowledge base (FFs) also failed. The failure to utilize library fragments was due to the following issues:

- Fragments appropriate for a given problem are missing in the knowledge base.
- Appropriate fragments are found, but the number of possible variants of carbon atom assignments in these fragments is so huge (more than 100 million) that the computer runs out of resources attempting to sort out all possible permutations (see Section 9.2.2.2).
- The number of MCDs built up by the program is so huge that the completion of structure generation is simply too long and human intervention halts the process.

Among the structures we failed to determine in the common mode and using library fragments were the following three alkaloids from the cryptolepine series:[29] cryptolepicarboline (A), cryptospirolepine (B), and cryptoquindolinone (C). The structures of these natural products are shown in Figure 9.10.

These molecules are relatively large, highly unsaturated, have four condensed benzene rings and double bonds contained in other cycles. All the molecules have large fragments (displayed in bold) containing no hydrogen atoms. These fragments account for half of the skeleton of the molecule, thus limiting the utility of COSY data. For structures A–C in Figure 9.10, COSY correlations are observed only for the contiguous protons in the 1,2-disubstituted benzene

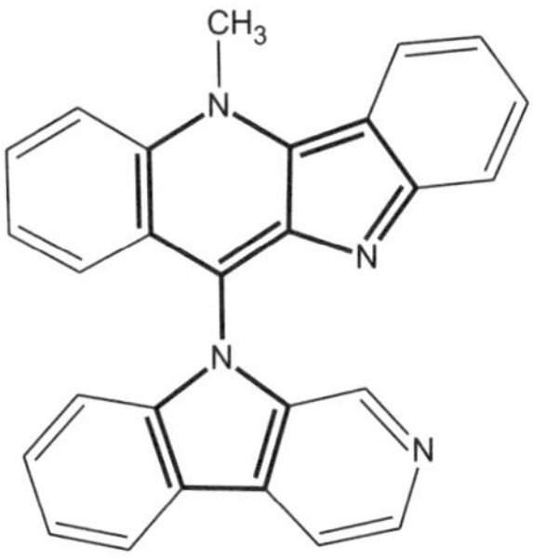

**A**

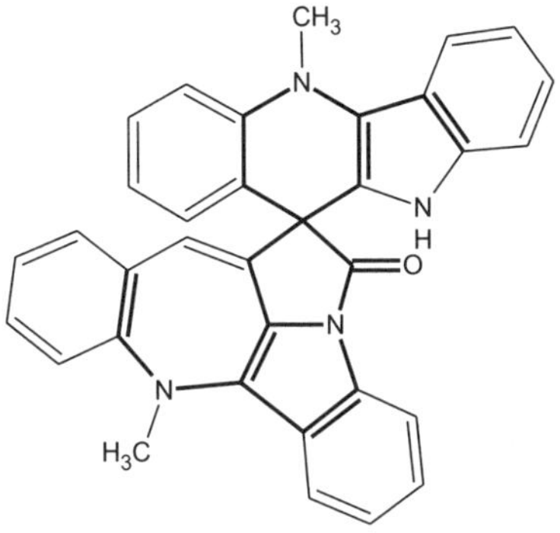

**B**

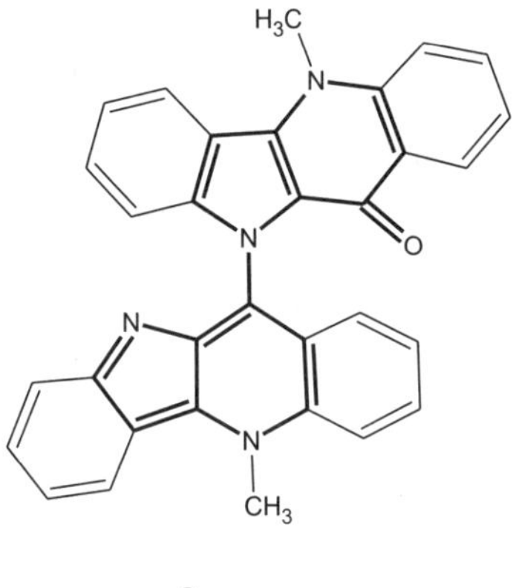

**C**

**Figure 9.10** Three challenging molecules from the cryptolepine series.

rings. Cryptospirolepine (B) has an especially complex structure consisting of two planar fragments linked through a carbon spiro-center and therefore lying in perpendicular planes. Each fragment makes up a system of conjugated bonds. One can expect long-range correlations ($^{n}J_{CH}$, $^{n}J_{HH}$, n ≥ 4) inside the planar fragments and little or no spin–spin coupling information transfer between the fragments. An unexpected chemical shift for the carbonyl amide group in the $^{13}$C NMR spectrum is observed at 188.4 ppm. It is more likely to correspond to a ketone carbon resonance since it is more representative of this type of structure. The examination of 2250 structures found in the system knowledge base and containing an amide group in a five-membered cycle confirmed that no such structures had a similar carbonyl chemical shift. In the IR spectrum of cryptospirolepine the frequency of the carbonyl absorption is also unusually low at 1611 cm$^{-1}$. These factors all contribute to a very challenging elucidation process for cryptospirolepine. The structure determination of compound C in Figure 9.10, a degradation product of cryptospirolepine, is also a complex analytical problem. In addition to fused benzene rings, the molecule contains a large "silent" fragment containing 19 carbon atoms and two N–CH$_3$ groups. The ketone carbonyl resonance also has an uncommonly low chemical shift value of 167.1 ppm characteristic of esters and amides, although representative of pyridones and quinolones.

The StrucEluc system contains algorithms and programmatic features that allow the user to create their own knowledge base. This knowledge base generally contains structures referring to chemical classes that are of particular interest to a given researcher or laboratory, as well as fragments excised from these structures using specific algorithms. As in the main system knowledge base, the carbon atoms of fragments and molecules are attributed the corresponding chemical shifts in the $^{13}$C NMR subspectra. The system therefore can be adjusted to expedite the structural characterization of new members related to an existing class(es) of compounds for which some data are available and upon which specialized knowledge bases can be built. A search for fragments consistent with the $^{13}$C NMR spectrum can be performed in each database separately or in any combination.

For this reason this approach was considered for determining the alkaloid structures A–C with the aid of a user database[29,30] adjusted for the structure analysis of members of the cryptolepine family. In this Section we will describe the structure elucidation of compounds A–B, whereas details of elucidating structure C will be discussed in Chapter 11. Assuming that the unknown compounds were members of this series, the information for earlier published members of the series was introduced. The compounds shown in Figure 9.11 were selected to create a user database.

The spectral data referring to these compounds were obtained from the literature.[31–37] 2D NMR spectra for compounds #4–#8 were found in the literature while others only provided 1D NMR spectra.

## 9.7.1   Creating a User Database

Alkaloids #1–#8 of the cryptolepine series along with the assigned $^{13}$C NMR spectra were included in the user library of full structures. The algorithm used to create a UF library is similar to the algorithm used to create the StrucEluc

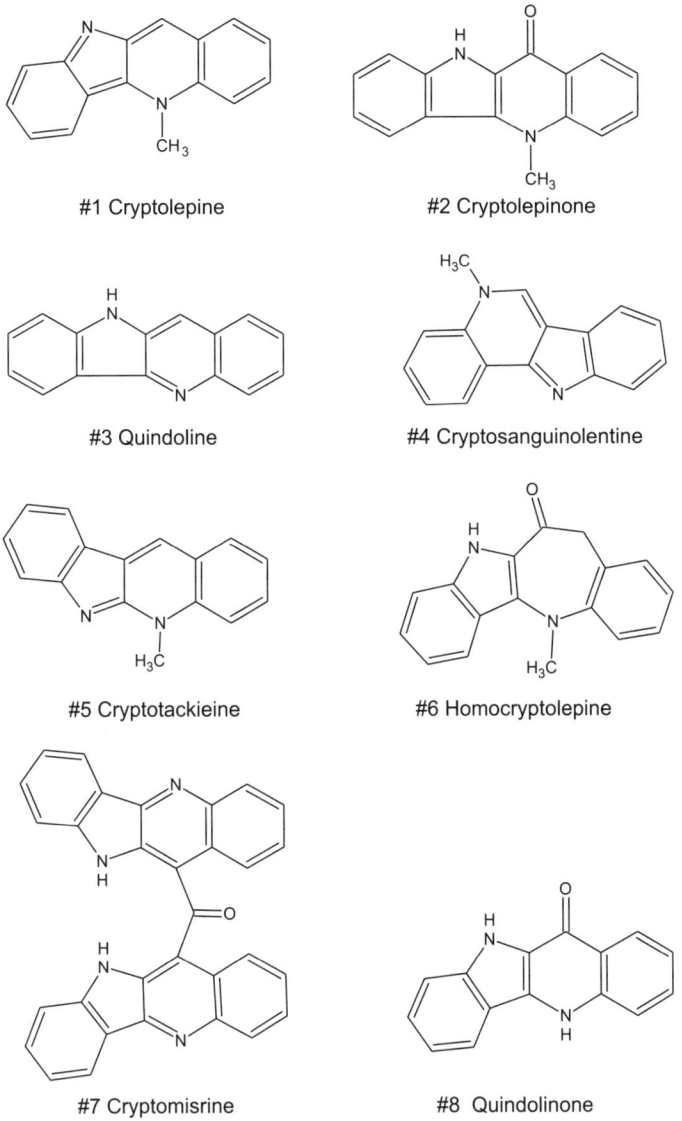

**Figure 9.11** Compounds from cryptolepine series that were used to create a UF Library.

system knowledge base and has been described in Section 7.4.1. The two basic steps are:

- The program generates as complete as possible a set of fragments from all the structures which are incorporated into the structural file. In so doing, the program is guided by a collection of rules providing for the generation of "chemically reasonable" fragments.

- Atoms in the fragments are assigned chemical shift values that they have in the corresponding structure.

The procedure described above produced a UF Library containing 342 fragments from the cryptolepine series. This database was then used to elucidate the challenging structures A–C.

## 9.7.2 Solving Challenging Problems

### 9.7.2.1 Cryptolepicarboline

Searching the $^{13}$C NMR spectrum of cryptolepicarboline through the UF library yielded 68 fragments. To create the MCDs from $^1$H–$^{13}$C HMBC data the following options were specified: (a) the minimal number of fragments to be included into each MCD is 3 ($q_{fr} = 3$); and (b) the minimal percentage of the total number of skeletal atoms absorbed by the fragments of each MCD is set to 50 ($p_{at} = 50\%$). These options are specified to minimize the number of free skeletal atoms in each MCD. 50 MCDs were created for $E = 1.5$ ppm, 25 of these survived the test for contradictions. One of the MCDs shown below (Figure 9.12) displays three fragments accounting for approx. 70% of the skeletal atoms.

The low $E$ value indicates that among the 68 selected fragments there were many carbon atoms with assignments very close to the experimental values. Structure generation resulted in: $k = 808 \rightarrow 160 \rightarrow 64$, $t = 1$ s. NMR spectra prediction positioned the cryptolepicarboline molecule first in the ranked file ($r_{all} = 1$), where $d_A(2) - d_A(1) = 2.9$ ppm.

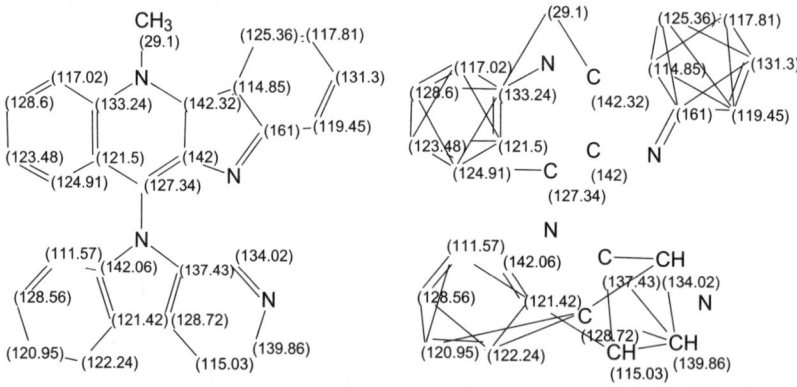

**Figure 9.12**   One of the MCDs created during the elucidation of cryptolepicarboline whose structure is shown, with the assigned chemical shifts, on the left-hand side of the Figure to assist in understanding of the MCD.

## 9.7.2.2 *Cryptospirolepine*

Searching the $^{13}$C NMR spectrum of cryptospirolepine in the UF library yielded 60 fragments. These produced 180 MCDs using a value of $E = 4$ ppm. Attempts to generate a structure failed since the calculations were too time-consuming. The conclusion was made that it was necessary to find the most perspective fragments among the 60 selected fragments. For this purpose the "generalized portrait" approach discussed earlier[1] (see Chapter 5) was utilized. The full list of fragments was "refined" and fragments corresponding to the nature of the structure analyzed were selected using the facilities contained within the program. For the "generalized portrait" of the cryptospirolepine structure, most functional groups in the selected fragments were identified.

The system fragment library was searched using the $^{13}$C NMR spectrum of cryptospirolepine as input. A list of FFs was produced ($L = 1437$). All ion-containing structures were removed (it was assumed that ionic structures were not expected) to provide 1287 fragments. The functional group library (see Chapter 5) was "sifted" through the FF set, and the first 12 functional groups along with the number of parent fragments are shown in Figure 9.13.

703 fragments (55 per cent) contain a benzene ring and almost half of these (338) show a 1,2-disubstituted aromatic pattern (1,2-Ar). Due to the atypical downfield chemical shift value of the carbonyl group (188.4 ppm) none of the FFs contained a tertiary amide. The hypothesis was that the molecule should contain at least one 1,2-Ar fragment. In the next step, the following ten 1,2-Ar containing fragments (Figure 9.14) were automatically sorted out of the 60 fragments extracted from the user database in the first step.

We see that some identical fragments have different chemical shift assignments related to the corresponding parent molecules that are included into the user library. MCDs were generated from these fragments using the following parameter values: $q_{fr} = 3$, $p_{at} = 50\%$, $E = 6$ ppm. As a result, 320 MCDs contained four fragments each were created. One of these is shown in Figure 9.15 (only HMBC connectivities are visible).

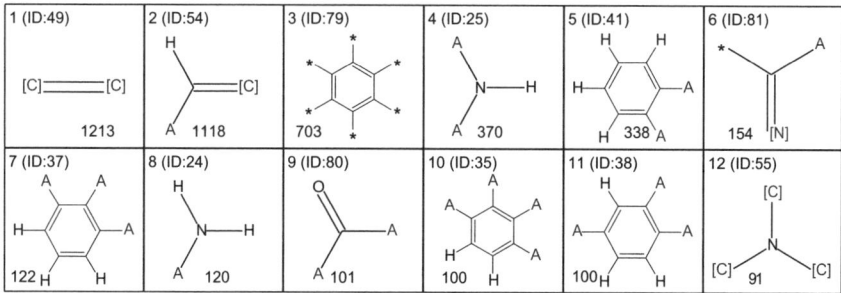

**Figure 9.13** Functional groups found in structures of fragments. Functional groups are ranked in descending values of the number of fragments containing a given functional group. The numbers are shown below the functional group structures.

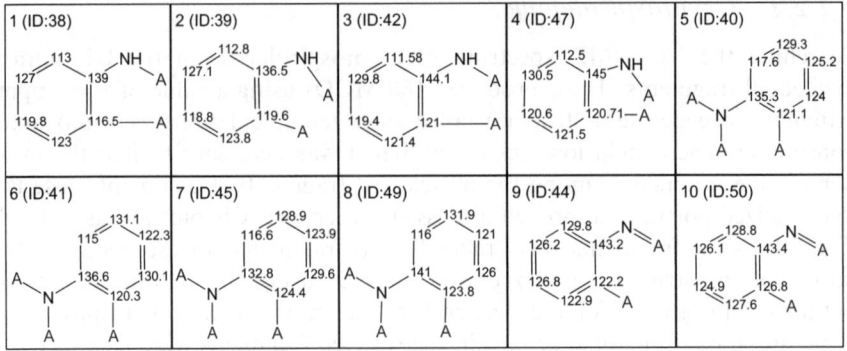

**Figure 9.14**   1,2-Ar-containing fragments sorted out of the 60 fragments found in the
UF library.

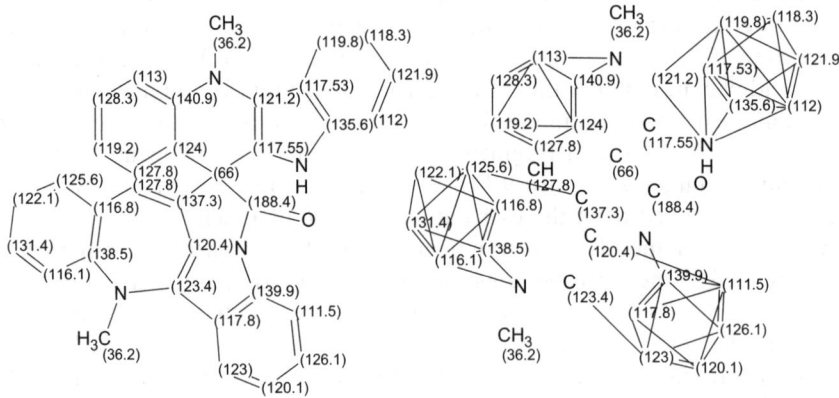

**Figure 9.15**   One of the MCDs created during the elucidation of cryptospirolepine
whose structure with assigned chemical shifts is shown to the left-hand
side to assist in understanding the MCD.

In these MCDs the fragments account for more than 70% of the skeletal
atoms suggesting they are likely good fragments. It was proven by IR spectral
experiments[37] that the analyzed molecule contains an amide group, so the
ketone functional group was added to the BADLIST to produce $k = 192 \rightarrow 1$,
$t = 18$ min 30 s under the single constraint $R_c = 5$–7. The only structure
consistent with the data was the structure of cryptospirolepine.

## 9.8   Elucidation of Symmetric and Ionic Structures

In the course of performing the structure elucidation of natural products a
number of unknown molecules with symmetry were encountered. The presence

or absence of molecular symmetry does not influence structure generation within 1D expert systems, but it was revealed that an attempt to generate symmetric structures from 2D NMR data led to unmanageable processing times. The cause of this failure was carefully investigated and the peculiarities of symmetric structure generation were discovered. An enhanced algorithm was developed to reveal symmetry features in 2D NMR data, and then it is automatically adjusted to the generation of symmetric molecules. As a result the processing time necessary for the generation of symmetric molecules is now of the same order as for molecules without symmetry. This algorithm is still not optimized and improvements are being researched to allow further acceleration of the structure generation process by *using symmetry* for this purpose.

A specific example will be considered. A new dimeric natural product, ashwagandhanolide (Figure 9.16, Structure a), was isolated by Subbaraju *et al.*[38] Its molecular formula was determined as $C_{56}H_{78}O_{12}S$ on the basis of the molecular ion observed at $m/z$ 975.5285. The structure of this compound was determined using 2D NMR data, as well as additional information obtained from comparison of experimental spectra with the structures and spectra of related molecules. In the article[38] only 35 HMBC correlations are reported (no COSY data were available). The number of correlations is small due to severe overlap in the $^1$H NMR spectrum. An attempt to solve the problem using StrucEluc in the common mode showed that the processing time and the number of generated structures would be unmanageable. Therefore a fragment search using the $^{13}$C NMR spectrum was performed and 5524 fragments

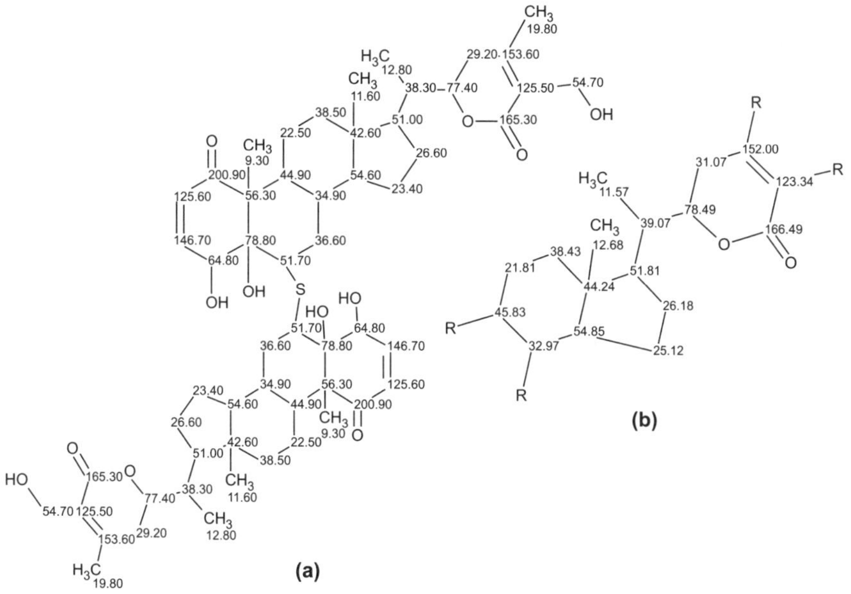

**Figure 9.16**  The structure of ashwagandhanolide (a) and a FF (b).

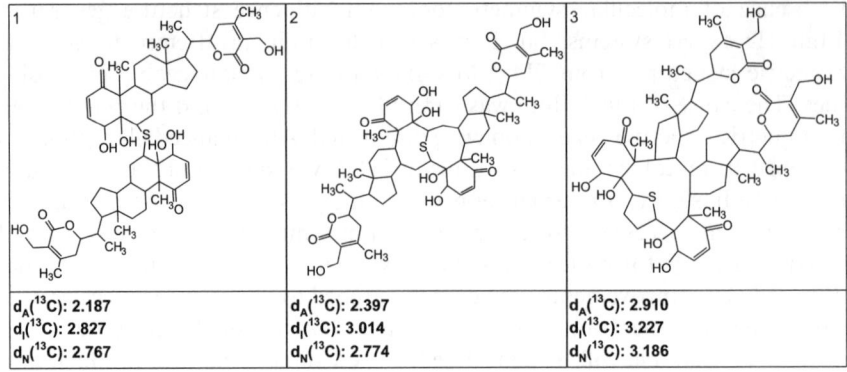

| $d_A(^{13}C)$: 2.187 | $d_A(^{13}C)$: 2.397 | $d_A(^{13}C)$: 2.910 |
| $d_I(^{13}C)$: 2.827 | $d_I(^{13}C)$: 3.014 | $d_I(^{13}C)$: 3.227 |
| $d_N(^{13}C)$: 2.767 | $d_N(^{13}C)$: 2.774 | $d_N(^{13}C)$: 3.186 |

**Figure 9.17**   The three top structures in the ranked file obtained as a result of the ashwagandhanolide structure elucidation.

were found in the fragment library. The first ranked fragment is shown in Figure 9.16, structure b. Visual comparison of the ashwagandhanolide structure with the structure of the fragment confirms that the fragment can be used as a substructure of the molecule analyzed and its carbon chemical shifts are close to the experimental values. The procedure of creating MCDs from the FFs was initiated and the program produced 960 MCDs. Checking MCDs for contradictions took 28 min, and structure generation resulted in $k = 960 \rightarrow 24 \rightarrow 6$, $t_g = 22$ s.

The three top structures in the ranked file are shown in Figure 9.17. The most probable structure, #1, coincides with the structure determined in the publication.[38]

From the very beginning, the StrucEluc system was intended for the identification of standard organic molecules. Since ionic structures are possible for both natural products and synthesized organic molecules, the structure generation algorithm was also adjusted to allow for the generation of *ionic* structures. Thus, in principle, the system is now capable of elucidating structures of any organic molecules regardless of their topological and electronic properties.

## 9.9   Challenges for Empirical and GIAO DFT NMR Prediction

In Chapter 3, empirical and quantum mechanical (QM) methods of NMR chemical shift prediction were compared as tools for correct structure selection during the process of molecular structure elucidation. In this examination it was interesting to learn whether empirical methods can be useful under the conditions when $d_A > 5$ ppm [*i.e.* MAE(HOSE) $> 5$ ppm] and how these methods perform in regard to structures in the literature (Bagno and co-workers[39]) deemed to be challenging.

Structure **9.17**, daphnipaxinin, is a structure suggested by Bagno *et al.*[39] to be an example of an unusual molecule which may not be properly treated using empirical approaches for NMR spectrum prediction. The assignment for structure **9.17** was performed by Yang *et al.*[40] who were the first to elucidate the structure.

**9.17**

This molecule provided an interesting example to test and challenge empirical methods of $^{13}$C chemical shift prediction. For structure **9.17**, the MAE(HOSE) and MAE(NN) values were $\sim 6.3$ ppm and displayed maximum deviations of $d_{max}$(HOSE) $= 14.29$, $d_{max}$(NN) $= 17.12$ ppm, whereas the QM calculations predicted the $^{13}$C NMR shifts more accurately giving MAE(QM) $= 3.92$ ppm. Using the facilities of ACD\CNMR Predictor to examine the calculation protocol we determined that the HOSE code algorithm failed to accurately predict the chemical shifts for two of the carbon atoms (those resonating at 179.5 and 113.8 ppm), because the database has no reference structures containing the atoms with the necessary environments. Nevertheless, the program offered chemical shift values of 166.2 and 115. ppm corresponding to these atoms using as an approximation the NN algorithms.

As mentioned earlier, the main application of chemical shift prediction is to confirm the correct structural hypothesis during the process of molecular structure elucidation. Therefore it was interesting to investigate whether an empirical approach could be applicable to the identification of structure **9.17**, in spite of the low prediction accuracy. The HMQC, HMBC, and COSY data of structure **9.17** presented in the work[40] were input into the StrucEluc[29] software. The program automatically detected the presence of NSCs.[12] Because of the presence of these NSCs, fuzzy structure generation[41] (see details in the next Chapter) was initiated. Structure generation options were set which assume the presence of an unknown number, $m$, of NSCs having an unknown length in the COSY and HMBC data. The following solution was found at a value of $m = 5$, $k = 10456 \rightarrow 5056 \rightarrow 2017$, $t_g = 2$ min 58 s.

According to the general CASE strategy,[29] the final structures were then ranked by $d_N$ values and HOSE code-based chemical shift predictions were performed for the first 20 structures of the ranked file. Then the structures were sorted based on increasing $d_{HOSE}$ values. The first three structures ranked in

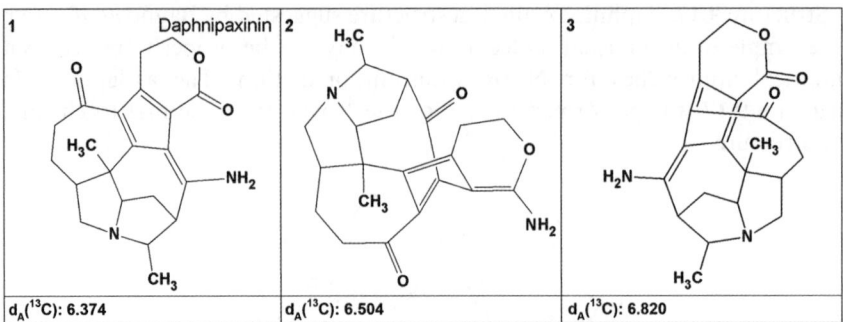

**Figure 9.18**   The first three structures of the output file ordered in ascending order of $d_{HOSE}$ values. The structure of daphnipaxinin is listed in first position.

ascending order of $d_{HOSE}$ values are shown in Figure 9.18. As we see, the suggested structure of daphnipaxinin was distinguished by the program to be the most probable. At the same time, automated [13]C NMR chemical shift assignment agreed with that suggested by the authors.[39,40] The next two structures have slightly larger deviations and, in addition, they contain strained and somewhat "exotic" fragments, which make them questionable.

Therefore, in spite of the unusual character of the structure and the large values of the deviations, an "engineering approach" (see Chapter 3) allows the program to correctly select this challenging structure from among 2000 candidates, although with very little preference over the closest members of the output file.

Bagno *et al.*[39] also tested the method of QM-based [13]C chemical shift prediction with other unusual structures which might seem challenging for empirical methods, namely buletunone (**9.18**) and corianlactone (**9.19**). It has been shown[42] that the application of StrucEluc allowed one to confidently identify the buletunone molecule from the 2D NMR data. The solution of this problem will be discussed in Section 13.2.

9.18                                              9.19

It was determined that the uncommon nature of the corianlactone structure, **9.19**, did not prevent the problem being solved using empirical methods of [13]C chemical shift prediction within the StrucEluc system. The 2D NMR data of

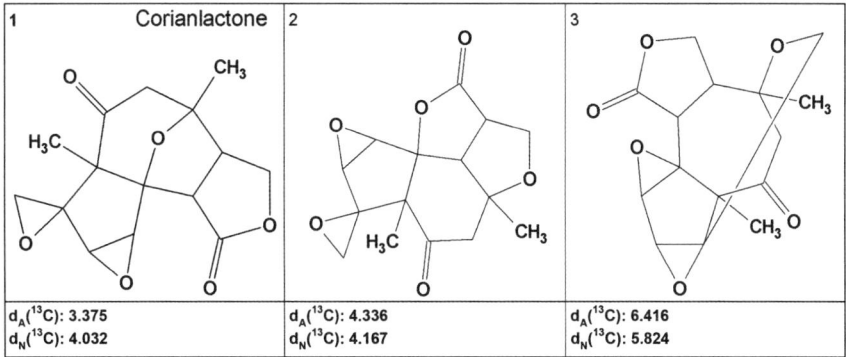

| 1 Corianlactone | 2 | 3 |
|---|---|---|
| $d_A(^{13}C)$: 3.375 | $d_A(^{13}C)$: 4.336 | $d_A(^{13}C)$: 6.416 |
| $d_N(^{13}C)$: 4.032 | $d_N(^{13}C)$: 4.167 | $d_N(^{13}C)$: 5.824 |

**Figure 9.19** The first three structures of the ordered output file resulting from the structure elucidation of the corianlactone molecule (**9.19**) using StrucEluc.

this compound were taken from the original publication[43] and input into the StrucEluc software. The following results were obtained: $k = 83 \rightarrow 72 \rightarrow 65$, $t_g = 4.7$ s. The three best structures in the ordered output file are shown in Figure 9.19.

The structure of corianlactone was confidently identified with the aid of the StrucEluc software in combination with ACD/CNMR Predictor. As has been demonstrated,[44] (see Chapter 4) empirical methods of $^{13}C$ chemical shift prediction can also be used for selecting the preferable configurations from a full set of stereoisomers associated with a given molecular structure. StrucEluc generated all of the 256 stereoisomers of corianlactone and the most probable relative configuration, as shown by structure **9.19**, was determined using HOSE- and NN-based $^{13}C$ NMR spectrum prediction. Stereoisomer **9.19** was ranked as the most likely isomer with MAE(HOSE) = 2.93 ppm and MAE(NN) = 3.89 ppm, whereas the MAE(QM) value found for structure **9.27** using the GIAO approach was 5.3 ppm.[39]

# References

1. M. E. Elyashberg, K. A. Blinov and E. R. Martirosian, *Lab. Autom. Inf. Manag.*, 1999, **34**, 15.
2. M. E. Elyashberg, A. J. Williams and G. E. Martin, *Prog. NMR Spectrosc.*, 2008, **53**, 1.
3. M. E. Elyashberg, K. A. Blinov, A. J. Williams, S. G. Molodtsov and E. R. Martirosian, *J. Nat. Prod.*, 2002, **65**, 693.
4. M. E. Elyashberg, E. R. Martirosian, Y. Z. Karasev, H. Thiele and H. Somberg, *Anal. Chim. Acta*, 1997, **337**, 265.
5. K. Funatsu and S. Sasaki, *J. Chem. Inf. Comp. Sci.*, 1996, **36**, 190.
6. B. D. Christie and M. E. Munk, *J. Chem. Inf. Comput. Sci.*, 1988, 28.

7.  M. Will, W. Fachinger and J. R. Richert, *J. Chem. Inf. Comput. Sci.*, 1996, **36**, 221.

8.  T. E. Elyashberg, Y. Z. Karasev and E. R. Martirosian, *Anal. Chim. Acta*, 1999, **388**, 353.

9.  S. G. Molodtsov, *Commun. Math. Chem. (MATCH)*, 1994, **30**, 213.

10. S. G. Molodtsov, *Commun. Math. Chem. (MATCH)*, 1994, **30**, 203.

11. S. G. Molodtsov, *Commun. Math. Chem. (MATCH)*, 1998, **37**, 157.

12. S. G. Molodtsov, M. E. Elyashberg, K. A. Blinov, A. J. Williams, G. E. Martin and B. Lefebvre, *J. Chem. Inf. Comput. Sci.*, 2004, **44**, 1737.

13. W. Bremser, *Anal. Chim. Acta*, 1978, **103**, 355.

14. P. Willett, J. Barnard and G. Downs, *J. Chem. Inf. Comput. Sci.*, 1998, **38**, 983.

15. A. G. Gonzales, F. Leon, L. Sanchez-Pinto, J. I. Pardon and J. Bermejo, *J. Nat. Prod.*, 2000, **63**, 1297.

16. H. M. Ge, B. Huang, S. H. Tan, D. H. Shi, Y. C. Song and R. X. Tan, *J. Nat. Prod.*, 2006, **69**, 1800.

17. H. Shigemori, S. Shimamoto, M. Sekiguchi, A. Ohsaki and J. Kobayashi, *J. Nat. Prod.*, 2002, **65**, 82.

18. J. Junker, W. Maier, T. Lindel and M. Köck, *Org. Lett.*, 1999, **1**, 737.

19. S. Ley, K. Doherty, G. Massiot and J.-M. Nuzillard, *Tetrahedron*, 1994, **50**, 12267.

20. J.-M. Nuzillard and G. Massiot, *Tetrahedron*, 1991, **47**, 3655.

21. C. Koren-Goldshlager, M. Aknin and Y. Kashman, *J. Nat. Prod.*, 2000, **63**, 830.

22. C. Koren-Goldshlager, M. Aknin, E. Gaydou and Y. Kashman, *J. Org. Chem.*, 1998, **63**, 4601.

23. A. Rudi and Y. Kashman, *J. Org. Chem.*, 1989, **54**, 5331.

24. H. Luesch, W. Y. Yoshida, R. E. Moore, V. J. Paul and S. L. Mooberry, *J. Nat. Prod.*, 2000, **63**, 611.

25. T. Ohta, Maruyama, M. Nagahashi, Y. Miyamoto, S. Hosoi, F. Kiuchi, Y. Yamazoe and S. Tsukamoto, *Tetrahedron*, 2002, **58**, 6631.

26. H. Luesch, W. Y. Yoshida, R. E. Moore and V. J. J. Paul, *J. Nat. Prod.*, 2000, **63**, 1106.

27. B. Marquez, K. S. Watts, A. Yokochi, M. A. Roberts, P. Verdier-Pinard, J. I. Jimenez, E. Hamel, P. J. Scheuer and W. H. Gerwick, *J. Nat. Prod.*, 2002, **65**, 866.

28. R. J. Capon, J. Ford, E. Lacey, J. H. Gill, K. Heiland and T. Friedel, *J. Nat. Prod.*, 2002, **65**, 358.

29. M. E. Elyashberg, K. A. Blinov, S. G. Molodtsov, A. J. Williams and G. E. Martin, *J. Chem. Inf. Comput. Sci.*, 2004, **44**, 771.

30. K. A. Blinov, D. Carlson, M. E. Elyashberg, E. R. Martirosian, S. G. Molodtsov and A. J. Williams, *J. Magn. Reson. Chem.*, 2003, **41**, 359.

31. S. Y. Ablordeppey, D.-B. Hufford, R. F. Borne and D. Dwumu-Badu, *Planta*, 1990, **56**(4), 416.

32. C. E. Hadden, W. K. Duholke, J. E. Guido, R. H. Robins, G. E. Martin, M. H. M. Sharaf and P. L. Schiff Jr., *J. Heterocycl. Chem.*, 1999, **36**, 525.

33. T. D. Spitzer, R. C. Crouch, G. E. Martin, M. H. M. Sharaf, P. L. Schiff Jr., A. N. Tackie and G. L. Boye, *J. Heterocycl. Chem.*, 1991, **28**, 2065.

34. M. H. M. Sharaf, P. L. Schiff Jr., G. E. Martin, C. H. Phoebe Jr. and A. N. Tackie, *J. Heterocycl. Chem.*, 1996, **33**, 239.

35. M. H. M. Sharaf, P. L. Schiff Jr., R. C. Crouch, A. Davies, C. W. Andrews, G. E. Martin, C. H. Phoebe Jr. and A. N. Tackie, *J. Heterocycl. Chem.*, 1995, **32**, 1631.

36. R. Crouch, A. Davies, T. D. Spitzer, G. E. Martin, C. H. Phoebe Jr, M. H. M. Sharaf, P. L. Schiff Jr. and A. N. Tackie, *J. Heterocycl. Chem.*, 1995, **32**, 1077.

37. A. N. Tackie, G. L. Boye, M. H. M. Sharaf, P. L. Schiff, R. C. Crouch, T. D. Spitzer, R. L. Johnson, J. Dunn, D. Minick and G. E. Martin, *J. Nat. Prod.*, 1993, **54**, 653.

38. G. V. Subbaraju, M. Vanisree, C. V. Rao, C. Sivaramakrishna, P. Sridhar, B. Jayprakasam and M. G. Nair, *J. Nat. Prod.*, 2006, **69**, 1790.

39. A. Bagno, F. Rastrelli and G. Saielli, *Chemistry*, 2006, **12**, 5514.

40. S.-P. Yang and J.-M. Yue, *Org. Lett.*, 2004, **6**, 1401.

41. M. E. Elyashberg, K. A. Blinov, S. G. Molodtsov, A. J. Williams and G. E. Martin, *J. Chem. Inf. Model.*, 2007, **47**, 1053.

42. M. E. Elyashberg, K. A. Blinov and A. W. Williams, *Magn. Reson. Chem.*, 2009, **47**, 371.

43. Y.-H. Shen, S.-H. Li, R.-T. Li, Q.-B. Han, Q.-S. Zhao, L. Liang, H.-D. Sun, Y. Lu, P. Cao and Q.-T. Zheng, *Org. Lett.*, 2004, **6**(10), 1593.

44. M. E. Elyashberg, K. A. Blinov and A. W. Williams, *Magn. Reson. Chem.*, 2009, **47**, 333.

CHAPTER 10

# The Challenge of Non-Standard Spectral Responses and the Role of Fuzzy Structure Generation

Examples discussed in Chapters 6 and 9 lead us to conclude that the CASE 2D NMR methodology can provide solutions for computer-assisted structure elucidation tasks in a reasonable time if the 2D NMR and MS data are *true, consistent* and *complete* (*i.e.* the number of observed NMR correlations is large enough to sufficiently define the connectivities within a structure). If at least one of these conditions is violated the possibility to somehow find a correct solution to the problem decreases significantly. As was shown earlier in Chapter 8, methods to overcome the presence of contradictions in experimental data have been suggested and the corresponding algorithms have been developed and implemented in the StrucEluc system. In this Chapter we will consider the different StrucEluc-based strategies for molecular structure elucidation in those situations when 2D NMR spectra contain non-standard correlations (NSCs). These strategies were developed and tested using a series of more than 60 separate structure elucidations with deliberately contradictory 2D NMR data. 2D NMR literature data, as well as raw 2D NMR spectra, were used in this study. The spectral data were entered into the program and the structures were assumed to be unknown. Later, an attempt was made to elucidate the structures on the assumption that no information existed regarding the presence of contradictions in the 2D NMR data. These problems were used to perform numerous experiments to determine and remove the contradictions. This was feasible, since the 2D NMR data related to these problems were known to contain COSY and/or HMBC connectivities of non-standard length.

It was shown that the program was capable of determining the *presence* of connectivities of non-standard length in 90% of all cases using the MCD

New Developments in NMR No. 1
Contemporary Computer-Assisted Approaches to Molecular Structure Elucidation
By Mikhail Elyashberg, Antony Williams and Kirill Blinov
© Royal Society of Chemistry 2012
Published by the Royal Society of Chemistry, www.rsc.org

checking procedure described in Chapter 8. These results are very encouraging, since experimental methods guaranteeing the precise determination of COSY and HMBC connectivity lengths are not available. Knowledge of the presence of contradictions in 2D NMR data gives the investigator valuable information that can determine the strategy of structure elucidation with these data. An invalid suggestion regarding the presence of non-standard connectivities may appear if there are contradictions between the atom properties (the kinds of hybridization) and the atom property correlation table (APCT, see Chapter 7) used during the data analysis. The validity of this statement can be checked by repeated data verification with the APCT disabled. A false statement can also appear in those rare cases when an unknown under study contains a pair of bonded heteroatoms if the absence of such atomic pairs is set in the program options as the default. In these cases a program message regarding the existence of contradictions can help the chemist to reveal the presence of bonded heteroatoms. For instance, it allowed the detection of the O-O group from only the published shift data during the elucidation of mycaperoxide H.[1] With this in mind, there were no cases of incorrect detection of non-standard connectivities when solving approx. 200 tasks where contradictions in 2D NMR data were not present.

For about 50% of the cases studied the program not only identified the contradictions in the data correctly, but was able to successfully remove them automatically to allow determination of the correct structure. Examples were encountered where the program resolved contradictions caused by the presence of a large number of NSCs, up to a total of eight.

In six out of 50 cases the program was unable to determine the presence of connectivities of non-standard length. In five of those six cases the 2D NMR data contained only one non-standard HMBC connectivity. This occurrence can be explained by the fact that if there are only one or two HMBC non-standard connectivities in the data, the atoms in the structure may be arranged so that their arrangement complies with the standard length of all connectivities. If the number of NSCs is large, such an arrangement of atoms is unlikely. The presence of implicit non-standard connectivities can become apparent as a result of structure generation and their subsequent filtration with the use of spectral libraries: if all of the generated structures obviously contradict the spectral data, the program produces an empty results file. Indirect evidence of the possibility that contradictions were not detected may not only be an empty result file but large values, more than 3.5–4.5 ppm, of the chemical shift deviations, $d_A$ and $d_N$, calculated for the first ranked structure.[2,3] Investigations have shown that non-standard connectivities were detected by both direct and indirect methods for 95% of the analyzed tasks containing contradictory data. If there are reasons to assume that the program did not detect contradictions in the initial data it would be highly likely that the problem could be solved with the use of "fuzzy structure generation" (FSG), as will be described in Section 10.3.

Since it is possible that 2D NMR data can contain implicit non-standard connectivities, the most probable structure generally requires additional verification by independent methods. It has been shown[4] that incorrect structures are often rejected on the basis of predicted chemical shifts and multiplets in the

¹H NMR spectrum (see the examples below). However, the most effective method, as we will see, is application of a process known as FSG. If the structure is generated after automated removal of contradictions then it is still desirable to check for the presence of non-standard connectivities. The connectivities can be verified with appropriate experimental parameter optimization to probe the values of the spin couplings.[5,6]

Therefore it is not always possible to find non-standard connectivities and to automatically resolve the contradictions in 2D NMR data sets. In practice, the following difficult situations may typically arise:

1. The program detects the presence of NSCs and makes an attempt to remove the contradictions in the data, but then reports that contradictions cannot be removed. Frequently, FSG can help to solve the problem. Generally, additional experiments are required in an effort to detect NSCs.
2. The program fails to detect non-standard connectivities and displays a message informing the researcher about the absence of contradictions. In this case, structure generation is initiated. The following outcomes are possible: (a) no structure is generated; and (b) the wrong structure(s) is generated, which can generally be recognized because of the large values of the ¹³C experimental *vs.* predicted deviations. Again, FSG can help in this situation.
3. The program detects the presence of non-standard connectivities, makes an attempt to remove the contradictions in the data, and displays a message that the contradictions were removed, although, in fact, some contradictions still remain. This is due to the fact that not all non-standard connectivities are lengthened. Possible non-desirable consequences, and the methods to overcome them, are similar to those listed for point 2 above.

It should be obvious that the most dangerous situations are when incorrect solutions are produced and these can occur for points 2 and 3 above and will be considered later in more details. Even in those cases when the program is not able to remove detected contradictions, specifically case 1, the fact that contradictions are identified is still extremely important.

We will first examine a series of examples of the successful automatic removal of contradictions from 2D NMR data and then will consider examples that demonstrate how a correct result can be obtained in those cases where the program does not remove all contradictions.

## 10.1   Examples of Successful Removal of Contradictions

*Example 1*
In the work of Leone *et al.*[7] five compounds related to a class of polyoxygenated sterols were identified and examined. For three of these compounds

the authors presented 1D $^1$H (600 MHz) and $^{13}$C (150 MHz) NMR data, as well as HMBC and COSY correlations, in the form of peak tables. Automated structure elucidation for two of the three natural products was performed without difficulty, since the related data were rather complete and coherent. The third compound (10.1) was more challenging and the analysis is presented here in detail.

**10.1**

The given molecular formula was $C_{29}H_{46}O_5$ and MW = 474. The molecular formula and peak lists for the HMQC, COSY, and HMBC spectra were fed as inputs into the program. These formed tables containing 43 COSY cross-peaks and 57 HMBC correlations. The default settings of the StrucEluc program were, as usual, set so that the COSY and HMBC cross-peaks were defined to result from coupling interactions within 2 or 3 bonds. However, the comparison of cross-peaks with the molecular structure, a process which can in reality only be performed after the structure of the unknown compound is determined, showed that there are four long-range correlations in the COSY spectrum corresponding to the following couplings (marked on structure 10.1 by arrows): $^5J_{HH}$ (H20–H25), $^4J_{HH}$ (H19–H21), $^4J_{HH}$ (H12–H26), and $^4J_{HH}$ (H7–H14). It is evident that these correlations contradict the default options and consequently the correct structure will not be generated. As mentioned previously, 2D NMR cross-peak intensity information can be useful, although not definitive, since the correlation response intensity is a function of both the size of the coupling, and the experimental optimization, in helping to define which correlations are "non-standard". Intensity information was, however, omitted in the peak table presented in the reference.[7]

The MCD was automatically formed by the program on the basis of the peak tables. A part of the MCD related to the COSY spectrum of compound 10.1 is

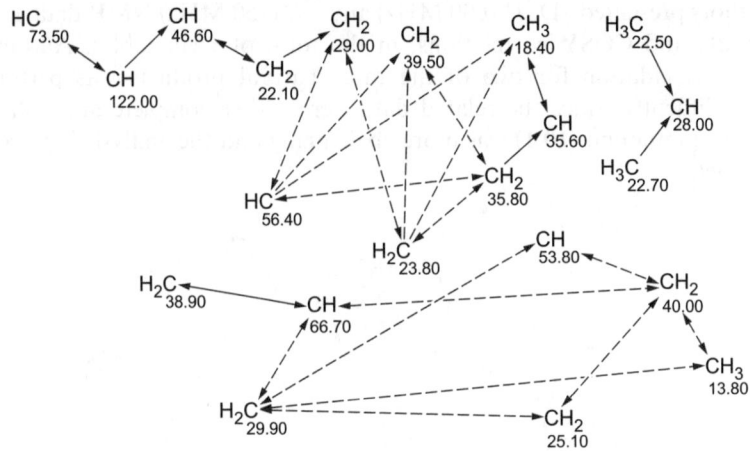

**Figure 10.1** A part of the MCD related to the COSY spectrum of compound **10.1**. Ambiguous COSY connectivities are drawn with dotted lines.

shown in Figure 10.1. For convenience, connectivities having a length of one C–C bond, typical for a COSY correlation, will be called α-connectivities and those corresponding to a length of one to two C–C bonds (typical for an HMBC correlation) will be referred to as β-connectivities. Correlations spreading over three C–C bonds will be referred to as γ-connectivities. For the example, as shown in Figure 10.1, the methylene groups associated with C(23.8), C(29.0), C(35.8), C(39.5), C(40.0) have three or more α-connectivities (some of them are ambiguous), which can lead to contradictions regarding the valence of the carbon atom. It is obvious that if three or more unambiguous connectivities go from a $CH_2$ group, then at least one of them is definitely of non-standard length (see Chapter 8). In the case under consideration visually, it can expected that the COSY data may have at least one NSC.

When the MCD was checked for contradictions some α-correlations were suggested to contradict the standard values. The procedure for resolving contradictions required four iterations and 14 s to resolve all contradictions.

An important feature of an advanced expert system is the ability to create an audit trail of the actions and decision-making process. This is particularly useful when a scientist needs to review, repeat or tweak the elucidation. To support this function a "contradiction resolution protocol" is generated by the StrucEluc system. It is worth noting that the minimal number of NSCs which can be present in the 2D NMR data is determined and shown in the protocol. In addition, connectivities that were edited by the program appear in the MCD marked by colors corresponding to the new connectivity lengths. In this case, the protocol indicated that the program detected and lengthened connectivities by one bond in the following order (see structure **10.1**): (C-20–C-25), (C-19–C-21), (C-12–C-26), and (C-7–C-14). The protocol

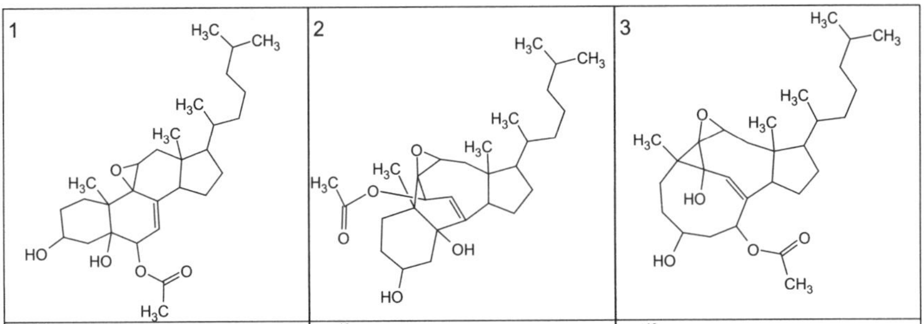

| 1 | 2 | 3 |
|---|---|---|
| $d_A(^{13}C)$: 1.693 | $d_A(^{13}C)$: 2.577 | $d_A(^{13}C)$: 2.851 |
| $d_I(^{13}C)$: 2.202 | $d_I(^{13}C)$: 2.649 | $d_I(^{13}C)$: 2.490 |
| $d_N(^{13}C)$: 1.766 | $d_N(^{13}C)$: 2.341 | $d_N(^{13}C)$: 2.528 |

**Figure 10.2** The results of the structure elucidation of compound **10.1**. The top three structures of the file are arranged in order of the increasing average deviation of the calculated spectra.

also summarized the steps producing a complete list of the new lengths of connectivities introduced to resolve the contradictions.

On the basis of the MCD with the corrected connectivity lengths, the generation and filtration of structures using spectral and structural libraries was performed. No changes were made to the default atom properties (hybridization, possible neighborhood with heteroatoms, *etc.*). The program generated 18 molecules ($k = 18$) in 5 s ($t_g = 5$ s). To determine the most preferred chemical structure, the $^{13}C$ NMR chemical shifts of the structures in the output file were calculated using methods described earlier (see Chapter 3). The top three structures of the file, arranged in order of increasing average deviation of the calculated spectra, are shown in Figure 10.2.

Structure **10.1** was listed as the first entry of the prioritized output file by all three methods of spectrum prediction. This is denoted as $r_{all} = 1$. The value of the deviation $d_A(1) = 1.69$ ppm and was markedly different from the value of that for the second structure [$d_A(2) - d_A(1) = 0.8$ ppm]. A significant difference in $d_A$ between the first and second structure is generally a good indication of the validity of the solution as discussed in Chapter 9. In this case the program successfully dealt with identifying and resolving all contradictions and elucidating the correct structure. Unfortunately, the program is not always able to resolve such issues because the algorithm for searching and resolving contradictions (Chapter 8) is based on heuristic principles.

*Example 2*

In the work of Shen *et al.*[8] a natural product (**10.2**), related to the class of taxoids, was identified. For this compound the authors presented $^1H$ and $^{13}C$ NMR data, as well as COSY and HMBC correlations, in the form of peak tables. These data were used as inputs to the StrucEluc system. The given molecular formula was $C_{28}H_{42}O_{12}$ with MW $= 570$ and the number of skeletal atoms was $n_{sk} = 40$:

**10.2**

The molecular formula and the cross-peaks from the COSY, HMQC, and HMBC spectra were fed as inputs into the program. These formed tables containing 22 HMQC direct correlations, 21 COSY cross-peaks and 47 HMBC correlations. Comparison of the cross-peaks with the molecular structure showed that one long-range correlation $^4J_{HH}$ (H-2–H-18) exists in the COSY spectrum and one $^4J_{CH}$ exists in the HMBC (H-16–C-22). These long-range COSY and HMBC connectivities are shown on structure **10.2**, with the corresponding two-sided blue and one-sided green arrows. Information regarding experiments utilized to check the correlation lengths was not available in the reference.[8]

The MCD was created by the program, and when the "check MCD for contradictions" command was applied the program displayed the appropriate message that some correlations contradicted the standard values. The algorithms attempted to resolve the contradictions. The procedure required three iterations and 3 s to successfully elongate contradictory connectivities. For this example, one COSY connectivity and all HMBC connectivities belonging to C-16, as well as one COSY connectivity at C-2, were lengthened by one bond. The protocol indicates the results of each iterative step of the program and produces a complete list of the new lengths of connectivities introduced.

The structure generation was run based on the corrected MCD with the following result: $k = 168 \rightarrow 150 \rightarrow 19$, $t_g = 2$ s). The deviation values ($d_A = 1.44$ ppm, $d_I = 1.54$ ppm, $d_N = 1.33$ ppm) are appropriate for a correct structure determination.

*Example 3*
The program was applied to the structure elucidation of betulinic acid, **10.3** (Peng *et al.*[9]), $C_{30}H_{48}O_3$, whose COSY spectrum contained only one non-standard connectivity (C-64)–(C-66) indicated by the blue arrow.

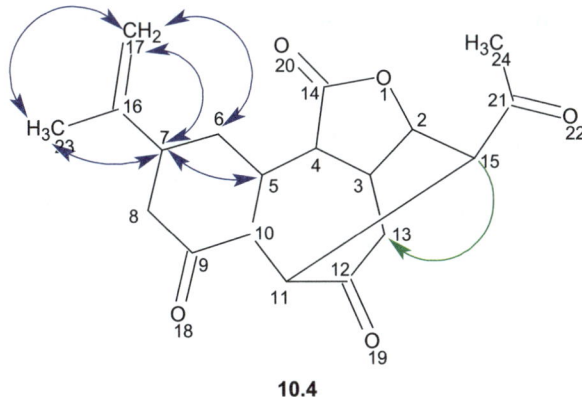

**10.3**

Contradictions in the 2D NMR data were searched and the one contradiction identified was removed by the program and, as a result, only one correct structure was generated in 1 s.

In the relatively simple cases described above, the program successfully overcame contradictions in the presence of several non-standard connectivities. However, the number of NSCs may be much larger, up to 15 or more.

*Example 4*

**10.4**

In the case of the analysis of horiolide, **10.4** (Radhika *et al.*[10]), whose 2D NMR spectra contain five NSCs in the COSY data and one in the HMBC spectrum, as indicated with arrows on the structure, the program found and automatically removed all contradictions. As a result, only one and correct structure was generated; $t_g = 0.066$ s, $d_A = 2.74$, $d_I = 2.78$, $d_N = 2.51$ ppm.

*Example 5*

The 2D NMR spectra of triterpene E, $C_{30}H_{48}O_2$ (**10.5**), which was isolated and characterized by Reynolds *et al.*,[11] contained a total of eight non-standard connectivities: seven in the COSY spectrum and one in the HMBC.

**10.5**

All contradictions were resolved automatically in two iterations, and one, correct structure, was generated in 0.5 s with $d_A = 1.67$ ppm.

In the next section we will consider some more complicated cases when the problem of contradiction presence in 2D NMR data was solved with the aids of StrucEluc.

## 10.2   Resolving Contradictions in More Complicated Cases

*Example 1*
This example demonstrates the ability of the program to verify the validity of the 2D NMR data and the results of structure elucidation reported in the literature. Horgen *et al.*[12] reported the structure of a novel natural product (**10.6**), isolated and identified as malevamide-D on the basis of spectroscopic data.

**10.6**

The high-resolution ESI mass spectrum of **10.6** gave a molecular ion $m/z$ 733.5114, suggesting the molecular formula $C_{40}H_{68}N_4O_8$. The $^1$H, $^{13}$C, COSY,

HMQC, and HMBC NMR spectra were recorded and presented in a peak table without any indication of the presence of correlations of non-standard lengths. The information captured in this table was used to determine the structure using the StrucEluc program. Checking for contradictions did not reveal the presence of any connectivities of non-standard length, and structure generation was completed in the automated mode without any constraints imposed. The results were: $k = 40 \rightarrow 10$, $t_g = 4$ s. Ranking the obtained structures suggested the following structure as the most likely:

**10.6′**

This solution does not formally allow one to question the validity of the generated structure, since the deviations of the three different types of calculated spectra from the experimental ones are well within the statistical limits: $d_A(1) = 1.64$, $d_I(1) = 1.86$, and $d_N(1) = 1.83$ ppm. Comparison of the resultant structure with that obtained by the authors[12] shows clearly that they differ in the functional groups highlighted in the central part of the molecule. The differing fragments are marked with bold lines in structures **10.6** and **10.6′**. It can be seen that the differences are fairly subtle. The calculation of the Tanimoto similarity match factor between the structures gave a value of 0.96. The validity of the generated structure becomes doubtful when the difference between the experimental and calculated $^1$H NMR spectra is considered. Using the ACD/HNMR prediction component of the program, comparison of the spectra indicated differences in the region 0.5–1.5 ppm: the predicted doublet of the methyl group (C-6) at $\delta = 1.4$ ppm is absent in the experimental spectrum. There are two suggestions that will explain this observation: either the authors[12] elucidated the wrong structure, or the 2D NMR data contains non-standard connectivities that were not revealed by the program.

Comparison of the HMBC correlations with the chemical shift assignment performed for structure **10.6** identified the presence of the $^6J_{CH}$ correlation (C-7–C-21), as indicated with the dotted arrow in structure **10.6**. However, the search for contradictions in the 2D NMR data produced no results. This indicates that the program identified an arrangement(s) of atoms under which the corresponding topological distances between the carbon nuclei, represented by their chemical shifts, are in agreement with the default values of 2D NMR correlations. In other words, the program found that it was possible to generate a chemical structure that meets both the constraints and the atom properties assigned by the program with the help of the APCT. As can be seen in structure **10.6′** atoms C-7 and C-21 are separated by two C–C bonds that correspond to the standard HMBC connectivity.

This observation was communicated to the authors,[12] with the result of a misprint in the published tables being identified. The experimental data actually displayed a HMBC connectivity (C-6–C-7) of standard length (see structure **10.6**). After the initial data were corrected, the generation was repeated with the following results: $k = 156 \rightarrow 44$, $t_g = 4$ s, $d_A(1) = 0.721$ ppm, $d_I(1) = 1.22$ ppm, $d_N(1) = 1.18$ ppm, $r_A = r_I = r_N = 1$. The most highly ranked structure that was generated coincided with that reported by the authors, structure **10.6**. The deviation values $d_A(1)$, $d_I(1)$, and $d_N(1)$ were also less than those calculated initially for the wrong structure. Note that the HMBC correlation between $CH_3$-6 and $CH_3$-7 would generally not be possible in structure **10.6′** or, if a correlation were observed, it would be very weak.

The conclusion of this study is that if the program is not able to reveal non-standard connectivities, it is possible that the program can generate a structure similar to the correct one. Therefore it is always desirable to verify the solutions suggested by the software for their correctness and stability. It can be very effective to check the structural validity by comparing the experimental and predicted $^1$H NMR spectral patterns visually. The stability of the result can be verified by lengthening some connectivities to which the weakest peaks of the 2D NMR spectra correspond, by loosening the limitations on the atom properties or by using the fragments from the knowledge base.

*Example 2*

The work of Habtemariam *et al.*[13] reported the structure of a novel steroid (**10.7**) isolated and identified as 17-epiacnistin A on the basis of both spectroscopic and X-ray diffraction data. In order to aid the discussion here the diagram includes all chemical shifts assigned to the carbon atoms.

**10.7**

The high-resolution electron impact mass spectrum of **10.7** indicated a molecular ion at $m/z$ 470, giving the molecular formula $C_{28}H_{38}O_6$. $^1H$, $^{13}C$, $^1H$-$^1H$ COSY, $^1H$-$^1H$ NOESY, HMQC, and HMBC (optimized for 7 Hz long-range couplings) NMR spectra were recorded using a Bruker AMX-400 spectrometer. However, only the $^1H$, $^{13}C$, and HMBC data were listed in the article[13] without any indication of the observation of HMBC correlations of non-standard lengths. Only this information was used to determine the structure using StrucEluc. Comparison of the HMBC correlations with the chemical structure (**10.7**) allowed identification of the presence of four HMBC connectivities of non-standard length, as indicated with arrows.

In addition to NSCs in the HMBC data, two proton shifts, 1.75 and 2.5 ppm, were both associated with two $^{13}C$ resonances each: CH(50.5; 1.75), $CH_2$ (23.5; 1.75) and CH(53.2; 2.5), CH2(34.6; 2.5) (see structure **10.7**). Superposition of the $^1H$ NMR signals listed increased ambiguity in the 2D NMR data.

The 2D NMR data were converted into connectivities to create a MCD. The MCD showed that $CH_2$(23.5) and CH(53.2) groups had no correlations with other carbon atoms. According to the original article,[13] the carbon atom with $\delta$23.5 ppm, in fact, does not correlate with any other atom, whereas the CH $\delta$53.2 ppm was deprived of connectivities by the program due to accidental degeneracy of signals in the $^1H$ NMR spectrum as mentioned above. The presence of two non-correlating carbon atoms in the MCD usually results in a long structure generation time and the number of potential structures will correspondingly be large, even in the case when the 2D NMR data are consistent.

The search for contradictions in the 2D NMR data produced no results. With the absence of contradictions, the structure generation procedure was initiated with no additional restrictions. As expected, the structure generation process was rather long. The program had produced no resulting chemical structures within 45 min and the generation process was aborted. Following the general strategy of problem solving using the StrucEluc system, an attempt to use fragments stored in the system database was made.

The program selected $L = 1846$ library fragments from the database that were consistent with the molecular formula of the analyzed compound, taking into account both the chemical shifts and the multiplicities in the 1D $^{13}C$ NMR spectrum. The first 35 records from the resulting file, with the selected fragments ranked in decreasing order of their molecular formula, are shown in Figure 10.3.

Among the selected fragments there are a number that are present in the "unknown" structure. The high probability that there will be fragments consistent with a truly unknown structure is critically important if the structure is to be successfully solved by StrucEluc. Obviously, however, when dealing with a true unknown, there is also no way to know if, in fact, any of the fragments are really consistent with the structure of the molecule. The MCD creation from all of the found fragments (FFs) was started using an initial value of $E = 0.5$ ppm (see Section 9.2.2.3). With $E = 0.5$ ppm, no MCDs were created, but with a value of $E = 2$ ppm the program created one molecular connectivity diagram (see Figure 10.4). It can be seen that the fragment of the MCD denoted

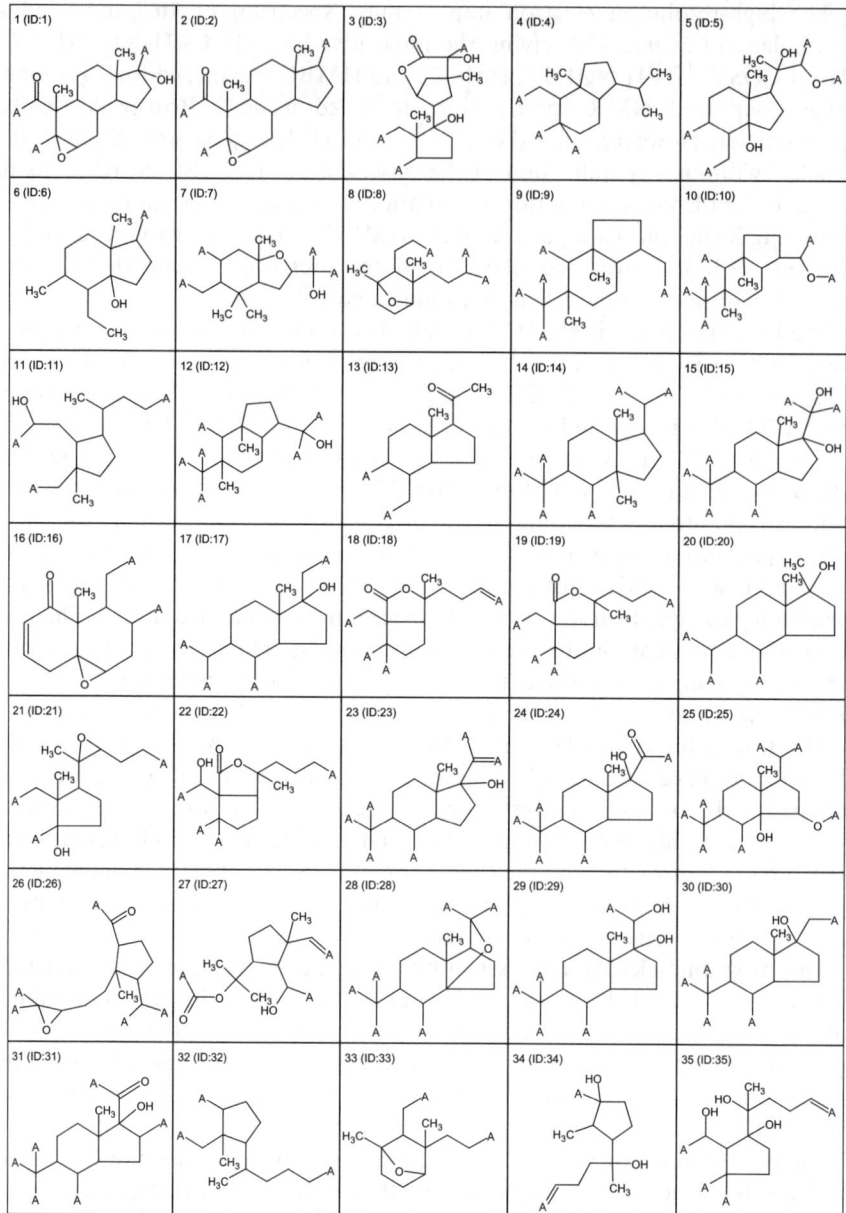

**Figure 10.3** The initial structures in the FF list obtained as a result of a fragment search using the $^{13}C$ NMR spectrum of compound **10.7** as input.

by bold lines corresponds to a portion of the structure **10.7**. Two oxygen atoms from the original molecular formula were incorporated but the "free" carbons resonating at δ 23.5 and 53.2 were not present in the fragment

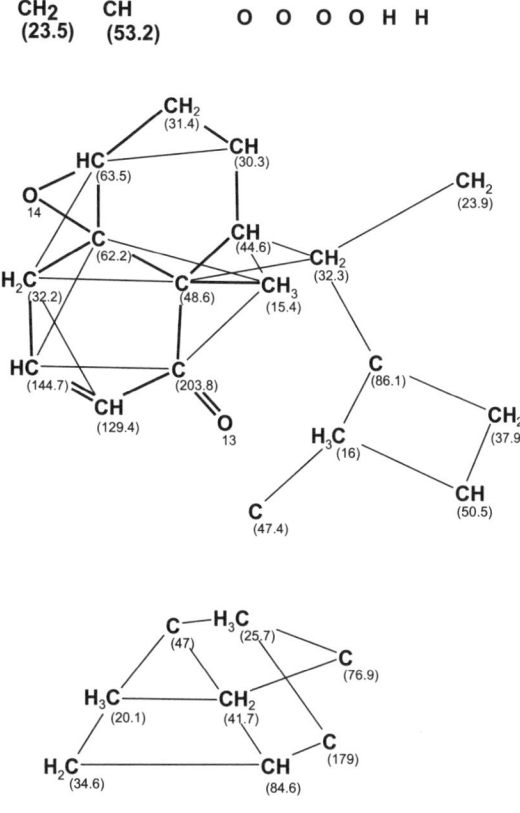

CH$_2$  CH  O  O  O  O  H  H
(23.5)  (53.2)

**Figure 10.4**  The molecular connectivity diagram created from the HMBC spectrum of compound **10.7** using FFs.

structure. The check for contradictions was re-run with the result that this MCD passed all algorithmic tests and the program again was unable to recognize any contradictions.

In reality, the user would not be aware of the presence of contradictions when working with a true unknown structure, so the common process of starting structure generation from the MCD was followed to mimic the actions of an investigator with no prior knowledge to work from. Due to the presence of two free carbon atoms, and even with the presence of a large fragment which had consumed 13 skeletal atoms, the generation time was relatively long: $t_g = 5$ min 40 s, with a resulting structure set of $k = 36$. After removing duplicates, 15 structures were stored ($k = 36 \rightarrow 15$). The first three structures from the ranked file are shown in Figure 10.5.

A value of $d_A = 2.80$ ppm was determined for the preferred structure within the limits of allowed values[2] and, consequently, the identified solution was not rejected. Comparison of the favored structure with structure **10.7** shows that they differ from each other only by the arrangement of both the OH and CH$_3$

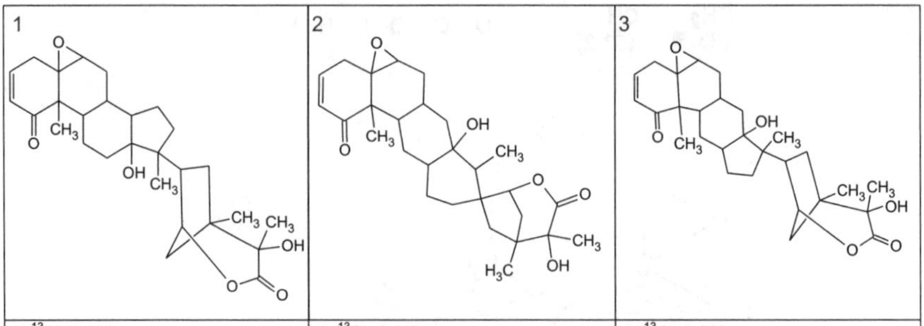

| $d_A(^{13}C)$: 2.797 | $d_A(^{13}C)$: 3.118 | $d_A(^{13}C)$: 3.206 |
| $d_i(^{13}C)$: 2.683 | $d_i(^{13}C)$: 3.176 | $d_i(^{13}C)$: 3.201 |
| $d_N(^{13}C)$: 3.234 | $d_N(^{13}C)$: 3.287 | $d_N(^{13}C)$: 3.574 |

**Figure 10.5**   The initial structures in the ranked output structural file produced from the MCD shown in Figure 10.4.

groups attached to the five-membered ring, confirming that the structural formula of the unknown compound obtained in the presence of contradictions was very similar to the actual structure. The calculation of the Tanimoto similarity match factor[14] for the suggested structure with the real structure gave a value of 0.975. Therefore in the presence of undetected contradictions in the 2D NMR data, StrucEluc can provide an "approximate solution" to the problem in definite cases. Understanding why the program failed to identify the non-standard connectivities can be explained by comparing the assignments of the chemical shifts in the real and preferred chemical structures. As shown in Figure 10.6, all connectivity lengths in the preferred structure match the default values. The values of the carbon chemical shifts and their properties assigned by the software are in agreement with the range of intervals postulated in the axiomatic knowledge of the system.

To ascertain the correctness of a particular solution, a check of its convergence may be of value. In order to check convergence for this particular example, the fragments found in the database were used as a basis for structure generation by assembling structures from fragments *having common atoms* (see Chapter 5). It should be noted that the $^{13}C$ NMR spectrum and the structure of the analyzed molecule were not present in the system's knowledge base.

The first 100 fragments from the list were selected and the atom overlapping mode of structure generation was initiated on the basis of the 1D $^{13}C$ NMR spectrum. Within 1 min 20 s one structure was obtained and was consistent with structure **10.7**. Visual comparison of the experimental and calculated spectra showed a very high degree of similarity with a deviation equal to 0.93 ppm, a value which supports the correctness of the found solution. A high degree of flexibility in the software design, the presence of structure generators based on different algorithms and ability for the user to take active part in the process, allows the correct structure to be identified even in the presence of hidden contradictions in HMBC data.

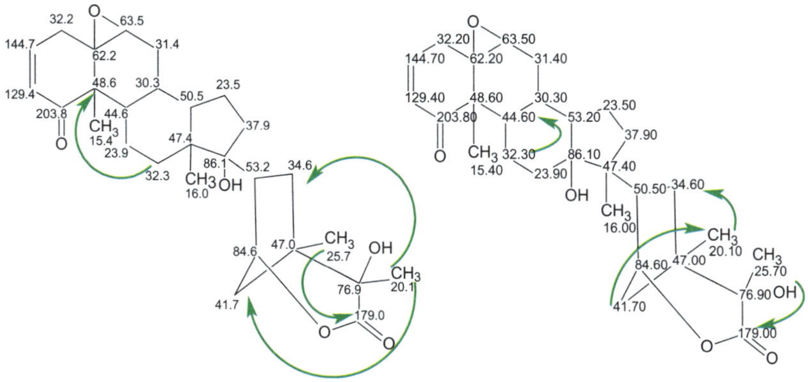

Structure **10.7**                                                    Preferred structure

**Figure 10.6** The comparison of assignments based on chemical shifts in the published (**10.7**) and StrucEluc preferred structures.

## 10.3 Solving Problems Using FSG

In those cases when correlations are present in the 2D NMR data with $^nJ$ where $n > 4$, the method of automatic removal of contradictions, unfortunately, does not work. The augmentation of the path between two intervening nuclei by one bond obviously cannot lead to the generation of a correct structure in this case. Moreover, due to a lack of constraints that are to be logically analyzed, even in those cases when $n = 4$ the algorithm gives no guarantee that all NSCs will be found and corrected. For example, the greater the number of carbon atoms with properly defined properties (in regards to the type of hybridization and different heteroatom neighborhoods) and/or the higher the total number of available 2D NMR connectivities, the higher the probability of successfully performing logical analysis to arrive at the correct structure. In contrast, severely proton-deficient molecules can be among the most challenging. Obviously, the problem becomes more computationally complicated as the complexity of a molecule increases.

Experience has shown that the number of NSCs contained within the 2D NMR data associated with a molecule, *m*, can be rather large – up to about 20 correlations. At the same time the augmentation of standard correlation lengths, *a*, could be 1–3. As an example of such situations several structures taken from literature[15–18] are used to demonstrate those examples with a large number of NSCs including $^5J$ and $^6J$ coupling constants (see Figure 10.7).

To overcome the described shortcomings a computational approach was suggested in Chapter 8 that has been defined as "fuzzy structure generation" or "FSG".[19] During the process of FSG the number of non-standard connectivities is restricted to a parameter *m* and their lengths can be augmented by a number of bonds equaling *a*. A strategy of determining the values of the parameters *m* and *a* was elaborated and methodology was developed that

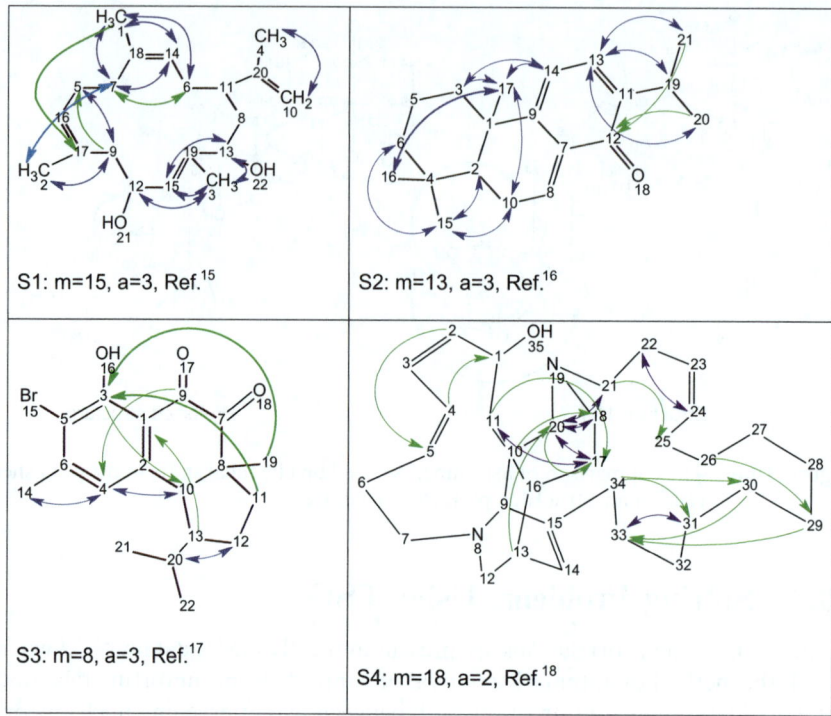

**Figure 10.7**   An illustration of a number of structures containing multiple NSCs. The non-standard COSY correlations are shown as blue arrows and the HMBC correlations by green arrows. In the legends for the structures *m* is the total number of NSCs, and *a* is the value of correlation lengthening allowed during the process of FSG (see below).

allows the program to identify both the number and length of the NSCs in 2D NMR data.[20] This was achieved by FSG on the basis of *m* and *a* parameters whose actual values become known during the process of problem solving. Further, we will discuss the methodology of applying FSG and show a series of examples which demonstrate its efficiency.

## 10.3.1   Modes of FSG

Numerous computational experiments have allowed us to conclude that if the program detects the presence of NSCs, but fails to resolve contradictions in the 2D NMR data using algorithms described in Chapter 8, then FSG should be used to solve the problem. Moreover, it is quite probable that structure elucidation from 2D NMR data on the basis of FSG can be considered as a general CASE strategy, because it is almost independent of the presence or absence of NSCs in the 2D NMR data.

FSG can easily be controlled by parameters that make up a set of options.[20] The two main parameters are: $m$, number of nonstandard connectivities; and $a$, the number of bonds by which some connectivity lengths should be augmented. Unfortunately, 2D NMR spectral data cannot deliver definitive information regarding the values of these variables and, as a matter of fact, both of them can be determined only during the process of structure elucidation. It has been concluded that in many cases the risk of choosing an erroneous value for $a$ can be avoided and the solution of a problem can be considerably simplified if the lengthening of the $m$ connectivities is replaced by their *deletion*. When set in the options, the program can ignore by deletion connectivity responses that have to be augmented (by convention, the parameter $a$ is set to a value of 16 in these cases). Such an approach can be successful in those cases when the number of 2D NMR connectivities is in some sense optimal. In this sense, we mean that the total number of connectivities (structural constraints), $N$, must be large enough to facilitate description of the chemical structure. In many instances, there are sufficient numbers of correlations in the ensemble of 2D NMR data acquired to essentially over determine the structure – in other words there is redundancy in some of the connectivity information. It can then be expected that deletion of $m$ of the connectivities will not dramatically influence either the generation time or the size of the output file. On the other hand, the number of combinations of $N$ connectivities taken $m$ at a time can be very large. This can dramatically impede problem solving to a point that it is not feasible to solve the problem. Indeed, some researchers have commented that some of the accordion-optimized long-range heteronuclear shift correlation experiments actually provide too many long-range correlations of the type $^nJ_{XH}$ where $n \geq 4$.

If the number of connectivities, $N$, is small then further decreasing $N$ by $m$ in a connectivity combination can lead to an excessive decrease in the number of structural constraints required for solving the problem. In such a case, the problem may be difficult to solve because the 2D NMR data structural constraints will only reduce the total number of possible isomers very slightly.

Independent of the use of augmentation or removal of connectivities, the crucial point in application of FSG is the number of connectivity combinations that should be checked during structure generation. For instance, if $N = 60$ and $m = 5$, then the number of connectivity combinations, $n_{math} = C_N^m$, is equal to $\sim 5.5$ million. Any attempt at structure generation has to be performed using each of these combinations. It is necessary to perform generation of structures from each of the $C_N^m$ data sets and obtain the output file as a unification of all of the intermediate results. Although the StrucEluc structure generator is fast, the productivity is certainly insufficient in terms of coping with a combinatorial problem as outlined here.

To overcome this difficulty the system is delivered with an algorithm capable of reducing the number of combinations without the risk of losing the correct solution. The first step is to reduce the total number of connectivities $N$ down to

$N_0$, where $N_0$ is the number of connectivities used to form the connectivity combinations. The data are pre-processed according to the following rules: (1) ambiguous connectivities are excluded from consideration; (2) if two connectivities C-1 to C-2 and C-2 to C-1 are present, then only one of them is included in a data set. One of the two equivalent correlations is redundant and corresponds to over determination of the data needed for solution of the structure. The second and most important step is based on the results of logical analysis of the initial 2D NMR data. If connectivity sets containing NSCs are identified (see Chapter 8), then groups of these connectivities are utilized to produce connectivity combinations. As a consequence connectivities that are suspected to be non-standard are included in all of the resulting combinations and the initial number of combinations reduces (as will be shown later this number can be reduced by many factors). In addition, the algorithm is capable of immediately detecting combinations of connectivities from which structure generation is impossible – a connectivity combination of this kind still contains at least one NSC. These combinations are skipped during the structure generation process. As a result FSG can be performed in a reasonable time, even in those cases when $n_{math}$ is very large. If the MCD checking process fails to detect NSCs in the 2D NMR data (according to our studies the probability of failure is approx. 10%), the program is forced to try all $C_N^m$ connectivity combinations. This can drastically increase the time to solve the problem and the described approach is inefficient. In these cases UFs and FFs[3,21] can frequently be helpful. The ability of the program to calculate and display the real number of connectivity combinations to be validated during FSG allows the user to approximately evaluate the complexity of a given task, even at the first stage of the structure elucidation process.

When option parameters are combined in a different way it is possible to initiate the following modes of FSG:

- *Mode 1*. Structures are generated such that the number of correlations that are extended is specified ($m = m_0$) and connectivity augmentation is also assigned ($a = a_0$). In this case for a HMBC correlation having a length of one or two skeletal bonds both the lower and upper length limits are updated and the connectivity length is extended to three bonds.
- *Mode 2*. Structure generation is performed using the following options: it is assumed that the number of extendable (or ignored) connectivities cannot exceed $m_{max}$, ($m = 1, 2, \ldots m_{max}$), whereas $a$ is equal to $a_0$. The $m_{max}$ value is defined as the *maximum* allowed number of NSCs in the 2D NMR data. Typically, $m_{max}$ value is set equal to 20, thereby covering a wide range of non-standard connectivities (see Figure 10.7). The program initially performs structure generation with a value of $m = 1$. If the attempt is unsuccessful, then the $m$ value is *automatically* incremented by 1 and a new run is made with $m = 2$ and so on. An iteration is declared unsuccessful if either no structure is stored after structure generation and spectral filtration or if an *unacceptable* solution was found. When $m$ reaches the $m_g$ value the program considers the 2D NMR data to be

consistent, then FSG is initiated with $m = m_g$. The program stops after completing structure generation with $m = m_g$ if the output structural file is not empty and if an *acceptable* solution is provided.

- *Mode 3*. The number of connectivities $m$ is allowed to vary between $m_{min}$ and $m_{max}$ values ($m_{min} \leq m \leq m_{max}$), whereas the fixed number of bonds $a_0$ is set. The minimum number $m_{min}$ is usually derived as a result of checking the 2D NMR data for consistency. The program stops when similar conditions as described for *Mode 2* are achieved.

- *Mode 4*. This mode is a generalization of *Mode 3* where the interval for $m$ value variation is defined by the condition $m_{min} \leq m \leq m_{max}$ at $m_{min} = 0$. The peculiarity of this mode is that it is a "generalized" mode of structure generation and can be initiated with $m = 0$. In this mode, the program starts by checking the hypothesis that NSCs are absent in a given 2D NMR dataset. If the dataset does not contain non-standard connectivities, then the program completes the process of structure generation and the further solution of the problem is carried out as described previously.[3,21,22] If an attempt with $m = 0$ proves to be unsuccessful then the program automatically performs FSG starting with $m = 1$, $a = 16$, and continues problem solving in the manner described earlier for *Mode 3*. The merit of such an approach is that no assumption regarding the $a$ value is necessary.

- *Mode 5*. This mode is initiated if it is necessary to perform FSG iteratively covering all values of $m$ starting from $m_{min}$ to $m_{max}$ without exclusion. For example, if structure generation is successful at $m = m_g$, then the program automatically switches to $m = m_g + 1$ and so on until it reaches $m = m_{max}$. The structures generated at each step are added to those generated during the previous step. This mode is useful to check the solution for stability to make sure that the best structures found at steps $m = m_g$ and $m = m_g + 1$ or higher are equivalent.

- *Mode 6*. This mode resembles *Mode 5*, but the function of this mode is to generate all structures for which the number of non-standard connectivities is *less or equal to* $m$ at the given value of $a$. The corresponding options are denoted as $\{m \leq m_0, a = a_0\}$. The number of connectivity combinations from which FSG is performed depends only on the $N_0$ and $m$ values. In contrast to the "step-by-step" modes, some combinations of the connectivities are united by this approach and this in principle can speed up the calculations. When this procedure is performed, only the maximal lengths of HMBC connectivities (*i.e.* two skeletal bond lengths) are enlarged. For example, consider a HMBC connectivity between C-1 and C-2 atoms whose "standard" length is varied from one to two skeletal bonds. In this mode the updated connectivity length varies from one to three skeletal bonds. It is important to note that the number of non-isomorphic structures generated in this mode is equal to the total number of non-isomorphic structures generated during all steps of *Mode 5*. However, the total time necessary for completion of FSG can be significantly different between these modes.

- *Mode 7.* This approach gives the researcher a chance to solve a problem in a fully automated mode. To initiate this mode the commands "allow fuzzy structure generation" and "determine options automatically" are selected. The program analyses the 2D NMR data and, depending on the results, makes a corresponding decision on the choice of the generation parameters and the strategy of their application. If a problem can be solved in the common mode (without using fragments), FSG with automatically determined parameters is very effective. For example, all problems described in Section 10.1 are automatically solved using this approach.

*Mode 6* with the parameter $a$ set equal to 16 can be considered as the most comprehensive mode, since, in principle, it will solve a problem in which the 2D NMR data contain an *unknown* number of non-standard connectivities of an *unknown* length. Experience has shown that depending on the complexity of the problem the $m$ value is typically set to be equal to 5, 10, or 15. If the problem is successfully solved with a given set of options, then the real $m$ and $a$ values are simply determined visually from the resulting structure, which is displayed along with all COSY and HMBC connectivities. In other cases these parameters can be estimated only by the trial method.

In addition to the approaches mentioned for controlling FSG, there is also a possibility to exclude the COSY data from the process of FSG as a user option. In some cases, especially those when the COSY data contain many NSCs requiring $a > 1$ while the HMBC data are rich enough, the exclusion of the COSY data both simplifies and accelerates the solution of the problem.

## 10.3.2  The Strategy of Applying FSG

The possibility of employing several different modes of FSG proves to be a very flexible analytical tool. However, the diversity of modes available is also a source of complexity, since the user has to choose the optimal mode when solving a specific problem. Before starting the calculations it is unclear which mode will lead to a solution in a reasonable time.

An attempt was made to answer the question of whether there is a general strategy of structure elucidation using FSG that works best. A set of more than 100 problems were selected in which either the HMBC or COSY spectra, or both, contained a total of 1–18 non-standard connectivities corresponding to a range of coupling constants $^nJ_{HH,CH}$ where $n = 4$–6. The structures under investigation were all natural products and the number of skeletal atoms in the molecules varied between 15 and 75 skeletal atoms. The experimental data were obtained from articles published mainly in the *Journal of Natural Products* or from collaborations with various laboratories.

For each problem, the NMR spectral data were entered into the program and graphically represented as MCDs. The procedure for checking the 2D NMR data for contradictions[19] was then applied to every problem. If the presence of NSCs was revealed then the program displayed the minimum number of nonstandard connectivities and made an attempt to automatically resolve the contradictions as described above. In successful cases, the updated MCDs were displayed with modified connectivities marked by a specific color.

As a result of these studies all problems were classified into three sets as follows:

1. 53 problems were identified where NSCs were detected and the initial MCDs were updated.
2. 34 problems were identified where the program revealed the presence of NSCs but failed to update the MCDs.
3. 13 problems were identified where the program failed to detect NSCs.

This classification describes all conceivable results of checking the MCDs. Depending on the results of checking the MCD, various modes or combinations of modes can lead to solution of the problem. Attempts to solve each problem were made using different FSG modes to investigate possible approaches. The problems for which valid solutions could not be found during the first attempt were eventually solved after utilizing different fuzzy generation options. Logical data pre-processing frequently allowed significant reduction of the number of connectivity combinations to be tested during the FSG. Figure 10.8 shows the ratio $\rho$ describing the number, $n_{real}$, of tested connectivity combinations to the theoretically calculated number of combinations, $n_{math} = C_{N_0}^m$ for the entire problem set. Figure 10.9 examines these combinations in greater detail.

The Figures demonstrate that the theoretical number of combinations can be hundreds of billions, but the real numbers reduce down to manageable dimensions. For instance, in 20 problems the theoretical number dropped by

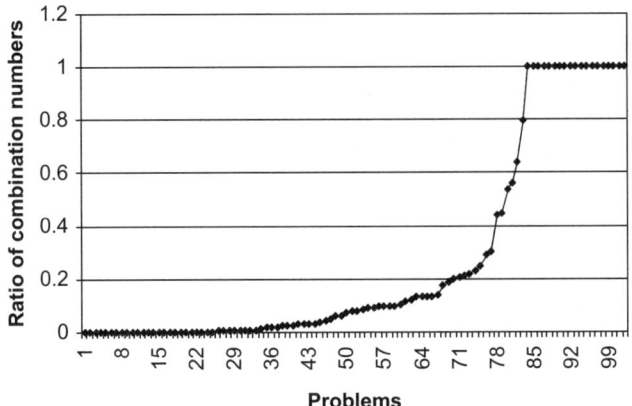

**Figure 10.8**  The ratio of numbers of real connectivity combinations to the numbers of theoretically possible combinations for the problems solved using FSG. The program failed to reduce the number of combinations mainly in those cases when non-standard connectivities were not detected during checking of the MCD.

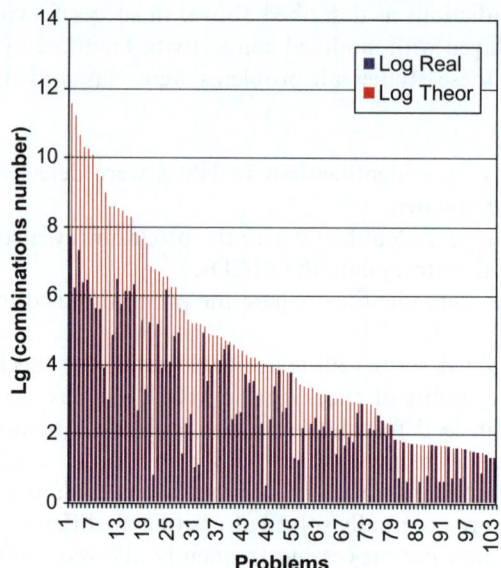

**Figure 10.9** A plot of the logarithms of the theoretical (red) and real (blue) numbers of connectivity combinations.

$10^4$–$10^6$ times, but the real numbers of combinations still remained rather large. Nevertheless, the speed of the structure generator algorithm was fast enough to solve almost all problems.

FSG did, however, fail for the elucidation of structure **S4** ($C_{32}H_{50}NO_2$) in Figure 10.7. The 2D NMR data contain 18 non-standard connectivities (12 HMBC and six COSY non-standard connectivities; five connectivities are of the type $^5J$). The theoretical number $n_{math}$ of connectivity combinations is equal to $\sim 43.10^{12}$ for this case. The difficulty could be circumvented by using the "fragment mode", but no large appropriate fragment was found in the database during the $^{13}C$ NMR search. The application of a large user fragment led to an extremely large set of MCDs with each containing the UF with different distributions of the carbon chemical shifts.[23] As a result these two combinatorial "explosions" hampered problem solving. The solution of such computationally difficult problems will hopefully be eased by further development of the algorithm providing fragment "implementation" in MCDs.

As a result of the studies described, general traits were identified that could help to find appropriate ways to solve a problem. These strategies, as applied to the three problem subsets mentioned above, are described in the following subsections.

### 10.3.2.1 NSCs were Identified and the MCD was Updated

Assuming that the MCD updating process was performed correctly (with the lengths of all NSCs increased), then *strict* structure generation is performed. If an acceptable solution is obtained then it should be checked for stability. FSG with the options $\{m = m_{min}/20$; stop at $m = m_g$, $a = 16\}$ is started from the initial MCD, not the updated MCD. The previously found solution will be confirmed if the first ranked structures for both strict and fuzzy solutions coincide. When an inequality $d^{st}_A(1) > d^{fuz}_A(1,m_g)$ is observed [$d^{st}_A(1)$, the deviation calculated for the first ranked structure of the solution found by strict structure generation; $d^{fuz}_A(1,m_g)$, the same found by FSG at $m = m_g$], then it is concluded that not all NSCs were lengthened and FSG should be repeated with $m_g + 1$ and so on until the minimum value of $d^{fuz}_A(1,m_g + v)$ and a valid solution is achieved at $m = m_g + v$. The corresponding structure is then considered as the most probable.

An unacceptable solution can be obtained as a result of strict structure generation from the updated MCD, *i.e.* a solution will be found for either $d^{st}_A(1) > D_A$, where $D_A$ is a threshold value or an empty structural file is obtained ($k = 0$). In both cases the program is automatically switched to the mode where $\{m = m_{min}/20$, stop at $m = m_g$, $a = 16\}$. Depending on the $m_g$ values and the complexity of the problem (the size of $m_{real}$ and the calculation time) evaluated during the first stages of solving the problem, the user can initiate FSG with the options $\{m \leq m_0, a = 16\}$, $m_0 = 5$, 10 or 15 to obtain the most reliable solution.

## 10.3.2.2 NSCs were Identified but the MCD Failed to be Updated

If the program identified NSCs but failed to update the MCD, then FSG is one manner by which to solve such a problem. Since the software application only displays the minimum number of NSCs while their associated lengths remain unknown, FSG with the options $\{m = m_{min} - 20$, stop at $m = m_g$, $a = 16\}$ should be used. The real numbers of the connectivity combinations, $n_{real}$, are displayed, as well as the number of combinations for a given $m = m_g$, and the predicted time for structure generation allows the user to easily evaluate the complexity of the problem and the suggested time for execution. If *Mode 6* can be applied based on acceptable time estimates then it should be used.

## 10.3.2.3 NSCs were not Detected

If non-standard connectivities were not revealed by checking the MCDs, then there are two ways to interpret this result: either the 2D NMR data are free of non-standard connectivities or the NSCs are present, but the program failed to detect them. Both of these situations are covered by FSG with the options $\{m = 0/20$, stop at $m = m_g$, $a = 16\}$. If NSCs are, indeed, absent from the 2D NMR data, then structure generation is performed with $m = 0$ with a non-zero output file and the values of deviation allow the user to determine whether the solution determined is acceptable. Obtaining deviation values that exceed the threshold $d_A$, or deriving an empty output file after spectral filtering, both serve as hints to the presence of latent non-standard connectivities.

When NSCs are not detected by logical data analysis then the number of connectivity combinations that must be tested during FSG cannot be reduced and it is equal to $C^m_{N_0}$, $m = 1, 2, 3,...$ at each $m^{th}$ step of the FSG process. This situation can cause significant difficulties due to an unmanageable number of connectivity combinations needing to be processed; as discussed previously, both FFs and UFs can assist in this situation.

It is difficult to describe the myriad of nuances associated with FSG, since these depend on each 2D NMR data set associated with a given problem. A series of examples illustrating the strategies leading to valid solutions with the minimum number of user assumptions will be presented in the next Section. Example problems were chosen where automatic updating of the MCD to resolve contradictions was inefficient. Particularly, the structures shown in Figure 10.7 were used as examples, but first a couple of much simpler problems will be considered.

## 10.3.3 Problem Solution using FSG in the Common Mode

*Example 1*
When the structure of *hermitamide* (**10.8**) was elucidated from the HMBC data published by Tan et al.,[24] the program did not reveal the presence of the HMBC (C-27–C-23) connectivity and, as a result, three structures were generated in 2 min 15 s ($k = 810 \rightarrow 14 \rightarrow 3$).

**10.8**

The deviations $d_A(1) = 2.47$, $d_I(1) = 2.80$, $d_N = 2{,}75$ ppm characterizing the best structure fall into the limits of admitted values, but there was one reason making the structure inacceptable: all of the three structures contain a $NH_2$ group, whereas two NH groups were distinctively identified in the $^1H$ NMR (600 MHz) spectrum at 5.55 and 8.22 ppm. For this reason, the generation was repeated in fuzzy mode with $\{m = 1–10, a = 16\}$. The result was $k = 1038 \rightarrow 354 \rightarrow 352$, $t_g = 20$ min, $d_A(1) = 0.71$ ppm, $d_I(1) = 0.95$ ppm and $d_N = 0.94$ ppm. The resulting preferred structure and the chemical shifts assignment of the $^{13}C$ NMR spectrum coincided with those given by the authors.[24]

*Example 2*
The HMBC spectrum of structure **10.9** in the study by McGarvey et al.[25] contains one $^5J_{CH}$ correlation indicated with an arrow. The presence of this non-standard connectivity between the C-5–C-20 atoms was detected by the program and the MCD was automatically corrected by lengthening by one bond. This was not sufficient, of course. On the basis of the corrected MCD, two structures were instantly generated and then rejected by the spectral filtering. The best, structure **10.10**, had deviations of 5–6 ppm. Such large deviation values can be interpreted as

evidence of a wrong solution. With this in mind FSG was performed from the initial MCD using the mode "determine options automatically". A single and correct structure was generated in 0.024 s and $d_A = 2.13$ ppm.

**10.9**                    **10.10**

*Example 3*

In the analysis of cleospinol A (Collins *et al.*[15]), with the molecular formula $C_{20}H_{32}O_2$ (**10.11**), the 2D NMR data are comprised of 21 COSY and 55 HMBC correlations. These data were used to evaluate the possibility of solving a problem in those cases when a large number of NSCs were present. In this case, the 2D NMR data contained the following combination of NSCs: 3 HMBC [2a(1), 1a(3)] + 12 COSY[8a(1), 3a(2), 1a(3)] = 15. This nomenclature describes the fact that there are three HMBC NSCs, two of which must be lengthened by one bond and one by three bonds; the information about the 12 COSY correlations is interpreted analogously. The total number of NSCs is 15. Such expressions are used to provide short and unambiguous designations of the numbers and lengths of the NSCs contained within the 2D NMR data.

The COSY connectivities are represented below on the structure by blue double-headed arrows, whereas the HMBC correlations are defined by green unidirectional arrows from the proton to the carbon to which it is long-range coupled.

**10.11**

The COSY, HMQC, and HMBC spectral data associated with the compound were fed into the program and the MCD was generated. A check of the MCD was accompanied by the automatic removal of contradictions. The software program displayed a message declaring that the contradictions had been detected and resolved, while the minimum number of NSCs was estimated to be equal to 7. Unfortunately, strict structure generation from the automatically edited MCD resulted in an empty output file. This result was interpreted as evidence of the presence of either undetected additional NSCs or those whose lengths must be augmented by more than one bond.

There are two possible trajectories from this point to solve the problem. Since, in general, there is no information about the number of NSCs and their lengths these values can be determined using a trial and error method. If it turns out that $a = 1$ then there is a chance to find a solution in a short time. The second approach is more general (see Section 10.3) and allows the user to ignore the problem of determining the maximal $a$ value. The cost, however, may be longer structure generation times and a consequently larger output file. Both approaches are described in detail below to illustrate different ways of solving the problem with FSG.

With the first approach FSG was initiated from the initial (not updated) MCD using *Mode 3* with the options $\{m = 7\text{--}20$, stop at $m = m_g$, $a = 1\}$. The program started the generation process automatically with $m_g = 14$ ($m = 7\text{--}13$ were immediately rejected) and the process was aborted by the operator at $m = 16$ with a zero result. An attempt to repeat FSG with $a = 2$ again gave an empty result file.

The possibility that one or more of the non-standard connectivities needed to be augmented by three bonds was assumed. When the options $\{m_{max} = 20$, stop at $m = m_g$, $a = 3\}$ were set then the program automatically started with $m_g = 10$ and completed the fuzzy generation process in 9 min 15 s (Pentium IV, 2.8 MHz) with $m = 14$. Three molecules were generated and two were stored in the results file after the removal of duplicate structures ($k = 3 \rightarrow 2$). The highest ranked structure coincided with the actual structure, **10.11**. The second structure gave a $d_A(2)$ value with $\Delta_{2-1} = d_A(2) - d_A(1) = 1.5$ ppm. A solution was found at $m = 14$ and not at $m = 15$ since, the COSY and HMBC connectivities between C-5 and C-9 carbons are coincidental. The program displays the final structure where all connectivities and their associated lengths can be visualized.

The second more universal and systematic approach was applied, and FSG was initiated assuming only that the number of nonstandard connectivities is not more than 15 (*Mode 6*), *i.e.* options $\{m \leq 15$, $a = 16\}$ were set. In this case 18 281 379 connectivity combinations from 40 225 345 056 theoretically possible combinations were used for structure generation. The following result was obtained: $k = 769 \rightarrow 430 \rightarrow 245$, $t_g = 29$ min 9 s, and the correct structure was ranked first by all methods of spectrum prediction, $r_{all} = 1$.

The program therefore identified the correct solution, even when 15 non-standard connectivities existed in the 2D NMR data and especially in the presence of HMBC and COSY connectivities representing both $^6J_{CH}$ and $^6J_{HH}$ correlations. Note that only $\sim 10^{-4}$ of the theoretically possible connectivity combinations were processed, although the real number of processed connectivity combinations $n_{real}$ is more than 18 million. Nevertheless, the high-speed structure generator present in the StrucEluc program completed the process in a reasonable time.

*Example 4*

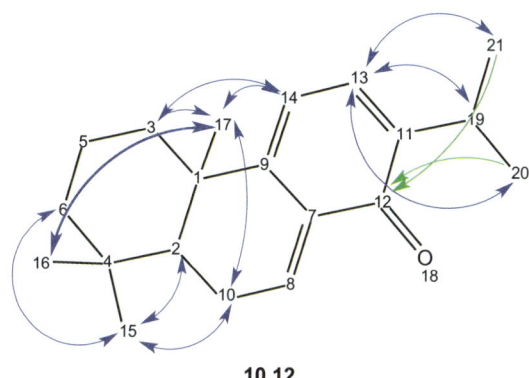

**10.12**

The 2D NMR data associated with a natural product with molecular formula $C_{20}H_{28}O$ isolated and identified by Mensah *et al.*[16] contains 14 NSCS: 2 HMBC[2a(1)] + 12 COSY [4a(1), 6a(2), 1a(3)] = 13. The presence of a minimum of six NSCs was detected and the program displayed a message indicating that the MCD was successfully updated. Strict structure generation of the updated MCD led to an empty output file, indicating that FSG must be used. As experience shows that the maximum value of $a$ usually does not exceed 3, it was suggested that a solution to this problem could be obtained with the options {$m = 6$–15, stop at $m = m_g$, $a = 3$}. The program initiated generation at a value of $m_g = 11$ and stopped when four structures were generated. The large deviations calculated for the best structure suggested that the solution was invalid (see Table 10.1). The

**Table 10.1** The results of four subsequent steps of FSG.

| m | No | $n_{math}$ (approx.) | $n_{real}$ | $\rho$ | Results |
|---|---|---|---|---|---|
| 11 | 41 | $3.1 \times 10^9$ | 17 629 | $5.6 \times 10^{-6}$ | $k = 4 \rightarrow 4$, $t_g = 1$ min 35 s, $d_A(1) = 6.22$ |
| 12 | 41 | $7.9 \times 10^9$ | 240 001 | $3 \times 10^{-5}$ | $k = 25 \rightarrow 19$, $t_g = 2$ min 40 s, $d_A(1) = 5.15$ |
| 13 | 41 | $17.6 \times 10^9$ | 1 601 574 | $1.2 \times 10^{-4}$ | $k = 218 \rightarrow 176 \rightarrow 110$, $t_g = 12$ min, $d_A(1) = 2.13$ |

results of subsequent structure generations with $m_g = 11$, 12, and 13 are presented in Table 10.1. All deviations calculated for the best structure achieved minimum values for the solution obtained at a value of $m_g = 13$: the best structure coincided with structure **10.12**. Carbon atom assignments were identical to those suggested by the authors.[16]

The practical approach was applied with the options $\{m \le 15, a = 16\}$. The following results were obtained: $k = 66\ 538 \rightarrow 38\ 407 \rightarrow 12\ 070$, $t_g = 2$ h 41 min, $n_{math} \approx 63 \times 10^9$, $n_{real} = 48\ 525\ 735$, $\rho = 7.6 \times 10^{-4}$, $r_{all} = 1$. It is worth noting that the $^{13}$C NMR spectra of 38 407 structures were calculated in 2 min 23 s for preliminary ranking using the incremental method.

*Example 5*

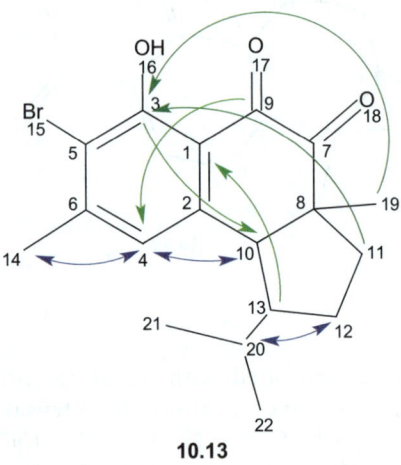

**10.13**

Natural product **10.13** was isolated and identified by Wellington *et al.*[17] using a combination of HMBC and COSY spectra. There are eight NSCs in the 2D NMR data: 5 *HMBC* [2*a*(3), 3*a*(1)] + 3 *COSY* [3*a*(1)] = 8. As a result of checking the MCD was updated and the minimum number of NSCs $m_{min}$ was determined to be 4. An attempt to perform strict structure generation was unsuccessful and the program produced an empty output file. The problem could be solved using "step-by-step" methods analogous to those described in the previous example, but we present here only the result of the practical approach.

For modeling a situation where nothing is known about the number and lengths of NSCs except that $m \ge 4$ the following FSG options to ensure minimization of the risk of losing the correct solution were set: $\{m \le 10, a = 16\}$. The result is defined by: $k = 114\ 638 \rightarrow 68\ 668 \rightarrow 23\ 213$; $t_g = 1$ h 52 min; $n_{real} = 52\ 427\ 715$, $n_{math} \approx 0.6 \times 10^9$; $\rho = 0.08$; $r_{all} = 1$. Using *Mode 6* the program performed an exhaustive search of all possibilities to generate structures (see Section 10.3.1), and the output file is large. The $^{13}$C NMR spectrum prediction for about 70 000 structures using the incremental method took about 5 min. The correct structure was distinguished among approx. 23 000 candidates by examining the values of the deviations calculated using all methods within the StrucEluc system.

*Example 6*
Kirsch and co-authors[26] reported the isolation and structure elucidation of a natural product with a molecular formula of $C_{25}H_{38}O_2$ (**10.14**) using 2D NMR data containing 23 COSY and 41 HMBC correlations. As illustrated by the arrows the following set of NSCs are observed in structure **10.14**: 2 *HMBC* [$2a(2)$] + 7 *COSY* [$4a(1)$, $3a(2)$] = 9.

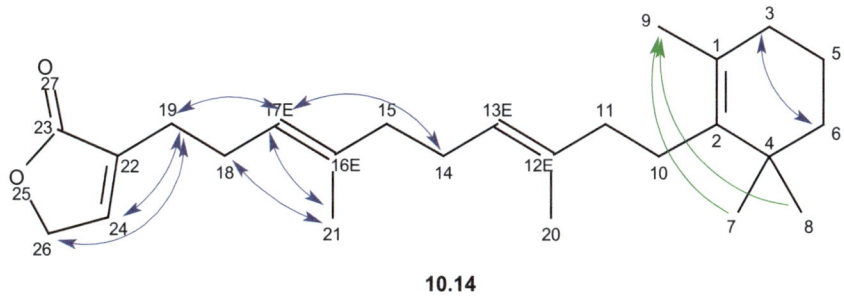

**10.14**

The following steps allowed identification of a correct solution to this problem:

1. When the MCD was checked, NSCs were detected but the program declared that the contradictions in the 2D NMR data could not be resolved. The minimum number of NSCs had a value of 6.
2. FSG was performed with the options {$m = 6$–10, stop at $m = m_g$, $a = 1$}, but resulted in an empty output file.
3. FSG with the options {$m = 6$–10, stop at $m = m_g$, $a = 2$} provided a valid solution in less than 7 min. The solution was obtained with $m = 9$, since $m = 1$–8 resulted in empty output files.

When the more general approach was applied with options {$m = 6$–20, stop at $m = m_g$, $a = 16$}, a single and correct structure was found in 2 min. With this, 370 950 potential connectivity combinations of $\approx 2.5 \times 10^9$ theoretically possible ($\rho \approx 10^{-4}$) were processed in this time.

The validity of the solution was checked by a time-consuming process using FSG with the options $m \leq 10$, $a = 16$ to give the following result: $k = 18 \rightarrow 17 \rightarrow 15$, $t_g = 28$ min 35 s, $n_{real} = 4\,830\,600$, $n_{math} \approx 10 \times 10^9$, $\rho = 4.7 \times 10^{-4}$, $r_{all} = 1$. The first ranked structure and associated atom assignment coincided with that determined by the authors.[26]

*Example 7*

Computational difficulties associated with structure generation arise with an increase in the complexity of the molecule, even in those cases when the 2D NMR data contain correlations of standard lengths only. The difficulties become especially serious if the NSCs exist in the COSY and HMBC spectra. The possibility of solving problems using FSG for a large molecule with a large number of NSCs in the 2D NMR data is illustrated in the following example.

Feller *et al.*[27] reported the isolation and structure determination of a new terpenoid **10.15** with a molecular formula of $C_{43}H_{66}O_{10}$ and therefore containing 53 skeletal atoms:

CH$_3$ 16.30   H$_3$C 18.60

46.20   37.30   29.20–CH$_3$ 20.60   CH$_3$ 51.00

31.70   213.40   51.50

23.50   O   212.80   173.70   H$_3$C 20.00

36.10   O   48.20   140.40

O   38.40

26.20   210.30   47.90   124.70

H$_3$C 21.70   48.90   46.40   40.70   26.60

32.60   125.00   CH$_3$ 25.50

126.40   28.20   85.00   69.70–OH

CH$_3$ 19.80   O   75.60   69.20   31.60

H$_3$C 20.30   170.20   75.80   20.10

O   H$_3$C 18.70   OH

**10.15**

The 2D NMR data were composed of 53 COSY and 94 HMBC correlations including 10 non-standard connectivities as shown on the representation of structure **10.15** and enumerated as follows: 5 *HMBC* [5a(1)] + 5 *COSY* [4a(1), 1a(2)] = 10. When the MCD was created, four carbon atoms, whose chemical shifts are marked in red, were not involved in any correlations. The presence of such "free" atoms introduces an additional obstacle to solving a problem and generally leads to an increase in structure generation time.

MCD checking accompanied by automated resolution of contradictions in the 2D NMR data produced a program message declaring that the minimum number of non-standard connectivities was seven and the MCD was updated by the program to resolve the contradictions. Structure generation from the updated MCD gave $k=56 \rightarrow 28$ and $t_g=21$ s. Structure **10.15'** was distinguished as the best one:

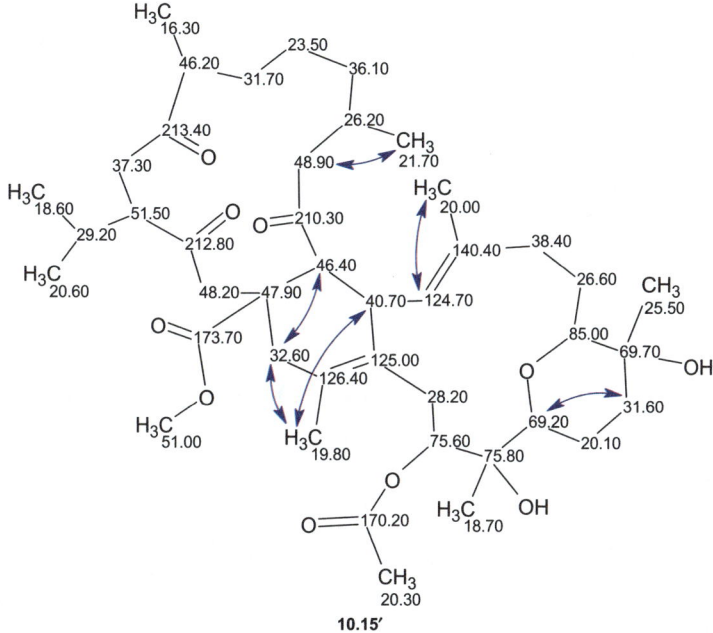

**10.15'**

The deviation value of $d_A(1)=2.74$ ppm is acceptable for a correctly recognized structure,[23] so there was no reason to reject the first ranked structure. However, when the number of detected NSCs is rather high (seven in our case), there exist, at least, the following two reasons to check the correctness of the best structure by FSG: (1) not all NSCs could be detected, and (2) not all detected NSCs should be lengthened only by one chemical bond.

FSG was performed using the options $\{m=7–15$, stop at $m=m_g$, $a=16\}$. A non-empty structural file was generated with $m_g=9$ ($n_{math} \approx 328 \times 10^9$, $n_{real} \approx 28 \times 10^6$, $\rho=8 \times 10^{-5}$) due to the accidental degeneracy of two COSY and HMBC NSCs, at C(40.7) and C(19.8) ppm, (otherwise only $m=10$ would be successful). The following results were obtained: $k=36 \rightarrow 21 \rightarrow 11$, $t_g=7$ h 50 min, $r_{all}=1$, $d_A(1)=2.62$ ppm, and the best structure coincided with that deduced by the authors.[27] The difference between the $d_A$ deviations for the correct (**10.15**) and incorrect (**10.15'**) structures is only 0.12 ppm. The calculated similarity coefficient was equal to 0.98 for these structures, and the structures differ only by the permutation of the carbons at C(47.9) and CH(46.4) ppm. The solution was fairly time consuming due a large number of connectivity combinations ($28 \times 10^6$), as well as the presence of four carbon atoms with no connectivities to other atoms. The considered problem provides evidence that the approach is valid, even in a situation when the analyzed molecule is large and the number of NSCs is big enough, including correlations corresponding to $a > 1$.

The examples presented in this Section demonstrate the high efficiency of the procedure suggested for logical analysis of 2D NMR data. The application of this procedure reduces the total number of connectivity combinations by approx. $10^4$–$10^6$ times and this allows the program to complete the FSG within a reasonable time. The examples given lead to the conclusion that the capability of FSG in the StrucEluc system enables complex tasks to be successfully resolved, even in those cases when the number of NSCs is large (10–15), and when correlation lengths exceed the default values of two to three bonds by two or even three additional bonds.

## 10.3.4   Problem Solution using FSG in the Fragment Mode

As shown in Chapter 9, the use of fragments stored in a content database, the so-called FFs selected as a result of a database search using $^{13}$C NMR spectra as inputs, as well as UFs, allows problems to be solved, even when common mode structure generation is time consuming and cannot be completed in a reasonable time. In this Section peculiarities of the fragment approach are investigated as a tool for solving problems in the presence of nonstandard connectivities.

As shown previously (see Chapter 9), the need to apply fragments to solve a problem arises when there is a deficit of hydrogen atoms in the molecular formula or when the number of connectivities in the 2D NMR data is simply not ample enough to produce a set of efficient structural constraints. One might assume that the larger the fragment involved in forming the MCD, then the quicker the time to arrive to a solution. However, in reality, as mentioned above, for a large fragment there are a huge number of permutations for different assignments of the experimental chemical $^{13}$C shifts to the fragment carbon atoms and this can lead to an extremely large number of MCDs for processing during structure generation. When FSG is applied to the set of MCDs containing fragments, the $m_g$ value can be different for each MCD and the $m_{min}$ value estimated during checking of the MCD generated in the common mode has low predictability. The specificity of FSG from a set of MCDs containing fragments will be demonstrated by examining a series of examples.

*Example 1*

Mdee and co-workers[28] isolated and identified using 2D NMR data (nine COSY and 44 HMBC correlations) a new bichalcone (**10.16**) of molecular formula $C_{30}H_{22}O_8$:

**10.16**

**10.17**

The COSY spectrum contained a set of NSCs 3 *COSY* [1*a*(1), 1*a*(2), 1*a*(3)], as illustrated by the blue arrows shown on structure **10.16**. Checking the MCD and using automatic contradiction removal detected a minimum number of two NSCs and resulted in updating of the initial MCD. However, no structures could be generated from the resulting MCD. An attempt to perform FSG from the initial MCD with the options {$m = 2$–$10$, stop at $m = m_g$, $a = 16$} was also unsuccessful; the FSG process turned out to be very time consuming, and tens of hours were predicted for the value $t_g$. A search of the fragment library using the $^{13}C$ NMR spectrum as an input was performed and 1138 fragments were selected. The first ranked fragment, **10.17**, showed good coincidence of its $^{13}C$ subspectrum with the chemical shifts of the unknown compound, and the program created four MCDs from this fragment.

FSG was initiated {$m = 2$–$15$, stop at $m = m_g$, $a = 16$} and was completed with the following result: $k = 156 \rightarrow 46$, $t_g = 6$ min 30 s, $r_{all} = 1$, *i.e.* the correct structure was unambiguously identified. Analysis of the conditions applied for FSG was performed from each MCD which delivered a result and gave:

MCD #1: $m_g = 10$, $n_{real} = 6060$, $n_{math} \approx 20 \times 10^6$, $\rho = 3 \times 10^{-4}$
MCD #2: $m_g = 6$, $n_{real} = 96$, $n_{math} \approx 0.3 \times 10^6$, $\rho = 3.2 \times 10^{-4}$
MCD #3: $m_g = 6$, $n_{real} = 3319$, $n_{math} \approx 0.5 \times 10^6$, $\rho = 7 \times 10^{-3}$
MCD #4: $m_g = 2$, $n_{real} = 27$, $n_{math} \approx 351$, $\rho = 8 \times 10^{-2}$

*Example 2*
Kehraus *et al.*[29] reported the separation and identification of a new natural product geometricin A, $C_{39}H_{63}N_2O_{12}P$ (**10.18**):

**10.18**

The reported 2D NMR data included 11 COSY and 99 HMBC correlations. Two NSCs characterized with $a = 1$ are present in the HMBC data set and are represented by green arrows. The program detected the presence of NSCs and determined $m_{min}$ to be equal to 2, but failed to resolve the contradictions. An attempt to apply FSG to the MCD showed that the problem would be too time consuming without the introduction of some key fragments. The following UFs were introduced to assist in solution of the problem:

**10.19**                      **10.20**

With the chemical shifts calculated for fragment **10.19** (as shown on the structure) the program produced 18 MCDs containing both UFs. With

$m_{min} = 2$ FSG was performed with $\{m \leq 3, a = 16\}$ to produce the following result: $k = 6180 \rightarrow 4900 \rightarrow 2450$; $t_g = 3$ min 30 s, $n_{math} = 57\ 154$, $n_{real} = 1056$, $\rho = 0.02$, $r_{all} = 1$. In reality, a result was obtained only from one of the 18 MCDs, since the other MCDs contained fragment **10.19** in a format whereby the carbon atom assignment did not correspond to that specific to the target structure. $^{13}$C NMR spectral prediction for the 4900 structures *via* the incremental method took only 27 s. As a result of prediction, the correct structure was ranked first in the output file.

*Example 3*
The next example indicates how the combined usage of different system tools can help in solving challenging problems. In this case (Feng *et al.*[30]), the HMQC, COSY and CIGAR-HMBC spectra were used to elucidate cytochalasin (**10.21**), $C_{29}H_{36}N_2O_5$. Comparison of the table of 2D NMR peaks cited in the original publication with the structure suggested by the authors revealed the presence of one HMBC non-standard connectivity C(17.4)–C(57.6) in the HMBC data.

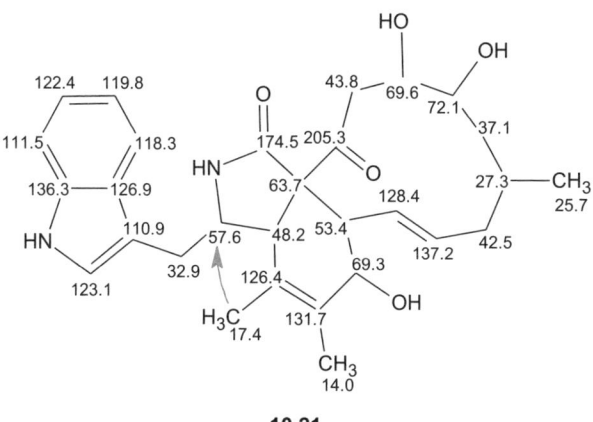

**10.21**

Solution of this problem proceeded using various system tools. Checking the MCD revealed the presence of contradictions and the program gave a message informing that the contradictions had been removed. The program made this conclusion because it eventually became possible to generate structures from the "corrected" MCD. Visual analysis of the corrected MCD showed that the connectivity (17.4–57.6) had not been lengthened, but other correct connectivities (126.4–14.0), (126.4–17.4), (126.4–57.6), (63.7–57.6) and (63.7–NH) had been lengthened. In reality, the program had not managed to remove the contradictions in the data. So, with the assumption that the program correctly adjusted the MCD, structures were generated with only one limitation, no four-membered cycles should be present, with the following results: $k = 3486$, $t_g = 32$ min. After the spectrum calculation and rank ordering the most preferable structure **10.21′** was characterized by a deviation of $d_A(1) = 3.8$ ppm. The structure with the assigned experimental chemical shifts is shown below:

**10.21'**

The $^{13}$C deviation value exceeding 3.5 ppm is rarely observed for a genuine structure; therefore the validity of obtained solution should be checked. The length of the connectivity 17.4–57.6 in this structure, marked with the solid arrow in the drawing, is "standard", whereas the structure had three HMBC $^4J_{CH}$ connectivities, as indicated with the dashed arrows. If the experimental spectra were available to the authors, a check for found non-standard connectivities using the potential relationship with the intensities of the corresponding peaks of HMBC spectrum would be made or alternatively additional experimental data would be generated.

In the absence of such opportunities the $^1$H NMR spectrum was calculated and compared with the experimental one. A considerable difference in the region 1–2 ppm was observed: instead of a singlet $\delta_H = 1.52$ ppm ($\delta_C = 17.4$ ppm) for the methyl group observed in the experimental spectrum, in the calculated spectrum a doublet at $\delta_H = 1$ ppm was assigned to this methyl group. A conclusion was made that the generated structure was incorrect and FSG may be an appropriate path forward from the initial, not corrected, MCD.

FSG was performed using $m = 1$ and a restriction on the presence of four-membered ring cycles. The result was $k = 4552 \rightarrow 112$, $t_g = 36$ min, $r_{all} = 1$ and the most preferable generated structure corresponded to that reported in the publication.[30]

When a $^{13}$C spectral search of the fragment library was performed, inspection of the first fragments from the beginning of the list showed that they were quite consistent with the $^{13}$C spectrum of the analyzed molecule. A solution was sought using these FFs. The following calculation parameters were selected: $l = 100$, $E = 2.5$ ppm, $q = 0$, $n_f = 1$ which resulted in $n_{MCD} = 11$. FSG using $m = 1$ gave $k = 16 \rightarrow 12$, $t_g = 8$ s, $r_{all} = 1$. Note that FSG using fragments elucidated the structure 250 times faster.

## 10.4  Is There an Alternative to FSG?

In Chapter 6 it was noted that some researchers suggested that it was possible to overcome the problem of NSCs by setting default values for $^4J_{CH,HH}$ for *all*

COSY and HMBC correlations observed in the 2D NMR spectra. It was important to answer the question: to what extent can the lengthening of *all* 2D NMR correlations act as a method for contradiction resolution in 2D NMR data? A study was undertaken[20] to answer this question. In this study[20] an attempt to identify the quantitative characteristics describing how structure generation time increases and the amount of structural information obtained decreases if all correlations belong in the $^{2-4}J$ range has been made. Here we will consider the results reported in the work.[20]

Computational experiments have demonstrated that it is possible to generate a full set of isomers corresponding to a given molecular formula for small molecules containing no more than 15–17 skeletal atoms[23] using the GENM[31–33] generator present in StrucEluc. It takes tens or even hundreds of hours when a PC Pentium IV, 2.8 GHz is used. For the work examining structural problems *via* 2D NMR data, only problems with less than 20 skeletal atoms were selected. The experimental data were borrowed from publications.

In Table 10.2 the structures, their molecular formulae, and the total number of structural isomers are given. The Table shows that all selected structures can be considered as small structures. For each problem the two sets of solution results obtained with two different options enabled are listed. In the first case all correlations were assumed to be $^{2-3}J$ by default. In the second case the option settings allowed the connectivity length to vary between 2 and 4. Also listed is the number of structures $k$ contained in the output file and the generation times $t_g^{(2-3)}$ and $t_g^{(2-4)}$ and their ratio $\tau = t_g^{(2-4)}/t_g^{(2-3)}$. The $\tau$ value demonstrates a slowing down of the generation process if generation is performed using default correlation lengths corresponding to $^{2-4}J$. The $\mu$ coefficient estimates the loss of structural information that takes place in this case.

Analysis of the Table shows that, even in the case of small molecules, the output file size increases considerably when the $^{2-4}J$ couplings are set as default. Under those conditions the portion of extracted structural information drops from 95–100% to 60–70%. At the same time the generation time increases by many times to hundreds or even tens of thousands times greater. Problem 6 is the most distinctive example where the size of the output file increased from 2 to 3036 structures, whereas the generation time increased by 6.5 million times!

For problems 2 and 5 in Table 10.2, the 2D NMR data contained NSCs and both tasks were solved automatically using FSG with the output file containing only one structure at the conclusion of the run. The solutions to these problems found by simply lengthening all correlations to $^{2-4}J$ resulted in an increase in the number of structures up to 25 (problem 2) and up to 1211 (problem 5) structures. The computing time grew by 25 times and 11 900 times, respectively.

Table 10.2 lists some examples that allow the examination of the dependence of the results on the default option settings. However, the most typical problems of the same structure size are presented in Table 10.3. In example 1,[43] eight out of 28 HMBC correlations were of non-standard length. The task was solved using the FSG mode and resulted in a single correct structure with a generation time of 41 min 35 s. Strict generation with the coupling constant values $^{2-4}J$ set by default gave $k = 4$ and $t_g = 24$ min. This example shows that if a molecule is small and the number of NSCs $^{2-4}J$ is large, then while the longer

**Table 10.2** The dependence of the amount of structural information extracted from 2D NMR data based on the nature of the coupling constant $^nJ$ value set as the default during the structure generation process.[a]

| No. | Structure, molecular formula, 2D NMR data available, number of possible isomers | Comparison of problem solutions obtained with $^{2-3}J$ and $^{2-4}J$ correlations | Structural information, $\mu$ and $\tau$ values |
|---|---|---|---|
| 1 | Ref.,[34] $C_{10}H_{17}Br_2ClO_2$, $N = 50\ 502\ 293$, HMBC (24) | $^{2-3}J$ <br> $k = 1$ <br> $t_g = 0.008$ s <br><br> $^{2-4}J$ <br> $k = 103\ (E = 6.69)$ <br> $t_g = 2.36$ s | $I_0 = 25.59$ <br> $\mu = 100\%$ <br><br> $E_0 = 25.59$ <br> $I = 18.90$ <br> $\mu = 74\%$ <br> $\tau = 295$ |
| 2 | Ref.,[35] $C_{13}H_{20}O_3$, $N = 14\ 431\ 269\ 166$ COSY (33), HMBC (48) | $^{2-3}J$ <br> 3 NSCs, COSY, <br> $a = 1$ <br> 2 NSCs HMBC, <br> $a = 1$ <br> FSG <br> $k = 1$ <br> $t_g = 0.014$ s <br><br> $^{2-4}J$ <br> $k = 25\ (E = 4.64)$ <br> $t_g = 0.342$ s | $I_0 = 33.75$ <br> $\mu = 100\%$ <br><br><br><br><br><br><br> $E_0 = 33.75$ <br> $I = 29.11$ <br> $\mu = 86\%$ <br> $\tau = 25$ |
| 3 | Ref.,[36] $C_{11}H_{12}O_3$, $N = 4\ 703\ 963\ 545$ HMBC (20) | $^{2-3}J$ <br> $k = 2\ (E = 1)$ <br> $t_g = 0.011$ s <br><br> $^{2-4}J$ <br> $k = 933\ (E = 9.86)$ <br> $t_g = 1$ min 46 s | $I_0 = 32.13$ <br> $\mu = 97\%$ <br><br> $E_0 = 32.13$ <br> $I = 22.27$ <br> $\mu = 69\%$ <br> $\tau = 9630$ |
| 4 | Ref.,[37] $C_{15}H_{22}O_2$, $N = 138\ 136\ 211\ 624$ HMBC (48) | $^{2-3}J$ <br> $k = 1$ <br> $t_g = 0.008$ s <br><br> $^{2-4}J$ <br> $k = 823\ (E = 9.69)$ <br> $t_g = 1$ min 28 s | $I_0 = 37$ <br> $\mu = 100\%$ <br><br> $E_0 = 37$ <br> $I = 27.31$ <br> $\mu = 74\%$ <br> $\tau = 11\ 100$ |

**Table 10.2**   (*Continued*).

| No. | Structure, molecular formula, 2D NMR data available, number of possible isomers | Comparison of problem solutions obtained with $^{2-3}J$ and $^{2-4}J$ correlations | Structural information, $\mu$ and $\tau$ values |
|---|---|---|---|
| 5 |  Ref.,[38] $C_{15}H_{26}O$, $N = 261\ 045\ 917$  COSY (12), HMBC (56) | $^{2-3}J$  $k = 1$  $t_g = 0.009$ s  2 NSC COSY, $a = 1$  FSG:  $^{2-4}J$  $k = 1211$  $(E = 10.23)$  $t_g = 1$ min 47 s | $I_0 = 27.95$  $\mu = 100\%$  $E_0 = 27.95$  $I = 17.72$  $\mu = 63\%$  $\tau = 11\ 900$ |
| 6 |  Ref.,[39] $C_{15}H_{20}O$, $N = 37\ 568\ 150\ 635$  HMBC (32) | $^{2-3}J$  $k = 2$ $(E=1)$  $t_g = 0.011$ s  $^{2-4}J$  $k = 3036$  $(E = 11.57)$  $t_g = 2$ h | $E_0 = 35.14$  $I = 34.14$  $\mu = 97\%$  $E_0 = 35.14$  $I = 23.57$  $\mu = 67\%$  $\tau = 6.5 \times 10^6$ |
| 7 |  Ref., 40 $C_{13}H_{28}N_2$, $N = 6\ 332\ 846$  HMBC (35) | $^{2-3}J$  $k = 2$ $(E=1)$  $t_g = 0.012$ s  $^{2-4}J$  $k = 2$ $(E=1)$  $t_g = 6$ s | $E_0 = 22.60$  $I = 21.60$  $\mu = 95\%$  $E_0 = 22.60$  $I = 21.60$  $\mu = 95\%$  $\tau = 500$ |
| 8 |  Ref.,[41] $C_{12}H_{12}O_3$, $N = 68\ 930\ 547\ 646$  HMBC (26) | $^{2-3}J$  $k = 2$ $(E=1)$  $t_g = 0.050$ s  $^{2-4}J$  $k = 22$ $(E=4.46)$  $t_g = 11$ s | $E_0 = 36$  $I = 35$  $\mu = 97\%$  $E_0 = 36$  $I = 31.54$  $\mu = 88\%$  $\tau = 220$ |

**Table 10.2** (*Continued*).

| No. | Structure, molecular formula, 2D NMR data available, number of possible isomers | Comparison of problem solutions obtained with $^{2-3}J$ and $^{2-4}J$ correlations | Structural information, $\mu$ and $\tau$ values |
|---|---|---|---|
| 9 | Ref.,[42] $C_{10}H_{14}O_2$, $N = 16\ 422\ 284$ HMBC (39) | $^{2-3}J$ $k = 1$ $t_g = 0.008$ s  $^{2-4}J$ $k = 132\ (E = 6.77)$ $t_g = 10$ s | $I_0 = 23.95$ $\mu = 100\%$  $E_0 = 23.95$ $I = 17.18$ $\mu = 72\%$ $\tau = 1250$ |

[a]Calculations were performed with a PC Pentium IV, 2.8 GHz with 1 GB of RAM. The abbreviation NSC represents non-standard connectivity.

correlations ($^5J$, $^6J$) are absent in 2D NMR data the application of FSG and the lengthening of all correlations both give comparable results in a case when the number of skeletal atoms is less than 20.

The problem solution time and the number of generated structures increase dramatically when the size of the molecule increases, and it is evident that the default setting of $^{2-4}J$ for the correlation length can only assist when the number of skeletal atoms is around 20 and the length of NSCs does not exceeds three skeletal bonds.

The solutions for problems 2 and 3 with $^{2-3}J$ coupling constants set as the default were identified in several seconds. However, when the coupling constants were set to $^{2-4}J$ as default, the program was aborted by the operator in about 15 h. At the same time the number of generated structures was around $10^5$, and it was impossible to predict the time left in order for structure generation to be completed.

Investigation of the results of problem 4 also provided interesting results. The correct solution was determined by two approaches: using automatic correction of the MCD ($k = 1$, $t_g = 0.009$ s) and using FSG with the options $\{m_{max} = 10$, stop at $m = m_g$, $a = 2\}$ and gave $k = 1$, $t_g = 0.031$ s. Structure generation with all correlations set to $^{2-4}J$ by default gave 61 structures in approx. 6 min. Rank ordering of the structural file according to the deviation values gave a structure with the deviations $d_A(1) = d_I(1) = 8.6$ ppm placed at the top of the ranked file. According to previously defined criteria, $d_A(1)$ should be less than 3.5–4.5 ppm for the correct structure and under this constraint the solution is likely incorrect. This implies that the generation process should be repeated with the coupling constants set to $^{2-5}J$ as the default.

It could be expected that the number of structures with $^{2-5}J$ set as default would be large so the structure generation process was executed in a mode

**Table 10.3** The dependence of structural information extracted based on the $J$ value range utilized for a series of fairly "large-sized" molecules.

| 1 | Ref.,[43] $C_{13}H_{12}O_4$ HMBC (28) | $^{2-3}J$ Eight non-standard HMBC connectivities ($a=1$). FSG was performed using options $\{m_{max}=10$, stop at $m=m_g$, $a=1\}$ Results: $k=1$; $t_g=41$ min 35 s <br> $^{2-4}J$ $k=4$; $t_g=24$ min |
|---|---|---|
| 2 | Ref.,[44] $C_{15}H_{19}NO$ HMBC (30) | $^{2-3}J$ $k=4$; $t_g=2.6$ s <br> $^{2-4}J$ Structure generation was aborted after 14 h with $k \sim 116\,000$ structures produced. An indicator predicted that at least double this number of structures would be produced. The problem could not be solved. |
| 3 | Ref.,[45] $C_{15}H_{14}O_4$ HMBC (31) | $^{2-3}J$ One non-standard HMBC correlation ($a=1$). FSG with options: $\{m_{max}=10$, stop at $m=m_g$, $a=1\}$ Results: $k=5$; $t_g=2$ s <br> $^{2-4}J$ Structure generation was aborted after 15 h with $k \sim 10^5$. |
| 4 | Ref.,[46] $C_{15}H_{20}O_2$ COSY (22), HMBC (62) | $^{2-3}J$ Two non-standard COSY correlations ($a=2$) Two non-standard HMBC correlations ($a=1$) Automatic MCD correction: $k=1$, $t_g=0.009$ s FSG $\{m_{max}=10$, stop at $m=m_g$, $a=2\}$: $k=1$, $t_g=0.031$ s <br> $^{2-4}J$ $k=61$; $t_g=5$ min 54 s Chemical shift deviations of the best structure: $d_A(1)=d_I(1)=8.6$ ppm; $d_H(1)=0.66$ ppm Wrong solution. <br> $^{2-5}J$ Run in the mode of isomer counting. The program was aborted in 6 min when $k$ reached $\sim 500\,000$ structures. |

whereby the structures were not written to disk. The program was later aborted by the operator after 6 min, since the number of resultant structures had already reached half a million.

The last example shows that even in the case of a molecule being fairly small, in this case with $n < 20$, and even when both the COSY and HMBC spectra contain a large number of standard correlations, the process of increasing the intervals allowed for the correlation lengths offered no solution. In such a situation, the most effective way to solve the problem appears to be FSG.

We assume that the results described in this Chapter strongly support the guideline that the lengthening of *all* correlations should be rejected as a general method of solving problems arising from the presence of NSCs in 2D NMR data. This conclusion is even more obvious if we take into account the distribution of problems solved using StrucEluc (see Chapter 14). The distribution of problems, as well as the data in Figure 7.3, show that molecules with less than 20 skeletal atoms are rarely found among natural products.

## 10.5   FSG as an Analytical Tool

The described examples have shown that FSG should be considered as the most general method for structure generation in expert systems based on 2D NMR spectra. It allows the process of structure elucidation to be initiated under conditions whereby the user has no idea about the presence or absence of NSCs or details regarding their real lengths. This is attained by setting appropriate generator options that provide a varying number of NSCs in the range of 0 up to $m_{max}$. Experience has shown that the value of $m_{max}$ is usually set equal to 10, 15 or 20 depending on the minimum number of NSCs detected by the program during the logical analysis of 2D NMR data. The real number of NSCs and their lengths can be determined by two approaches.

The first approach allows identification of these variables along with the structure of the unknown using a trial method. In so doing, the variables $m$ (true number of NSCs) and $a$ (augmentations of connectivity lengths) are varied.

The second approach assumes only that the $m$ variable does not exceed some $m_{max}$ value, while no information about $a$ value is necessary. This approach is advisable since it is fully automated and more universal. Its shortcoming is that FSG can in some cases be more time consuming and, in addition, the output file can be larger. This can be of little impact since calculation time can be relatively inexpensive in the present era of computational cost and structure generation can be performed in background mode anyway. At the very least computations can be left overnight, as is common for 2D NMR data acquisition. We suppose that $t_g = 5$–$15$ h is still acceptable because the separation and identification of a natural product by traditional methods usually takes weeks and even months. To elucidate the structure of a new compound without any danger associated with the presence of an unknown number of NSCs of unknown lengths is attractive enough to afford time-consuming structure generation.

As a result of dramatically increasing the speed of $^{13}$C NMR spectrum prediction (6000–10 000 shifts/s), a large output file no longer hampers fast candidate structure spectral prediction that is necessary for the optimal elimination of isomorphic structures and file ranking in order to select the most probable structure. For instance, the $^{13}$C spectrum prediction and average deviation calculation for 38 407 isomers generated from the molecular formula $C_{20}H_{28}O$ (structure **10.12**) took 2 min 23 s, for 4900 isomers of $C_{39}H_{63}N_2O_{12}P$ (structure **10.8**) – only 27 s. FSG can be concluded to be an appropriate analytical tool for application to the structure elucidation of organic molecules using 2D NMR spectra.

In each individual case a strategy for problem solving is chosen using an estimation of the problem complexity. This is possible on the basis of computing the number of connectivity combinations which can be processed during FSG. Preliminary logical analysis of the 2D NMR data allows the reduction of the calculated number of combinations by $10^3$–$10^5$ times. In spite of a large enough number of remaining combinations to be processed (thousands to millions), the speed of the structure generator in StrucEluc is such that it copes with a task in a reasonable time.

The strategy of FSG has been illustrated using a series of real-world examples, including molecules whose 2D NMR spectra contained up to 15 NSCs with lengths varying between four and six bonds. When a data set lacked 2D NMR correlations but was accompanied by the presence of NSCs, then the problem can be solved utilizing the fragments found in a fragment library and employing user-proposed substructures. However, the solution of such problems can be difficult, since a large number of MCDs can be created from the fragments. To circumvent this difficulty in the future, algorithms for fragment "implementation" in a MCD, as well as FSG from fragment-containing MCDs, needs to be further improved. This work is underway.

In the literature, examples are described where lengthening *all* correlations by one bond circumvented the problem of NSCs in 2D NMR data. This possibility has been investigated here and has shown that even the structures of small molecules with less than 20 carbon atoms, in general, cannot be elucidated using this approach. The lengthening of connectivities by one bond is evidently in vain when the NSCs are characterized by $^{5-6}J$ coupling constants present in the 2D NMR data.

# References

1. P. Phuwapraisirisan, S. Matsunaga, N. Fusetani, N. Chaitanawisuti and H. Mycaperoxide, *J. Nat. Prod.*, 2003, **66**, 289.
2. M. E. Elyashberg, K. A. Blinov, A. J. Williams, S. G. Molodtsov and E. R. Martirosian, *J. Nat. Prod.*, 2002, **65**, 693.
3. M. E. Elyashberg, K. A. Blinov, S. G. Molodtsov, A. J. Williams and G. E. Martin, *J. Chem. Inf. Comput. Sci.*, 2004, **44**, 771.

4.  M. E. Elyashberg, K. A. Blinov and E. R. Martirosian, *Lab. Autom. Inf. Manag.*, 1999, **34**, 15.
5.  B. L. Marquez, W. H. Gerwick and R. T. Williamson, *Magn. Reson. Chem.*, 2001, **39**, 499.
6.  G. E. Martin, in *Ann. Rep. NMR Spectrsosc.*, ed. G. A. Webb, Academic Press, New York, 2002, vol. 46, pp. 37.
7.  P. Leone, J. Redburn, J. Hooper and R. Quinn, *J. Nat. Prod.*, 2000, **63**, 694.
8.  Y. C. Shen, K.-L. Lo, C.-Y. Chen, Y.-H. Kuo and M.-C. Hung, *J. Nat. Prod.*, 2000, **63**, 720.
9.  C. Peng, G. Bodenhausen, S. Qiu, H. H. S. Fong, N. R. Farnsworth, S. Yuan and C. Zheng, *Magn. Reson. Chem.*, 1998, **36**, 267.
10. P. Radhika, P. V. S. Rao, V. Anjaneyulu and R. N. Asolkar, *J. Nat. Prod.*, 2002, **65**, 737.
11. W. F. Reynolds, S. McLean, S. J. Burke and H. Jacobs, *Magn. Reson. Chem.*, 2001, **39**, 757.
12. F. D. Horgen, E. B. Kazmierski, H. E. Westenburg, W. Y. Yoshida, P. J. Scheuer and D. Malevamide, *J. Nat. Prod.*, 2002, **65**, 487.
13. S. Habtemariam, B. Skeleton, P. Waterman and A. White, *J. Nat. Prod.*, 2000, **63**, 512.
14. P. Willett, J. Barnard and G. J. Downs, *Chem. Inf. Comput. Sci.*, 1998, **38**, 983.
15. D. O. Collins, W. F. Reynolds and P. B. Reese, *J. Nat. Prod.*, 2004, **67**, 179.
16. A. Y. Mensah, P. J. Houghton, S. Bloomfield, A. Vlietinck and D. V. Berghe, *J. Nat. Prod.*, 2000, **63**, 1210.
17. K. D. Wellington, R. C. Cambie, P. S. Rutledge and P. R. Bergquist, *J. Nat. Prod.*, 2000, **63**, 79.
18. J. H. H. L. Oliveira, A. Grube, M. Köck, R. G. S. Berlinck, M. L. Macedo, A. G. Ferreira and E. Hajdu, *J. Nat. Prod.*, 2004, **67**, 1685.
19. S. G. Molodtsov, M. E. Elyashberg, K. A. Blinov, A. J. Williams, G. E. Martin and B. Lefebvre, *J. Chem. Inf. Comput. Sci.*, 2004, **44**, 1737.
20. M. E. Elyashberg, K. A. Blinov, S. G. Molodtsov, A. J. Williams and G. E. Martin, *J. Chem. Inf. Model.*, 2007, **47**, 1053.
21. K. A. Blinov, D. Carlson, M. E. Elyashberg, G. E. Martin, E. R. Martirosian, S. G. Molodtsov and A. J. Williams, *J. Magn. Reson. Chem.* 2003, **41**, 359.
22. M. E. Elyashberg, K. A. Blinov, E. R. Martirosian, S. G. Molodtsov, A. J. Williams and G. E. Martin, *J. Heterocycl. Chem.*, 2003, **40**, 1017.
23. M. E. Elyashberg, K. A. Blinov, A. J. Williams, S. G. Molodtsov and G. E. Martin, *J. Chem. Inf. Model.*, 2006, **46**, 1643.
24. L. T. Tan, T. Okino and W. H. Gerwick, *J. Nat. Prod.*, 2000, **63**, 952.
25. B. D. McGarvey, A. B. Attygalle, A. N. Starratt, B. Xiang, F. C. Schroeder, J. E. Brandle and J. Meinwald, *J. Nat. Prod.*, 2003, **66**, 1395.
26. G. Kirsch, G. M. Kong, A. D. Wright and R. Kaminsky, *J. Nat. Prod.*, 2000, **63**, 825.
27. M. Feller, A. Rudi, N. Berer, I. Goldberg, Z. Stein, Y. Benayahu, M. Schleyer and Y. Kashman, *J. Nat. Prod.*, 2004, **67**, 1303.

28. L. K. Mdee, S. O. Yeboah and B. M. Abegaz, *J. Nat. Prod.*, 2003, **66**, 599.
29. S. Kehraus, G. M. Köning and A. D. Wright, *J. Nat. Prod.*, 2002, **65**, 1056.
30. Y. Feng, J. W. Blunt, A. L. J. Cole and M. H. G. Munro, *J. Nat. Prod.*, 2002, **65**, 1274.
31. S. G. Molodtsov, *Commun. Math. Chem. (MATCH)*, 1994, **30**, 213.
32. S. G. Molodtsov, *Commun. Math. Chem. (MATCH)*, 1994, **30**, 203.
33. S. G. Molodtsov, *Commun. Math. Chem. (MATCH)*, 1998, **37**, 157.
34. A. R. Diaz-Marrero, J. Rovirosa, J. Darias, A. San-Martin, M. Cueto and A.-C. Plocamenols, *J. Nat. Prod.*, 2002, **65**, 585.
35. C. Klemke, S. Kehraus, A. D. Wright and G. M. König, *J. Nat. Prod.*, 2004, **67**, 1058.
36. J.-Q. Gu, T. N. Graf, D. Lee, H.-B. Chai, Q. Mi, L. B. S. Kardono, F. M. S. Setyowati, R. Ismail, S. Riswan, N. R. Farnsworth, G. A. Cordell, J. M. Pezzuto, S. M. Swanson, D. J. Kroll, I. Falkinham, J. O., M. E. Wall, M. C. Wani, A. D. Kinghorn and N. H. Oberlies, *J. Nat. Prod.*, 2004, **67**, 1156.
37. K. B. Iken and B. J. Baker, *J. Nat. Prod.*, 2003, **66**, 888.
38. Y. Fukushi, C. Yajima, J. Mizutani and S. Tahara, *Phytochemistry*, 1998, **49**, 593.
39. M. Gavagnin, E. Mollo, F. Castelluccio, A. Crispino and G. Cimino, *J. Nat. Prod.*, 2003, **66**, 1517.
40. M. R. Heinrich, Y. Kashman, P. Spiteller and W. Steglich, *Tetrahedron*, 2001, **57**, 9973.
41. D. G. Nagle, Y.-D. Zhou, P. U. Park, V. J. Paul and I. Rajbhandary, *J. Nat. Prod.*, 2000, **63**, 1431.
42. S.-S. Moon, J.-Y. Lee and S.-C. Cho, *J. Nat. Prod.*, 2004, **67**, 889.
43. B. S. Joshi, K. L. Singh and R. Roy, *Magn. Reson. Chem.*, 2001, **39**, 771.
44. J. M. S. López, M. M. Insua, J. P. Baz, J. L. F. Puentes and L. M. C. Hernández, *J. Nat. Prod.*, 2003, **66**, 863.
45. Y. Hernández-Romero, J.-I. Rojas, R. Castillo, A. Rojas and R. Mata, *J. Nat. Prod.*, 2004, **67**, 160.
46. D. Stærk, B. Skole, F. S. Jørgensen, B. A. Budnik, P. Ekpe and J. W. Jaroszewski, *J. Nat. Prod.*, 2004, **67**, 799.

# CHAPTER 11
# *Challenging Structure Elucidator*

In this Chapter we will give examples of challenges solved with the aid of StrucEluc. The first two examples are related to the elucidation of complex alkaloids belonging to the cryptolepine (**11.1**) series[1,2] (see Chapter 9).

**11.1**

## 11.1 Structure Elucidation of a Degradant of Cryptospirolepine

Martin *et al.*[1] utilized cryogenic NMR probe technology[3] and the StrucEluc system to characterize degradants of a complex spiro monocyclic alkaloid, the cryptospirolepine molecule, **11.2**. A 2.5 mg sample of this compound had been stored in a sealed 5 mm NMR tube in $d_6$-DMSO for ~10 years and the compound **11.2** had degraded.

**11.2**

New Developments in NMR No. 1
Contemporary Computer-Assisted Approaches to Molecular Structure Elucidation
By Mikhail Elyashberg, Antony Williams and Kirill Blinov
© Royal Society of Chemistry 2012
Published by the Royal Society of Chemistry, www.rsc.org

The two major degradation products of cryptospirolepine, DP-1 and DP-2, $\sim 35$ and $\sim 16\%$ of the total sample respectively, were isolated by reversed-phase, semi-preparative HPLC. NMR samples of approx. $\sim 0.5$ mg and $\sim 200$ μg, respectively, were used for the structural characterization effort.

The major component DP-1 was quickly identified by a $^{13}$C NMR search in the ACD/CNMR database to identify a known natural product, cryptolepinone **11.3**.

CH₃

**11.3**

MS on the second isolate (DP-2) gave a molecular ion, $MH^+ = 479$, which suggested a molecular formula for DP-2 of $C_{32}H_{22}N_4O$. A 1D $^{13}$C NMR spectrum was not available as is very common in natural product structure elucidation when very small samples are involved. The $^{13}$C shift inputs were thus created from the HSQC and HMBC spectra. Eighteen peaks were identified in the HSQC (two $CH_3$ and 16 CH) data and 13 peaks were extracted from the HMBC to give a total of 31 peaks. According to the molecular formula, the molecule contains 32 carbon atoms. It was concluded that one quaternary carbon atom did not show an HMBC peak and one was added to the spectrum with a chemical shift of 130 ppm, in the middle of the aromatic interval (an "axiom"). The overnight HMBC spectrum contained 32 readily assigned responses. To avoid contradictions caused by the presence of non-standard correlations (NSCs), the extra peaks observed in the second HMBC experiment were attributed to a range of potential couplings and allowed to be $^{2-4}J_{CH}$ (this is a specific "axiom" for the system).

Attempts to solve this problem in both the common and fragment modes quickly showed that structure generation would be extremely time consuming. This was interpreted as a hint to apply a user database (342 fragments) formed from the known structures of the cryptolepine series as described in Chapter 9. Searching the $^{13}$C NMR spectrum in the user database resulted in 44 fragments. 776 molecular connectivity diagrams (MCDs) were created and each MCD contained four found fragments. No constraints on the generated structures were imposed. The result of structure generation was $k = 1572 \rightarrow 228 \rightarrow 8$, $t_g = 12$ s. As the structures were ranked by the deviation values, the best structure was found in first position as shown in Figure 11.1.

All three methods of $^{13}$C NMR prediction pointed to structure #1 as the best one. This allowed Martin *et al.*[1] to conclude that the structure of compound DP-2 is cryptoquindolinone **11.4**.

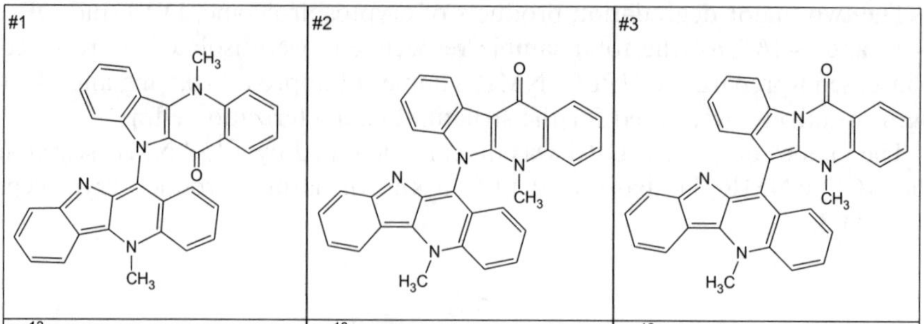

| #1 | #2 | #3 |
|---|---|---|
| $d_A(^{13}C)$: 1.967 | $d_A(^{13}C)$: 2.635 | $d_A(^{13}C)$: 3.973 |
| $d_I(^{13}C)$: 2.924 | $d_I(^{13}C)$: 4.741 | $d_I(^{13}C)$: 5.116 |
| $d_N(^{13}C)$: 2.667 | $d_N(^{13}C)$: 4.042 | $d_N(^{13}C)$: 4.354 |

**Figure 11.1** The first three structures of the ranked file which was deduced as the solution to the elucidation of DP-2.

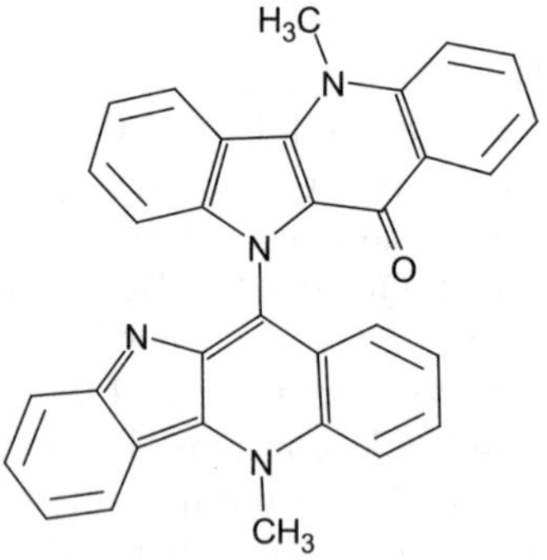

**11.4**

The authors[1] also considered how StrucEluc could assist in solving this problem when both traditional and computer-based approaches are combined. It is common for an experienced spectroscopist to detect molecular fragments simply by visual analysis of 1D and 2D NMR data. The approach is based on experience, knowledge, and insight of a highly qualified researcher, and the structural information extracted can therefore be invaluable. Providing spectroscopists with software tools that can facilitate the assembly of the molecular structure in an interactive mode, while allowing them to modify their hypotheses, is of obvious value. This approach was expected to have a synergistic effect.

The ability of the StrucEluc system to act as an assistant to the elucidation process was tested for this example. Visual analysis of the MCD produced for compound DP-2 allowed the experts to clearly see three 1,2-Ar fragments ("hexagonal stars" produced by HMBC connectivities inside of rings) and suggest the presence of the fourth one. HMBC connectivities identify the connection of the two aromatic fragments via the N-CH$_3$ group and binding of

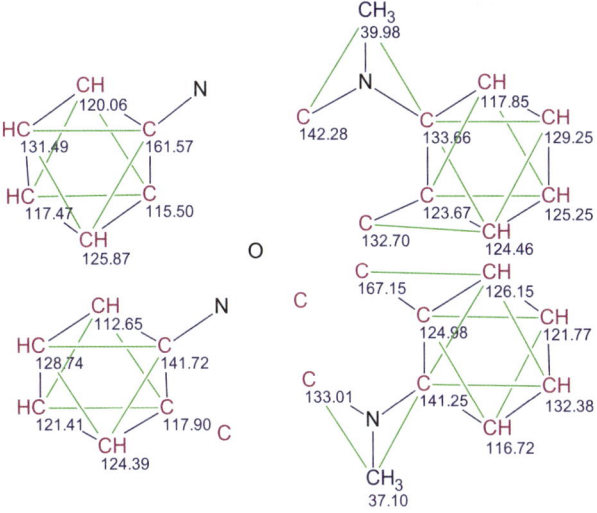

**Figure 11.2** MCD of DP-2 displaying fragments deduced by the expert. Ambiguous connectivities are not shown.

another N-CH$_3$ group to the third benzene fragment. The resulting MCD is shown in Figure 11.2.

The results of structure generation from this manually created MCD were: $k = 496,528 \rightarrow 26$, $t_g = 13$ min 15 s. The correct structure was again identified as the most probable one using all three methods of $^{13}$C NMR prediction. The application of the spectroscopist's insight had a beneficial effect and allowed progress without the user database. The above example indicates that a highly qualified expert is capable of determining very complex structures relying on his or her knowledge and the capacity of the system for deducing all, without any exception, logical consequences following from postulated suggestions (axioms and hypotheses).

## 11.2 Solution of a Cryptolepine Family "Puzzle"

When the 2D NMR data were being acquired for cryptospirolepine (**11.2**) by Martin and co-workers,[4] data were also accumulated in late 1991 for another alkaloid fraction from *Crocothemis sanguinolenta*, which was given the notebook designation TC-6. A data set consisting of proton and carbon reference spectra, COSY, ROESY, $^1$H-$^{13}$C HMQC, and HMBC spectra in MeOD was acquired. A structure consistent with all of the available data was not assembled in 1991/1992 when these data were first examined. It was associated with a small sample amount, very scant knowledge about the cryptolepine family and the accuracy of the instruments available in that time. Therefore TC-6 was reinvestigated ten years later by Blinov *et al.*[2] using new instrumentation, and StrucEluc was applied again in a mode of tight interaction with the spectroscopist.

The retained reference sample of this alkaloid was 95% pure with a molecular mass of 448 Da. Major fragmentation was simple with the molecule essentially splitting into two "halves" producing fragment ions at 217 and 232 Da. The accurate mass was measured as 448.1683 Da, which is within 1.2 ppm of the theoretical mass of the empirical formula of $C_{31}H_{21}N_4$.

Despite a relatively congested proton NMR spectrum at 400 MHz, the COSY spectrum still readily allowed the protons of the four individual four-spin systems to be identified and ordered. These included ordered sets of resonances as follows:

8.88 – 8.23 – 7.79 – 7.86
8.86 – 7.59 – 7.85 – 7.57
8.68 – 7.52 – 7.58 – 7.11
8.31 – 7.76 – 7.53 – 7.80

In addition, a $CH_3$ singlet was observed at 5.28 ppm in the $^1H$ NMR spectrum which can be attributed to an indoloquinoline *N*-methyl group. A singlet resonating at 7.90 ppm was plausibly interpretable as an isolated aromatic proton. Using the HMQC correlation data, directly bonded carbons were associated with their respective protons, which suggested the substructural fragments **11.5–11.8**. The HMQC data, obviously, also correlated the N-CH$_3$ singlet at 5.28 ppm, with a carbon resonating at 43.1 ppm and the isolated aromatic proton resonance at 7.90 ppm correlated with a carbon resonating at 115.8 ppm.

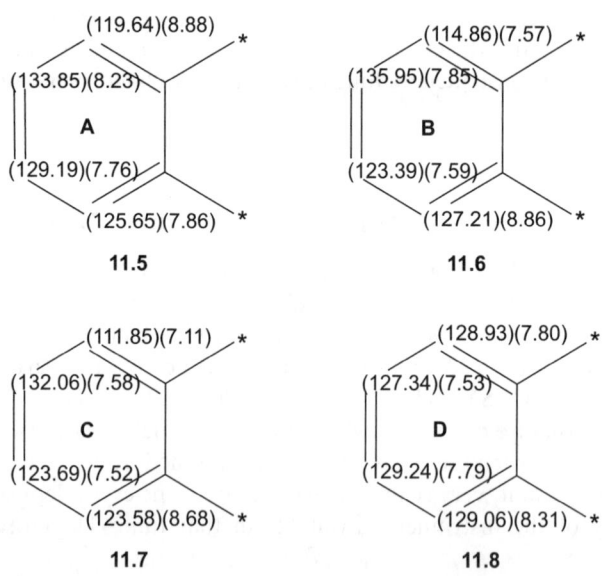

When the molecular formula and all NMR data were fed into StrucEluc, the MCD shown in Figure 11.3 was created. Because of the highly congested region in the vicinity of 129 ppm in the $^{13}$C spectrum, there was some potential for ambiguity in the assignments. All ambiguous correlations are displayed with dotted lines, which allow a spectroscopist to visually analyze the whole picture and edit correlations step-by-step in accord with chemical common sense.

To remove ambiguous correlations a heuristic approach suggested by spectroscopic common sense was used. We will describe in detail the first step of the process. Figure 11.3 shows that there are two very close chemical shifts in the $^1$H NMR at 7.86 and 7.85 ppm. The first of them was assigned to ring A, with the second to ring B. It is obvious that these chemical shifts may be exchanged because they are observed in a very congested area of the $^1$H spectrum. Furthermore the following reasoning was conducted.

In ring A a distinct HMBC correlation of standard length from H(7.86) to C(133.85) was revealed. This fact allows one to suggest that the chemical shift 7.86 is really related to the ring A, whereas the chemical shift 7.85 is related to ring B. The latter suggestion is confirmed by the distinct HMBC correlation of standard length from H(7.85) to C(127.21). With this in mind, researchers removed ambiguous correlations from H(7.86) that are directed to carbons

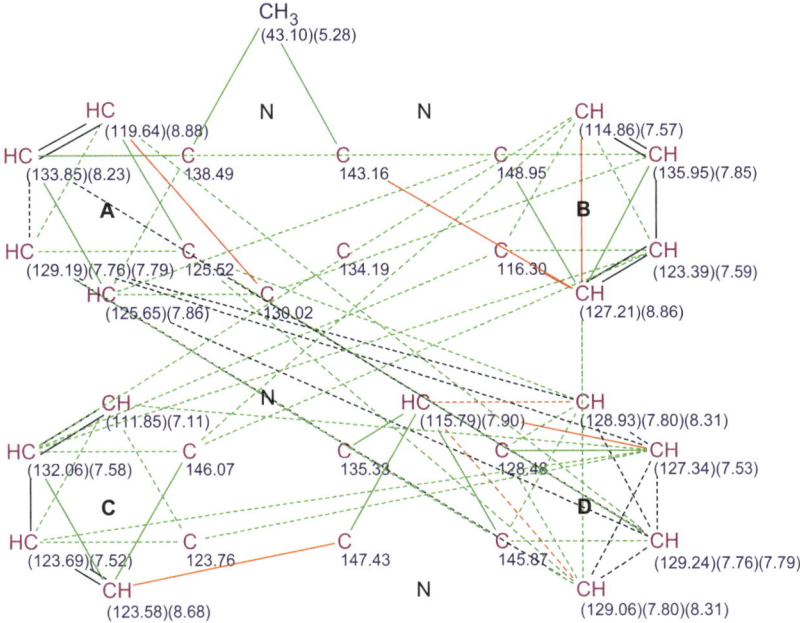

**Figure 11.3** The MCD showing all potentially ambiguous correlation pathways as dashed lines. Solid lines denote correlations that were initially thought to be correct. Vicinal connectivities are denoted by solid black lines. Two- and three-bond heteronuclear correlations are presented using solid or dashed green lines (the latter possibly ambiguous correlations); suggested longer-range correlations ($^nJ_{CH}$, n ≥ 4) are shown in orange.

situated in rings B and D, as well as the removal of ambiguous correlations from H(7.85) directed to carbons situated in ring A was made. Working in an analogous fashion, some minor revisions of proton/carbon pairings involving resonances with closely similar chemical shifts were performed and ambiguous connectivities were removed for all four-spin systems.

A ROESY correlation was observed from the *N*-methyl resonance at 5.28 ppm to the aromatic proton resonating at 8.88 ppm. The *single* ROE correlation observed from the N-CH$_3$ is important in that it excludes cryptolepine **(11.1)** from consideration as a possible fragment of the TC-6 structure. The reason is that the *N*-methyl group in systems containing cryptolepine exhibits correlations to *both* peri-aromatic protons, which would in turn be required again if one of these indoloquinoline systems were a constituent of the TC-6 structure. Consequently, the presence of a single ROESY correlation from N-CH$_3$ can be used as a very distinctive feature of the target molecule. An important cross-ring ROESY correlation was observed between the isolated aromatic proton resonating at 7.90 ppm (singlet for this proton) and the proton resonating at 7.80 ppm consistent with the MCD shown by Figure 11.3. Thus, the ROESY data can be quite important as an internal check for the consistency of the elucidation process when dealing with condensed polynuclear heteroaromatic systems such as this present example.

The finally revised proton-carbon chemical shift pairings are shown in the MCD represented by Figure 11.4. Approx. 48 h of spectroscopist interaction

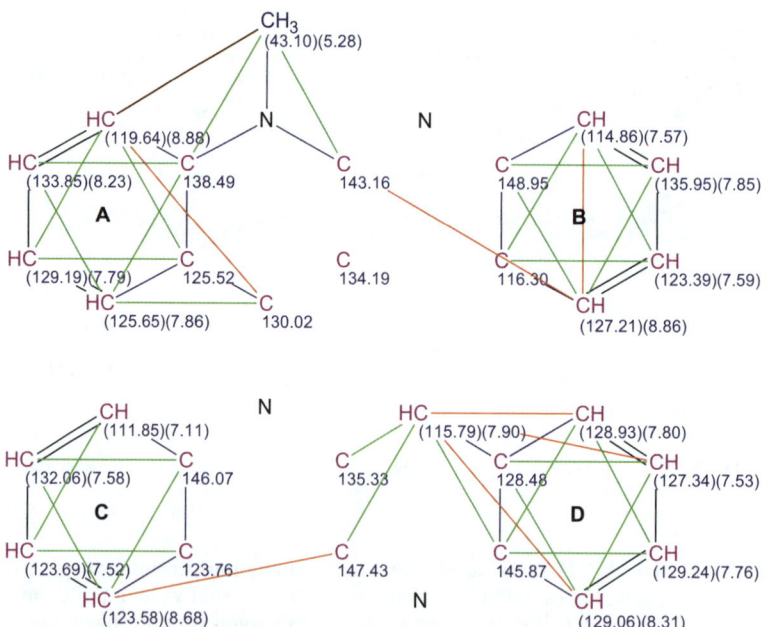

**Figure 11.4**  The final MCD obtained by continued pairwise successive removal of ambiguities associated with all four ring systems.

with the StrucEluc program package was required to reach this point in the structure elucidation process from the initial extraction of the four-spin systems represented by structures **11.5–11.8** from the COSY and HMQC data.

At this stage, one of the significant advantages of StrucEluc was demonstrated. It is the ability of the spectroscopist to work with the MCD family to successfully resolve ambiguities of this type which gives the benefit resulting from the synergistic interaction between a spectroscopist and the program.

In contrast, a spectroscopist working alone when faced with these sorts of entangled, closely spaced proton and carbon chemical shifts could spend a vast amount of time without success, particularly once correlations from the various protons to their respective long-range coupled carbons are added when the HMBC data are considered in attempting to solve the structure. In part, this sort of confusion was probably responsible for the frustrated initial attempts to manually elucidate the structure of this molecule.

From the MCD shown in Figure 11.4, the structure generation process was initiated and the following result was obtained: $k = 353 \rightarrow 266$, $t_g = 10$ s. $^{13}$C chemical shift calculation with subsequent file sorting allowed the program to distinguish the set of top ranked structures presented in Figure 11.5.

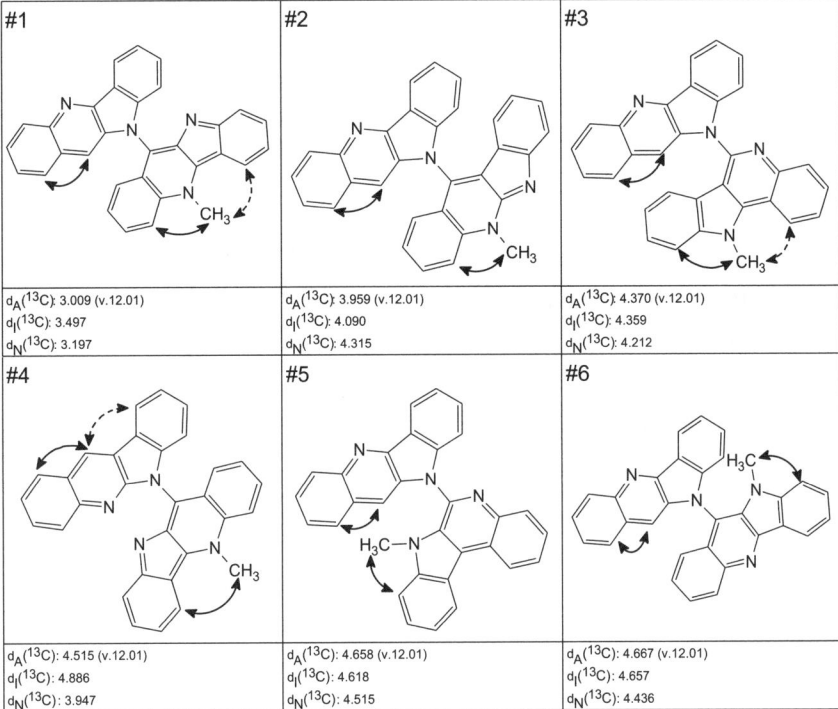

| #1 | #2 | #3 |
|---|---|---|
| $d_A(^{13}C)$: 3.009 (v.12.01) | $d_A(^{13}C)$: 3.959 (v.12.01) | $d_A(^{13}C)$: 4.370 (v.12.01) |
| $d_I(^{13}C)$: 3.497 | $d_I(^{13}C)$: 4.090 | $d_I(^{13}C)$: 4.359 |
| $d_N(^{13}C)$: 3.197 | $d_N(^{13}C)$: 4.315 | $d_N(^{13}C)$: 4.212 |

| #4 | #5 | #6 |
|---|---|---|
| $d_A(^{13}C)$: 4.515 (v.12.01) | $d_A(^{13}C)$: 4.658 (v.12.01) | $d_A(^{13}C)$: 4.667 (v.12.01) |
| $d_I(^{13}C)$: 4.886 | $d_I(^{13}C)$: 4.618 | $d_I(^{13}C)$: 4.657 |
| $d_N(^{13}C)$: 3.947 | $d_N(^{13}C)$: 4.515 | $d_N(^{13}C)$: 4.436 |

**Figure 11.5** The first six of 266 non-identical structures generated by StrucEluc and sorted on the basis of $d_A(^{13}C)$. Arrows show experimental (solid) and expected (dotted) ROESY correlations from CH$_3$ group and from the isolated aromatic proton 7.90 ppm.

Taking into consideration the *single* observed ROE correlation in the ROESY spectrum from the N-CH$_3$ group, structures #1 and #3 may be eliminated from consideration. Structure #4 was ruled out on the basis of the observed single ROE correlation from the isolated aromatic proton resonating at 7.90 ppm. Structure #2, with the more favorable $d_A(^{13}C)$ value, is consistent with this observation from the ROESY data, whereas structures #5 and #6 can be rejected due to deviation values. Based on these arguments, the structure of TC-6 is finally assigned as shown by **11.9** to which the name quindolino-cryptotackieine was given.

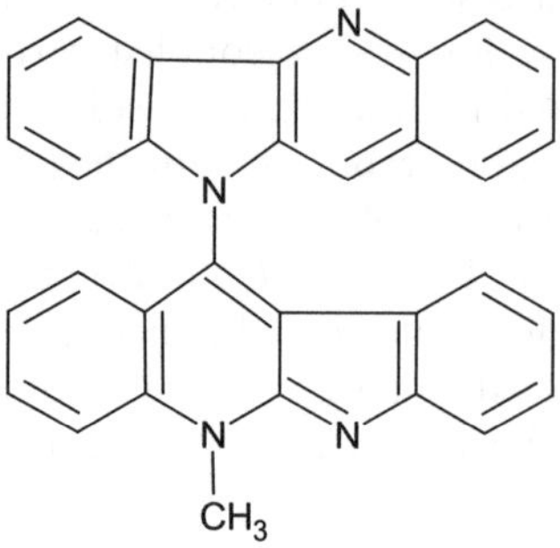

**11.9**

Thus, application of StrucEluc allowed the authors[2] to solve a problem which had remained unsolved for 10 years. Other examples of the application of CASE to natural product structure elucidation have been published elsewhere.[5–9]

# 11.3   Structure Elucidation of Two Unexpected Reaction Products in a Reaction of an α,β-Unsaturated Pyruvates

In the work by Sharman *et al.*[5] structure elucidation of two unexpected reaction products in a reaction of an α,β-unsaturated pyruvates (**11.10**) was performed.

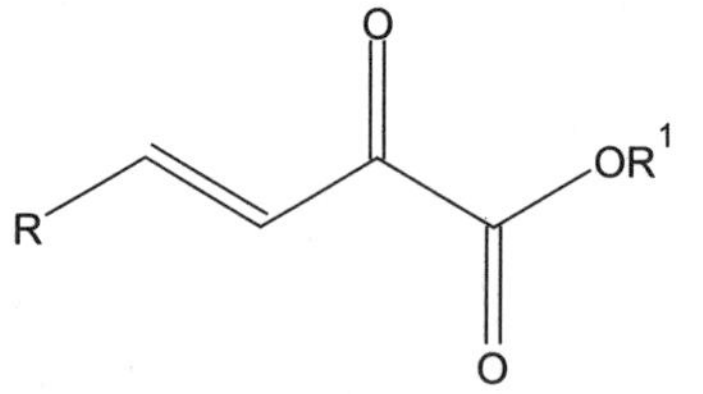

**11.10**

The reaction between one of these pyruvates (R = napthyl, R$^1$ = ethyl) and ethyl diazoacetate, using ZrCl$_4$ as the Lewis Acid, yielded two products. Initial evaluation of the $^1$H NMR and MS data indicated that neither of these products was among those expected from this reaction.

For the first product, double-quantum filtered (DQF)-COSY (20 correlations), HMQC (14 correlations), and HMBC (47 correlations) experiments were carried out and a molecular formula of $C_{20}H_{20}O_5$ was determined from the low-resolution MS and the carbon count extracted from the HMBC and HMQC spectra. An initial survey of the 2D data did not reveal a trivial solution to the structural problem.

The experimental data were input into StrucEluc and approx. 1700 structures were generated in approx. 1 min. This was reduced to 121 after filtering. Inspection of many of the hypothetical structures revealed highly improbable ring junctions. The absence of reasonable structures was considered as a hint to the presence of NSCs.

In this work[5] an early version of StrucEluc was used where fuzzy structure generation was still not implemented and as a result the following heuristic approach was employed: increase the lengths of correlations for those atoms which show a high number of correlations. In this case, the default lengths were increased to 2–4 and 2–5 bonds for HMBC and COSY, respectively, for the two protons showing the highest number of correlations. The results of the structure generation were $k = 5000 \rightarrow 347$, $t_g = 4$ h 30 min. The standard methodology allowed selection of the structure **11.11** as the most probable.

**11.11**

This structure markedly differed from the next ranked structure which showed $d_A = 6.75$ ppm. Structure **11.11** was later confirmed to be the correct structure by X-ray crystallography.

An alternative method of problem solving was also tested using this example. The naphthyl carbon skeleton found in the starting material was introduced as a user fragment for which [13]C chemical shifts were predicted. This assumption is reasonable based on the nature of the reaction. A corresponding MCD was created, and the structure generator was run again with the following results: $k = 92 \rightarrow 36 \rightarrow 9$, $t_g = 1$ s. The reduction in time and the number of resultant structures was substantial (in this case, introduction of a user fragment also removed the problem associated with the presence of NSCs), whereas, again, the correct structure was ranked as number one. This clearly demonstrates the large gains in efficiency, which can be achieved by using all of the available data.

Because the spectra of the second unknown were similar to spectra of structure **11.11**, it was suggested that the corresponding structures are also similar. With this in mind, the authors[5] automatically generated a user database from the structure **11.11** and performed the fragment search in it with the [13]C

NMR spectrum of the unknown. As a result of structure generation from MCDs created from selected fragments, a single structure **11.12** was produced:

**11.12**

This was also subsequently confirmed by X-ray crystallography. Because the molecules were small, the gains due to training seem trivial. As was shown above, for much larger molecules, it can mean the difference between days and minutes.

## 11.4   CASE Application for Identification of Drug Impurities in Mixture

The previous examples demonstrated the efficiency of the application of CASE approaches to solve difficult molecular structure elucidation problems where the unknowns are complex molecular structures and where the experimental NMR data are hard to interpret. Recently, Codina and co-workers[10] employed StrucEluc to identify drug impurities, generally small molecules existing in a mixture at very low concentrations. They described the combined use of pre-parative gas chromatography (prep-GC), GC-atmospheric pressure chemical ionization-mass spectrometry (APCI-MS), 1D and 2D NMR, and CASE for isolation and rapid elucidation of impurities at major and minor percentage levels in a pharmaceutical matrix. The total amount of one of the isolated and identified impurities was less than 60 nmol. The structure elucidation from 1D and 2D NMR data became possible due to recent advances in small cold probe technology[11] which have pushed the limits of sample concentration to the microgram (nanomolar) range. The use of instrumentation equipped with these probes leads to a dramatic reduction in either the NMR acquisition time (compared to when room temperature standard probes are used) or the amount of material needed. The results obtained in this work[10] are very interesting and they will be described here in more details.

The sample preparation preceding the NMR spectra acquisition was per-formed very thoroughly. GC with a preparative fraction collector (PFC) has been used to facilitate the identification of four volatile impurities in a phar-maceutical matrix. The trapping process was optimized using liquid sorbents, and the impurities were trapped directly into a deuterated solvent. Challenges related to the pharmaceutical matrix were overcome by derivatization with boron trifluoride in methanol and extraction with heptane, producing the methyl esters of the carboxylic acid impurities and main component. GC coupled to APCI-MS with a time-of-flight (TOF) detector was used to acquire accurate mass and isotopic data for the impurities, leading to the determination of their molecular formulae (MF).

APCI mass spectra were recorded on a Bruker MicroTOFQ mass spectrometer equipped with a Bruker GC-APCI interface. Data were acquired at a resolution of 6000 ($m/z$ 205), and accurate mass calibration was achieved by external calibration. The chemical formulas of the [M + H]+ ions were determined using Bruker SmartFormula software. Formulas with mass measurement errors of less than 2 mDa were considered, and the quality of the isotopic match, represented by a millisigma value, was used to determine a unique formula.

NMR spectra were recorded on Bruker AVANCE III 600 spectrometer, equipped with a 1.7 mm triple-resonance ($^1$H/$^{15}$N/$^{13}$C) single-axis gradient cryogenic probe. Homonuclear experiments included 1D $^1$H and 2D $^1$H-$^1$H-DQF-COSY. Heteronuclear experiments included HSQC and HMBC.

## 11.4.1 Structure Elucidation and Quantification

The APCI mass spectra of the derivatized known compound KC (**11.13**)

**11.13**

and derivatized impurities A, B, C, and D are shown in Figure 11.6. Data for the [M + H]$^+$ ions for all the components were found and the following molecular formulae were assigned with high confidence: KC′ ($C_{13}H_{16}O_2$), A′($C_9H_{12}O_2$), B′( $C_{13}H_{18}O_2$), C′($C_{13}H_{16}O_2$), and D′($C_{14}H_{20}O_3$). It is obvious that the C′ impurity is an isomer of KC′.

The MS data for the derivatized impurity A′ showed a tropylium ion at $m/z$ 91 (C7H7+, error 2.1 mDa) and strongly suggested the structure **11.14**:

**11.14**

$^1$H and HSQC NMR data were also acquired and confirmed the structure.

The MS data were not sufficient to unambiguously determine the structure of the derivatized impurities B, C, and D. Therefore 1D ($^1$H) and 2D COSY, HSQC and HMBC NMR experiments were acquired. The quality of the data was such that it allowed the elucidation of the structures in a semi-automatic fashion by using the StrucEluc system. The experiments were run with high

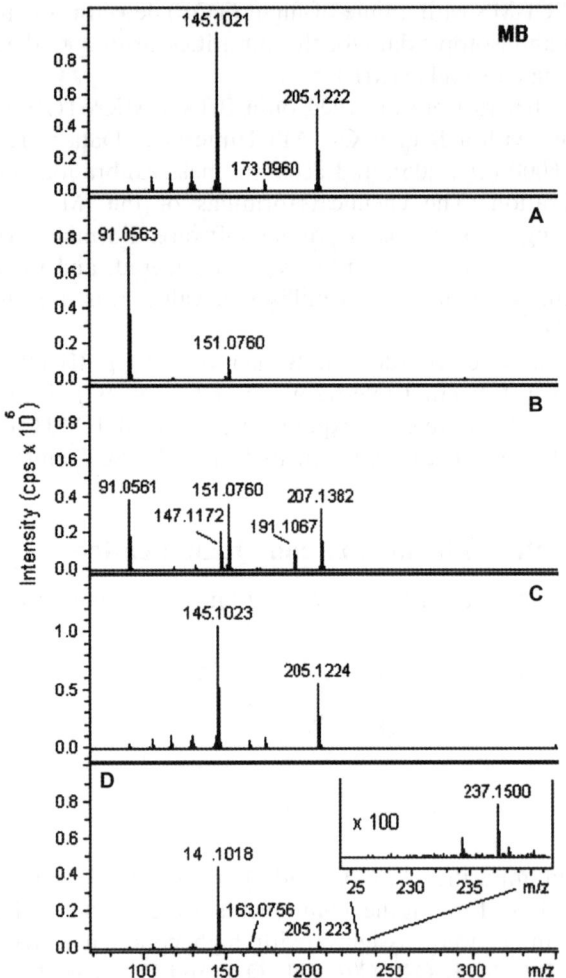

**Figure 11.6**   APCI+ mass spectra of the derivatized MB and impurities A, B, C, and D.

resolution in the indirect dimension in order to avoid peak overlapping, decrease the number of ambiguous correlations, and therefore facilitate the CASE analysis. Figure 11.7 shows the HSQC for the derivatized impurity D' (10 μg), acquired in 8 h and 5 min.

1D and 2D NMR data for the derivatized impurities were automatically processed with NMR Workbook (ACD/Labs). The $^1$H spectrum was manually peak picked, and the peaks automatically transferred into the 2D spectra. The automatic 2D peak picking was visually inspected for peaks the program failed to pick. In most cases, the peak picking was good and required minimal human optimization.

The NMR Workbook project containing the peak-picked NMR data was then opened in StrucEluc and the obtained molecular formula was introduced.

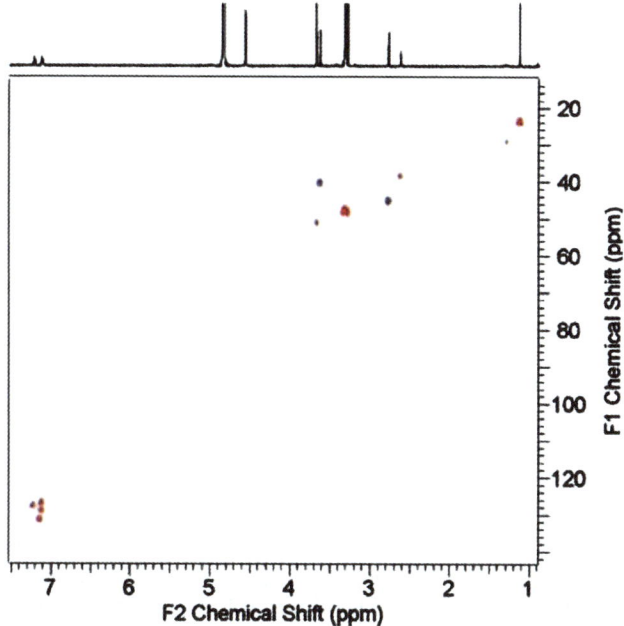

**Figure 11.7**    The 600 MHz HSQC NMR spectrum of the derivatized impurity D in methanol-d4.

The data were checked, and some structural information was added in a step called by authors[10] "data grooming". For impurities B, C, and D, this step included:

1. Completing the $^{13}$C chemical shift table (merged) by adding information about the number of carbon atoms for each different carbon chemical shift and the number of protons directly attached to each of those carbon atoms;
2. Completing the $^{1}$H chemical shift table (merged) by adding the number of protons for each different proton chemical shift;
3. Calibrating the HMBC correlations so that correlations were mainly (>80%) 2 or 3 bonds, except for the case of impurity B where *all* of the correlations were set up to be 2 or 3 bonds purposefully.

After the data were groomed, the MCD was generated and examined for contradictions. No contradictions were found, and, therefore, structures were calculated based on the MCD. No connectivities were extended during the calculations. The generated structures were filtered by carbon assignment during the generation step (maximum match factor 4 ppm and shift difference 20 ppm). For the structures that passed through the filter, $^{13}$C and $^{1}$H chemical shifts were predicted using a neural network (NN), and the average deviations between predicted and experimental chemical shifts $d_N(^{13}C)$ and $d_N(^{1}H)$ were obtained. Then the most probable structure ("best structure") was selected by the program.

For the derivatized impurity B', it was found that $k = 65 \rightarrow 13$, $t_g = 1$ s. These 13 structures were ranked by ascending $d_N(^{13}C)$. The second ranked structure **11.15** (#2) was found to be the best.

**11.15**

#1, #3 and #4 also had very good and similar deviations [$d_N(^{13}C) < 1$ ppm, $d_N(^1H) < 0.2$ ppm] and were duplicates of the same structure #2, but with different assignments for the aromatic ring. The difference in $d_N(^{13}C)$ between the best structure and the others obtained was large enough ($>0.5$ ppm) to conclude that the correct structure was #2. The APCI ion source fragmentation was also consistent with this structure. Substituted tropylium fragments were observed at $m/z$ 191 ([M + H − CH$_4$]$^+$, error 0.0 mDa) and $m/z$ 147 ([M + H − C$_2$H$_4$O$_2$]$^+$, error 0.4 mDa). Loss of C$_4$H$_8$ from [M + H]$^+$ to give the $m/z$ 151 fragment (C$_9$H$_{11}$O$_2$, error 0.6 mDa) was consistent with the *tert*-butyl group and further loss of C$_2$H$_4$O$_2$ from this ion gave the $m/z$ 91 fragment.

Since the authors[10] were not only interested in determining the structure for the derivatized impurity B', but also its $^{13}C$ and $^1H$ chemical shift assignment for adding to the database and reuse to assist with further analyses, they carried out a detailed examination of the results obtained with StrucEluc. The assignments for #1 and #3 were ruled out based on the multiplicity of the ring aromatic protons. Chemical shift assignments for #2 and #4 were both possible, and they could not easily be differentiated because of the overlapping $^1H$ and $^{13}C$ chemical shifts.

In the case of the derivatized impurity C', an isomer of KC', the structure generation gave the following result: $k = 10 \rightarrow 6$, $t_g < 1$ s. StrucEluc found **11.16** (#1) to be the most probable structure:

**11.16**

while #2 was identical to it and had slightly different chemical shift assignment (the assignment of two atoms could not be unambiguously determined because the $^1H$ and $^{13}C$ chemical shifts are overlapping). Structures #1 and #2 were significantly different from the rest of the structures [$d_N(^{13}C)$ and $d_N(^1H) < 1$ and $<0.2$ ppm, respectively, and $\Delta = d(\#3) - d(\#1) > 2.5$ ppm]. The APCI fragmentation observed was very similar to that observed for the KC' and, therefore, was consistent with the assigned structure.

For the derivatized impurity D′, 15 structures were generated and six passed through the filter. The calculation time was 2 s. Structure #1 was also found to be the most probable one by StrucEluc with $d_N(^{13}C) = 0.9$, $d_N(^1H) = 0.1$, and $\Delta d_N(^{13}C)$ between this structure and the others was bigger than 3 ppm, strongly suggesting the most probable structure **11.17** was the correct one:

**11.17**

The $^1H$ and $^{13}C$ chemical shift assignments obtained were *all* unambiguous. The APCI mass spectrum gave only a very weak intensity $[M + H]^+$ ion, which rapidly loses $CH_3OH$ to give $m/z$ 205. Further fragmentation was very similar to that observed for the derivatized KC, except for a considerably more intense $m/z$ 163 ($[M + H - C_4H_{10}O]^+$, error 0.2 mDa), which was fully consistent with the absence of a C=C double bond in the parent ion structure. This peak was determined to be formed from KC during the derivatization process, because of the absence of a corresponding peak in the GC-EI chromatogram of the pre-derivatized sample.

The derivatized impurities were quantified by $^1H$ NMR using the derivatized KC′ as an external reference of known concentration.[12,13] The concentration of the impurities were calculated by direct comparison of the values obtained from the integration of the singlet between 3.61 and 3.65 ppm, methylene adjacent to the methyl ester group. The results were remarkably similar to those estimated by GC-EI-MS and are presented in Table 11.1.

The structures of the impurities A, B, and C before derivatization are shown in Figure 11.8.

Codina and co-workers[10] have presented one of the first examples of computer-assisted structure elucidation of volatile impurities isolated by prep-GC and demonstrated that the combination of the proposed *in silico* methods and instrumentation is successful, even in cases where the total amount of material is of the order of 60 nmol. The CASE approach is not only strikingly fast,

**Table 11.1**  Impurity quantification by $^1H$ NMR (qNMR); comparison with EI-MS estimated values.

| Compound | qNMR | | Estimated by MS | |
|---|---|---|---|---|
| | *nmol* | *μg* | *nmol* | *μg* |
| KC′ | 1578 | 322 | 1578 | 322 |
| A′ | 143 | 21 | 98 | 15 |
| B′ | 176 | 36 | 192 | 40 |
| C′ | 355 | 72 | 338 | 69 |
| D′ | 58 | 14 | 42 | 10 |

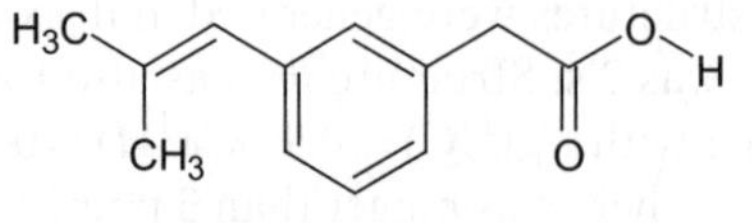

KC, N = 3,518,314,219 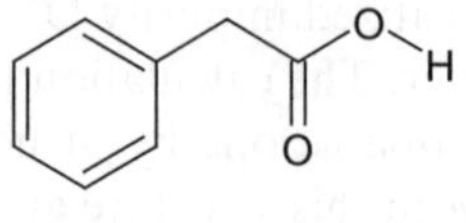                               A, N = 989,647

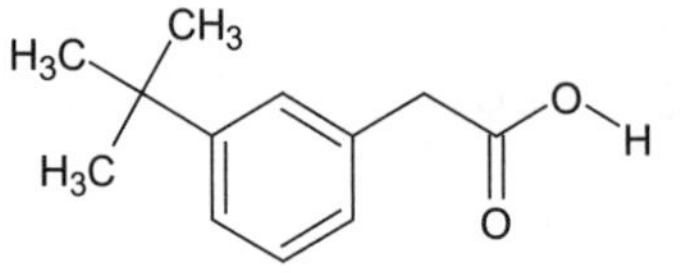

B, N = 1,264,165,511 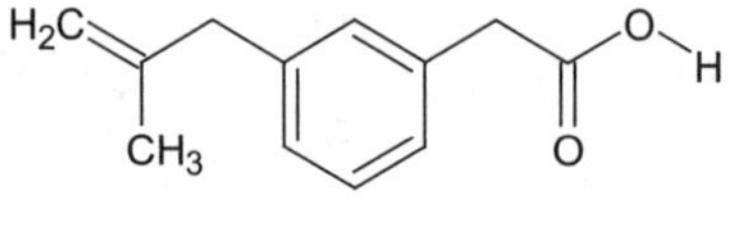                               C, N = 3,518,314,219

**Figure 11.8**   The structures of identified impurities and numbers of possible structural isomers calculated for these structures.

enabling the elucidation of the structures from raw MS and NMR data in minutes, but also is very reliable giving unique structures for each of the impurities with high confidence.

The sample quality achieved in this work by prep-GC combined with MS and NMR analyses allowed CASE to succeed with little human intervention. The authors envisage a wider use of prep-GC to facilitate the structure identification of impurities in the pharmaceutical industry. CASE can be applied to identify samples obtained by means other than prep-GC provided that the MF can be determined and the NMR data have sufficient quality in terms of S/N and resolution. Further advances in sample isolation/preparation and data quality together with software refinements have the potential to closely approach the fully automated structure identification of small molecules.

## 11.5   Structure Elucidation of Crizotinib

Crizotinib (**11.18**) is a rather challenging molecular structure. In spite of its modest size, the molecular formula $C_{21}H_{22}Cl_2FN_5O$ contains nine heteroatoms of four chemical types (including F and Cl) and the DBE is equal to 12.

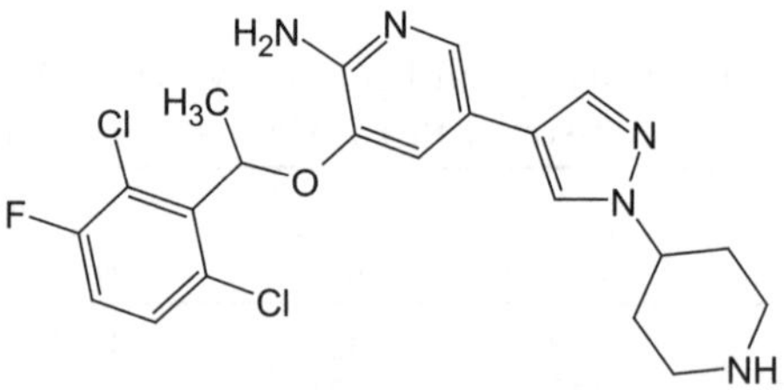

**11.18**

Codina[14] examined whether the StrucEluc system would be capable of elucidating this structure from 2D NMR data. The following spectral information

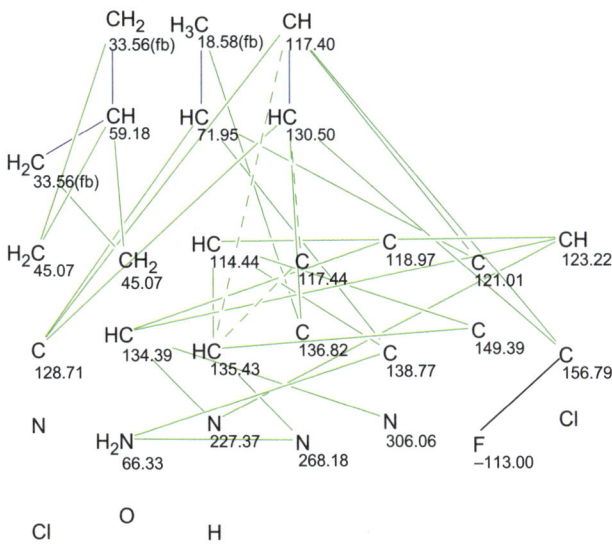

**Figure 11.9** The MCD created from 2D NMR spectral data acquired for crizotinib molecule. HMBC connectivities are colored in green, and COSY Connectivities are colored in blue.

was acquired: $^1$H and $^{13}$C NMR, $^1$H-$^1$H COSY, $^1$H-$^{13}$C HSQC, $^1$H -$^{15}$N HSQC, $^{19}$F-$^{13}$C HSQC, $^1$H-$^{13}$C HMBC, $^1$H-$^{15}$N HMBC, and $^{19}$F-$^{13}$C HMBC. The data were input into the program and the MCD in Figure 11.9 was generated.

Checking the MCD for the presence of contradictions showed that the data are free of NSCs. To speed up the structure generation, the following constraints were introduced manually: carbon atoms C(59.18, 4.12), C(71.95, 6.08), and C(156.79) were marked as having at least one heteroatom as a neighbor, while the neighborhood with heteroatoms was forbidden for carbons C(114.44) and C(117.44). Structure generation was completed with the result: $k = 130183 \rightarrow 12 \rightarrow 3$, $t_g = 42$ h 48 min, and the ranked output file is presented in Figure 11.10.

Figure 11.10 shows that despite the long period of time required for the structure generation process, the correct structure of crizotinib was identified correctly and ranked first. Note that the time for structure generation, about 2 days, is of the same order in this case as was necessary for acquiring 1,1-ADEQUATE or INADEQUATE spectra from the small amount of sample. As the acquisition time for 2D NMR experiments is significantly reduced due to the application of cryogenic probes,[3] nowadays 1,1-ADEQUATE experiment can be completed in 2.5–10 h. To evaluate the influence of 1,1-ADEQUATE information being included with the HMBC and COSY data, the corresponding correlations were input into the program. A new MCD was created and fuzzy structure generation was performed in fully automated mode from the MCD without any user intervention. The result was: $2554 \rightarrow 4 \rightarrow 1$, $t_g = 5$ min.

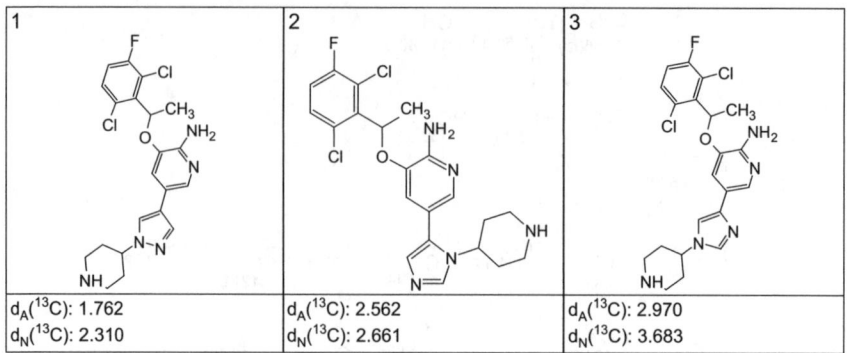

| 1 | 2 | 3 |
|---|---|---|
| $d_A(^{13}C)$: 1.762<br>$d_N(^{13}C)$: 2.310 | $d_A(^{13}C)$: 2.562<br>$d_N(^{13}C)$: 2.661 | $d_A(^{13}C)$: 2.970<br>$d_N(^{13}C)$: 3.683 |

**Figure 11.10** The ranked output file obtained as a result of crizotinib structure elucidation.

It can be concluded that in spite of the lack of 2D NMR structural constraints before using the 1,1-ADEQUATE data StrucEluc could elucidate the correct solution after long calculation times. The application of the 1,1-ADEQUATE experimental data increased the speed of the structure generation 500 times.

# References

1. G. E. Martin, B. D. Hadden, C. E. Russell, D. J. Kaluzny, J. E. Guido, W. K. Duholke, B. A. Stiemsma, T. J. Thamann, R. C. Crouch, K. A. Blinov, M. E. Elyashberg, E. R. Martirosian, S. G. Molodtsov, A. J. Williams and P. L. J. Schiff, *J. Heterocycl. Chem.*, 2002, **39**, 1241.

2. K. A. Blinov, M. E. Elyashberg, E. R. Martirosian, S. G. Molodtsov, A. J. Williams, M. M. H. Sharaf, P. L. J. Schiff, R. C. Crouch, G. E. Martin, C. E. Hadden, J. E. Guido and K. A. Mills, *Magn. Reson. Chem.*, 2003, **41**, 577.

3. G. E. Martin, D. J. Russell, K. A. Blinov, M. E. Elyashberg and A. J. Williams, *Ann. Rep. NMR Spectrosc.*, 2003, **1**, 1.

4. A. N. Tackie, G. L. Boye, M. H. M. Sharaf, P. L. J. Schiff, R. C. Crouch, T. D. Spitzer, R. L. Johnson, J. Dunn, D. Minick and G. E. Martin, *J. Nat. Prod.*, 1993, **56**, 653.

5. G. J. Sharman, I. C. Jones, M. P. Parnell, M. C. Willis, M. F. Mahon, D. V. Carlson, A. J. Williams, M. E. Elyashberg, K. A. Blinov and S. G. Molodtsov, *Magn. Reson. Chem.*, 2004, **42**, 567.

6. N. Lysek, E. Rachor and T. Lindel, *Z. Naturforsch.*, 2002, **57C**, 1056.

7. J.-P. Bouillon, B. Tinant, J.-M. Nuzillard and C. Portella, *Synthesis*, 2004, **5**, 711.

8. G. N. Belofsky, M. Anguera, P. R. Jensen, W. Fenical and M. Köck, *Chem. Eur. J.*, 2000, **6**, 1355.

9. C. Steinbeck, V. Spitzer, M. Starosta and G. von Poser, *J. Nat. Prod.*, 1997, **60**, 627.

10. A. Codina, R. W. Ryan, R. Joyce and D. S. Richards, *Anal. Chem.*, 2010, **82**, 9127.

11. T. F. Molinski, *Nat. Prod. Rep.*, 2010, **27**, 321.

12. I. W. Burton, M. A. Quilliam and J. A. Walter, *Anal. Chem.*, 2005, **77**, 3123.

13. G. Wider and L. Dreier, *J. Am. Chem. Soc.*, 2006, **128**, 2571.

14. A. Codina, *Private communication.*

# CHAPTER 12
# Structural Revisions of Natural Products with the Aid of the Structure Elucidator System

## 12.1 Introduction

In previous Chapters we have discussed the various benefits of a symbiotic interaction between the researcher and computer algorithms to facilitate computer-aided structure elucidation. Elyashberg and co-authors[1] considered a series of real-world examples in which incorrect structures were deduced by experienced scientists. In that work[1] it was shown that the application of a CASE system could help researchers to avoid such incorrect structural conclusions.

In 2005 Nicolaou and Snyder[2] published a review entitled "Chasing molecules that were never there: misassigned natural products and the role of chemical synthesis in modern structure elucidation" . The review posits that both imaginative detective work and chemical synthesis still have important roles to play in the process of solving nature's most intriguing molecular puzzles. Another review entitled "Structural revisions of natural products by total synthesis" was recently presented by Maier.[3]

According to Nicolaou and Snyder,[2] around 1000 articles were published between 1990 and 2004 in which the originally determined structures needed to be revised. Figuratively speaking, it means that 40–45 issues of the imaginary "*Journal of Erroneous Chemistry*" were published in which all articles contained only incorrectly elucidated structures and, consequently, at least the same number of articles were necessary to describe the revision of these structures. The associated labor costs necessary to correct structural misassignments, and subsequent reassignments, are very significant and, generally, are much higher than those associated with obtaining the initial solution. From

New Developments in NMR No. 1
Contemporary Computer-Assisted Approaches to Molecular Structure Elucidation
By Mikhail Elyashberg, Antony Williams and Kirill Blinov
© Royal Society of Chemistry 2012
Published by the Royal Society of Chemistry, www.rsc.org

these data, it is evident that the number of publications in which the structures of new natural products are incorrectly determined is quite large and reducing this stream of errors is clearly a valid challenge. The authors of the review[2] comment that "there is a long way to go before natural product characterization can be considered a process devoid of adventure, discovery, and, yes, even unavoidable pitfalls". The review of Maier[3] confirms this conclusion.

It can be expected that the application of modern CASE systems can frequently help the chemist to avoid pitfalls or, in those cases when the researcher is challenged, then the expert system can at least provide a cautionary warning. This supposition is based on the fact that molecular structure elucidation can be formally described as deducing all logical corollaries from a system of statements which ultimately form a *partial axiomatic theory* (see Chapter 2). These corollaries are all conceivable structures that meet the initial set of axioms.[4-6] The great potential of ES is due to the fact that these systems can be considered as an inference engine applicable to the knowledge presented the set of axioms. The expert system Structure Elucidator (StrucEluc) possesses these capabilities.

Nicolaou and Snyder[2] noted that the development of spectroscopic methods in the second half of the 20th century resulted in a revolution in the methodology of structure elucidation. We believe that the continued development of algorithms and accompanying software platforms and expert systems will further revolutionize structure elucidation. We also believe that the employment of expert systems will lead to significant acceleration in the progress of organic chemistry and natural products specifically as a result of reduced errors and increased efficiencies.

This Chapter considers the application of CASE systems to a series of examples in which the original structures were later revised. We demonstrate how the chemical structure could be correctly elucidated if 2D NMR data were available and the expert system StrucEluc was employed. We will also demonstrate that if only 1D NMR spectra from the published articles were used, then simply the empirical calculation of $^{13}C$ chemical shifts for the hypothetical structures frequently enables a researcher to realize that the structural hypothesis is likely incorrect. We also analyze a number of erroneous structural suggestions made by highly qualified and skilled chemists. The investigation of these mistakes is very instructive and has facilitated a deeper understanding of the complicated logical–combinatorial process for deducing chemical structures.

## 12.2 Examples of Structure Revision Using an Expert System

In this Section a series of articles are reviewed in which an incorrect structure was initially inferred from the MS and NMR data and then revised in later publications. In so doing, we will demonstrate how the problem would have been solved if the StrucEluc system was used to process the initial information from the very beginning. The partial axiomatic theories were formed by the system from the spectrum–structure data and suggestions from the researchers presented in the corresponding articles.

The number of new natural products separated and published in the literature each year is huge. Obviously, it is impossible for a scientific group to verify all structures presented in all articles. Therefore to choose the appropriate publications for consideration in this article we were forced to rely on those publications where the earlier identified structures were revised. Many references related to such structures were found in a review[2] covering the time period 1990–2005, whereas a series of later publications were revealed *via* an Internet search. As a result, we chose publications that were easily accessible. We then selected articles where the 2D NMR data were presented for the original structures (in the best cases – both for original and revised ones). With these data it was possible to analyze the full process of moving from the original spectra to the most probable structure and then clearly identify those points where questionable hypotheses led to the incorrect structures. If the 2D NMR data were not available within an article, then it was only possible to assess the quality of the suggested structure on the basis of $^{13}$C NMR spectrum prediction.

It was difficult to decide how the various types of structure revision cases could be classified. All of the problems were divided into four categories depending on the method or combination of methods which allowed reassignment of the original structure. We used the four following approaches that could be distinguished: reinterpretation of experimental data, re-examination of the 2D NMR data, application of chemical synthesis, and $^{13}$C NMR spectrum prediction. The reinterpretation of experimental data is required in those cases, for example, when an incorrect molecular formula is suggested, wrong fragments were suggested or artifacts in the 2D spectral data were taken as real signals, *etc*. In all of the cases, it is impossible to obtain the correct structure. The reinterpretation of 2D data is necessary when a human expert misinterpreted the data because they were unable to enumerate all possible structures corresponding to the data.

## 12.2.1 Revision of Structures by Reinterpretation of Experimental Data

Randazzo *et al.*[7] isolated two new compounds, named halipeptins A and B, from the marine sponge *Haliclona* sp. Their structures were determined by extensive use of 1D and 2D NMR (including $^1$H-$^{15}$N HMBC), MS, UV and IR spectroscopy assuming that these compounds belong to a class of materials with an elemental formula containing only CHNO, this assumption being an axiom. Halipeptin A showed a parent ion peak at $m/z$ 627.4073 [(M + H)$^+$] in the high-resolution fast-atom bombardment mass spectrum (HRFABMS) consistent with a molecular formula of $C_{31}H_{54}N_4O_9$ (calculated mass equals 627.3969 for $C_{31}H_{55}N_4O_9$ with $\Delta m = 0.0104$, *i.e.* 16.6 ppm). The following structure (**12.1**) was suggested for halipeptin A (the suggested chemical shift assignment for the carbon and nitrogen nuclei is shown to simplify the observation of changes in the shift assignment when the structure is revised):

**12.1**

A four-membered ring cycle is known to occur very seldom in natural products. The authors[7] commented that a four-membered ring containing an N–O bond appears to be a rather intriguing and unprecedented moiety. The presence of an N–O bond was inferred from an IR band at 1446 cm$^{-1}$ which was considered characteristic for an N–O bond as stretching in this range has already been observed in similar systems. Taking into account the axioms and accompanying examples described within the first group above, such a consideration, in our opinion, is not convincing. The occurrence of this band does not contradict the presence of this specific fragment, but it also does not provide absolute evidence for the presence of the fragment in the analyzed structure. Moreover, all compounds containing $CH_2$ groups also absorb in this region.[8] The unusual experimental chemical shift ($\delta_N = 290.9$ ppm, $NH_3$ as reference) of the nitrogen nucleus associated with the hypothetical four-membered ring (the typical experimental $\delta_N$ values in reference compounds used by Randazzo et al.[7] are 110–120 ppm) was explained in terms of the ring strain in the oxazetidine system. The large $^1J_{CH}$ values of 147.4 and 149.4 Hz observed for the two methylene protons, which is in excellent agreement with previously reported couplings for these ring systems, were considered as further support for the presence of this uncommon fragment.

To compare the suggested structure **12.1** with the results obtained from the StrucEluc software, the postulated molecular formula $C_{31}H_{55}N_4O_9$ and spectral data including $^{13}C$ and $^{15}N$ NMR spectra, HSQC, $^1H$-$^{13}C$ and $^1H$-$^{15}N$ HMBC were used as input for the program. It was assumed that all axioms and hypotheses are consistent, that the valences of all nitrogen atoms are equal to 3, and that C≡C and C≡N bonds were forbidden, whereas the N–O bond was permitted. No constraints on the ring cycle sizes were imposed. Molecular structure generation was run from the molecular connectivity diagram (MCD)[9] produced by the system and provided the result: $k = 6 \to 4 \to 4$, $t_g = 0.1$ s. This notation indicates that six structures were generated in 0.1 s, and two sequenced operations – spectral–structural filtering and the removal of duplicates – yielded four different structures. [13]C NMR spectrum prediction allowed us to select structure **12.2** as the most probable according to the minimal values of the mean average deviations ($d_A \cong d_N = 3.6$ ppm) of the experimental [13]C chemical shifts from calculated ones. These different approaches of NMR prediction have been discussed in more detail elsewhere[10,11] and characterized in Chapter 3. They are included in the ACD/NMR Predictors software[12] and implemented into StrucEluc.

H3C
22.00
56.40 / CH3
H3C 22.30
O
48.50–177.30
O
H3C—18.40
14.40
O
H3C 26.10 45.70
173.30
N
H
N
44.20
35.60 80.50
H3C 14.20
34.10 82.50
83.80 —CH3
23.10
31.20 31.90
O
172.40=O
H
169.60 N—O
49.50
N —CH3
64.50 30.70
169.20
28.10 35.10
60.80
H3C
18.00
O
H3C
18.40
OH

**12.2**

Structure **12.1** was not generated. The deviations obtained are twice as large as the value of the calculation accuracy (1.6–1.8 ppm), but in cases such as this a decision regarding the structure quality is taken after analyzing the maximum deviations. A linear regression plot obtained using both HOSE and NN chemical shift predictions is presented in Figure 12.1. The graph and prediction limits were calculated using options available within the graphing program (Microsoft Excel). The graph shows that there is a single point lying outside the prediction limits and that the difference between the experimental (83.8 ppm) and calculated (45 ppm) chemical shifts is equal to approx. 40 ppm. This suggests that: (i) structure **12.2** is certainly wrong, and (ii) it is probable that at least one non-standard correlation (NSC) is present in the 2D NMR data. According to the general methodology inherent to the StrucEluc system, fuzzy structure generation (FSG)[13] should be used in such a situation. FSG was therefore executed and the presence of one NSC of an unknown length was assumed. The results are: $k = 304 \to 284 \to 183$ and $t_g = 35$ s. Figure 12.2 shows the first three structures of the output file ranked in order of increasing deviations following [13]C spectrum prediction. Structure **12.1** as suggested by the authors[7] was ranked first, which means that they indeed

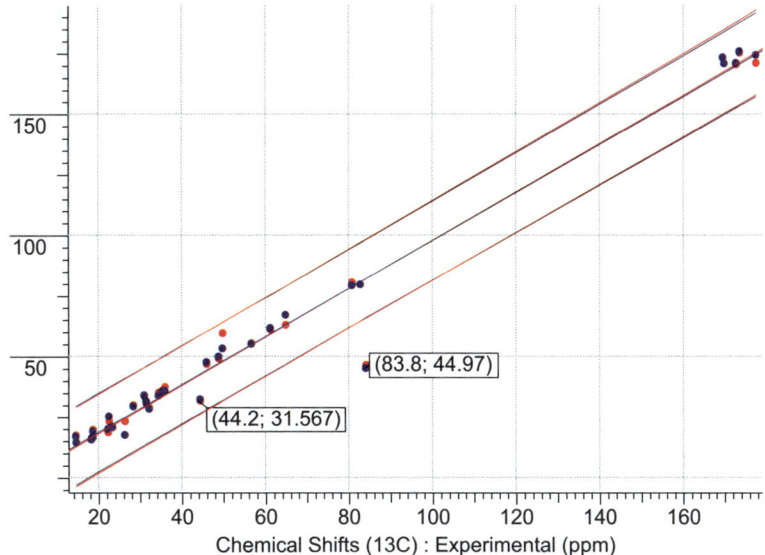

**Figure 12.1** Linear regression plots for structure **12.2** generated from both HOSE and NN methods of $^{13}C$ chemical shift prediction. The first number shown in a box denotes the experimental chemical shift, whereas the second is the calculated value. Both the HOSE and NN predictions practically coincide with the 45° line ($\delta_{calc} = \delta_{exp}$). Prediction limit lines are also shown.

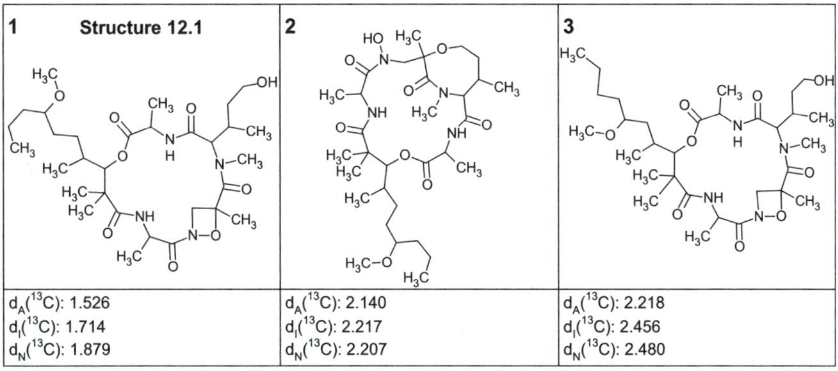

**Figure 12.2** The first three structures of the ranked structural file when a molecular formula of $C_{31}H_{55}N_4O_9$ was assumed. The numbers in the top left of each box are the rank numbers.

inferred the best structure among all possible structures from the initial data (axioms). The crucial axiom influencing the final solution is the assumed molecular formula.

In the next article[14] by the same group of authors, they reported that by using superior high resolution mass-spectrometer (HRMS) instrumentation capable of reaching a resolution of approx. 20 000 they had revised the molecular formula. A hint in regards to how to revise the structure was provided by the following finding: when a related natural product halipeptin C was isolated the presence of an unexpected sulfur atom in this compound was clearly detected by HRMS. The authors[14] suggested that the molecule halipeptin A also contained a sulfur atom instead of two oxygen atoms to give a molecular formula of $C_{31}H_{54}N_4SO_7$. In this case a pseudomolecular ion peak was found at $m/z$ 649.3628 ($M + Na^+$, $\Delta m = -0.0017$ or 2.6 ppm). For the original molecular formula $C_{31}H_{55}N_4O_9$, the difference between the measured and calculated molecular mass was much higher: 0.0160 or 24.6 ppm, so the wrong hypothesis about the elemental composition would probably be rejected if a more precise $m/z$ value was obtained in the earlier investigation.[7] With the revised molecular formula the following structure was deduced:[14]

**12.3**

We will now show how this problem would be solved using the StrucEluc software. The accurate molecular mass of 627.4073 determined in reference[7] was used as input for the molecular formula generator. Taking into account the number of signals in the $^{13}C$ NMR spectrum and the integrals in the $^1H$ NMR spectrum, the following admissible limits on atom numbers in molecular formula (the axioms of chemical composition) were set: C(31), H(52–56), O(0–10), N(0–10), S(0–2). For the initially determined mass of 627.4073±0.1, three possible molecular formulae were generated: $C_{31}H_{54}N_4O_9$ ($\Delta m = -0.0104$, 16.6 ppm), $C_{31}H_{54}N_4O_7S_1$ ($\Delta m = -0.0281$, 44 ppm), and $C_{31}H_{54}N_4O_5S_2$ ($\Delta m = -0.0459$, 73 ppm, where the mass differences are shown in brackets.

If high precision MS instruments are used, then a mass difference exceeding 10 ppm is commonly not acceptable. We suppose that in our case the value $\Delta m = 16$ ppm should suggest the presence of other elements or re-examination of the sample on a more advanced MS instrument.

We will show that if a CASE system is available, correct structure elucidation of an unknown compound is possible, even under non-ideal conditions. Although $C_{31}H_{54}N_4O_9$ is obviously the most probable molecular formula based on the calculated mass defect the closest related formula, $C_{31}H_{54}N_4O_7S_1$, can also be taken into account with the StrucEluc system.

Both the molecular formulae and the 2D NMR spectral data[7] were used to perform structure generation with the same axioms listed earlier. The valence of the sulfur atom was set equal to 2. An output file containing 303 structures was produced in 36 s. The three top structures of the output file ranked in ascending order of deviations are presented in Figure 12.3. The Figure shows that the revised structure **12.3** is placed in first position by the program, whereas the original structure is listed in second position. Application of the StrucEluc software would provide the correct structure from the molecular ion recorded, even at modest resolution MS. This example also illustrates the methodology[15] based on the application of an expert system which allows a user simultaneously to determine both the molecular and structural formula of an unknown compound.

For clarity, the differences between the original and revised structures are shown in Figure 12.4.

Sakuno *et al.*[16] isolated an aflatoxin biosynthesis enzyme *inhibitor* with molecular formula $C_{20}H_{18}O_6$. It is labeled as TAEMC161. The following structure for this alkaloid was suggested from the 1D NMR, HMBC, and NOE data (an experimental chemical shift assignment suggested by authors is displayed):

**12.4**

During the process of structure elucidation, the authors[16] postulated that the $^{13}C$ chemical shift at 173.50 ppm was associated with the resonance of the ester group carbon. The spectral data were input into the StrucEluc system and,

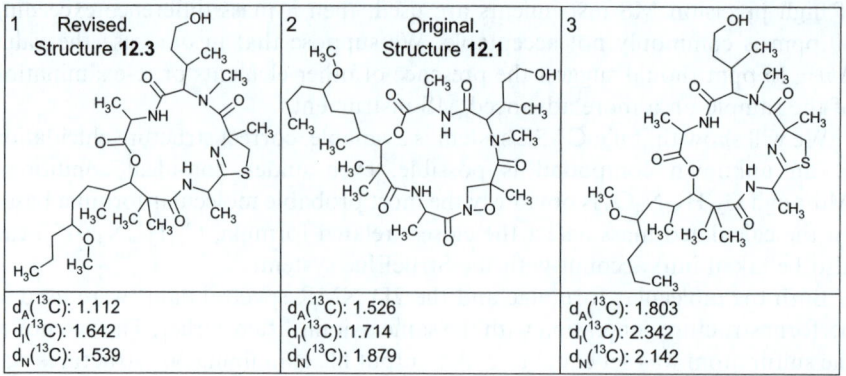

| 1    Revised. Structure **12.3** | 2    Original. Structure **12.1** | 3 |
|---|---|---|
| $d_A(^{13}C)$: 1.112 <br> $d_I(^{13}C)$: 1.642 <br> $d_N(^{13}C)$: 1.539 | $d_A(^{13}C)$: 1.526 <br> $d_I(^{13}C)$: 1.714 <br> $d_N(^{13}C)$: 1.879 | $d_A(^{13}C)$: 1.803 <br> $d_I(^{13}C)$: 2.342 <br> $d_N(^{13}C)$: 2.142 |

**Figure 12.3**   The top three structures of the output file generated from the two molecular formulae $C_{31}H_{54}N_4O_9$ and $C_{31}H_{54}N_4O_7S_1$. The numbers top left in each box are rank numbers.

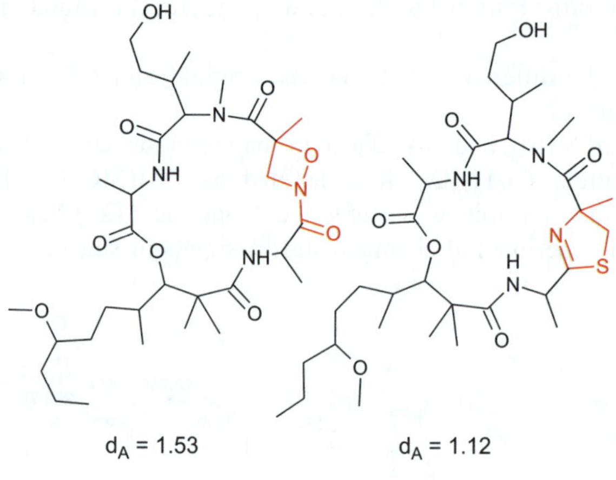

$$d_A = 1.53 \qquad\qquad\qquad d_A = 1.12$$

**Figure 12.4**   The original and revised structures of halipeptin A.

similar to Sakuno *et al.*,[16] the O=C–O group was involved into the process of FSG by manually adding to the MCD. The results gave: $k = 174 \rightarrow 80 \rightarrow 60$, $t_g = 30$ s. When the output file was ordered as described above, then structure **12.4** occupied the first position but with deviation values of approx. 4.5 ppm. Such large deviations suggest caution[9,17,18] and that close inspection of the data is required. It should be remembered that the accuracy of chemical shift calculation is approx. 1.6–1.8 ppm.

Wipf and Kerekes[19] compared the NMR and IR spectra of TAEMC161 with a number of spectra of its structural relatives and found close similarity between the spectra of TAEMC161 and viridol, **12.5**.

**12.5**

In this molecule both carbonyl groups are ketones and the structure is in accordance with the 2D NMR data used for deducing structure **12.4**. Density functional theory calculations of $^{13}$C chemical shifts were performed by the authors[19] for structures **12.4** and **12.5** using the GIAO approximation. It was proven that TAEMC161 is actually identical to **12.5**. We repeated structure generation without any constraints imposed on the carbonyl groups, with the following result: $k = 494 \rightarrow 398 \rightarrow 272$, $t_g = 1$ min 40 s. The three top structures in the ranked output file are presented in Figure 12.5.

The Figure shows that empirical prediction of $^{13}$C chemical shifts convincingly demonstrates the superiority of the revised structure over the original suggested for TAEMC161. The differences between the original and revised structures are shown in Figure 12.6.

In 1997 Cóbar *et al.*[20] isolated three new diterpenoid hexose-glycosides, calyculaglycodides A, B, and C, and their structures were determined from MS, 1D NMR, COSY, $^{1}$H-$^{13}$C HMBC, and NOE spectra. The following structure was suggested for calyculaglycodide B (molecular formula $C_{30}H_{48}O_8$):

**12.6**

In 2001 the same group[21] reinvestigated this natural product and discovered that structure **12.6** is incorrect. A hint to revision of the structure was obtained on the basis of the comparison of NMR spectra of similar compounds which were isolated from the same material. It was noticed that the NMR spectra of all compounds including an aglycone substructure contained indistinguishable

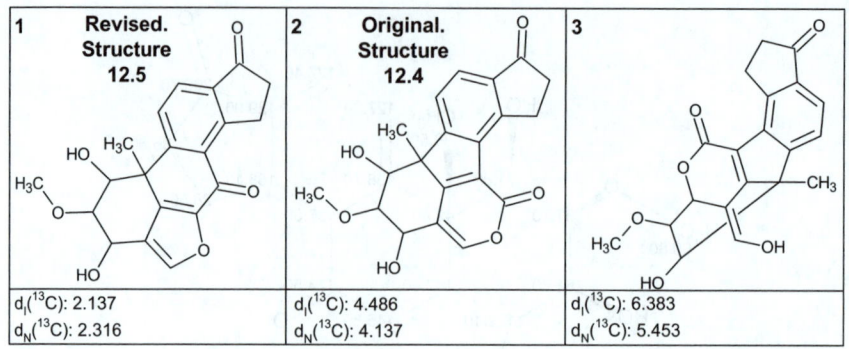

| 1 Revised. Structure 12.5 | 2 Original. Structure 12.4 | 3 |
|---|---|---|
| $d_I(^{13}C)$: 2.137 $d_N(^{13}C)$: 2.316 | $d_I(^{13}C)$: 4.486 $d_N(^{13}C)$: 4.137 | $d_I(^{13}C)$: 6.383 $d_N(^{13}C)$: 5.453 |

**Figure 12.5**  The top three structures of the output file generated for compound $C_{20}H_{18}O_6$ (viridol). The numbers top left in each box are rank numbers.

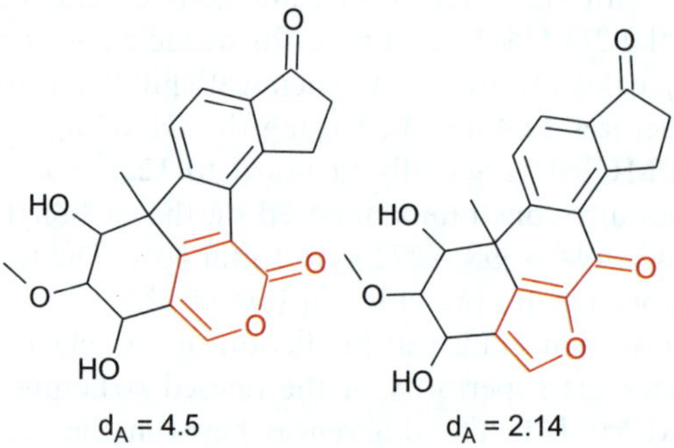

$d_A = 4.5$                                        $d_A = 2.14$

**Figure 12.6**  The original and revised structures of the inhibitor (viridol).

portions of the spectra. With this in mind, the NMR and mass spectra of calyculaglycodides A, B, and C were thoroughly reinvestigated and, as a result, the revised structure **12.7** was postulated for calyculaglycodide B:

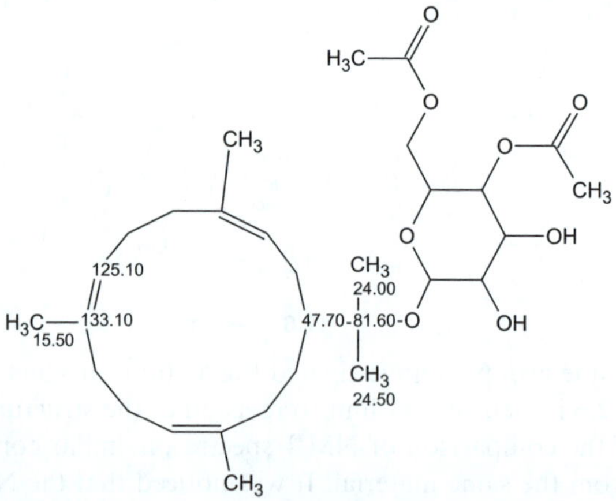

**12.7**

Freshly recorded NMR spectra showed that the HMBC connectivity $CH_3(15.5) \rightarrow C(47.7)$ was earlier identified as an artifact, while a strong correlation of the dimethyl group to $C(47.5)$ was missed. As a consequence, the initial set of axioms was false and inferring the correct structure was absolutely impossible. The $^{13}C$ chemical shifts predicted for structure **12.6** led to average deviations of values approx. 2 ppm, which are of an appropriate magnitude to not further question the correctness of structure.

When the corrected HMBC data were input into StrucEluc, the program detected the presence of NSCs, and FSG was carried out. During fuzzy generation, the program determined that there were two NSCs and provided the following results: $k = 10 \rightarrow 6$, $t_g = 1$ h 39 min. The time of structure generation is quite long, because, in this case, the program tried to generate structures from 861 different combinations of connectivities (see Section 10.3). The revised structure was selected using $^{13}C$ spectral prediction to choose the most probable one (see Figure 12.7). The difference between the structures is only in the positions of the double bond and methyl group on the large cycle (see Figure 12.8).

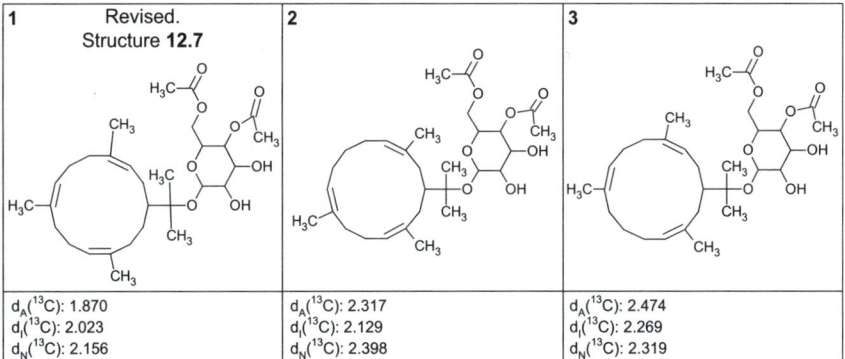

| 1 Revised. Structure **12.7** | 2 | 3 |
|---|---|---|
| $d_A(^{13}C)$: 1.870 $d_I(^{13}C)$: 2.023 $d_N(^{13}C)$: 2.156 | $d_A(^{13}C)$: 2.317 $d_I(^{13}C)$: 2.129 $d_N(^{13}C)$: 2.398 | $d_A(^{13}C)$: 2.474 $d_I(^{13}C)$: 2.269 $d_N(^{13}C)$: 2.319 |

**Figure 12.7**   The top structures of the output file generated by the StrucEluc software for the $C_{30}H_{48}O_8$ compound calyculaglycodide B.

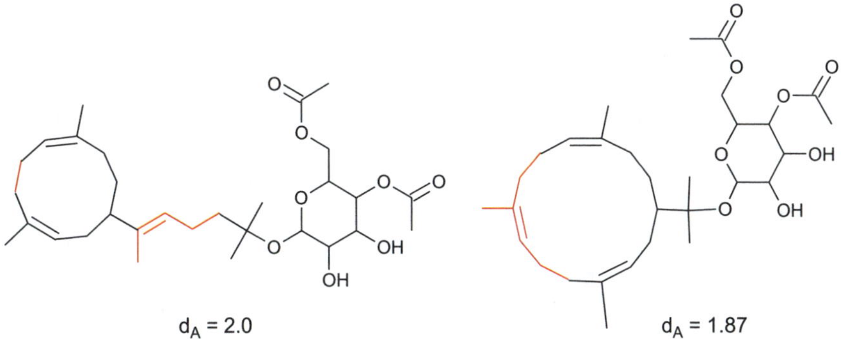

$d_A = 2.0$          $d_A = 1.87$

**Figure 12.8**   The original and revised structures of calyculaglycodide B.

Ralifo and Crews[22] reported on the separation (an isolated amount of approx. 3.2 mg) of (–)-spiroleucettadine, **12.8** ($C_{20}H_{23}N_3O_4$), the first natural product to contain a fused two-aminoimidazole oxalane ring. In spite of the modest size of this molecule, the high value of the double bond equivalent (DBE = 11) hints that the structure elucidation may be a very complicated problem.

**12.8**

The structure was inferred on the basis of the 2D NMR data, as well as by structural and spectral comparison between structure **12.8** and a series of known molecules of similar structure and origin. The authors[22] suggested the presence of a guanidine group ($\delta C$ 159.0) substituted with two methyls (axiom 1). This proposition was justified based on the characteristic $NCH_3$ signals at (29.3; 2.48) and (26.0; 2.91), along with the HMBC correlation from $NCH_3$(2.48) to C(159). The absence of an expected HMBC correlation from $NCH_3$(2.91) to C(159.0) was considered as acceptable and the possible reason for the absence of the correlation was not analyzed. The position of carbon C(48.8) was confirmed by HMBC from this nucleus to the hydrogen $\delta H$(1.97) attached to C(38.0) (axiom 2). The signal of the exchangeable hydrogen in the $^1H$ NMR spectrum was assigned to an OH group (axiom 3), but no attempt to confirm this postulate by IR spectroscopy was mentioned in the article. The relative stereochemistry of structure **12.8** was determined using a combination of ROESY data and molecular modeling. The absolute stereochemistry was determined using OED-CD spectroscopy.

As a result of utilizing the 1D and HMBC NMR data published by the authors[22] as an input to the StrucEluc system, the following result was obtained under the conditions of strict structure generation: $k = 117 \rightarrow 83 \rightarrow 79$, $t_g = 10$ s. Figure 12.9 presents the best ranked structures from the start of the output file. Note that structures containing too "exotic" fragments were deleted. The postulated axioms led to a preferred structure that differs from the original structure **12.8**, which was also generated: instead of the C=NH fragment this structure contains a C=O group, whereas the OH group is replaced by an $NH_2$ group. The third and fourth structures also contain a carbonyl group at the same position. There is no doubt that if the computer-based solution presented in Figure 12.10 was available to Crews's group, one of the leading groups in the chemistry of natural products, then their elucidated structure for **12.8** would be questioned and a different and likely correct structure would be found after appropriate revision of the experimental data and set of axioms.

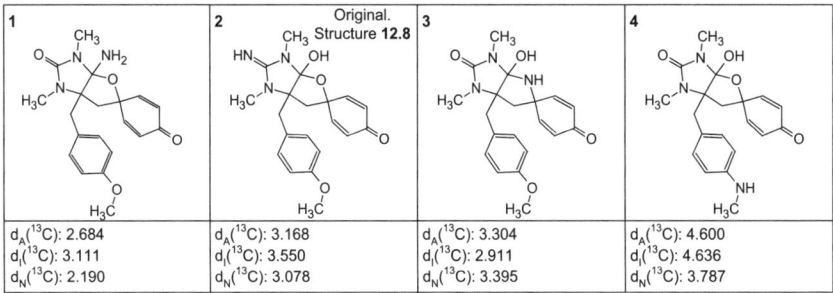

| 1 | 2 Original. Structure **12.8** | 3 | 4 |
|---|---|---|---|
| $d_A$($^{13}$C): 2.684 | $d_A$($^{13}$C): 3.168 | $d_A$($^{13}$C): 3.304 | $d_A$($^{13}$C): 4.600 |
| $d_I$($^{13}$C): 3.111 | $d_I$($^{13}$C): 3.550 | $d_I$($^{13}$C): 2.911 | $d_I$($^{13}$C): 4.636 |
| $d_N$($^{13}$C): 2.190 | $d_N$($^{13}$C): 3.078 | $d_N$($^{13}$C): 3.395 | $d_N$($^{13}$C): 3.787 |

**Figure 12.9**   The top ranked structures of the output file generated by the StrucEluc software for the $C_{20}H_{23}N_3O_4$ compound (–)-spiroleucettadine elucidated from the 2D NMR data obtained by Ralifo and Crews.[22] The numbers in the top left of each box are rank numbers.

Structure **12.8** was met with keen interest by the natural products and synthetic communities and several attempts to synthesize it were undertaken but without any success. Questions regarding the original structure elucidation process therefore arose. Aberle *et al.*[23] suggested structures **12.9** and **12.10** as alternatives, but density functional theory (DFT) calculations of chemical shifts performed by Crews and co-workers[24] showed that both of them should be declined.

**12.9**          **12.10**

With this in mind, Crews and co-workers[24] fulfilled a successful re-isolation of spiroleucettadine, and X-ray analysis established the correct structure of spiroleucettadine, shown as **12.11** below.

**12.11**

Fresh 2D NMR data on spiroleucettadine were obtained and verified.[24] It was revealed that the connectivity from C48.8 to C38.0 for structure **12.8** in methanol-$d_4$ was actually due to a solvent $J_{CH}$ peak. In this case axiom 2 was false. An inconsistency in axiom 1 became evident due to the lack of parity displayed between the two *N*-methyl groups as follows from structure **12.11**. The relative stereochemistry was also revised as shown in structure **12.11** and its superiority over structures **12.8–12.10** was proven by DFT calculations.

When the new 2D NMR data were input into the StrucEluc system, the structure generation was performed with very "liberal" atom properties: no constraints for heteroatom neighboring for carbons with chemical shifts in the interval range of 113.7–158.6 ppm. The following solution was obtained: $k = 342 \rightarrow 256$, $t_g = 8$ h 2 m. The reason for the long generation time, the so-called "overnight mode", was the high DBE value and the lack of structural restrictions. The best structures are presented in Figure 12.10.

The revised structure **12.11** was selected as the most probable one by the program in accord with the results of crystallographic analysis and the conclusions of the researchers.[24] The differences between the original and revised structures are shown in Figure 12.11.

Since three isomeric structures (**12.8–12.10**) and the first ranked structure in Figure 12.10 were considered as potential candidates for the genuine structure, the authors[24] carried out DFT-based [13]C chemical shift calculations using the B3LYP/6-31G*//B3LYP/6-31G* protocol for all stereoisomers. This resulted in the examination of a total of 16 structures and their modifications where the oxygen atom in the five-membered ring was migrated either "up and down". It was found that the configuration of structure **12.11** corresponds to the minimum discrepancy between the experimental and calculated spectra, whereas structure **12.10** obtained a low rank.

We performed [13]C chemical shift prediction using HOSE code-based and NN algorithms[11,12] for the same structure set (see Table 12.1). Note that both

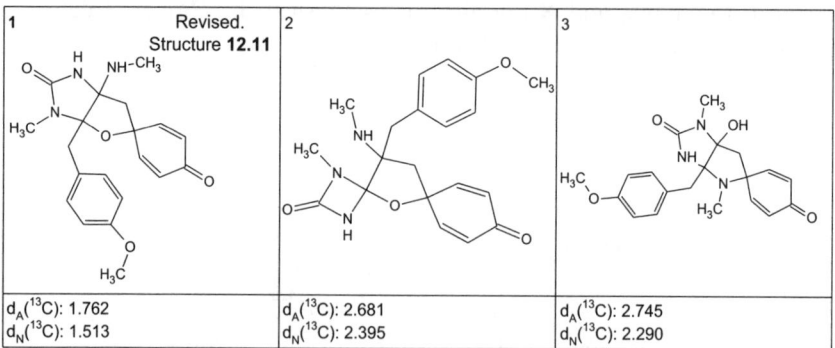

| 1 | 2 | 3 |
|---|---|---|
| Revised. Structure **12.11** | | |
| $d_A(^{13}C)$: 1.762 | $d_A(^{13}C)$: 2.681 | $d_A(^{13}C)$: 2.745 |
| $d_N(^{13}C)$: 1.513 | $d_N(^{13}C)$: 2.395 | $d_N(^{13}C)$: 2.290 |

**Figure 12.10**    The highest ranked structures of the output file generated by the StrucEluc software for the $C_{20}H_{23}N_3O_4$ compound from the new 2D NMR data obtained by White *et al.*[24] The numbers top left in each box are rank numbers.

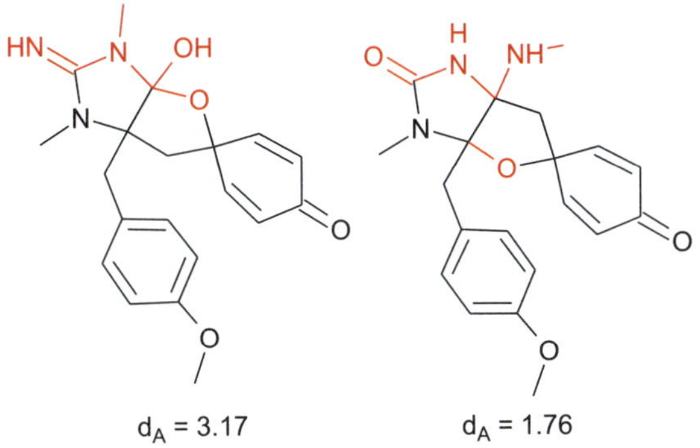

$d_A = 3.17$ $d_A = 1.76$

**Figure 12.11** The original and revised structures of (–)-spiroleucettadine.

methods take stereochemistry into account (see Section 12.3). As a result stereoisomer **12.11** was also distinguished as the best by empirical calculations. The total elapsed time was 7 min, with no geometry optimization being necessary.

Buske *et al.*[25] described the structural elucidation of antidesmone, **12.12**, a novel type tetrahydroisoquinoline alkaloid with the molecular formula $C_{19}H_{29}NO_3$:

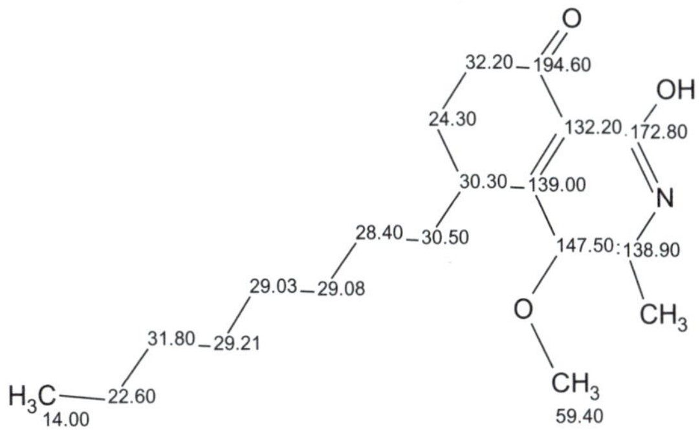

**12.12**

Antidesmone was identified as an unprecedented and novel alkaloid where the nitrogen is located in the aromatic ring and the substitution pattern, in particular the unusual *n*-octyl residue on the isocyclic ring, is also unique. The authors[25] reported that no HMBC correlations to carbon 172.8 could be found, but from the chemical shift and molecular formula they deduced the presence of an OH group attached to this carbon. This axiom crucially influenced the solution of the problem. The absolute configuration of antidesmone was determined using its methyl ether for which quantum chemical (QM) calculations of CD and UV spectra were performed.

The NMR data presented[25] were used to determine which structure would be deduced by StrucEluc from the published spectral data as the best structure if the assumptions of the researchers[25] were included into the initial data of the

**Table 12.1** Selection of the correct structure and the best stereoisomer of spir-oleucettadine. Structures are labeled as described by White *et al.*[24]

| 1 (ID:9) I, Revised | 2 (ID:12) L | 3 (ID:11) K | 4 (ID:10) J |
|---|---|---|---|
| | | | |
| $d_A(^{13}C)$: 1.799 | $d_A(^{13}C)$: 1.811 | $d_A(^{13}C)$: 1.853 | $d_A(^{13}C)$: 1.952 |
| 5 (ID:14) N | 6 (ID:13) M | 7 (ID:15) O | 8 (ID:16) P |
| | | | |
| $d_A(^{13}C)$: 2.152 | $d_A(^{13}C)$: 2.178 | $d_A(^{13}C)$: 2.412 | $d_A(^{13}C)$: 2.506 |
| 9 (ID:1) A | 10 (ID:7) G | 11 (ID:5) E | 12 (ID:3) C |
| | | | |
| $d_A(^{13}C)$: 2.666 | $d_A(^{13}C)$: 2.783 | $d_A(^{13}C)$: 2.783 | $d_A(^{13}C)$: 2.877 |
| 13 (ID:8) H | 14 (ID:6) F | 15 (ID:4) D | 16 (ID:17) B, Original |
| | | | |
| $d_A(^{13}C)$: 2.937 | $d_A(^{13}C)$: 2.937 | $d_A(^{13}C)$: 2.970 | $d_A(^{13}C)$: 3.039 |

program. The attachment of an OH group at carbon 172.8 was accepted as an axiom. The first run was performed in strict generation mode with the result $k = 13092 \rightarrow 12636 \rightarrow 1031$, $t_g = 1$ min 13 s. The first ranked structure gave deviations with values between 3.5–4.7 ppm. This hinted at the presence of at least one NSC. At the same time structure **12.12** was not generated. FSG was initiated with the following result: $k = 144228 \rightarrow 116496 \rightarrow 6604$, $t_g = 19$ min 28 s. The best structure was identical to that in the previous run, but structure **12.12** was generated this time and ranked in 113th position by NN-based chemical shift calculation. It is very convincing that structure **12.12** is incorrect. It is obvious that some incorrect restrictions (axioms) were included into the initial set of statements.

The problem was solved using StrucEluc to analyze the 2D NMR data. Our common methodology was used: no user-defined constraints were imposed on the generated structures and the fragment =C–O–H remained disconnected in the MCD. Strict structure generation gave the following result: $k = 59916 \rightarrow 51888 \rightarrow 4274$, $t_g = 6$ min 5 s. Chemical shift calculations using all three methods promoted the structure **12.13** to first position in the ranked output file with the following average deviations: $d_A = 1.437$, $d_N = 2.767$, $d_I = 1.964$.

**12.13**

**12.13A**

Structure **12.12** was also generated, but it was ranked $342^{nd}$ by NN prediction and $183^{rd}$ using HOSE-based prediction. Application of StrucEluc allowed us to establish the most probable structure and reject the original structural suggestion by the authors.[25]

In the next article published by the same group[26] it was reported that structure **12.12** was mistaken due to the poor quality of the 2D NMR spectral data obtained from a small amount of sample. The correct structure, **12.13**, was inferred for antidesmone from fresh 2D NMR data including HSQC, HMBC, COSY, and NOESY. When the new HMBC data were used as input for the StrucEluc system the program produced the following results: $k = 3972 \rightarrow 3876 \rightarrow 323$, $t_g = 1$ min 13 s. The best structure **12.13A** ($d_A = 0.974$, $d_N = 2.056$, $d_I = 1.572$) coincided with structure **12.13**, but the chemical shift assignment was refined according to the improved 2D NMR data and the chemical shifts at 147.5 and 138.9 were exchanged. For clarity, the differences between the original and revised structures are shown in Figure 12.12.

This example shows that even in those cases when the spectral data are of low quality the correct structure can still be determined in certain cases. It was possible because when the StrucEluc system is utilized the chemist can afford to avoid subjective suggestions such as those postulated by the authors.[25]

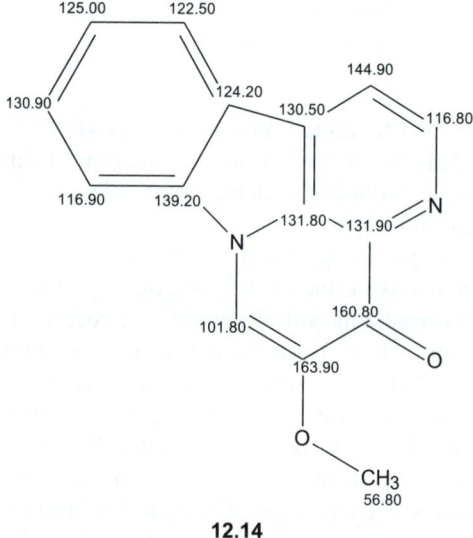

$d_A = 3.5$                    $d_A = 1.44$

**Figure 12.12**   The original and revised structures of antidesmone.

## 12.2.2   Revision of Structures with Application of Chemical Synthesis

In 2004 Hsieh *et al.*[27] isolated a new alkaloid with molecular formula $C_{15}H_{10}N_2O_2$ (DBE = 12) and named as drymarietin (5-methoxycanthin-4-one). Using a combination of $^1H$-$^{13}C$ HMBC and $^1H$-$^{15}N$ HMBC 2D NMR data, they hypothesized the drymarietin structure to be as shown in **12.14** with the chemical shift assignment shown.

**12.14**

This alkaloid showed interesting anti-HIV activity and has been mentioned in a series of review articles dealing with bioactive natural products.[28]

In 2009 Wetzel *et al.*[28] revised structure **12.14**. They synthesized 5-methoxycanthin-4-one and discovered that the synthetic product displayed spectroscopic data significantly different from those of the drymarietin alkaloid. Extensive re-evaluation of the spectroscopic data published for this and related alkaloids led them to conclude that drymarietin is identical to the known alkaloid cordatanine **12.15** (4-methoxycanthin-6-one):

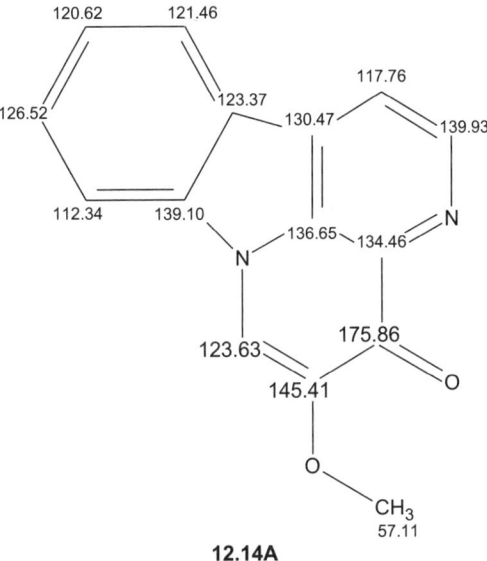

**12.15**

To investigate whether CASE methods could help researchers to avoid a pitfall in this case, we first predicted the $^{13}$C chemical shifts of structure **12.14** and determined that all average deviations were 8–9 ppm. This unambiguously demonstrated that the structure does not correspond to the $^{13}$C NMR spectrum. The calculated shifts are shown in structure **12.14A** where the shifts with the largest differences are shown in a larger font:

**12.14A**

Figure 12.13 shows a linear regression plot for the experimental *versus* calculated shifts for structure **12.14**.

Figure 12.14 provides convincing evidence that the structure and chemical shift assignment are wrong. We posited the question what structure would be

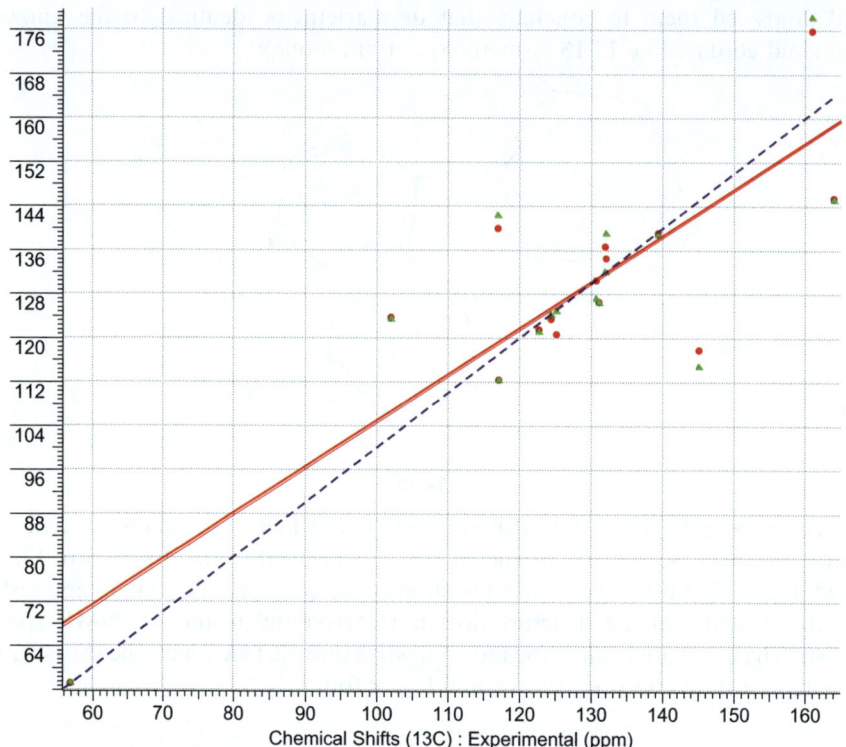

Database: Proposed Structures
- • Chemical Shifts (13C) : HOSE Calc. (ppm) (Current Record) (15 pts)
- ▲ Chemical Shifts (13C) : NN Calc. (ppm) (Current Record) (15 pts)

**Figure 12.13**   Linear regression plots for structure **12.14** generated using both HOSE and NN methods of $^{13}$C chemical shift prediction. The linear regression parameters are: $R^2(\text{HOSE}) = 0.742$, $\delta_{\text{HOSE}} = 0.843\delta_{\text{exp}} + 20.3$; $R^2(\text{NN}) = 0.710$, $\delta_{\text{NN}} = 0.841\delta_{\text{exp}} + 20.9$. The intersection angle between the regression plot and the 45° line is equal to –4°.

inferred by the StrucEluc program if the data of Hsieh *et al.*[27] were used as input for the system?

The program created an MCD which clearly showed the presence of a benzene ring. The corresponding atoms were therefore connected by chemical bonds. Structure generation quickly identified the presence of three NSCs in the 2D NMR data, and FSG performed using $m = 3$ and $a = 1$ ($a$ is the number of bonds by which the connectivity length should be augmented) gave the following result: $k = 3149 \rightarrow 1463 \rightarrow 146$, $t_g = 56$ s.

The best ranked structures are presented in Figure 12.14 where the correct structure **12.15** was ranked first. Application of $^{13}$C spectrum prediction, therefore, showed that structure **12.16** was wrong. The correct solution **12.15** was then obtained without any synthesis of the suggested structure **12.14**. If the authors[27] had used fast $^{13}$C chemical shift prediction to verify their hypothesis

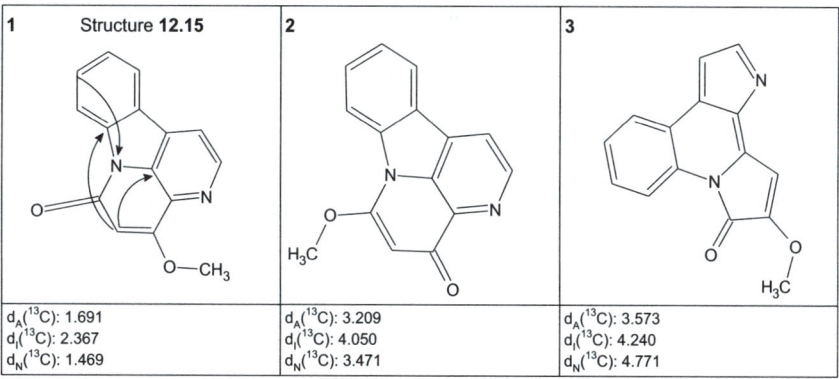

| 1 Structure **12.15** | 2 | 3 |
|---|---|---|
| $d_A(^{13}C)$: 1.691<br>$d_I(^{13}C)$: 2.367<br>$d_N(^{13}C)$: 1.469 | $d_A(^{13}C)$: 3.209<br>$d_I(^{13}C)$: 4.050<br>$d_N(^{13}C)$: 3.471 | $d_A(^{13}C)$: 3.573<br>$d_I(^{13}C)$: 4.240<br>$d_N(^{13}C)$: 4.771 |

**Figure 12.14** The top ranked structures inferred by the StrucEluc system from the spectral data obtained by Hsieh *et al.*[27] The numbers in the top left of each box are the rank ordered numbers.

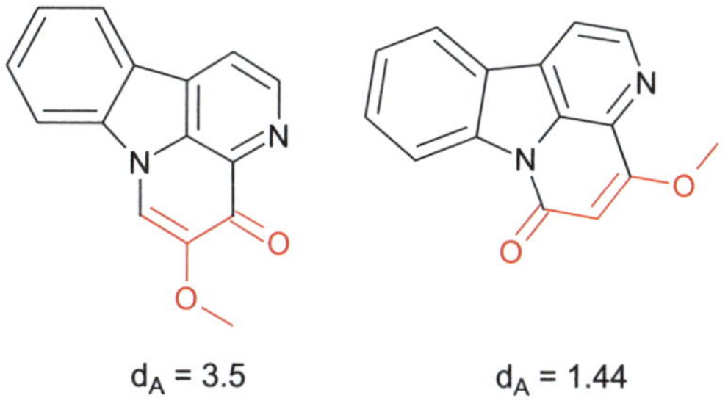

$d_A = 3.5$        $d_A = 1.44$

**Figure 12.15** The original and revised structures of cordatanine.

(structure **12.14**), then it would allow them to detect the wrong structural suggestion. In this case, no chemical synthesis would be necessary to disprove structure **12.14**.

Structure **12.14**, which was synthesized by Wetzel *et al.*,[28] was also confirmed by strict structure generation (no NSCs) from the 2D NMR data[28] with the following results: $k = 4083 \rightarrow 3874 \rightarrow 1439$, $t_g = 12$ min 6 s. The first ranked structure coincided with structure **12.14**.

The structure of cordatanine (**12.15**) was ranked first by the system. Non-standard HMBC correlations are shown using arrows. For clarity the differences between the original and revised structures are shown in Figure 12.15.

Wetzel *et al.*[28] comment in the conclusion of their article that their results "demonstrate that structure elucidations based only on spectroscopic data bear some risks of misinterpretation" and that "efforts regarding the total synthesis of alkaloids (performed *sine ira et studio*) helped to identify an erroneous structure assignment". We agree with the authors,[28] but our results show that when a software program such as the StrucEluc system is utilized the risks of misinterpretation can be minimized and laborious total synthesis can theoretically be avoided. This example also convincingly shows that $^{13}$C chemical shift calculation and dereplication of any isolated natural product are very useful as the first steps towards structure identification. Spectrum prediction frequently allows researcher to recognize if the suggested structure is reliable, while dereplication can help to identify the unknown if its structure is already present in a database.

In 2006 Wu *et al.*[29] isolated a new series of alkaloids, particularly cephalandole A, **12.16**. Using 2D NMR data (not tabulated in the article) they performed a full $^{13}$C NMR chemical shift assignment as shown on structure **12.16**.

Mason *et al.*[30] synthesized compound **12.16** and after inspection of the associated $^1$H and $^{13}$C NMR data concluded that the original structure assigned to cephalandole A was incorrect. The synthetic compound displayed significantly different data from those given by Wu *et al.*[29] The $^{13}$C chemical shifts of the synthetic compound are shown on structure **12.16A**.

**12.16**                                    **12.16A**

Cephalandole A was clearly a closely related structure with the same elemental composition as **12.16**, and structure **12.17** was hypothesized as the most likely candidate. Compound **12.17** was described in the mid 1960s and this structure was synthesized by Mason *et al.*[30] The spectral data of the reaction product fully coincided with those reported by Wu *et al.*[29] The true chemical shift assignment is shown in structure **12.17**.

**12.17**

For clarity the differences between the original and revised structures are shown in Figure 12.16.

We expect that $^{13}$C chemical shift prediction, if originally performed for structure **12.16**, would encourage caution by the researchers (we found $d_A = 3.02$ ppm). Figure 12.17 presents the correlation plots of the $^{13}$C chemical shift values predicted for structure **12.16** by both the HOSE and NN methods *versus* experimental shift values obtained by Wu *et al.*[29] The large point

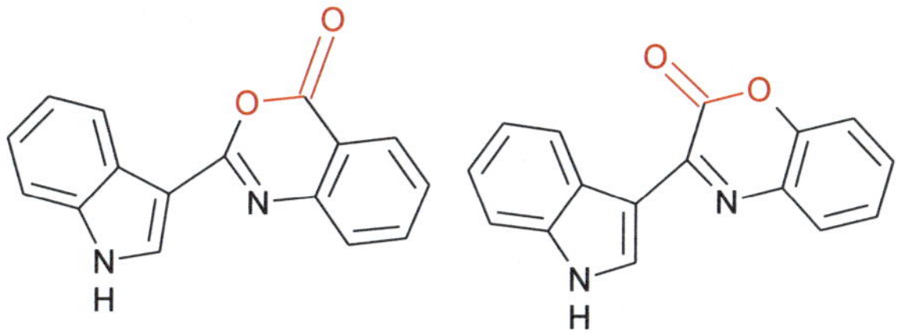

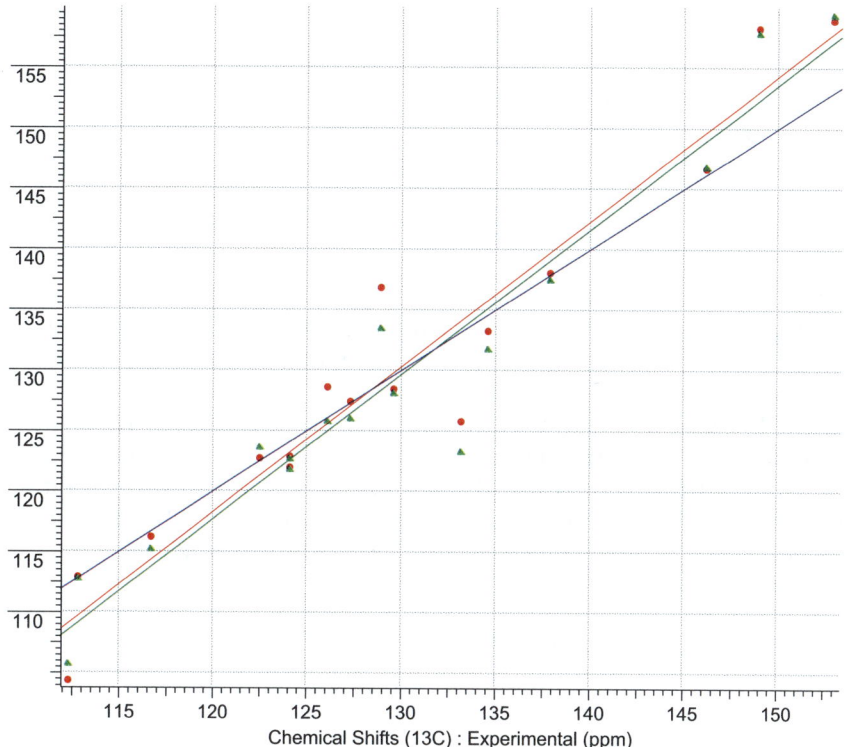

$d_A = 3.0$                    $d_A = 1.38$

**Figure 12.16**  The original and revised structures of cephalandole A.

Database: Proposed Structures
- Chemical Shifts (13C) : HOSE Calc. (ppm) (Current Record) (16 pts)
- Chemical Shifts (13C) : NN Calc. (ppm) (Current Record) (16 pts)

**Figure 12.17**  Correlation plots of the $^{13}$C chemical shift values predicted for structure **12.16** by HOSE and NN methods *versus* experimental shift values obtained by Wu *et al.*[29] Extracted statistical parameters: $R^2(\text{HOSE}) = 0.932$, $\delta_{\text{HOSE}} = 1.20_{\delta\text{exp}}-25.6$.

scattering, the regression equation, the low $R^2 = 0.932$ value (an acceptable value is usually $R2 \geq 0.995$) and the significant magnitude of the $\gamma$-angle between the correlation plot, and the 45° line (a visual indication for disagreement between the experiment and model) could indicate inconsistencies with the proposed structure and should encourage close consideration of the structure. Our experience has demonstrated that a combination of warning attributes can serve to detect questionable structures, even in those cases when the 2D NMR data and StrucEluc system are not used for structure elucidation. In Chapter 13 we will return to the cephalandole A story and reinvestigate this case under conditions when 2D NMR data are available.

In 1988 Sharma *et al.*[31] isolated two natural products sclerophytins A and B (structures **12.18** and **12.19** correspondingly):

**12.18**                                                    **12.19**

The novel structural features of these oxygen-bridged heterocycles and the significant cytotoxic properties of **12.18** have attracted the attention of chemists. At the same time, the relative stereochemistry at C-2, C-3, C-6, and C-7 were dubious, and a series of syntheses were undertaken to verify these structures.[32] In consideration of the fact that the synthetic analogs of **12.18** differed significantly from the originally isolated marine metabolites, an extensive NMR analysis of sclerophytins A and B was undertaken.[33,34] The real structures of these natural products were revealed to be **12.18A** and **12.19B** which are characterized by molecular masses and molecular formulae differing from those found by Sharma *et al.*[31]

**12.18A**                                              **12.19B**

Since the MS and tabulated 2D NMR data of the original structure **12.18** were not available to us, we carried out [13]C chemical shift predictions for structures **12.18** and **12.18A**. The following deviations were obtained:

Structure **12.18** - $d_A = 3.01$, $d_N = 2.52$, $d_A(max) = 9.57$, $R^2(HOSE) = 0.985$

Structure **12.18A** - $d_A = 1.37$, $d_N = 1.89$, $d_A(max) = 4.95$, $R^2(HOSE) = 0.996$

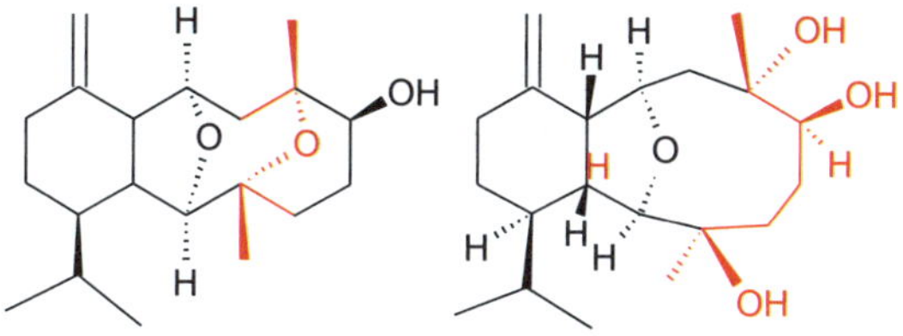

$d_A = 3.0$             $d_A = 1.37$

**Figure 12.18**  The original and revised structures of sclerophytin A.

The data can be used to reject structure **12.18**. The superiority of structure **12.18A** is convincingly confirmed by comparison of both deviations and $R^2$ values calculated for structures **12.18** and **12.18A**. For clarity, the differences between the original and revised structures are shown in Figure 12.18.

For revision of the structure of sclerophytin B, Friederich *et al.*[33] synthesized the compound and determined the structure of the reaction product using a combination of MS and 2D NMR. When the 1D, HMQC, and HMBC data published by the authors[33] were input into StrucEluc, the system automatically detected the presence of two NSCs in the HMBC data and generated a unique structure, **12.19B**, in 0.17 s with $d_A = 1.59$ ppm. The solution obtained is evidence that structure **12.19** is incorrect and could not have been inferred as a candidate from the MS and NMR data presented in the work.[33]

Sakano *et al.*[35] reported the isolation of the novel lanosterol synthase inhibitors epohelmins A (**12.20**) and B (**12.21**). The structures were determined by detailed spectroscopic analysis and proposed to be novel 9-oxa-4-azabicyclo [6.1.0]-nonanes. These structure assignments gave rise to doubts based on both chemical and spectroscopic grounds.[36]

Snider and Gao[36] comprehensively analyzed both the spectral and chemical aspects of the study of epohelmins A and B. They observed that the originally suggested bicyclo-[6.1.0]nonane structures could cyclize readily to give pyrrolizidin-1-ol structures and pointed to the observed chemical shifts as being more consistent with the rearranged product. They suggested structures **12.22** and **12.23** correspondingly as being more appropriate hypotheses.

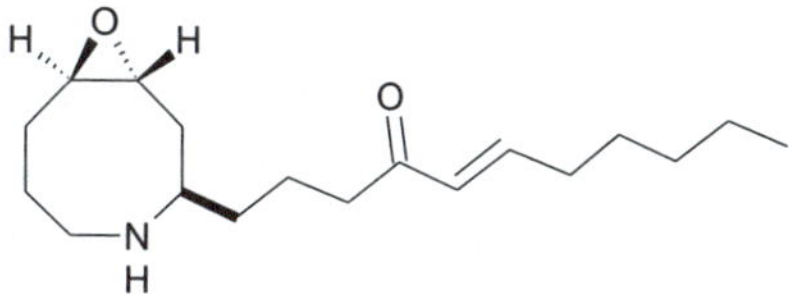

**12.20** 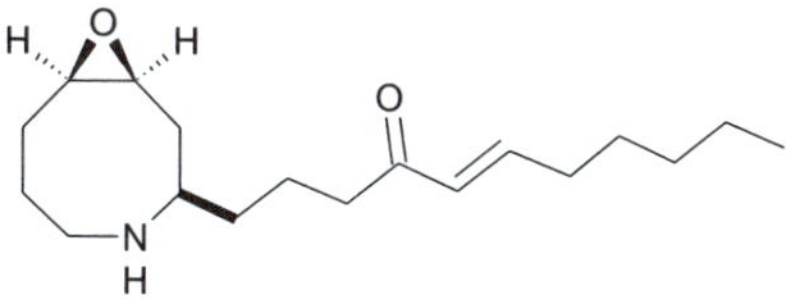                  **12.21**

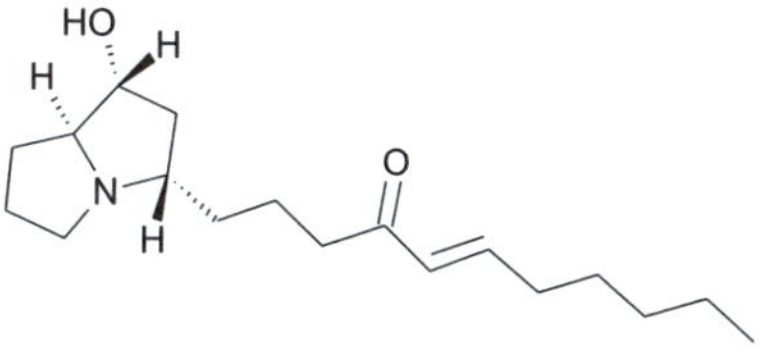

**12.22** 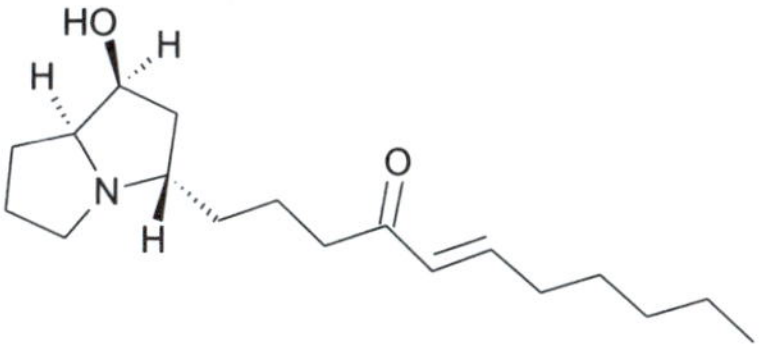                  **12.23**

To validate their suggestions, the authors[36] developed an eight-step synthesis of epohelmin A (12.22) and an 11-step synthesis of epohelmin B (12.23). The $^1$H and $^{13}$C NMR spectra of 12.22 and 12.23 were identical to those reported for epohelmin A (12.20) and epohelmin B (12.21), and the revised structures of these compounds were therefore unambiguously established via chemical synthesis.

2D NMR spectra of the investigated compounds were not available to us, so only the prediction and comparison of the $^{13}$C NMR spectra of competing structures 12.21 and 12.23 was possible, together with review of the discrepancies between the predicted and experimental data (see Table 12.2).

Table 12.2 unambiguously shows that structure 12.23 is superior over structure 12.21. For clarity, the differences between the original and revised structures are shown in Figure 12.19.

It is likely that if 2D NMR data were available to the researchers, then application of StrucEluc would deliver the correct structure very quickly and structure 12.21 would immediately be rejected by the program due to the very large deviations, especially with a $d_{A(max)}$ value of 21.4 ppm. Multi-step syntheses would also not be necessary to resolve the structural problem. However, at the same time the method of synthesizing epohelmin A and epohelmin B would not be developed! This contradictory peculiarity of the reassignment problem was strongly underlined in the review by Nicolaou and Snyder,[2] in which a number of striking examples were given.

In 2000 Hardt et al.[37] isolated a new cytotoxic marinone derivative neomarinone, molecular formula $C_{26}H_{32}O_5$, for which structure 12.24 was determined from the 1D and 2D NMR data:

12.24                                                    12.25

The authors noted that the connectivity of the sesquiterpenoid side-chain, and the presence of a methylated cyclopentane ring, were established by $^1$H NMR, HMBC, and COSY data. It is worth noting that all HMBC connectivities

**Table 12.2** Comparison of deviations and $R^2$ values calculated for competing structures 12.21 and 12.23.

| Structure | $d_A$ (ppm) | $d_N$ (ppm) | $d_{A(max)}$ | $R^2_{(HOSE)}$ | $R^2_{(N)}$ |
|-----------|-------------|-------------|--------------|----------------|-------------|
| 12.21     | 4.00        | 4.17        | 21.4         | 0.978          | 0.980       |
| 12.23     | 1.23        | 1.25        | 4.84         | 0.999          | 0.999       |

between the atoms forming a five-membered cycle are always of standard length: all combinations of connectivities meet the 2D NMR axioms. This results in difficulties in the unambiguous determination of the atom arrangement in the ring cycle from the HMBC data. The chemical shift assignment for the mentioned fragments is displayed in structure **12.24**.

On the basis of the novel structure of the sesquiterpenoid unit in neomarinone, in 2003 Kalaitzis *et al.*[38] attempted to investigate its biosynthesis via labeling studies with [13]C-labeled intermediates. The feeding experiments unexpectedly resulted in the modification of the earlier published structure **12.24** of neomarinone. The labeling studies and 2D NMR data, including an INADEQUATE experiment, allowed the researcher to obtain evidence that the true structure of neomarinone is **12.25**. The crucial observation disproving structure **12.24** was the INADEQUATE connectivity between carbons resonating at 25.10 and 123.90 ppm.

Tabulated 2D NMR data were not available from the original references,[37,38] and it was not possible to apply StrucEluc to this problem. Instead [13]C NMR chemical shift prediction was applied to structures **12.24** and **12.25**. The results obtained were:

Structure **12.24** - $d_A = 3.22$, $d_N = 3.43$, $R^2_{HOSE} = 0.995$, $d_{A(max)} = 9.0$

Structure **12.25** - $d_A = 1.08$, $d_N = 2.01$, $R^2_{HOSE} = 0.999$, $d_{A(max)} = 5.20$

For clarity the differences between the original and revised structures are shown in Figure 12.20.

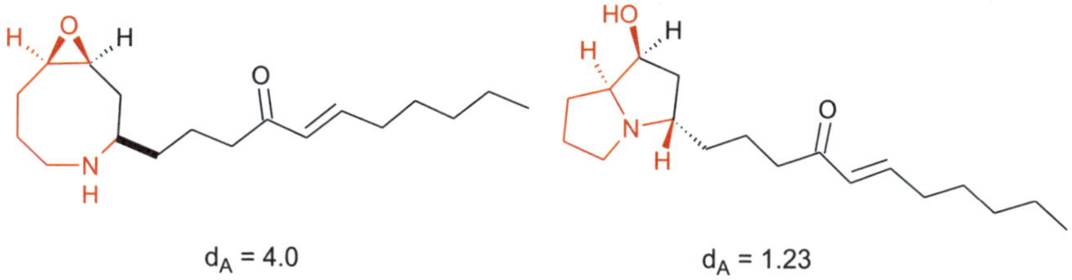

$d_A = 4.0$      $d_A = 1.23$

**Figure 12.19** The original and revised structures of epohelmin B.

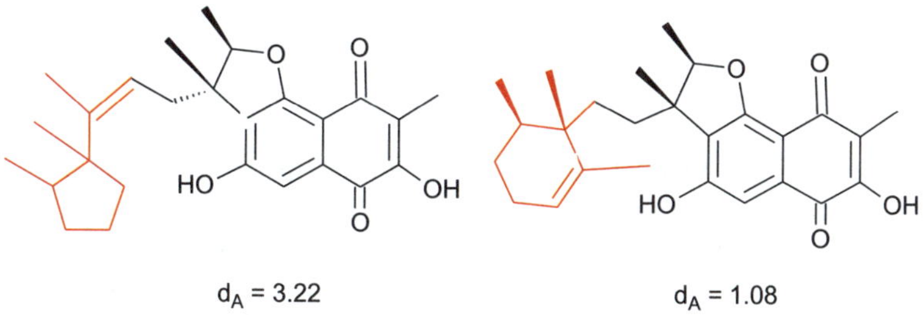

$d_A = 3.22$      $d_A = 1.08$

**Figure 12.20** The original and revised structures of neomarinone.

It is likely that the application of StrucEluc would allow the correct structure to be recognized by its small deviation values in the ranked output file.

## 12.3 Revision of Structures by the Re-examination of 2D NMR Data

In 1992 Suemitsu and co-workers[39] isolated a new natural product, porritoxin, with molecular formula $C_{17}H_{23}NO_4$, for which the following structure (**12.26**) was determined from the NMR data:

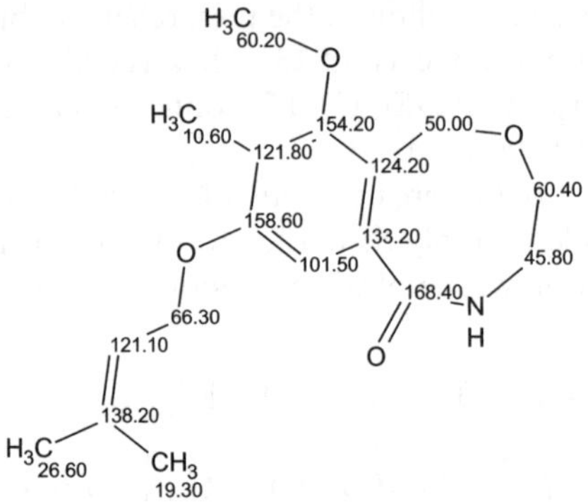

**12.26**

In 2002 the same group[40] reinvestigated the structure of porritoxin by detailed analysis of 2D NMR data, including COSY, $^1H$-$^{13}C$ HMBC, and $^1H$-$^{15}N$ HMBC experiments. This led to the revised structure **12.27**:

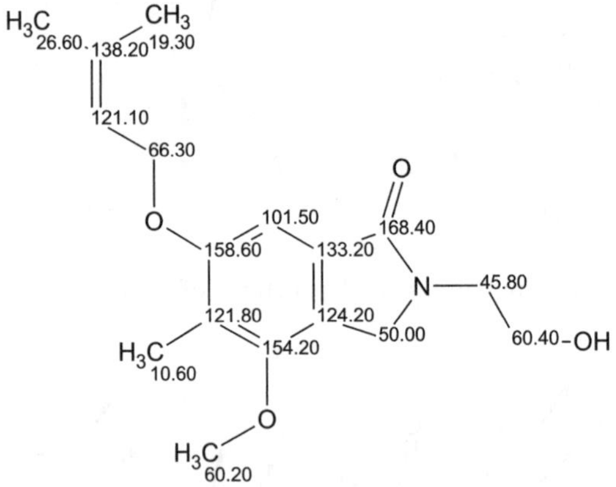

**12.27**

Only the $^1H$-$^{13}C$ HMBC data were used with the StrucEluc system to produce two structures in 1 s (see Figure 12.21) in FSG mode (one NSC was detected). The correct structure was reliably distinguished using $^{13}C$ chemical shift prediction. The original structure **12.26** was not generated, because the presence of three NSCs must be permitted to allow its generation. For completeness, FSG was restarted with $m = 3$, $a = x$ option ($m$ is the number of NSCs and $a = x$

means that the lengths of the NSCs are unknown). The results are as follows: $k = 52\,998 \rightarrow 20\,163 \rightarrow 12\,573$, $t_g = 6$ min 50 s. NN-based $^{13}$C chemical shift prediction was performed for the output file (calculations took 50 s). The correct structure was ranked in first place based on deviations, whereas the original structure was placed only in 59$^{th}$ position with $d_A = 3.71$ ppm. The suggested structure for **12.26** would be immediately rejected if $^{13}$C spectrum prediction was performed to check the reliability of the structure assignment.

For clarity the differences between the original and revised structures are shown in Figure 12.22.

Komoda *et al.*[41] isolated a new lipoxygenase inhibitor tetrapetalone A (20 mg of material), structure **12.28**, with a molecular formula of $C_{26}H_{33}O_7N$. The chemical structure was determined using a combination of IR, $^1$H, $^{13}$C NMR and

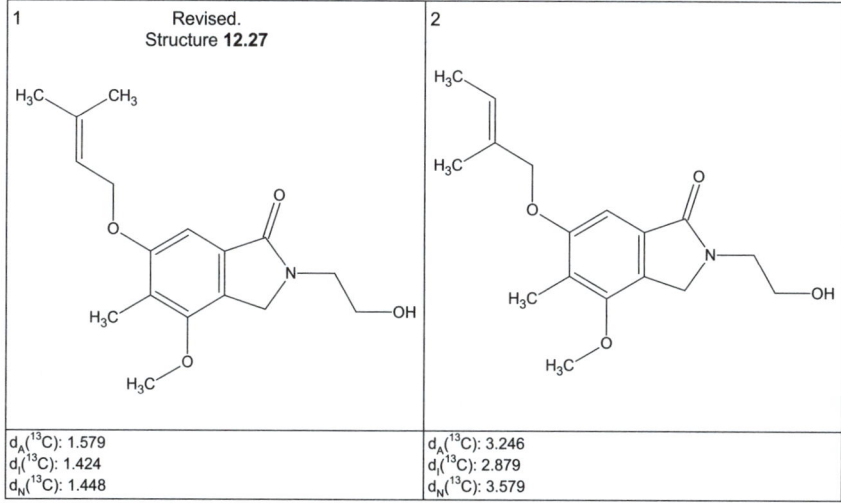

**Figure 12.21** The structures of the output file generated by StrucEluc software for the $C_{17}H_{23}NO_4$ compound (porritoxin). The numbers top left in each box are rank numbers.

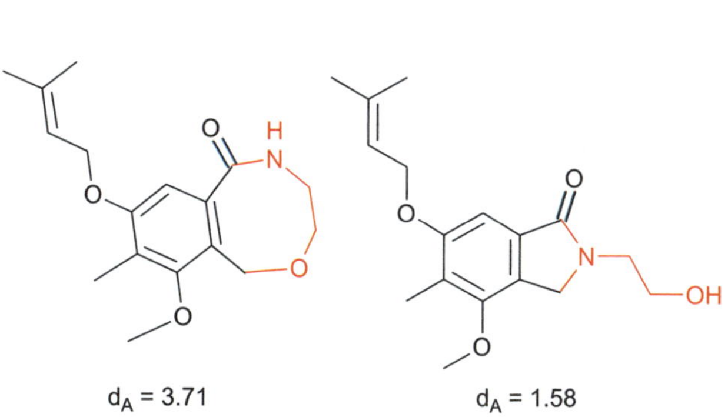

$d_A = 3.71$          $d_A = 1.58$

**Figure 12.22** The original and revised structures of porritoxin.

DEPT spectra, and HMQC, $^1$H-$^1$H COSY, HMBC and 2D-INADEQUATE data, and by methylation with diazomethane. The authors[41] inferred structure **12.28** using a common approach for organic chemists: four fragments were constructed on the basis of the 2D NMR correlations and then the fragments were joined taking into account the HMBC data. The set of mentioned fragments that should be present in the analyzed structure can be considered as a set of structural axioms. The stereochemistry was investigated by the coupling constants in $^1$H NMR and NOESY data, and the modified Mosher's method.[42]

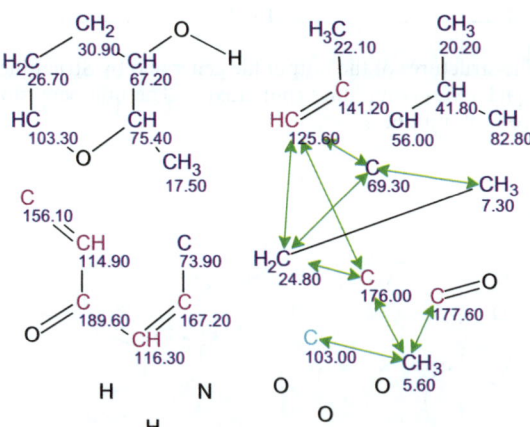

**12.28**

All available spectral data and the associated postulated fragments were input into StrucEluc. The fragments were drawn into the MCD window,[9,43] MCD, as shown in Figure 12.23. The chemical bonds are denoted by black lines and the HMBC correlations by green lines.

**Figure 12.23**   The MCD which shows the fragments suggested by the authors[41] and used by the StrucEluc software for the purpose of structure generation. The green arrows denote the HMBC correlations and the black lines the chemical bonds. The following colors are used to denote the atom hybridizations: violet, sp$^2$; dark blue, sp$^3$; light blue, not sp.

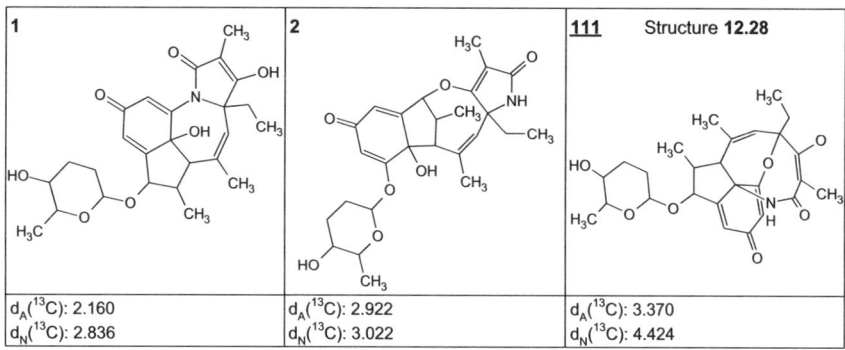

| 1 | 2 | 111   Structure **12.28** |
|---|---|---|
| $d_A(^{13}C)$: 2.160<br>$d_N(^{13}C)$: 2.836 | $d_A(^{13}C)$: 2.922<br>$d_N(^{13}C)$: 3.022 | $d_A(^{13}C)$: 3.370<br>$d_N(^{13}C)$: 4.424 |

**Figure 12.24**   The first, second, and 111$^{th}$ structures in the ranked output file produced by StrucEluc as a solution to the problem of tetrapetalone A
structure elucidation. The 111$^{th}$ structure is equivalent to structure
**12.28** of tetrapetalone A suggested by other authors.[41] The numbers in
the top left of each box are the rank numbers.

Structure generation from the MCD led to the following results: $k = 16\,465 \rightarrow$
$13\,672 \rightarrow 9203$ and $t_g = 61$ s. Ranking the output file in ascending order of mean
average error values placed structure **12.28** into 111$^{th}$ position. The first two
structures and the structure occupying position 111 are shown in Figure 12.24.

The automatically obtained solution to the problem delivered the best structure from among almost 10 000 candidates. The structure was characterized by
deviation values that were significantly smaller than those found for structure
**12.28**. It should be obvious that structure **12.28** cannot be the correct structure.

The same group[44] undertook a reinvestigation of the structure of tetrapetalone A. In this study the $^{1}H$-$^{15}N$ HMBC data were used to provide more
convincing evidence of the structural conclusions. As a result structure **12.28**
was revised and the following structure (**12.29**) was assigned to tetrapetalone A.
The stereochemistry was determined as shown below:

**12.29**

Comparison of structure **12.29** with the first structure in Figure 12.25 leads to the conclusion that the StrucEluc system has generated and automatically selected the true structure of tetrapetalone A without using any additional information. The structure could therefore be correctly identified in several minutes if the StrucEluc system was used for solving this problem. Moreover, all 256 stereoisomers of structure **12.29** were generated and HOSE-code based $^{13}$C chemical shift calculation was performed to select the most probable stereochemistry, which also coincided with the stereoconfiguration shown in structure **12.29**.

For clarity the differences between the original and revised structures are shown in Figure 12.25.

In 1990 Cáceres *et al.*[45] isolated the dolabellane diterpenoid palominol with molecular formula $C_{20}H_{32}O$, for which structure **12.30** was suggested with the $^{13}$C shift assignment shown:

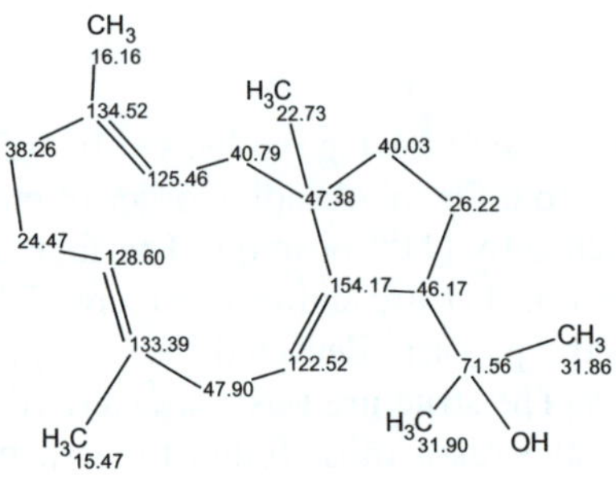

**12.30** 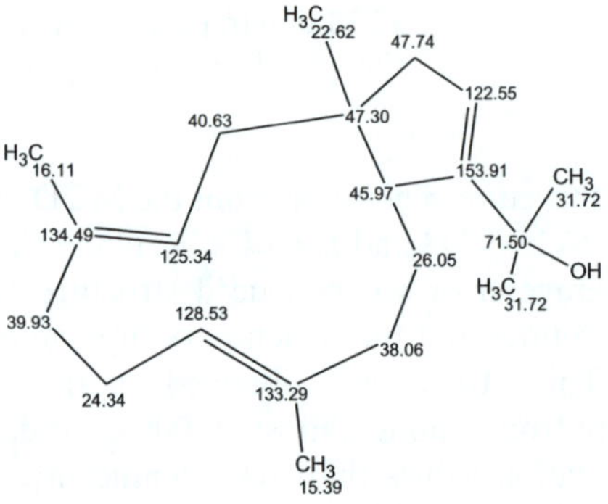                                                              **12.31**

In 1993 the same group[46] reinvestigated structure **12.30** using HMQC, HMBC, COSY, INADEQUATE, and ROESY data, and established that structure **12.31** was the actual structure. Using the StrucEluc system and utilizing 1D NMR, HMQC, and HMBC data we obtained four structures in 1 s in fuzzy generation mode with one NSC detected by the program. Structure **12.30** was not generated at all. Our studies showed that many NSCs, around eight,

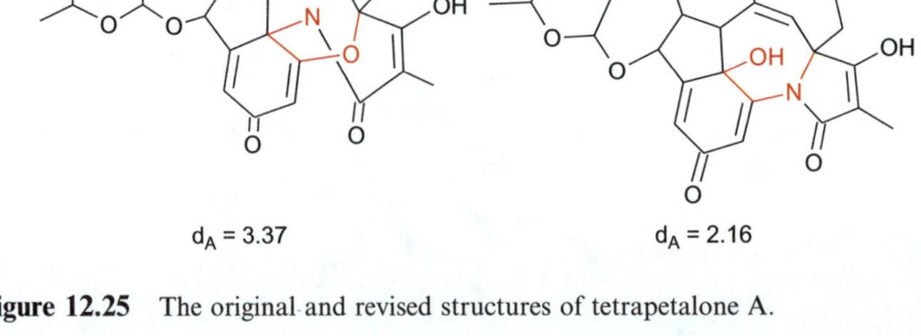

$d_A = 3.37$                                                        $d_A = 2.16$

**Figure 12.25**   The original and revised structures of tetrapetalone A.

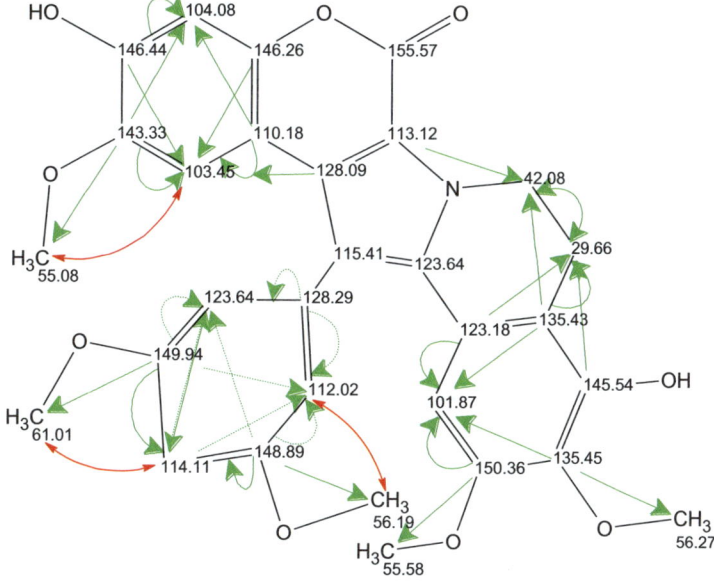

**Figure 12.26** The original and revised structures of palominol.

would need to be present in the HMBC data to allow it to be generated. $^{13}$C chemical shift prediction was performed for the four candidate structures. In so doing, both the *cis-* and *trans-*configurations of the double bonds included into the 11-membered ring were taken into account. The smallest deviations ($d_A = 2.18$ ppm) were found for the *trans-*configurations and the priority of structure **12.31** was confirmed (for double-bond *trans-*configurations in structure **12.30**, the value $d_A = 2.56$ ppm was found). For clarity the differences between the original and revised structures are shown in Figure 12.26.

Further testing of the StrucEluc system used the experimental data of Krishnaiah *et al.*[47] for the structure elucidation of a newly separated alkaloid lamellarin γ. The following structure (**12.32**) was deduced by the authors[47] from the molecular formula $C_{30}H_{27}O_9N$, $^1$H, and $^{13}$C NMR spectra, and 2D NMR data (HMQC, HMBC and NOESY):

**12.32**

The chemical shift assignments suggested in the original work[47] are shown on the chemical structure **12.32**. The green arrows indicate HMBC correlations, while the double-sided red arrows show the NOESY correlations. The dotted green lines are used to denote ambiguous connectivities. It is obvious that the structure is in agreement with the suggestion that all HMBC correlations are of a standard length (2 or 3 bonds, $^{2-3}J_{CH}$), whereas the NOESY correlations support the structure only in those cases when the methoxy groups at 61.01 and 56.19 ppm are asymmetrically oriented on the 1,3,5-trisubstituted benzene ring. The chemical shift assignment of structure **12.32** shows that the chemical shifts of the 1,3,5-trisubstituted benzene ring and the methoxy groups do not meet the local symmetry of this fragment. There is no reason that the theoretically symmetric carbons at 112.0 and 123.6 ppm should be so distinct.

Considering this observation, we performed $^{13}C$ chemical shift prediction for structure **12.32** using ACD/NMR Predictors[12] based on both the HOSE code and NNs algorithmic approaches.[48] The following results were obtained: $d_A = 4.70$ ppm, $d_N = 5.29$ ppm. It is obvious that the calculated deviations are extremely high in terms of providing confirmation of structure **12.32**. The correlation plots of the $^{13}C$ chemical shift values predicted for structure **12.32** *via* both prediction approaches are presented in Figure 12.27.

The data shown in Figure 12.27 and represented by the statistical parameters indicates that the calculated $^{13}C$ NMR chemical shifts differ significantly from the experimental values. This observation encouraged us to apply StrucEluc to validate the assignment.

The molecular formula and associated spectral data[47] were input into StrucEluc and a MCD was created. An attempt to perform structure generation in common mode,[9] where possible structures are assembled from "free" atoms, indicated that solving the problem would be extremely time consuming. This is accounted for by a deficit in the number of hydrogen atoms in the molecular formula in which the DBE = 18. A lack of HMBC correlations can be observed in structure **12.32**. According to a general methodology described elsewhere,[9] in such a situation the application of fragments stored into the system fragment library can be helpful. A fragment search using $^{13}C$ NMR chemical shifts resulted in the selection of 2318 fragments whose $^{13}C$ chemical shifts agreed with the experimental spectrum. The found fragments, ranked in descending order of carbon atom numbers, are displayed in the software program and fragments placed at the top of the ranked file are considered as the most likely, since they consume a large number of skeletal atoms. For instance, in the case described here the first fragment had the molecular formula $C_{17}H_{10}NO_4$ and the $^{13}C$ chemical shifts of the fragment were close to those observed experimentally.

The MCD creation procedure was applied to the top ten ranked found fragments, and 192 MCDs were produced. Each MCD contained only one fragment – the first ranked one, and the observed difference between the MCDs was in regards to the chemical shift assignments of fragment carbons performed automatically by the software program. Consequently, the lengths of the HMBC correlations corresponding to different pairs of associated chemical shifts in the different MCDs are different. FSG[13] was initiated with the following options: $m = 0$–20; $a = x$ (the augmentations of the connectivities are unknown) and was completed in 11 min with the following results: $k = 133\ 504 \rightarrow 120\ 816 \rightarrow 1530$. The chemical shift prediction for approx. 121 000 molecules took 11 min.

Structure **12.33**, characterized by $d_A = 1.26$ ppm and $d_N = 2.55$ ppm, was distinguished as the best structure:

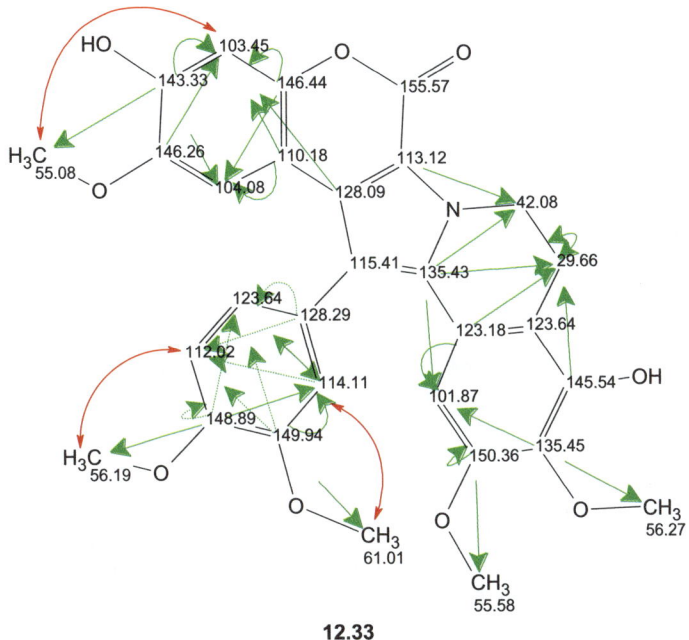

**12.33**

A comparison of deviations calculated for structures **12.32** and **12.33** shows that structure **12.33** is much more probable. However, structure **12.33** possesses an attribute which suggests that there may be a need for chemical shift reassignment: one of the four NOESY correlations (see the left portion of structure **12.33**) does not make sense chemically. At the same time structure **12.32**, suggested by the authors,[47] was also generated by the program and placed in 21st position by the ranking procedure. This also confirms the superiority of structure **12.33** over structure **12.32**.

The next step was to automatically find the chemical shift assignments of structure **12.33**, which are in accord with both the HMBC and NOESY correlations. As shown above, there are a lot of identical structures among the >120 000 structures generated from the 192 MCDs. For our purpose, we collected all isomorphic structures for structure **12.33**, 384 in total, in a separate file. We then performed NMR spectral predictions and ranked the file.

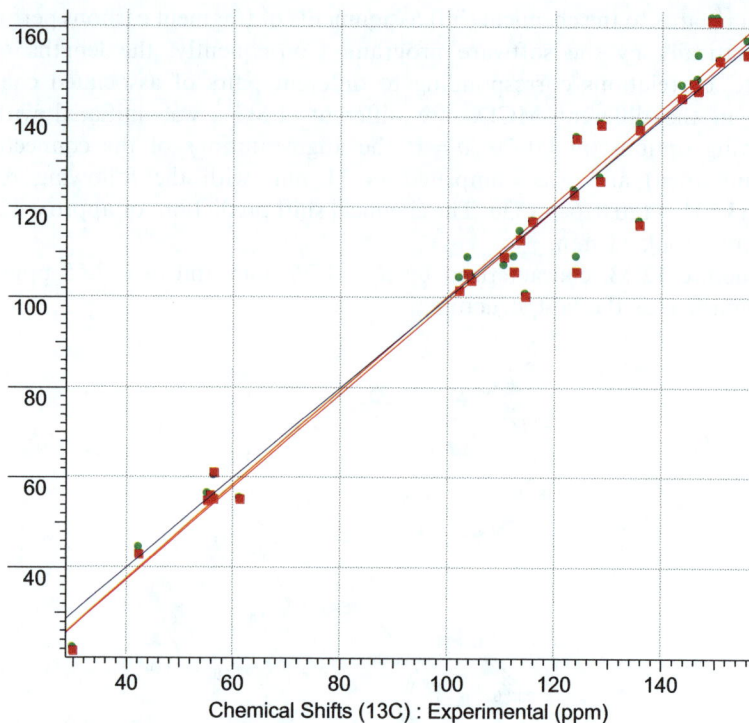

Database: Proposed Structures
- Chemical Shifts (13C) : NN Calc. (ppm) (Current Record) (30 pts)
- Chemical Shifts (13C) : HOSE Calc. (ppm) (Current Record) (30 pts)

**Figure 12.27**   Correlation plots of $^{13}$C chemical shift values predicted for structure **12.32** by HOSE (red points) and NN (green points) prediction methods *versus* experimental shift values. The target line Y = X is colored in blue. The $R^2$ value calculated by the HOSE-based method is equal to 0.965.

**Table 12.3**   Comparison of deviations and $R^2$ values calculated for competing structures **12.32** and **12.33A**.

| Structure | $d_A$ (ppm) | $d_N$ (ppm) | $d_A(max)$ | $R^2_{(HOSE)}$ | $R^2_{(NN)}$ |
|---|---|---|---|---|---|
| **12.32** | 4.70 | 5.29 | 18 | 0.965 | 0.967 |
| **12.33A** | 1.26 | 2.55 | 5 | 0.997 | 0.993 |

The structure ranked first fit both the HMBC and NOESY spectra and structure **12.33A** was finally selected.

Deviations and $R^2$ values calculated for structures **12.32** and **12.33A** are presented in Table 12.3.

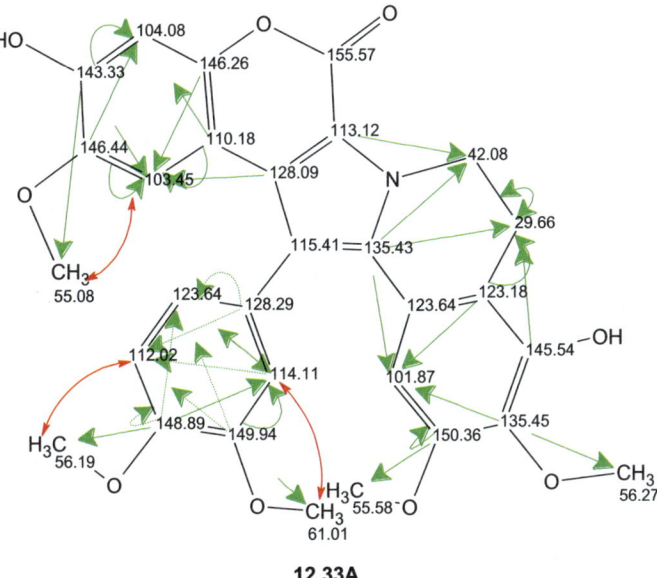

$d_A = 4.7$ $d_A = 1.26$

**Figure 12.28** The original and revised structures of lamellarin γ.

Table 12.3 shows the evident superiority of structure **12.33A** over structure **12.32**. For clarity, the differences between the original and revised structures are shown in Figure 12.28.

**12.33A**

In 2004 Hiort *et al.*[49] isolated from the Mediterranean sponge *Axinella damicornis* seven new natural products including four pyranonigrins featuring a novel pyrano[3,2-*b*]pyrrole skeleton previously unknown in nature. All structures were elucidated on the basis of extensive 1D and 2D NMR spectroscopic studies ($^{1}$H, $^{13}$C, COSY, HMQC, HMBC, NOE difference spectra) and MS analysis. For the two chiral pyranonigrin molecules, particularly for pyranonigrin A ($C_9H_{10}NO_5$, DBE = 7) **12.34**, the absolute configurations were established by quantum mechanical calculation of their circular dichroism (CD) spectra.

**12.34a**

In 2007 Schlingmann *et al.*[50] isolated from the marine fungus *Aspergillus niger* a compound of molecular formula $C_9H_{10}NO_5$ whose physical data were identical to those published by Hiort *et al.*[49] for pyranonigrin A. Interpretation of the NMR data did not permit the authors[50] to assign structure **12.34a** to pyranonigrin A. They suggested that the correct structure is one of the following three possible candidates:

**12.34b**          **12.34c**          **12.34d**

Similar to the previous report[49] the structure determination of the pyranonigrin A was based on the interpretation of spectroscopic data, especially MS and NMR data, which included HSQC, COSY, ROESY, HMBC, and an essential $^1$H-$^{15}$N HMBC. Comprehensive analysis of the experimental 1D and 2D NMR spectra allowed the authors[50] to reject hypotheses **12.34b** and **12.34c**. It was concluded that pyranonigrin A was consistent with structure **12.34d**. To further prove this finding the researchers produced hydrophobic derivatives of the analyzed compound suitable for comparison of experimental UV/CD spectra with that of *ab initio* predicted data (*in vacuo*), since the substance itself was soluble only in polar solvents. As a result of extensive experimental and theoretical investigations, the structure of pyranonigrin A was unambiguously elucidated, and its absolute configuration was determined.

The initial spectral data presented for pyranonigrin A by Hiort *et al.*[49] were input into the StrucEluc system, and strict structure generation was performed excluding any NSCs as the authors had suggested (an axiom). The results were as follows: $k = 109 \rightarrow 81 \rightarrow 72$, $t_g = 0.3$ s. The first and sixth ranked structures are presented in Figure 12.29.

The first ranked structure, similar to **12.34a**, is characterized by unacceptably large deviations, while the suggested original structure **12.34a** should be

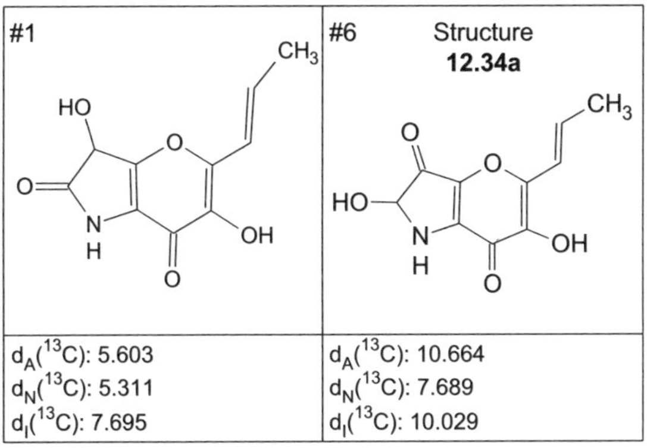

| #1 | #6 Structure **12.34a** |
|---|---|
| $d_A(^{13}C)$: 5.603 | $d_A(^{13}C)$: 10.664 |
| $d_N(^{13}C)$: 5.311 | $d_N(^{13}C)$: 7.689 |
| $d_I(^{13}C)$: 7.695 | $d_I(^{13}C)$: 10.029 |

**Figure 12.29** The first and sixth ranked structures of the output file produced using strict structure generation for pyranonigrin A. The numbers in the top left of each box are rank numbers.

immediately rejected as it had a large deviation of $d_A = 10.6$ ppm. The hypothesized structures **12.34b**–**12.34d** were not generated at all. As mentioned earlier, large deviations found for the first ranked structure should be considered as an indication to the possible presence of non-standard correlations in the 2D NMR data. The next step was FSG with options $m = 1$, $a = x$ to provide the result: $k = 3024 \rightarrow 2130 \rightarrow 1144$, $t_g = 14$ s. The correct structure **12.34d** was generated and ranked first ($d_A = 2.03$), structure **12.34c** was ranked fifth ($d_A = 5.26$) and structure **12.34a** was placed in $31^{st}$ position. Structure **12.34b** was not generated.

To check the solution for stability we performed FSG using $m = 2$ and $a = x$ as options to provide the following results: $k = 18275 \rightarrow 10725 \rightarrow 3506$, $t_g = 2$ min 23 s. Under the condition that two NSCs may be present in a structure, all structures (**12.34a**, **12.34b**–**34d**) considered by the authors[50] were generated. During this run, the program produced a full set of structures containing all six possible rearrangements of OH, NH and C=O groups on the five-membered cycle. These structures along with their rank ordered positions in the output file are presented in Figure 12.30.

Figure 12.30 convincingly demonstrates the priority of the correct structure, **12.34d**, whereas the original structure, **12.34a,** was placed in $95^{th}$ position by the program. Note that the structure ranked as #7 was the best one in the file obtained by strict structure generation (see Figure 12.30), because only this structure and structure **12.34a** meet the authors[49] restrictive suggestion (axiom) regarding the absence of non-standard correlations in the 2D NMR data. Structure **12.34b** could be considered only using the suggestion that it contains two NSCs. For clarity the differences between the original and revised structures are shown in Figure 12.31.

The example shows that even small molecules with a deficit of hydrogen atoms can become a structure elucidation challenge using traditional

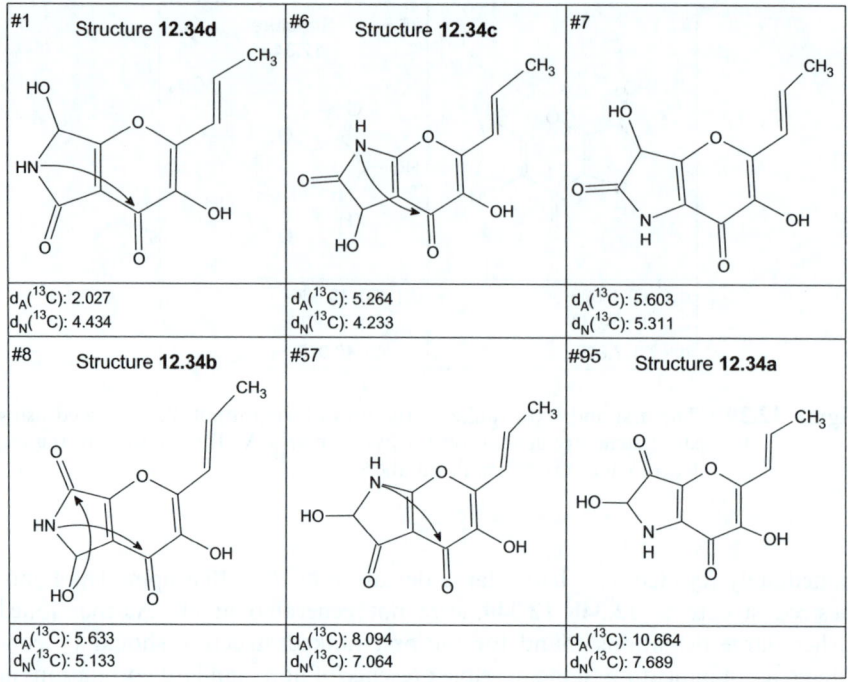

**Figure 12.30** The full set of structures containing all arrangements of OH, NH, and C=O groups on a five-membered cycle. The numbers in the left upper corner denote the rank of the corresponding structure and the arrows show non-standard HMBC correlations. The numbers in the top left of each box are rank numbers.

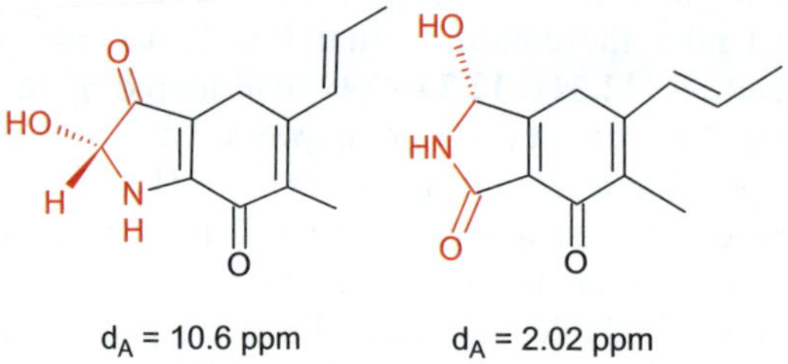

$d_A$ = 10.6 ppm $\qquad$ $d_A$ = 2.02 ppm

**Figure 12.31** The original and revised structures of pyranonigrin A.

approaches. The application of the StrucEluc program would allow Hiort *et al.*[49] to automatically generate all conceivable candidate structures and select the correct molecule in a much reduced time. If only $^{13}$C chemical shift prediction was performed for the original structure, then it would immediately show that the structure is incorrect since $d_A = 10.66$ ppm. New hypotheses would need to be examined.

## 12.4 Structure Selection on the Basis of Spectrum Prediction

Johnson *et al.*[51] reported the unexpected isolation of a novel thiopyrone CTP-431 with molecular formula $C_{23}H_{29}NO_5S$. On the basis of both mass spectrometry and 2D NMR data (HMQC, HMBC, COSY, and NOESY) two possible structures were suggested:

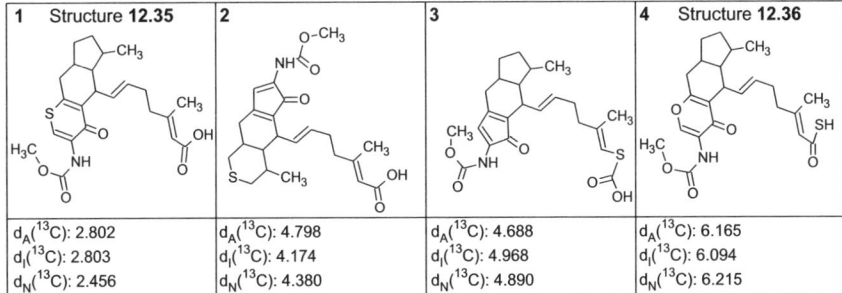

**12.35**     **12.36**

To choose between these two structures, the authors[51] performed DFT GIAO $^{13}C$ chemical shift calculations allowing them to select structure **12.35** as the most probable. The conclusion was supported by the results of X-ray crystallography.

StrucEluc was applied to solve this problem by taking into account all possible valences of the sulphur atom: 2,4,6. Fuzzy Structure Generation delivered the following solution from the HMBC and COSY data: $k = 622 \rightarrow 374$, $t_g = 4$ s. The top four structures in the ranked output file are presented in Figure 12.32.

The Figure shows that the correct structure, **12.35**, was reliably distinguished, while the alternative structure, **12.36**, was placed only in fourth position in the ranked file. We have previously shown[9,17,18] that large deviations ($>6$ ppm) indicate without doubt that structure **12.36** should be rejected. For clarity the differences between the two competing structures are shown in Figure 12.33.

| 1    Structure **12.35** | 2 | 3 | 4    Structure **12.36** |
|---|---|---|---|
| $d_A(^{13}C)$: 2.802 | $d_A(^{13}C)$: 4.798 | $d_A(^{13}C)$: 4.688 | $d_A(^{13}C)$: 6.165 |
| $d_I(^{13}C)$: 2.803 | $d_I(^{13}C)$: 4.174 | $d_I(^{13}C)$: 4.968 | $d_I(^{13}C)$: 6.094 |
| $d_N(^{13}C)$: 2.456 | $d_N(^{13}C)$: 4.380 | $d_N(^{13}C)$: 4.890 | $d_N(^{13}C)$: 6.215 |

**Figure 12.32**   The top ranked structures inferred by the StrucEluc system when the spectral data obtained by Johnson *et al.*[51] were used. The structure of thiopyrone (**12.35**) was ranked first by the system. The numbers top left in each box are rank numbers.

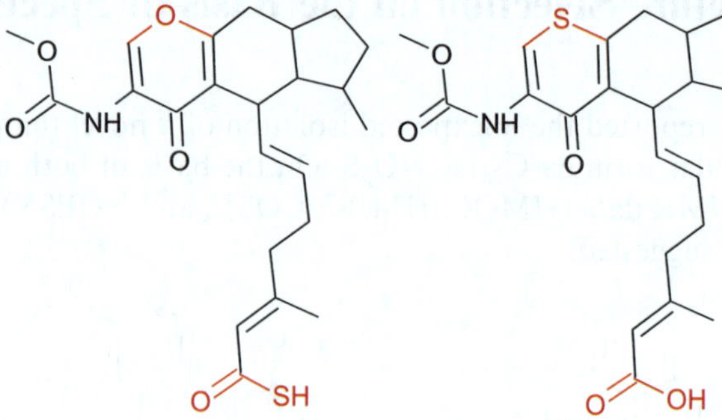

$d_A = 6.15$                                       $d_A = 2.8$

**Figure 12.33**    The rejected and real structures of thiopyrone CTP-431.

This study indicates that the StrucEluc system can identify the correct structure almost instantly. In connection with this example it should be noted that by using only HMBC it is not possible to detect the position of the S atom. However, when HMBC is used within StrucEluc in combination with structure generation and $^{13}$C NMR spectrum prediction new possibilities arise: the position of the S atom in the molecule was correctly and quickly detected without time-consuming QM calculations. This demonstrates the strength of the CASE approach.

Takashima *et al.*[52] isolated a component from tree bark for which structure **12.37** (*Brosium allene*) was elucidated. The structure assignment was based on high-resolution MS, $^{1}$H, $^{13}$C and 2D NMR data. The 2D NMR data were not disclosed.

Hu *et al.*[53] recognized that the $^{13}$C NMR signal at 139 ppm was assigned to the central allenic carbon in **12.37**, even though the central carbon signal of allenes normally appears near 200 ppm. This discrepancy served as an impetus for reinvestigation of this compound.

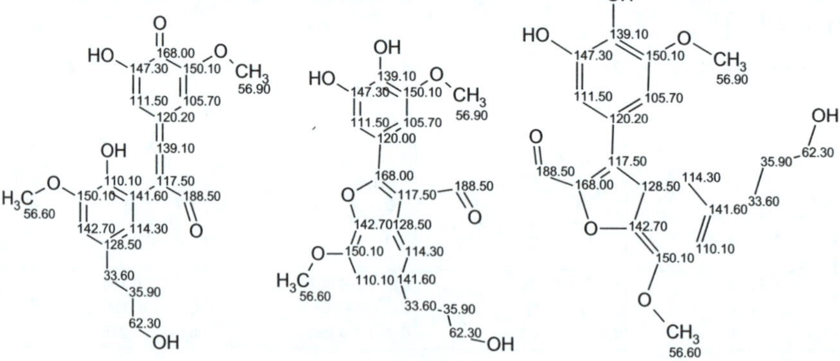

**12.37**                          **12.38**                          **12.39**

The authors[53] performed QM computational modeling of the $^{13}$C chemical shifts expected for **12.37**. Geometry optimizations were performed with B3LYP

[6-31G (2d,2p)] and with Hartree-Fock method, HF[6-31G (2d,2p)]. The spectral data were calculated using DFT functionals B3LYP and mPW1PW91, as well as the HF approach. None of the data sets matched well. For the signal assigned as 139 ppm the calculated value was found to consistently be equal to ~230 ppm. Although QM-based NMR signal prediction is only approximate, a deviation value of 90 ppm is extreme. This observation was considered as evidence that structure **12.37** is not correctly assigned. The authors[53] also doubted that **12.37** represents a molecular arrangement isolable under standard conditions.

To verify their suggestion the authors[53] evaluated the reactivity of structure **12.37** and, taking into account the results of the chemical shift predictions, suggested two alternative structures, **12.38** and **12.39**, as possibilities. QM-based [13]C chemical shift prediction for both proposed structures led the researchers to conclude that structure **12.38** provided the best match between the experimental and calculated values. Finally, the authors showed that structure **12.38** was identical to a known compound mururin C.[54]

We also performed [13]C chemical shift prediction using our empirical prediction methods[12] for all three structures. The deviations resulting from the empirical and QM predictions are presented in Figure 12.34. The Figure shows that structure **12.37** is rejected by all methods and that structure **12.38** is indeed the most probable. It is evident that the StrucEluc system would reject structure **12.37** if it was generated from 2D NMR data. At the same time Figure 12.34 demonstrates that the choice of **12.38** as the best structure relative to **12.37** could be made almost instantly using empirical methods of chemical shift prediction and without the application of time consuming QM calculations.

The Figure also confirms our previous conclusion[55] that the accuracy of empirical methods of rapid chemical shift predictions is about two times higher than QM-based predictions. For clarity the differences between the original and revised structures are shown in Figure 12.35.

| Structure **12.37** | Structure **12.38** | Structure **12.39** |
|---|---|---|
| $d_A$([13]C): 12.891 | $d_A$([13]C): 1.801 | $d_A$([13]C): 3.735 |
| $d_N$([13]C): 12.003 | $d_N$([13]C): 1.561 | $d_N$([13]C): 4.212 |
| maxd$_A$([13]C): 67.840 | maxd$_A$([13]C): 12.430 | maxd$_A$([13]C): 25.350 |
| $d_Q$([13]C): 9.395 | $d_Q$([13]C): 3.900 | $d_Q$([13]C): 5.655 |

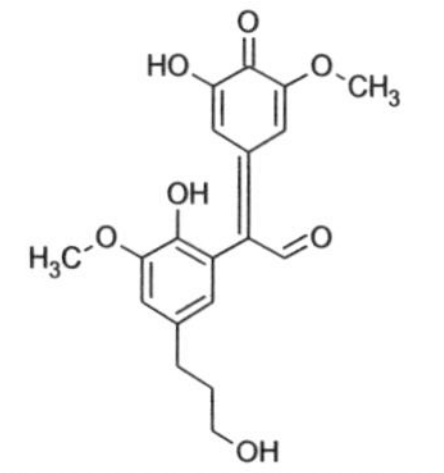

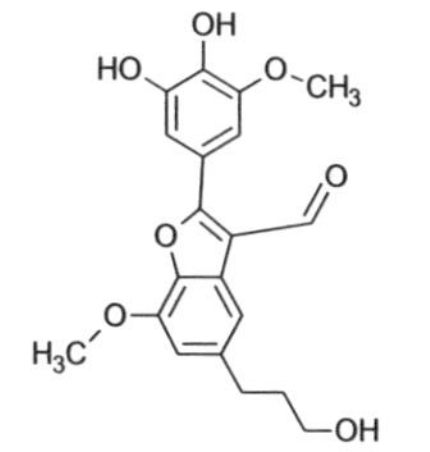

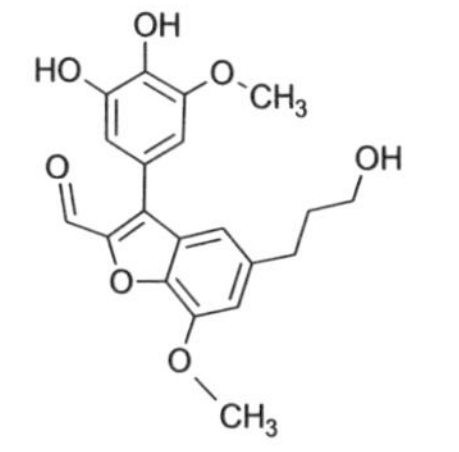

**Figure 12.34** Comparison of discrepancies between experimental and calculated [13]C chemical shift for structures **12.37**, **12.38** and **12.39**. $d_Q$ is the mean average deviation (MAE) found as a result of QM calculations.

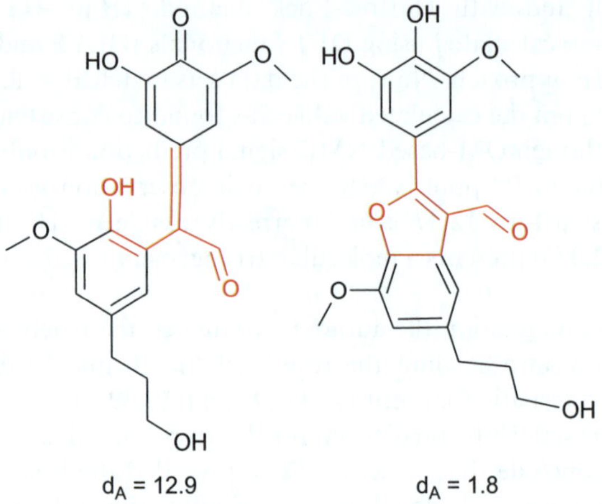

**Figure 12.35** The original and revised structures of *Brosium allene*.

## 12.5 Discussion

In this Chapter we have tried to provide answers to the following important questions. (i) Are the pitfalls arising during molecular structure elucidation unavoidable? (ii) Can modern computer-aided methods of molecular structure elucidation be used to minimize the probability of inferring incorrect structures from spectral data?

To investigate these questions we analyzed a large number of examples for which the originally determined structures of novel natural products were revised in later publications. In all cases, when the 2D NMR data were available the expert system StrucEluc[56] was used to determine whether the correct structure could be inferred from the experimental spectra and assumptions or "axioms" suggested by the researcher.

To make the process of structure elucidation more transparent, in Chapter 2 we expounded the main statements of the common methodology describing this process in the form of an axiomatic theory. The examples considered here show that this theory not only adequately reflects the nature of the problem, but it is also a very important and effective analytical tool which can, and should, be employed routinely in the practice of spectroscopic analysis. This approach appears to be unique for the natural sciences and we failed to find another example of a problem where the initial knowledge could be so clearly and explicitly represented in the form of a set of axioms (hypotheses) and then all logical corollaries, in our case a set of structures, would be automatically inferred then, with subsequent selection, to provide the most probable corollary, in theory the correct chemical structure.

It is also necessary to underline a very important general property of the problem of structure elucidation from *spectral* data. It was noted (see Chapter 1) that this problem was related to the class of so-called "inverse problems".[57]

The consequence of this is that a unique and correct solution can be deduced only as a result of using additional information taken from different sources. Therefore, the chance of fully replacing human intellect with a computational algorithm is unlikely at best. Moreover, in accordance with the Bohr principle of complementarity,[58] the methodology of computer-assisted structure elucidation includes two major elements that complement each other. They are deterministic logic (enhanced with combinatorial analysis) of the computer and the knowledge and intuition of the investigator. The interaction of these elements in the process of solving the problem is what gives rise to the synergistic effect to allow the elucidation of complex molecules. It is therefore necessary to find a rational way of combining connectivities deduced algorithmically from experimental 2D NMR data with additional information such as chemical considerations, hints based on visual spectrum analysis, *etc.*, provided by a scientist in order to obtain a solution to the problem in a reasonable time.

The effectiveness of this relationship between a researcher and a computer accounts for the possibility of the program to produce all consequences, without exception, following from the axiomatic set provided by the researcher. The many examples presented in this Chapter show that if a researcher's assumptions are incorrect then the solution to a problem is invalid – it does not contain the correct structure.

It has been shown that if the initial NMR data did not contain artifacts and misinterpreted peaks then, in the majority of cases, the software allows the chemist to choose the correct structure. Errors in suggestions made by the researchers or incorrectly interpreted spectral data input into the system leads to output structures whose unlikelihood is easily revealed simply by the application of $^{13}$C NMR chemical shift prediction. This allows the researcher to immediately recognize that a particular structural suggestion is not correct or is at least questionable. Figuratively speaking, an expert system can play the role of a "polygraph detector" helping to identify whether a structural hypothesis corresponds to a genuine structure.

As well as $^{13}$C chemical shift prediction, the dereplication of the structure of any isolated natural product is very useful as a first step towards structure identification. The dereplication process can help to identify the unknown if its structure is already present in a database.

The analysis of the examples considered in this Chapter allows us to distinguish the following types of errors which are quite commonly made by researchers in the process of forming their initial hypotheses and then in the further deduction of the structure from MS and NMR data:

- The elemental composition is incorrectly identified to provide the wrong molecular formula.
- Due to insufficient resolution of the mass-spectrometer, the *m/z* value is determined incorrectly. This also leads to an incorrect molecular formula.
- The observation of a spectral feature characteristic for a fragment is erroneously interpreted as evidence of the presence of a particular

fragment in a molecule. It should be kept in mind that if the implication $A_i \rightarrow X_j$ is true, then the inverse implication $X_j \rightarrow A_i$ can be true or not true.

- Some 2D NMR peaks resulting from a solvent artifact can be erroneously interpreted as part of the 2D NMR spectrum of the unknown compound. As a result the correct structure cannot be inferred. Recording spectra in at least two different solvents can be helpful to detect such issues.

- Some important 2D NMR signals can be missed in the peak-picking process and this can certainly prevent generation of the correct structure in certain cases.

- Suggested structures are not checked using the most significant characteristic spectral features in either IR or Raman spectra. For instance, the absence of any absorption in the IR area 3200–3700 cm$^{-1}$ will reject any hypothetical structure containing an alcohol group.

- The absence of peaks corresponding to expected correlations in an experimental 2D NMR spectrum may be ignored. As the spectroscopist is an integral part of a symbiotic partnership between a human and a program, the best chosen 1–3 structures (not thousands of possible ones!) should be carefully analyzed regarding their correspondence to the experimental spectra. If the expert, using his knowledge and experience, determines that some 2D NMR correlation that was expected in the 2D NMR of a given concrete structure was not observed, then this fact should be considered as a warning of the plausibility of the structure.

- All 2D NMR correlations are assumed to have only standard lengths. As a result, a correct structure whose HMBC or COSY spectra contain non-standard correlations will be lost.

- The number of non-standard correlations allowed in 2D NMR data may be incorrectly estimated by the researcher and as a result the correct structure is missed.

- $^{13}C$ chemical shift prediction might not be performed for the suggested structure. Almost all of the original structures that were identified to be incorrect in this article would have been either rejected or declared suspicious if $^{13}C$ NMR spectrum calculations were performed. There are, of course, various NMR prediction algorithms and based on our experience and expertise we recommend HOSE code or NN algorithms over rules-based approaches.

- When several fragments are deduced from the 2D NMR data by a researcher then the human expert frequently is unable to take into account all possible ways of combining fragments to complete assembly of the structure using, as a rule, HMBC correlations. Many thousands of structures should be checked and as a result the wrong structure may be selected.

When an expert system is employed for the purpose of structure elucidation the overwhelming majority of subjective errors made by the human expert can be either avoided or detected during the process of solving the problem or as a result of validating the most probable structure by NMR spectrum prediction. Some methodological guidelines given below can be helpful.

In general, the process of structure elucidation is known[17] to be reduced to the superposition of constraints on a finite number of isomers that correspond to the molecular formula of an unknown (see Chapter 1). The number of isomers can be very large, even for relatively small molecules[17] (see Figure 1.1 in Chapter 1). For instance, structure generation using the modest molecular formula $C_{11}H_{12}N_4$ produced 2 258 672 147 012 isomers.[59] Researchers try to introduce as many as possible constraints to provide a manageable number of suggested structures. As was shown above, the issue is that some constraints introduced by user assumptions can be erroneous. The application of an expert system can minimize the number of user assumptions as a result of the high speeds of both structure generation and spectrum prediction: a great number of isomers can be generated in a reasonable time and then fast spectrum prediction allows the program to quickly select the most probable structure. We advise great care when postulating the presence of some fragments and setting atom properties. At the same time, the fast NMR prediction algorithms discussed in Chapter 3 give the user an opportunity to solve the problem repeatedly trying different constraints (spectral and structural hypotheses). Such a solution (structural set) containing a structure characterized by the minimum deviations is considered as at the most preferable one. An expert system also allows the researcher to utilize two or three possible molecular formulae if the elemental composition of the unknown is not clear or the resolving power of the MS instrument is insufficient.

The most challenging part of the structure elucidation process using 2D NMR data is establishing the presence of NSCs, as well as their number and length. To overcome the serious difficulties associated with NSCs, the FSG algorithm[13,60] was implemented into StrucEluc. This algorithm is capable of solving a problem under the conditions that neither the number of the NSCs nor their lengths are known. Due to the nature of the sophisticated FSG algorithm not all possible combinations of connectivities are tried, only a small number of them, and this dramatically reduces the generation time. The following recommendation is given: if the $d_A(1) > 3.5$ ppm was found for the highest ranked structure then it is likely incorrect and must be examined further. FSG should initially be performed with $m = 1$, $a = x$ parameters, and if the new $d_A(1)$ value reduces in value then there is likely at least one NSC. The typical value of $d_A$ acceptable for the correct structure is 1.0–2.5 ppm.

In those rare cases when an unknown molecule is classed as "exotic", then the correct structure may be characterized by deviations which are close to or exceeding a threshold of 3.5 ppm. The reason is that empirical methods are known to exhibit at least one principal drawback: if the database created for the purpose of HOSE prediction, or the training set for the NN algorithm does not contain specific atoms representing the atom environments in the molecule under investigation, then the empirical methods can fail to predict the chemical shift of such atoms with sufficient accuracy.

Examples of such "exotic" structures are corianlactone (**12.40**), hexacyclinol (**12.41**), and daphnipaxinin (**12.42**), for which $d_A$ values were 2.93, 3.65 and 6.34 ppm correspondingly.

**12.40**            **12.41**            **12.42**

It has been shown[55,61] (see also Section 9.9) that in spite of the unusual character of these structures and the large values of the deviations, the application of StrucEluc allows the program to correctly select these challenging structures from many candidates while using the structure ranking methodology described above. The intriguing story about the structure elucidation of hexacyclinol was described in a series of publications.[61–65]

$^{13}$C chemical shift calculation should be considered as the most severe filter to reject all invalid structures and to select the most probable one. However, the average deviations between experimental and predicted spectra that serve as effective criteria for structure assessment are calculable only if chemical shift assignment is completed. The series of examples considered in this Chapter confirm the usefulness of creating linear regression plots of calculated $^{13}$C chemical shifts against experimental shifts. These graphs allow visual inspection of the point scattering along the full chemical shift scale, whereas the regression equation and accompanying statistical parameters give numerical criteria for comparing the different suggested structures. A regression plot can also help to detect a small incorrect feature within a molecule when the remaining structure is very close to the correct one (see the halipeptins case).

It has also been shown[6] that if shift assignment is not available, which can happen when CASE methods are not used, then a visual comparison of the graph bars depicted for the experimental and calculated spectra for a series of suggested structures frequently allows the researcher to identify which structure is the most probable: structures characterized by large outliers should be treated as suspicious.

It would be very attractive to determine some quantitative criteria to allow preliminary estimation of the complexity of a problem. We have failed to find such criteria, so far, because there are a great number of factors influencing the complexity of the problem and, unfortunately, all of them become known only after a structure is elucidated. Nevertheless, the following properties of the initial data have been identified as factors making solving a problem more difficult:

- A deficit of hydrogen atoms in the molecular formula, and therefore a large value of DBE;
- When the number of experimentally available 2D NMR correlations is markedly less than the number of theoretical correlations for a given structure (discovered *a posteriori*);

- When there is severe signal overlap in the $^1H$ 2D NMR spectra;
- When the 2D NMR data contain nonstandard correlations;
- When the unknown is very large and contains many heteroatoms.

As mentioned earlier the size of the molecule is not a crucial factor: sufficient 2D NMR correlations allow the system to routinely identify large and complex molecules.[9,18] At the same time even molecules of modest size (<15 skeletal atoms) become difficult to identify when there is a high degree of unsaturation. The histogram of molecular masses of the molecules discussed in this article is presented in Figure 12.36. The histogram shows that the majority of structures initially elucidated incorrectly are of modest sizes with molecular masses between 200 and 400 Da.

We conclude that the application of expert systems, such as StrucEluc, could dramatically accelerate the structure elucidation of novel natural products, improve the reliability of identification and reduce the number of publications containing erroneous structures. The examples considered in this Chapter clearly demonstrate that an expert system, previously referred to as an *artificial intelligence system*, is no more than a powerful amplifier of the human intellect. We may expect that as expert system algorithms improve, and computers become faster, then more complex problems will be solvable (as the "gain factor" of the "amplifier" will become higher). We expect that in the near future the further development of expert systems will make such software applications versatile analytical tools that will ultimately become indispensable, not only for structure elucidation but also for the determination of the most probable relative stereochemistry of a newly isolated or synthesized natural product. We also believe that the teaching of CASE methods in universities will help a new generation of chemists to work more efficiently. It will eventually lead to such expert systems becoming routine tools available in the majority of organic and analytical chemistry laboratories.

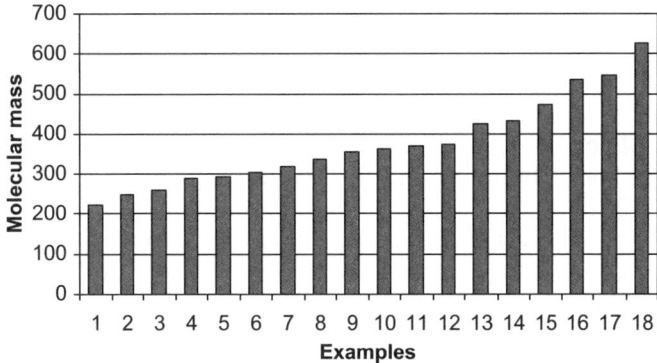

**Figure 12.36** Histogram of molecular masses of examples discussed in this article.

# References

1. M. E. Elyashberg, A. J. Williams and K. A. Blinov, *Nat. Prod. Rep.*, 2010, **27**, 1296–1328.
2. K. C. Nicolaou and S. A. Snyder, *Angew. Chem. Int. Ed.*, 2005, **44**, 1012–1044.
3. M. E. Maier, *Nat. Prod. Rep.*, 2009, **26**, 1105–1124.
4. L. A. Gribov, M. E. Elyashberg and L. A. Moscovkina, *J. Mol. Struct.*, 1971, **9**, 357–371.
5. M. E. Elyashberg, L. A. Gribov and V. V. Serov, *Molecular spectral analysis and computers*, Nauka, Moscow, 1980.
6. M. E. Elyashberg, K. A. Blinov and A. J. Williams, *Magn. Reson. Chem.*, 2009, **47**, 371–389.
7. A. Randazzo, G. Bifulco, C. Giannini, M. Bucci, C. Debitus, G. Cirino and L. Gomez-Paloma, *J. Am. Chem. Soc.*, 2001, **123**, 10870–10876.
8. G. Socrates, *Infrared and Raman Characteristic Group Frequencies: Tables and Charts*, Wiley, Chichester, 2004.
9. M. E. Elyashberg, K. A. Blinov, S. G. Molodtsov, A. J. Williams and G. E. Martin, *J. Chem. Inf. Comput. Sci.*, 2004, **44**, 771–792.
10. M. E. Elyashberg, A. J. Williams and G. E. Martin, *Prog. Nucl. Magn. Reson. Spectrosc.*, 2008, **53**, 1–104.
11. K. A. Blinov, Y. D. Smurnyy, T. S. Churanova, M. E. Elyashberg and A. J. Williams, *Chemometr. Intell. Lab. Syst.*, 2009, **97**, 91–97.
12. *Advanced Chemistry Development. ACD/NMR Predictors. Prediction suite includes 1H, 13C, 15N, 19F, 31P NMR prediction*, http://www.acdlabs.com.
13. M. E. Elyashberg, K. A. Blinov, A. J. Williams, S. G. Molodtsov and G. E. Martin, *J. Chem. Inf. Model.*, 2007, **47**, 1053–1066.
14. C. D. Monica, A. Randazzo, G. Bifulco, P. Cimino, M. Aquino, I. Izzo, F. De Riccardisc and L. Gomez-Paloma, *Tetrahedron Lett.*, 2002, **43**, 5707–5710.
15. M. E. Elyashberg, Y. Z. Karasev and R. Martirosian, *Analyt. Chim. Acta*, 1999, **388**, 353–363.
16. E. Sakuno, K. Yabe, T. Hamasaki and H. Nakajima, *J. Nat. Prod.*, 2000, **63**, 1677–1678.
17. M. E. Elyashberg, K. A. Blinov, A. J. Williams, S. G. Molodtsov and G. E. Martin, *J. Chem. Inf. Model.*, 2006, **46**, 1643–1656.
18. K. A. Blinov, D. Carlson, M. E. Elyashberg, G. E. Martin, E. R. Martirosian, S. G. Molodtsov and A. J. Williams, *Magn. Reson. Chem.*, 2003, **41**, 359–372.
19. P. Wipf and A. D. Kerekes, *J. Nat. Prod.*, 2003, **66**, 716–718.
20. O. M. Cóbar, A. D. Rodriguez, O. L. Padilla and J. A. Sanchez, *J. Org. Chem.*, 1997, **62**, 7183–7188.
21. Y.-P. Shi, A. D. Rodriguez and O. L. Padilla, *J. Nat. Prod.*, 2001, **64**, 1439–1443.
22. P. Ralifo and P. Crews, *J. Org. Chem.*, 2004, **69**, 9025–9029.

23. N. Aberle, S. P. B. Ovenden, G. Lessene, K. G. Watson and B. J. Smith, *Tetrahedron Lett.*, 2007, **48**, 2199–2203.

24. K. N. White, T. Amagata, A. G. Oliver, K. Tenney, P. J. Wenzel and P. Crews, *J. Org. Chem.*, 2008, **73**, 8719–8722.

25. A. Buske, S. Busemann, J. Mühlbacher, J. Schmidt, A. Porzel, G. Bring-mann and G. Adam, *Tetrahedron*, 1999, **55**, 1079–1086.

26. G. Bringmann, J. Schlauer, H. Rischer, M. Wohlfarth, J. Mühlbacher, A. Buske, A. Porzel, J. Schmidt and G. Adam, *Tetrahedron*, 2000, **56**, 3691–3695.

27. P.-W. Hsieh, F.-R. Chang, K.-H. Lee, T.-L. Hwang, S.-M. Chang and Y.-C. Wu, *J. Nat. Prod.*, 2004, **67**, 1175–1177.

28. I. Wetzel, L. Allmendinger and F. Bracher, *J. Nat. Prod.*, 2009, **72**, 1908–1910.

29. P.-L. Wu, Y.-L. Hsu and C.-W. Jao, *Nat. Prod.*, 2006, **69**, 1467–1470.

30. J. J. Mason, J. Bergman and T. Janosik, *J. Nat. Prod.*, 2008, **71**, 1447–1450.

31. P. Sharma and M. J. Alam, *Chem. Soc., Perkin Trans.*, 1988, **1**, 2537.

32. L. A. Paquette, O. M. Moradei, P. Bernardelli and T. Lange, *Org. Lett.*, 2000, **2**, 1875–1878.

33. D. Friedrich, R. W. Doskotch and L. A. Paquette, *Org. Lett.*, 2000, **2**, 1879–1882.

34. D. Friedrich and L. A. Paquette, *J. Nat. Prod.*, 2002, **65**, 126–130.

35. Y. Sakano, M. Shibuya, Y. Yamaguchi, R. Masuma, H. Tomada, S. Omura and Y. Ebizuka, *J. Antibiot.*, 2004, **57**, 564–568.

36. B. B. Snider and X. Gao, *Org. Lett.*, 2005, **7**, 4419–4422.

37. I. H. Hardt, P. R. Jensen and W. Fenical, *Tetrahedron Lett.*, 2000, **41**, 2073–2076.

38. J. A. Kalaitzis, Y. Hamano, G. Nilsen and B. S. Moore, *Org. Lett.*, 2003, **5**, 4449–4452.

39. R. Suemitsu, K. Ohnishi, M. Horiuchi, A. Kitagichi and T. Odamura, *Phytochemistry*, 1992, **31**, 2325–2326.

40. M. Horiuchi, T. Maoka, N. Iwase and K. Ohnishi, *J. Nat. Prod.*, 2002, **65**, 1204–1205.

41. T. Komoda, Y. Sugiyama, N. Abe, M. Imachi, H. Hirota and A. Hirota, *Tetrahedron Lett.*, 2003, **44**, 1659–1661.

42. I. Otani, T. Kusumi, Y. Kashman and H. J. Kakisawa, *Am. Chem. Soc.*, 1991, **113**, 4092–4096.

43. M. E. Elyashberg, K. A. Blinov, A. J. Williams, S. G. Molodtsov and E. R. Martirosian, *J. Nat. Prod.*, 2002, **65**, 693–703.

44. T. Komoda, Y. Sugiyama, N. Abe, M. Imachi, H. Hirota, H. Koshinoe and A. Hirota, *Tetrahedron Lett.*, 2003, **44**, 7417–7419.

45. J. Cáceres, M. E. Rivera and A. D. Rodríguez, *Tetrahedron*, 1990, **46**, 341.

46. A. D. Rodríguez, A. L. Acosta and H. Dhasmana, *J. Nat. Prod.*, 1993, **56**, 1843–1849.

47. P. Krishnaiah, V. L. N. Reddy, G. Venkataramana, K. Ravinder, M. Srinivasulu, T. V. Raju, K. Ravikumar, D. Chandrasekar, S. Ramakrishna and Y. Venkateswarlu, *J. Nat. Prod.*, 2004, **67**, 1168–1171.

48. M. E. Elyashberg, K. A. Blinov, S. G. Molodtsov, T. S. Churanova and A. J. Williams, *ChemSpider Journal of Chemistry*, 2009, http://www.chemmantis.com/Article.aspx?id=889.

49. J. Hiort, K. Maksimenka, M. Reichert, S. Perović-Ottstadt, W. H. Lin, V. Wray, K. Steube, K. Schaumann, H. Weber, P. Proksch, R. Ebel, W. E. G. Müller and G. Bringmann, *J. Nat. Prod.*, 2004, **67**, 1532–1543.

50. G. Schlingmann, T. Taniguchi, H. He, R. Bigelis, H. Y. Yang, F. E. Koehn, G. T. Carter and N. Berova, *J. Nat. Prod.*, 2007, **70**, 1180–1187.

51. T. A. Johnson, T. Amagata, A. G. Oliver, K. Tenney, F. A. Valeriote and P. Crews, *J. Org. Chem.*, 2008, **73**, 7255–7259.

52. J. Takashima, S. Asano and A. Ohsaki, *Tennen Yuki Kagobutsu Toronkai Koen Yoshishu*, 2000, **42**, 487.

53. G. Hu, K. Liu and L. J. Williams, *Org. Lett.*, 2008, **10**, 5493–5496.

54. J. Takashima, S. Asano and A. Ohsaki, *Planta Med.*, 2002, **68**, 621.

55. M. E. Elyashberg, K. A. Blinov, Y. D. Smurnyy, T. S. Churanova and A. J. Williams, *Magn. Reson. Chem.*, 2010, **48**, 219–229.

56. M. E. Elyashberg, K. A. Blinov, S. G. Molodtsov, Y. D. Smurnyy, A. J. Williams and T. S. Churanova, *J. Cheminformatics*, 2009, http://www.jcheminf.com/content/1/1/3.

57. L. A. Gribov, M. E. Elyashberg and V. V. Serov, *J. Mol. Struct.*, 1978, **50**, 371–387.

58. N. Bohr, *Atomic Physics and Human Knowledge*, Wiley, New York, 1958.

59. K. A. Blinov, M. E. Elyashberg and A. J. Williams, unpublished results.

60. S. G. Molodtsov, M. E. Elyashberg, K. A. Blinov, A. J. Williams, G. E. Martin and B. Lefebvre, *J. Chem. Inf. Comput. Sci.*, 2004, **44**, 1737–1175.

61. A. J. Williams, M. E. Elyashberg, K. A. Blinov, D. C. Lankin, G. E. Martin, W. F. Reynolds, J. A. Porco, C. A. Singleton and S. Su, *J. Nat. Prod.*, 2008, **71**, 581–588.

62. G. Saielli and A. Bagno, *J. Org. Lett.*, 2009, **11**, 1409–1412.

63. J. A. J. Porco, S. Su, X. Lei, S. Bardhan and S. D. Rychnovsky, *Angew. Chem. Int. Ed.*, 2006, **45**, 1–4.

64. S. D. Rychnovsky, *Org. Lett.*, 2006, **8**, 2895–2898.

65. J. J. La Clair, *Angew. Chem., Int. Ed.*, 2006, **45**, 2769–2773.

# CHAPTER 13

# *Comparison of Systematic CASE Systems* versus *a Traditional Approach*

## 13.1 Introduction

As discussed in previous Chapters, molecular structure elucidation remains a major challenge and an important problem in the domain of organic chemistry and molecular spectroscopy. If a researcher is engaged in the investigation of a chemical reaction, then certainly one of the primary aims is to determine the structure of the resulting reaction product(s). In the domain of natural products the isolation and elucidation of chemical structures are major challenges. In both cases the structure determination process, in general, is reduced to forming some structural hypotheses and then their subsequent verification. The generation of structural hypotheses is the initial step in the process of structure elucidation.

Each hypothesis is the result of the comprehensive logical treatment of the available spectral and chemical information associated with the structure under analysis. Nowadays, the main source of spectral information is a combination of 1D and 2D NMR spectra complemented by the molecular formula and fragmentation ions determined from one of the many variants of mass spectrometry.[1] The information obtained from 2D NMR data is frequently rather complicated and fuzzy by nature (see Chapter 2). During the interpretation of 2D NMR data several structural hypotheses are often produced, each of them fitting the experimental data and other available information. Hypothesis generation by humans offers a series of obvious difficulties: (1) there is no guarantee that *all* possible hypotheses will be enumerated; and (2) there are no criteria allowing the selection of the most credible hypotheses. As a result, numerous articles revising previously reported chemical structures are quite

New Developments in NMR No. 1
Contemporary Computer-Assisted Approaches to Molecular Structure Elucidation
By Mikhail Elyashberg, Antony Williams and Kirill Blinov
© Royal Society of Chemistry 2012
Published by the Royal Society of Chemistry, www.rsc.org

common, and many examples can be found in the review published by Nicolaou and Snyder,[2] as well as reported in our work[3] and in Chapter 12.

The main objective of the article by Nicolaou and Snyder[2] was to demonstrate to the chemical community that molecular structure elucidation is not a
routine problem. For complete structure elucidation it is frequently necessary
not only to utilize both spectral and X-ray data but also to confirm the structure
*via* total synthesis. Different research teams can offer different conclusions for a
structure, since investigators working in a traditional manner have no access to
an exhaustive list of all possible candidate structures. Therefore it is necessary
to utilize a method that will systematically generate all possible structural
hypotheses, as well as identifying the most probable structure from this series.
Choosing several of the most probable structures using CASE methods (for
instance, see references[4–7]) can, as shown in previous Chapters, dramatically
reduce the number of hypotheses which should be examined and finally verified.
It can reduce the time consumed by highly qualified specialists, as well as reduce
the number of potential errors of misinterpretation.

When CASE systems are employed the relationships between spectral data
and structural information (the so-called "axioms", see Chapter 2) are used by
CASE programs to derive *all* logical consequences (resulting structures) from
the set of axioms to solve the task which the spectroscopist performs when
elucidating a structure without computational assistance. Any changes in the
set of axioms (varying assumptions, adding new assumptions, *etc.*) influence the
set of final structural hypotheses. Alternatively, if some new hypothetical
structures are added by the researcher to an earlier set of candidate structures,
then it is necessary to verify that they are in agreement with the initial axioms.
If a new candidate structure contradicts the initial axioms then either the new
suggestion is erroneous or the given axioms should be revised. For example, a
new candidate structure with a molecular formula or molecular mass differing
from the experimentally determined values is only feasible when the corresponding MS data are in doubt.

As discussed earlier a CASE expert system could produce an output file
containing many tens of thousands of structures. All of them may formally
meet the criteria and constraints provided by *all* of the 2D NMR correlations,
but the chemical shift assignments of the carbon and hydrogen nuclei for the
majority of structures usually contradicts a number of the spectrum–structure
correlations. A credible final structure or set of structures containing the correct
structure can be distinguished from the full set of candidates using NMR
spectrum prediction.

During the last decade, there has been a significant growth in the number of
publications devoted to the application of quantum mechanical (QM) chemical
shift calculations for identifying the most credible structure(s). It has been
shown[8–13] that if an adequate calculation protocol is chosen a QM approach
provides calculation accuracy that is, in general, enough for the successful
validation of candidate structures and, in particular, for the revision of structures which were originally determined incorrectly. One of the most recent and
striking examples of the successful application of the QM approach was

described by Rychnovsky[14] who refuted the incorrect structure of hexacyclinol proposed by Schlegel *et al.*[15] and suggested another structure which was then unambiguously proven by synthesis and X-ray analysis.[16] A systematic method to confirm the structure of hexacyclinol using Structure Elucidator (StrucEluc) was also described elsewhere.[17] QM chemical shift prediction is rather time consuming relative to other approaches. Different authors report different time costs for such calculations, depending on the molecular size, flexibility, and the number of possible conformers for which chemical shifts should be predicted. The processing time can vary from several hours to several tens of hours. It is therefore necessary to reduce the number of candidate structures to which QM calculations should be applied as much as possible before starting a series of calculations. It is natural to expect that prior to performing QM chemical shift calculations (if necessary) a minimum set of candidate structures should be chosen on the basis of fast chemical shift prediction by empirical methods. Structure generation and subsequent ranking of the candidate structures in descending order of their probability consumes only a few minutes on a modern processor using today's expert systems.[18] Obviously, the application of QM methods can play a decisive role in such cases when the analyzed structures contain "exotic" fragments that were absent from the training set (see Chapter 3). Another situation whereby QM prediction could help to identify a preferable structure may occur when the differences between the experimental NMR chemical shifts and those calculated by empirical methods for several "best structures" are too small to enable a certain choice.

In a series of publications the potential of QM methods as an analytical tool was evaluated on molecules for which the number of heavy atoms was most frequently around 20 and rarely reached 30 atoms. Meanwhile, for example, many natural products molecules contain 40–100 or more heavy atoms and the prediction of chemical shifts for such molecules is not attainable by QM methods as yet. For large molecules we can only rely on empirical methods for chemical shift prediction.

It is worthy to note that some publications (for example, see reference[19]) devoted to structure elucidation assisted by QM chemical shift prediction frequently do not mention empirical methods at all. In many cases it is seemingly alleged that the QM approach is the unique prediction method for proving or disproving a proposed structure. Other works (for example[20,21]) compare the accuracy of QM methods with the accuracy of older versions of empirical approaches[22] that do not appropriately represent the performance of contemporary programs.[23–27] The capabilities of the newer empirical methods for identifying the most probable structure within a set of proposed structures compared with results attained using QM methods was reported previously.[28] Some of the most interesting examples considered in the reported work will be discussed here.

The authors[28] selected a series of articles in which the QM approach was successfully used to distinguish the correct structure among a series of suggested molecules or for revising the originally hypothesized chemical structure. For each case, if 2D NMR data were available then an attempt to solve the problem systematically was made using the expert system StrucEluc.[29,30] It was found that the right structure was also assigned as the most probable one in the examples considered by both QM and fast empirical NMR chemical shift predictions,[23,31] whereas alternative and incorrect structures suggested by researchers were ranked lower. The examples studied confirm the rationale of a general approach (see Chapter 3) in which the most probable structure is established as a result of the joint application of a CASE expert system in combination with both empirical and QM methods for chemical shift prediction.

## 13.2 Problems Solved in the Common Mode of the StrucEluc Expert System

In this Section we will consider several examples which demonstrate how the problems could be solved using the StrucEluc system operating in the Common Mode.

*Example 1*
Kim *et al.*[32] separated two new natural products and attributed them to boletunone A (**13.1**) and boletunone B (**13.2**) using 1D and 2D NMR data. Both structures are presented with the $^{13}$C chemical shift assignment suggested by authors:[32]

**13.1**                                         **13.2**

Steglich and Hellwig[33] have shown that structures **13.1** and **13.2** are wrong, and the following alternative structural formulae for these compounds, **13.3** and **13.4**, were offered and proven:

**13.3**                                         **13.4**

For boletunone A, it has been shown[34] (see also Chapter 9) that application of StrucEluc allowed reliable determination of the right structure **13.3** in 0.15 s. It was also demonstrated how the problem of boletunone B could be solved very quickly and correctly in a systematic way. Since the boletunone B story was used by Bagno *et al.*[8] for challenging DFT chemical shift calculations considered as an analytical tool, we will explain here the solution of this problem in more detail.

Bagno *et al.*[8] showed that QM-based NMR chemical shift prediction was capable of distinguishing between the structural hypotheses of **13.2** and **13.4** and could determine the genuine structure. It was demonstrated that a quantum chemical approach could be used to resolve disputes regarding the acceptance or rejection of different isomeric structures. To calculate the $^1$H and $^{13}$C spectra for both the original proposal **13.2** and the revised structure **13.4,** the geometries were optimized at the B3LYP/6-31G(d,p) level in the gas phase, and the NMR properties were calculated with B3LYP/cc-pVTZ both in the gas phase and with a solvent reaction field of DMSO.

To make a choice as to the most favorable of the two structures, the following statistical characteristics were used: the parameters $a$ and $b$ for the linear regression equation $\delta_{calcd} = a + b\delta_{exptl}$; the correlation coefficient, $R^2$; the mean absolute error (MAE) defined as $\Sigma_n|\delta_{calcd} - \delta_{exptl}|/n$; the *corrected* mean absolute error (CMAE) defined as $\Sigma_n|\delta_{corr} - \delta_{exptl}|/n$, where $\delta_{corr} = (\delta_{calcd} - a)/b$ and therefore corrects for systematic errors. The corrected chemical shifts are referred to as the scaled shifts.

The authors[8] found that for the correct structure (**13.4**) MAE($^{13}$C) = 7.2 ppm, CMAE($^{13}$C) = 1.9 ppm and $R^2$ = 0.9984, whereas for the wrong structure, **13.2**, these values were equal to 6.0 ppm, 3.7 ppm and 0.9952 correspondingly. These parameters indicate that structure **13.4** is preferable. The calculation of the $^1$H chemical shifts and coupling constants confirmed that the revised structure **13.4** is in better agreement with the experimental spectrum than the originally proposed structure **13.2**. An attempt to improve prediction accuracy by optimizing the structures at the MP2/cc-pVDZ level indicated that the changes in geometry were very small and NMR properties were very similar.

For systematic elucidation of the boletunone B structure, the 1D and 2D NMR data associated with boletunone B were input into the StrucEluc software. The $^1$H signal multiplicities determined by Kim *et al.*[32] were introduced for *all* methyl groups and for two CH groups [$\delta C = 77.4$, $\delta H = 4.34$(d) and $\delta C = 71.9$, $\delta H = 3.77$(s)]. The multiplicities of these groups had been determined with sufficient reliability. During the first program run strict structure generation (see Chapters 9 and 10) was performed [*i.e.* no non-standard correlations (NSCs) are assumed in 2D NMR data]. The result was: $k = 142 \rightarrow 57 \rightarrow 57$, $t_g = 3.8$ s.

Neither the **13.2** nor **13.4** structures were found among the 57 structures since the presence of one NSC in the HMBC NMR data was postulated both in the original and revised structures. $^{13}$C chemical shift prediction was performed for all structures in the output file using the neural net (NN) approach and all structures were ranked in ascending order of the average chemical shift deviation $d_N(^{13}C)$ calculated for each structure. The top structures contained within the ranked structure file are presented in Figure 13.1.

The average deviation value of the first ranked structure is twice as large as the average deviation of the NN method (1.6–1.8 ppm), which indicates a need to repeat structure generation in fuzzy mode[18] (see Chapter 10). Structure generation was repeated in this mode allowing one NSC in the HMBC data as assumed by both groups of investigators.[32,33] This gave the following result: $k = 1211 \rightarrow 383 \rightarrow 374$, $t_g = 14$ s.

The top ranked structures in the file are shown in Figure 13.2.

Figure 13.2 shows that the revised structure is identical to the best structure and the difference between the average deviations calculated for the second and first structures is significant.[18] The original structure **13.2** was not found in the output file, since it contradicts the multiplicity of the $^1H$ signal at $\delta = 3.77$ ppm; a singlet is observed in the $^1H$ spectrum, whereas a doublet should be expected if structure **13.2** was correct. Fuzzy structure generation (FSG) performed without

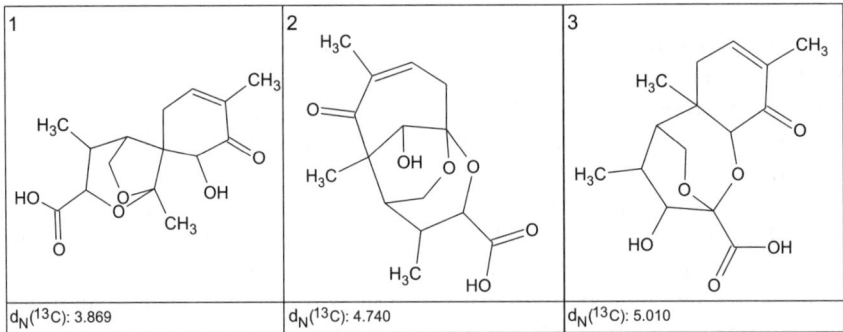

**Figure 13.1**   The three first structures of the file produced by strict structure generation from the NMR data measured for boletunone B. The structures are ranked in ascending order of average deviation $d_N(^{13}C)$.

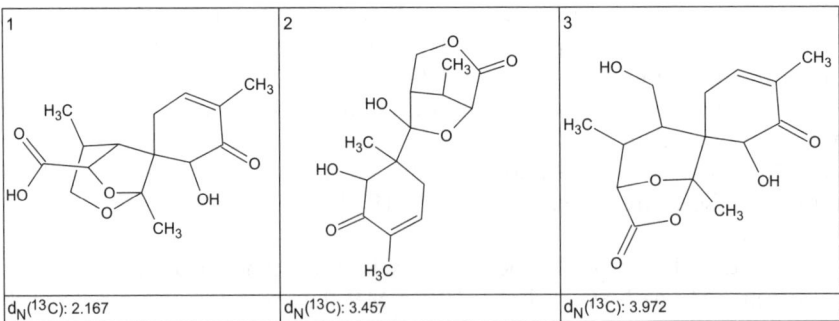

**Figure 13.2**   The top ranked structures in the output file resulted from fuzzy structure generation using the NMR data obtained for boletunone B. The top-ranked structure is identical to the revised structure **13.4**.

taking into account the multiplicities ($k = 32\,814 \rightarrow 8621 \rightarrow 7533$, $t_g = 1$ min 10 s) produced structure **13.2** and it was ranked in third position in the resulting structure file ($d_N{}^{13}C$ is equal to 3.08 ppm for this structure). Direct comparison of the linear regression parameters calculated for structures **13.2** and **13.4** shows that $R^2(\mathbf{13.4}) = 0.999$ and $R^2(\mathbf{13.2}) = 0.995$ and these parameters are practically the same as those found by Bagno *et al.*[8] The application of the systematic approach outlined here would allow researchers to immediately identify the correct structure that could then be further justified and investigated in more details by quantum chemical calculations if it were deemed necessary.

*Example 2*

As well as articles devoted to the evaluation of DFT chemical shift calculation as a potential tool for identification of a proposed structure, there are also publications in which this approach has been applied to solving real chemical problems. Recently, Sanz *et al.*[19] employed $^{13}C$ and $^{15}N$ chemical shift calculations by the GIAO approximation of the DFT method to choose between the structures forming two pairs of isomers (**13.5** or **13.6**) and (**13.7** or **13.8**):

13.5         13.6         13.7         13.8

The authors[19] synthesized and separated two samples, #1 and #2, with the molecular formulae of $C_{19}H_{12}F_3N_3$ (1) and $C_{17}H_{10}F_3N_3S$ (2). For the purpose of structure determination and confirmation, 1D NMR spectra ($^1H$, $^{13}C$, $^{15}N$ and $^{19}F$) in combination with gradient-selected (gs) $^1H$-$^{13}C$ gs-HMBC and $^1H$-$^{15}N$ gs-HMBC data were used. The authors[19] derived the two pairs of alternative isomers shown above and the problem was reduced to selection of the correct structure within each of the isomer pairs. For this purpose, the authors calculated both the $^{13}C$ and $^{15}N$ chemical shifts using the GIAO method of DFT approximation for two model compounds:

13.9         13.10

The model compounds **13.9** and **13.10** were used in order to simplify the QM computations. Linear regressions of the calculated *versus* experimental $^{13}C$ chemical shifts, the latter assigned for the proposed structures, allowed the authors[19] to confirm configuration **13.9** as the most preferable [$R^2(\mathbf{13.9}) = 0.998$; $R^2(\mathbf{13.10}) = 0.984$]. $^{15}N$ chemical shift prediction resulted in $R^2(\mathbf{13.9}) = 0.992$ and $R^2(\mathbf{13.10}) = 0.991$ values for the corresponding model structures and only slightly supports model **13.9** as the preferred structure. In order to show how the problem could be solved without utilizing model structures and QM calculations the structure elucidation process was repeated for both samples using the systematic approach.

The experimental 1D and 2D NMR data and the molecular formulae obtained for the two samples were input into the StrucEluc system and the two problems, 1 and 2, were solved. For the structure elucidation process both $^1H$-$^{13}C$ and $^1H$-$^{15}N$ HMBC correlations were used as initial data.

*Problem 1*
Strict structure generation gave the following result: $k = 106 \rightarrow 44 \rightarrow 20$, $t_g = 27$ s. For all of the structures, $^1H$, $^{13}C$, and $^{15}N$ chemical shifts were calculated using the NN approach and the resulting structure file was ranked in ascending order of $d_N(^{13}C)$. The first three structures of the ranked file are presented in Figure 13.3.

The Figure shows that all average deviations – $d_N(^{13}C)$, $d_N(^1H)$, $d_N(^{15}N)$ and $d_I(^{13}C)$-indicate that structure **13.5** is the most probable one. The values $R^2(\mathbf{13.5}) = 0.984$ and $R^2(\mathbf{13.6}) = 0.895$ calculated for $^{13}C$ prediction by the NN method supports the carbon atom assignment that was performed auto-matically and the conclusion of Sanz *et al.*[19]

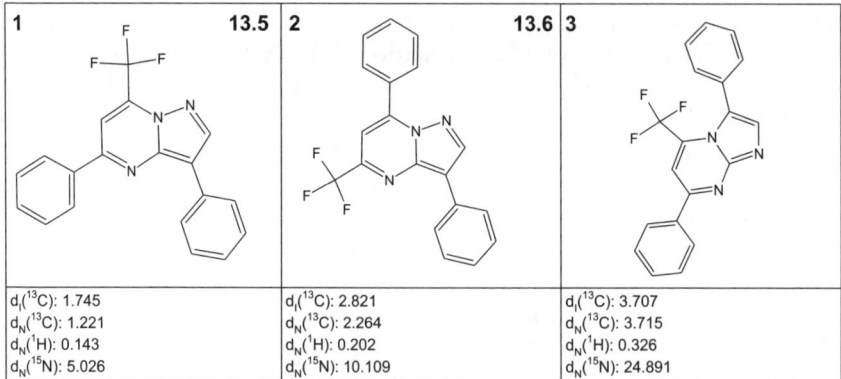

| 1 | 13.5 | 2 | 13.6 | 3 |
|---|---|---|---|---|
| $d_I(^{13}C)$: 1.745 | | $d_I(^{13}C)$: 2.821 | | $d_I(^{13}C)$: 3.707 |
| $d_N(^{13}C)$: 1.221 | | $d_N(^{13}C)$: 2.264 | | $d_N(^{13}C)$: 3.715 |
| $d_N(^1H)$: 0.143 | | $d_N(^1H)$: 0.202 | | $d_N(^1H)$: 0.326 |
| $d_N(^{15}N)$: 5.026 | | $d_N(^{15}N)$: 10.109 | | $d_N(^{15}N)$: 24.891 |

**Figure 13.3** The first three structures of the ranked structural file produced as a solution for Problem 1. The correct structure is confirmed by average deviations calculated for $^1H$, $^{13}C$ and $^{15}N$ spectra. $d_I(^{13}C)$ denotes the average deviation found when the incremental method of chemical shift prediction was employed.

*Problem 2*

The first program run was executed using the strict mode of structure generation. Only one structure, **13.11,** was produced in 7 s. The very large values of the calculated average deviations [$d_N(^{13}C) \sim 9$ ppm, $d_N(^{15}N) \sim 65$ ppm] suggested that structure **13.11** was likely incorrect and, consequently, non-standard HMBC connectivities might exist within the 2D NMR data.

**13.11**

Therefore, FSG was performed under the conditions $m = 1$–15, $a = x$, that is the possible number of NSCs is allowed to be between 1 and 15, and lengthening of connectivities is replaced by their removal during structure generation. This mode allows problems to be solved when 2D NMR data contain an unknown number of NSCs having unknown lengths (see Chapter 10). The following result was obtained: $k = 1481 \rightarrow 415 \rightarrow 164$, $t_g = 11$ min.

Chemical shift prediction and structure ranking promoted structure **13.7** to the first ranked position, whereas structure **13.8** was ranked as fourth. Both structures with average deviations are presented in Figure 13.4.

In this case the preferable structure was also indicated by all NMR shift predictions having the lowest deviations, as well as the calculated values of $R^2$**(13.7)** $= 0.974$ and $R^2$**(13.8)** $= 0.807$. Additional evidence for structure **13.7** is the need for a seven-bond NSC in structure **13.8**.

In the examples discussed in this Section empirical methods of chemical shift prediction are not only dramatically faster, but, in definite cases, are even more reliable because no structural models simplifying the calculations were utilized. Nevertheless, the coincidence of structural assignments made by both QM and empirical approaches shows that researchers can choose the method that is more attractive for them.

# 13.3   Problems Solved in Fragment Mode

The analytical process taken by experts for molecular structure elucidation from 2D NMR data has been described in a series of books (for example, see references[35–37]). The spectroscopist usually tries to assemble some fragments from atoms and their spectral signals, and then commonly combine the fragments using HMBC correlations until a complete molecular structure or set of plausible structures are constructed. As previously discussed, there is no

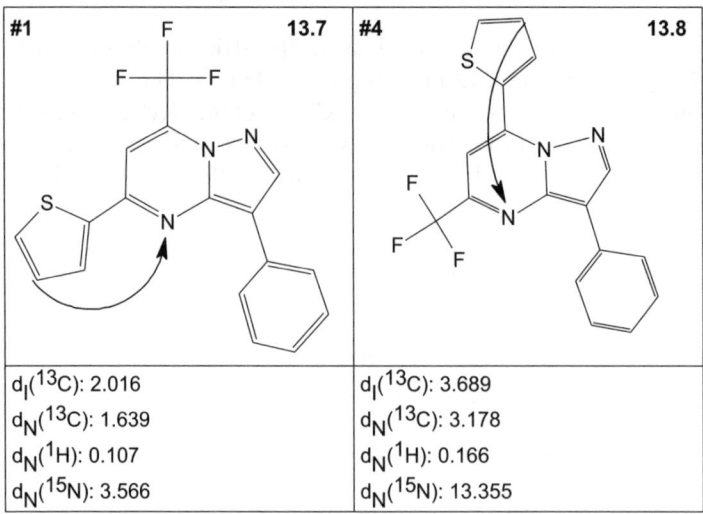

**Figure 13.4** Structures **13.7** and **13.8** proposed in the work of Sanz *et al.*[19] with their calculated average deviations. The arrows show long-range $^1$H-$^{15}$N HMBC correlations of $^5J_{NH}$ for structure **13.7** and $^7J_{NH}$ for structure **13.8**. For these structures values of $R^2$(**13.7**) = 0.974 and $R^2$(**13.8**) = 0.807 were calculated.

guarantee that all possibilities will be taken into account by the process of manual structure assembly. Moreover, some superfluous structural hypotheses may be suggested. If selection of the right structure is performed by QM chemical shift prediction then additional time will be consumed for checking these structures. In this Section we will demonstrate how an expert system supplied with new empirical chemical shift prediction methods can be used to assist the spectroscopist in obtaining the right solution in an optimal manner.

*Example 1*

Balandina *et al.*[10] synthesized a novel quinoxaline and determined its molecular formula $C_{16}H_{10}N_2O_2$ from the MS data [$m/z = 262(M+)$] combined with elemental analysis data. To elucidate the structure of this compound, the authors[10] used $^1$H, $^{13}$C and, $^{15}$N NMR spectra. The assignment of the $^1$H and $^{13}$C NMR spectra was accomplished using data derived from DEPT, 2D COSY-GP, HSQC, and HMBC experiments. Analysis of the NMR data provided two fragments containing H, C, and N atoms with assigned chemical shifts. Three quaternary carbons (151.04, 138.29 and 134.68) without HMBC correlations, one hydrogen atom and two oxygen atoms were not assigned to either of the fragments. The initial data for forming structural hypotheses is presented in Figure 13.5.

Using these data, and some additional chemical considerations, the authors suggested six structures which are presented in Figure 13.6.

To select the right structure, $^1$H, $^{13}$C and $^{15}$N chemical shifts were predicted for structures **13.12–13.17** using the DFT framework and using a hybrid

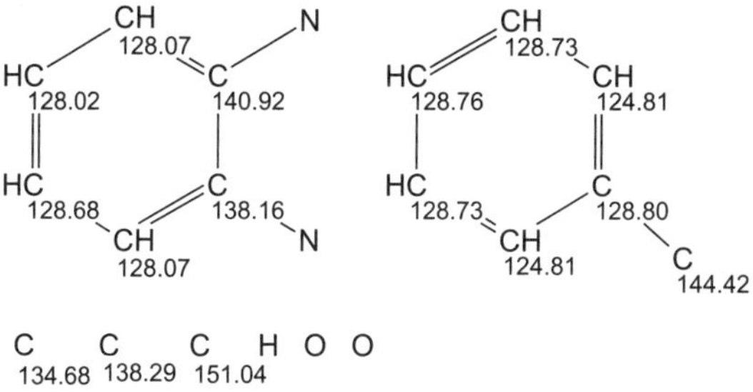

**Figure 13.5** Initial structure information for the generation of structural hypotheses.

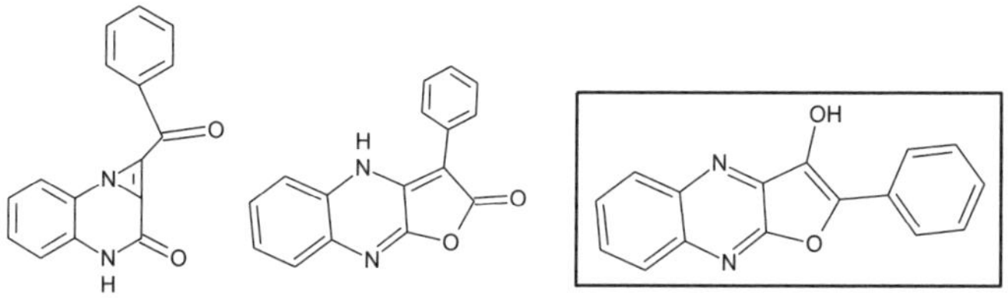

**13.12**          **13.13**          **13.14**

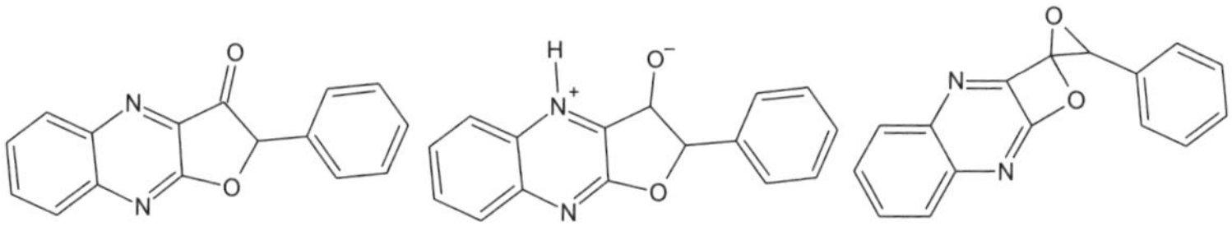

**13.15**          **13.16**          **13.17**

**Figure 13.6** Six suggested structures derived from the experimental data. Structure **13.14** corresponds to the correct structure.

exchange-correlation function, GIAO B3LYP, at the 6-31G(d) level. Full geometry optimizations were performed under *ab initio* RHF/6-31G conditions. Linear correlation coefficients of the experimental *versus* calculated $^{13}$C chemical shifts ($R^2$), root-mean-square errors (rms), slope (a), standard deviations (sd) and mean absolute deviations (MAD $= \Sigma[|\delta_{exp} - \delta_{calc}|]/n$) for structures **13.12–13.17** were computed. As a result, structure **13.14** was identified as the most probable ($R^2 = 0.9758$, rms $= 1.16$ ppm, sd $= 1.2$ ppm, MAD $= 7.03$ ppm). Other proposed structures were rejected by the authors due to smaller $R^2$ values ($R^2 = 0.01$–0.57) and larger deviations. It should be noted that the $R^2$ values have a reasonable interpretation only in those cases when *experimental* chemical shifts are assigned to the atoms of competing structures. Otherwise, selection of the preferable structure can be attained only by simple comparison of the experimental with the calculated spectrum and based on determining outliers. Obviously, the application of an expert system for structure elucidation provides chemical shift assignments that agree with the

2D NMR correlations and consequently the selection of the best structure occurs automatically.

To solve this problem using a CASE approach, spectral data presented in the work[10] were entered into the StrucEluc system. The fragments and atoms shown in Figure 13.5 were eventually transformed into a molecular connectivity diagram (MCD). The atom properties for three carbon atoms not included into the fragments were automatically set as $sp^2$/*not defined* (atom hybridization is $sp^2$, possibility of neighboring heteroatoms is not defined). Structure generation was performed in the automatic mode and FSG was allowed. The following result was obtained: $k = 247 \rightarrow 16 \rightarrow 4$, $t_g = 1$ s.

Empirical chemical shift prediction was performed for all nuclei. Subsequent structural ranking by $d_N(^{13}C)$ deviation resulted in the structure ordering shown in Figure 13.7.

Structure **13.14** is the best structure according to the shift predictions for all nuclei presented in Figure 13.7. Moreover, the deviations for structure **13.14** are dramatically smaller than those for the next (#2) ranked structure for all nuclei and suggests a high reliability for the solution.[18] Note that the $d_N(^{13}C)$ deviation for structure **13.14** is almost four times smaller than the average deviation calculated by the GIAO approach. Figure 13.8 shows the chemical shift assignments performed for structure **13.14** and deduced by the authors[10] and listed with deviations calculated by StrucEluc.

All deviation values calculated for the new assignment [including the deviations $d(^{15}N)$] are smaller than those found for the former assignment. The largest errors are related to the carbons at 151.4 and 138.29 ppm, differences of 11 and 16.5 ppm. It is also interesting to note that all suggested structures **13.12**–**13.17**, except structure **13.14** were not generated by the program, since the atom property correlation table (APCT) prevents the assembly of structures whose atoms would have chemical shifts differing dramatically from the experimental shifts.

For completeness, structure generation was repeated with both the APCT and filters (structural and spectral) switched off. Finally, 59 non-isomorphic structures were generated, including structures **13.12** and **13.13** characterized

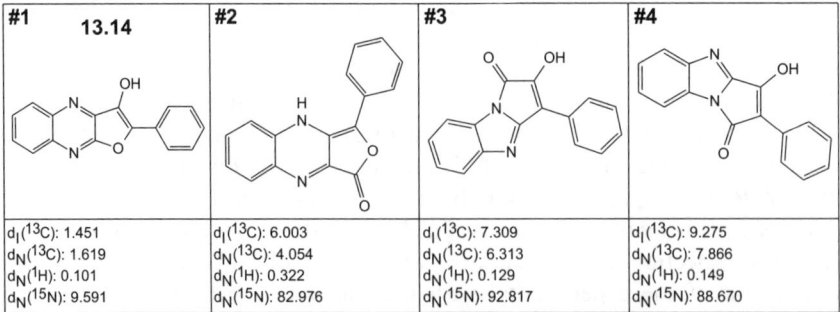

**Figure 13.7**   Output structural file ranked by $d_N(^{13}C)$ deviation.

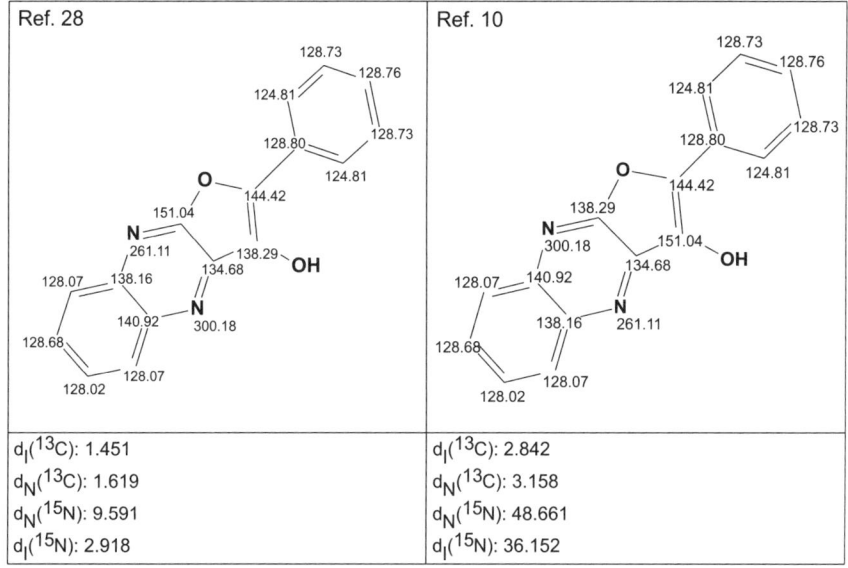

| | |
|---|---|
| $d_I(^{13}C)$: 1.451 | $d_I(^{13}C)$: 2.842 |
| $d_N(^{13}C)$: 1.619 | $d_N(^{13}C)$: 3.158 |
| $d_N(^{15}N)$: 9.591 | $d_N(^{15}N)$: 48.661 |
| $d_I(^{15}N)$: 2.918 | $d_I(^{15}N)$: 36.152 |

**Figure 13.8** The chemical shift assignments for structure **13.14** automatically determined in work[28] and deduced by the original article authors.[10] The deviation values indicate that some chemical shifts assigned on the basis of QM prediction should be exchanged.

by $^{13}C$ chemical shift deviations of 6–8 ppm. Structures **13.15** and **13.17** could not be generated because according to the initially postulated conditions the mono-substituted benzene ring is connected to the carbon atom with a shift of 144.42 ppm whose hybridization was assigned as $sp^2$ ($sp^3$ is necessary for **13.15** and **13.17**).

Using the $^{13}C$ chemical shift values predicted for structure **13.14** by DFT and NN methods linear regressions were calculated for the experimental shift values based on automatic shift assignments relative to the assignments suggested by the authors,[10] and as shown in Figure 13.9.

Figure 13.9 shows that for chemical shifts calculated by the NN algorithm the regression line is very close to the $Y = X$ line. The DFT line lies 7 ppm below the $Y = X$ line. We can observe that the value of $R^2$(DFT) suggests only an acceptable *linear correlation* between the experimental and predicted shifts. However, it does not characterize the true quality of prediction. Consequently, the $R^2$ criterion should not be considered as a measure of intrinsic prediction precision for a given method. As our experience shows, the average deviation is an effective and rather reliable criterion for selection of the most probable structure. The difference between the DFT- and NN-regression lines can be accounted for by systematic errors in the calculations performed with the GIAO approximation. These errors may be different for different molecules, different versions of the DFT based programs and for different ranges of $^{13}C$ NMR spectrum.

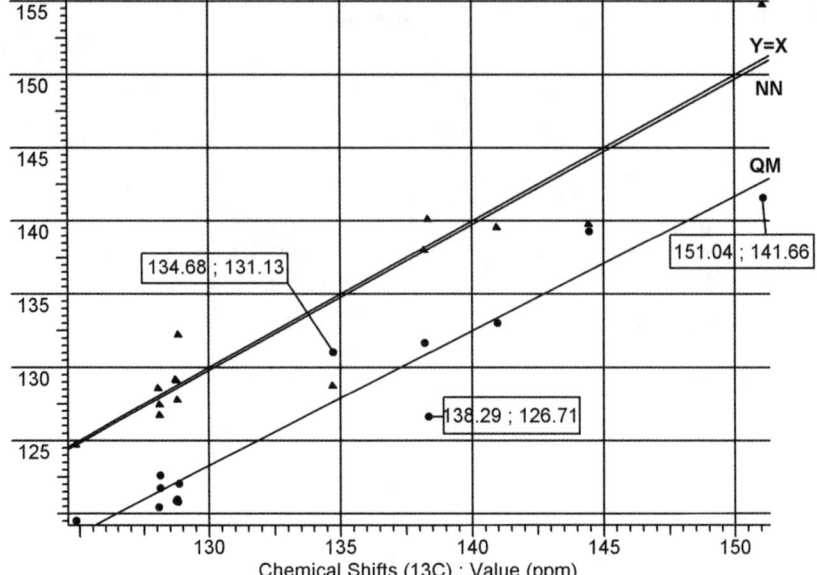

Chemical Shifts (13C) : Value (ppm)

Database: Proposed Structures
- ● Chemical Shifts (13C) : QM Calc. (ppm) (Current Record) (16 pts)
- ▲ Chemical Shifts (13C) : NN Calc. (ppm) (Current Record) (16 pts)

**Figure 13.9**  Correlation plots of $^{13}$C chemical shift values predicted for structure **13.14** (automatic assignment) by the DFT and NN methods *versus* the experimental shift values. The target line $Y = X$ is shown. The coordinates of some points are shown in frames where the *first* value designates the experimental shift and the *second* represents the calculated shift. The regression parameters are: $R^2(NN) = 0.905$, $R^2(QM) = 0.937$, rms(NN) = 2.54 ppm and rms(QM) = 1.89 ppm.

Figure 13.10 confirms the conclusion regarding the chemical shift assignment given in reference:[10] both NN and QM chemical shift predictions indicate that the assignment is incorrect. As the article does not contain an atom numbering scheme, perhaps some confusion in the carbon atom labeling arose when spectral information was listed in the article.[10] Finally, the $^{13}$C NMR chemical shifts for structures **13.12**–**13.17** were predicted and then graphically compared the predicted spectra with the experimental one as shown in Figure 13.11.

The difference between the experimental and NN-predicted spectra is dramatic for all structures, except structure **13.14**. All incorrect structures could be immediately rejected before performing QM calculations and the right structure would be quickly identified if hypotheses were offered by a human expert.

The authors[10] note that "an attempt to predict $^1$H and $^{13}$C NMR shifts values of structure **13.14** based on additivity rules ... would be totally unsuccessful" because ... "estimation of chemical shifts according to the additive scheme implemented in the 'estimate' utility of CambridgeSoft's ChemDraw Program[22] gives very poor prediction of $^{13}$C chemical shifts". With a more appropriate prediction algorithm,[23–25,27,38] this comment is obviously invalid.

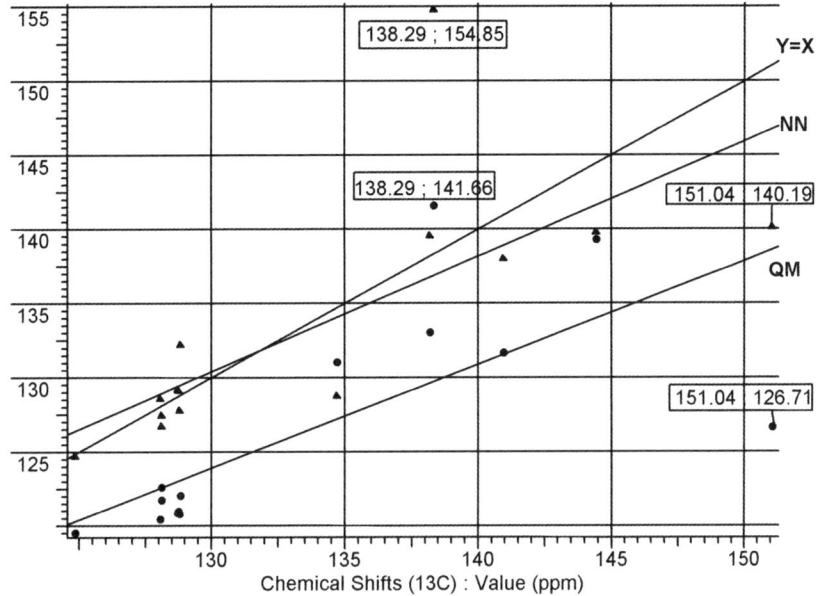

Database: Proposed Structures
- Chemical Shifts (13C) : QM Calc. (ppm) (Current Record) (16 pts)
- Chemical Shifts (13C) : NN Calc. (ppm) (Current Record) (16 pts)

**Figure 13.10**  The correlation plots of the $^{13}$C chemical shift values predicted for structure **13.14** (assignment given in reference[10]) by DFT and NN methods *versus* the experimental shift values. The target line $Y = X$ is shown. The coordinates of some points are shown in frames where the *first* value designates the experimental shift and the *second* represents the calculated value. The regression parameters are $R^2(NN) = 0.551$, $R^2(QM) = 0.538$, rms(NN) = 5.54 ppm and rms(QM) = 5.10 ppm.

We share the authors' enthusiasm regarding progress in QM methods applied to chemical shift prediction. With our results described above, however, the conclusion that "non-empirical calculations of chemical shifts are very cheap in the sense of computational costs and most of the researchers can run them easily on their desk computers (3–5 h per one isomer on a Pentium 4 2.8 GHz processor with 512 MB RAM)" provides a less than ideal situation relative to the use of other computational approaches for NMR prediction.

*Example 2*

The same group of authors[21] reported the application of a more complex molecule structure elucidation strategy that was described in their work.[10] A novel organic compound was investigated by 1D and 2D NMR experiments (DEPT, NOESY, COSY, HSQC, HMBC, and HMBC $^1$H-$^{15}$N). EI mass spectra were recorded on a TRACE MS instrument (Finnigan MAT) and a MALDI mass spectrum was also obtained. From the MS data a molecular formula of $C_{27}H_{22}N_4O_3$ was established. 2D NMR data analysis allowed the authors to assemble three fragments with chemical shifts assigned to $^1$H, $^{13}$C,

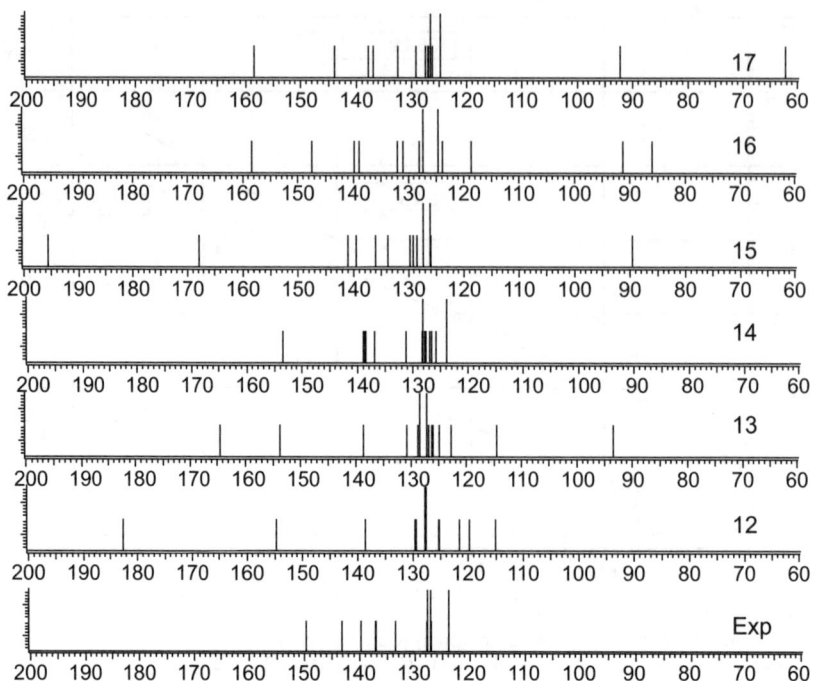

**Figure 13.11**  Experimental $^{13}$C chemical shifts in comparison with the chemical shifts predicted by the NN algorithm for proposed structures **13.12–13.17**.

and $^{15}$N nuclei. Two quaternary C atoms, one N atom, and three O atoms remained unassigned.

The authors[21] suggested two structural hypotheses (**13.19** and **13.20**), which are both in accordance with the molecular formula $C_{27}H_{22}N_4O_3$ and four other structures (**13.15–13.18**) with a molecular formula differing from that determined from the MS data (see Figure 13.12).

To identify the best structure from among the suggestions **13.18–13.23**, Balandina *et al.*[21] calculated the $^{13}$C chemical shifts for all candidate structures using QM methods. As a result of statistical processing of the data, structure **13.23** was selected as the most preferable structure since its correlation coefficient of $R^2 = 0.996$ was the highest in value and the MAD $= 5.65$ ppm was the minimum.

An attempt was made to solve this problem by employing the usual strategy for the StrucEluc system to elucidate unknown structures using the fragment mode. On the basis of spectral data presented in the article,[21] the program produced the MCD shown in Figure 13.13.

The atom corresponding to the chemical shift of 157.77 ppm was introduced with the property $sp^2/not\ defined$, whereas the atom corresponding to 117.38 ppm was assigned the property of *not defined/not defined*, as the indicated chemical shift can be observed either for $sp^2$ or $sp^3$ (O–C–O) hybridized carbons. HMBC connectivities are marked by green lines in Figure 13.13.

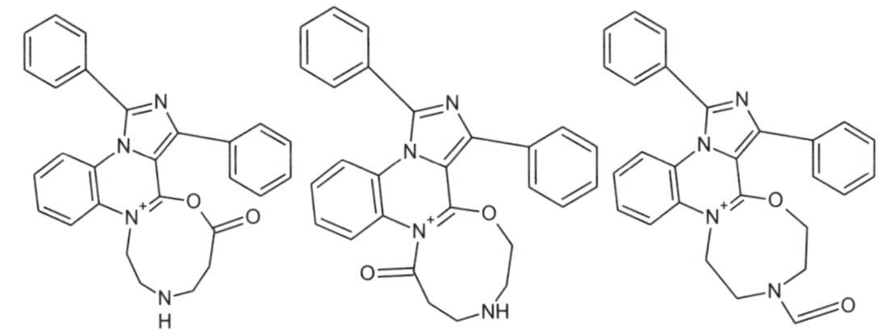

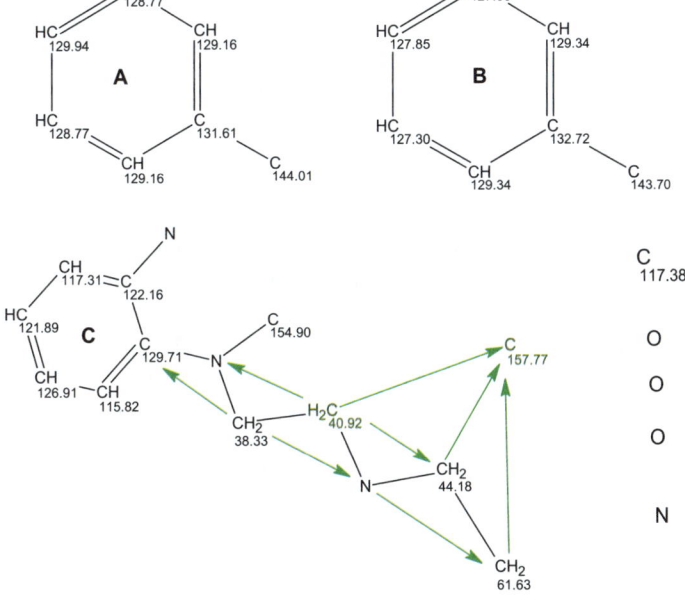

**13.18**: C$_{27}$H$_{23}$N$_4$O$_2$    **13.19**: C$_{27}$H$_{23}$N$_4$O$_2$    **13.20**: C$_{27}$H$_{23}$N$_4$O$_2$

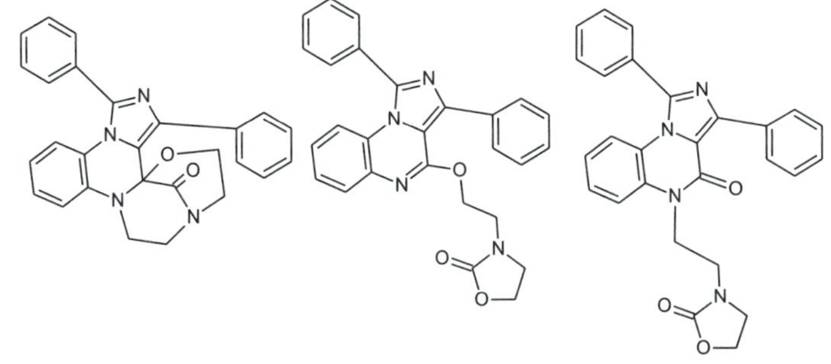

**13.21**: C$_{27}$H$_{22}$N$_4$O$_2$    **13.22**: C$_{27}$H$_{22}$N$_4$O$_3$    **13.23**: C$_{27}$H$_{22}$N$_4$O$_3$

**Figure 13.12**  Structures suggested in the work.[21]

**Figure 13.13**  The MCD containing the three fragments derived by Balandina *et al.*[21] The green arrows denote HMBC correlations.

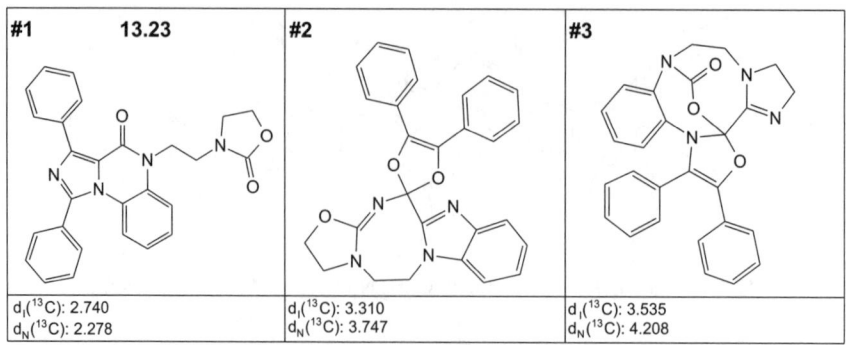

| #1     13.23 | #2 | #3 |
|---|---|---|
| $d_I(^{13}C)$: 2.740<br>$d_N(^{13}C)$: 2.278 | $d_I(^{13}C)$: 3.310<br>$d_N(^{13}C)$: 3.747 | $d_I(^{13}C)$: 3.535<br>$d_N(^{13}C)$: 4.208 |

**Figure 13.14**    The output structural file ranked by $d_N(^{13}C)$ deviation.

Structure generation accompanied by structural filtering was performed under standard conditions. To reduce the output file, three- and four-membered cycles were forbidden and a "Geometry" option was enabled to exclude deliberately "ugly" structures. The results gave: $411 \rightarrow 44 \rightarrow 25$, $t_g = 0.9$ s. For all structures, $^{13}C$ chemical shifts were calculated by both the NN and incremental approaches. The first three structures of the ranked file are shown in Figure 13.14.

Comparison of the best structure #1 of the ranked file with the suggested structures establishes structure #1 as identical to structure **13.23**. Therefore structure **13.23** corresponds to the molecular formula and the constraints graphically represented in the MCD. Structure **13.23** and the largest fragment C depicted in the MCD contains a chain $-N-CH_2CH_2-N-$, whereas this fragment is absent from structure **13.22**. Structure **13.22** is possible only if the initial set of constraints is changed. In this case it would mean that the chain part of the large fragment, which is well confirmed by $^1H-^{13}C$ and $^1H-^{15}N$ HMBC correlations, has been determined incorrectly. There are, however, no grounds for revision of the mentioned chain.

Structures **13.18**–**13.21** all have molecular formulae differing from that determined experimentally. Although possible, it is unlikely. In addition, structure **13.19** contains a fragment that contradicts the side chain of the core fragment C discussed previously. Structure **13.20** would show a doublet signal for the carbonyl group and, consequently, can be rejected for this reason. Finally, the carbon atom at 154.9 ppm in structure **13.21** has to be $sp^3$ hybridized and this is impossible. The structural suggestions are therefore not appropriate. The authors,[21] nevertheless, discussed all six structures as part of their methodological analysis in the publication.

Since structures **13.18** ($C_{27}H_{23}N_4O_2$) and **13.21** ($C_{27}H_{22}N_4O_2$) contain fragment C the structure elucidation process was performed using these molecular formulae to obtain all consequences of replacing these constraints. New MCDs were created and the following results were obtained.

Structure **13.18**: $k = 88$, $t_g = 2.5$ s, with structure **13.18** selected as the best with $d_N(^{13}C) = 4.24$ ppm. Structure **13.18** should be rejected as the deviation is almost twice as large as that obtained for structure **13.23**.

Structure **13.21**: FSG was run with the APCT and spectral filters switched off, since structure **13.21** would not be generated with these filters switched on. The results were: $k = 13826 \rightarrow 13569 \rightarrow 271$, $t_g = 61$ s. $^{13}$C chemical shifts for the resulting 13 569 structures were calculated using the incremental method in 77 s. Structure **13.21** was again selected as the best structure with $d_N(^{13}C) = 4.81$ ppm and $d_I(^{13}C) = 6.07$ ppm. The deviation values are large enough to suggest that, in this case, structure **13.21** has been "forcibly" derived from the available NMR experimental data. It is worthy to note that the predicted chemical shift for the carbon atom at 154.90 ppm differs from the experimental value by $> 70$ ppm.

Balandina *et al.*[21] again compared their results with NMR chemical shift prediction contained within the CambridgeSoft ChemOffice program. They stated that "information on the chemical shifts for the carbon atom of the $N^+=C—O$ fragment is lacking in the database of the ChemOffice program package. Hence, if these structures were formed in the course of the reaction, they could not be established in terms of this empirical approach". Based on the data obtained for structure **13.18**, the current study showed that this limitation was not present in the program.[23,26] In spite of the fact that the examples employed for justification of a methodology based on QM NMR calculations seems to be rather weak, we agree with the conclusion of the authors[21] that the "combined use of modern 2D NMR experiments and *ab initio* chemical shift calculations is efficient". This approach may be the only computational approach if a molecule contains exotic substructures that are unknown for a program based on empirical methods of spectrum prediction. At the same time, CASE expert systems supplied with fast and accurate algorithms for empirical chemical shift prediction can frequently help the researcher to avoid time consuming QM computations.

## 13.4 The Combined Application of Empirical and Non-Empirical Methods of Chemical Shift Assignment

Bagno and Saielli[9] published an interesting overview of advances in quantum chemical prediction of NMR spectra and their parameters. As an example of NMR spectrum prediction performed for natural products, they[9] reported a computational study of the molecule of nimbosodione **13.24**.

**13.24**

Compound **13.24** was originally isolated and identified by Ara *et al.*[39] from spectral data, but, as a result of the total synthesis of nimbosodione and careful analysis of its NMR spectra, Li *et al.*[40] showed that the original NMR spectral data and their assignment were incorrect. Even though Li *et al.*[40] measured the ${}^1$H and ${}^{13}$C NMR spectra of nimbosodione accurately, they failed to complete a new spectral assignment. Bagno *et al.*[9] calculated the ${}^1$H and ${}^{13}$C NMR chemical shifts of structure **13.24** using a DFT approach and clearly showed that the spectral assignment and chemical shifts of several definite atoms reported by Ara *et al.*[39] were incorrect. The absence of a correct spectral assignment prevented the authors[9] from comparing the predicted and experimental chemical shifts.

In order to allow this a search in the ACD/Labs NMR Database[23] for structures having a skeletal framework similar to that of nimbosodione was performed. As mentioned in Chapter 7, the database contains more than 400 000 structures with chemical shifts assigned to both carbon and hydrogen atoms. Five structures were found as a result of the search. These structures (ID 2–6) and their assigned ${}^{13}$C chemical shifts are presented in Figure 13.15. Comparison of the found structures with the nimbosodione structure allowed one to assign the chemical shifts of the compound under investigation (ID 1) as shown in Figure 13.15. To check the assignment, both the ${}^1$H and ${}^{13}$C chemical shifts of nimbosodione were calculated using a NN approach.[26]

The two sets of both old and new assigned experimental ${}^{13}$C chemical shifts, and the corresponding values calculated by DFT and NN approaches, are collected in Table 13.1.

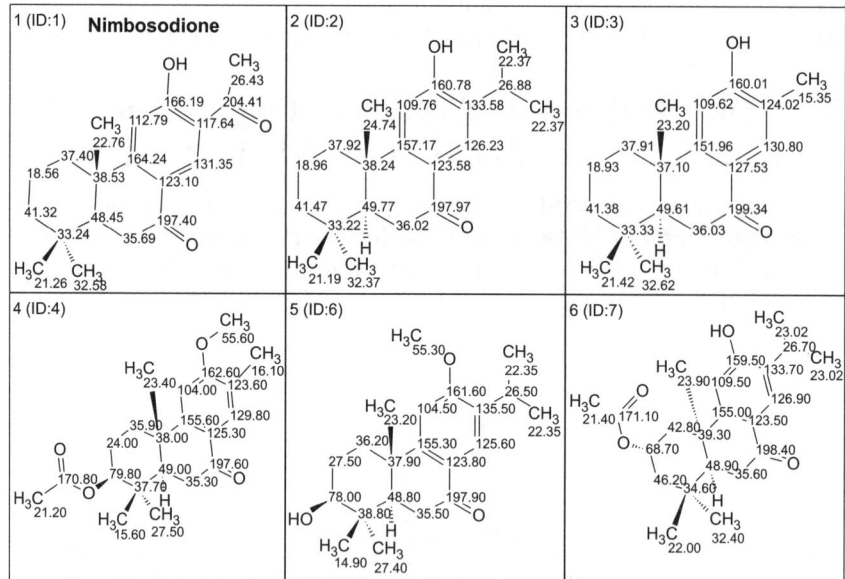

**Figure 13.15**   The structure of nimbosodione (ID 1) assigned in accordance with the experimental NMR spectra reported by Li *et al.*[40] The assignments of carbon atoms of the reference structures (ID 2–6) found in the ACD/ NMR DB were used.

**Table 13.1**    A comparison of experimental (Exp.) and calculated (Calc.) $^{13}$C chemical shifts for nimbosodione.

| Atom | Exp.[39] | Exp.[40] assignment[28] | Calc. QM[9] | Calc. NN |
|------|------|------|------|------|
| C1 | 37.95 | 37.4 | 41.75 | 36.8 |
| C2 | 18.90 | 18.56 | 23.78 | 19.38 |
| C3 | 41.37 | 41.32 | 45.41 | 40.92 |
| C4 | 33.31 | 33.24 | 41.19 | 33.87 |
| C5 | 36.03 | 48.45 | 39.53 | 49.38 |
| C6 | 49.58 | 35.69 | 53.63 | 37.25 |
| C7 | 198.60 | 197.4 | 201.75 | 197.77 |
| C8 | 157.08 | 123.1 | 128.25 | 125.46 |
| C9 | 159.08 | 164.24 | 173.26 | 161.31 |
| C10 | 33.31 | 38.53 | 46.34 | 38.89 |
| C11 | 109.62 | 112.79 | 117.47 | 111.48 |
| C12 | 159.15 | 166.19 | 177.21 | 165.79 |
| C13 | 157.42 | 117.64 | 122.09 | 119.57 |
| C14 | 130.78 | 131.35 | 139.49 | 131.98 |
| C18 | 15.10 | 32.58 | 34.35 | 29.17 |
| C19 | 23.22 | 21.26 | 23.06 | 26.49 |
| C20 | 21.31 | 22.76 | 25.07 | 27.53 |
| COCH3 | 32.59 | 26.43 | 28.07 | 29.48 |
| COCH3 | 198.62 | 204.41 | 212.84 | 205.42 |

Statistical analysis of the data gave the following average chemical shift deviations and $R^2$ values for the old and new assignments:

1. Old assignment – $R^2$(DFT) = 0.979; $R^2$(NN) = 0.981; $d$(DFT) = 10.48 ppm, $d$(NN) = 7.87 ppm.
2. New assignment – $R^2$(DFT) = 0.994; $R^2$(NN) = 0.999; $d$(DFT) = 6.26 ppm, $d$(NN) = 1.72 ppm.

Both methods of chemical shift calculation provide evidence that the old nimbosodione assignment is incorrect. The old (**13.25**) and new (**13.26**) measured chemical shifts, and their assignments, are shown on structures **13.25** and **13.26**.

**13.25**                      **13.26**

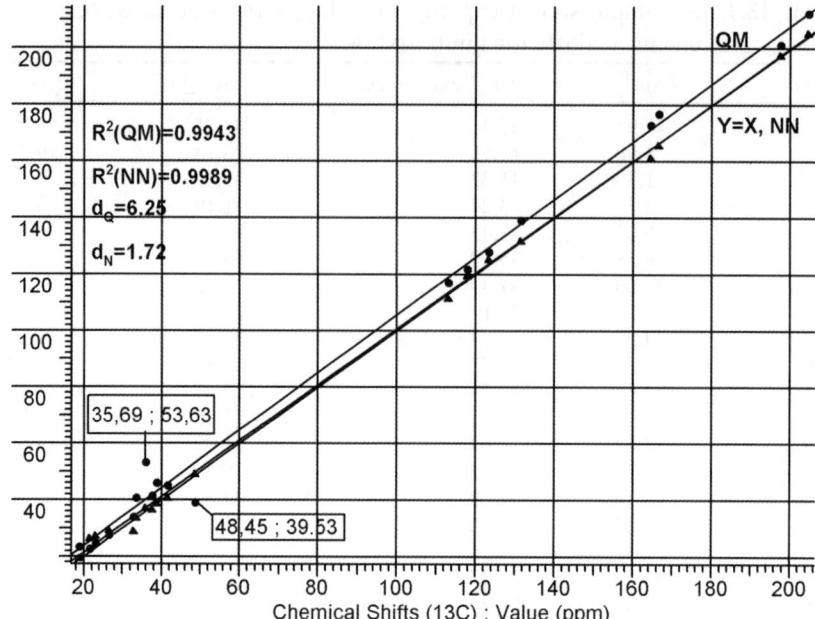

Database: Proposed Structures
• Chemical Shifts (13C) : QM Calc. (ppm) (Current Record) (19 pts)
▲ Chemical Shifts (13C) : NN Calc. (ppm) (Current Record) (19 pts)

**Figure 13.16**  Correlation plots of $^{13}$C chemical shift values predicted for the new nimbosodione assignment by DFT (circles) and NN (triangles) methods *versus* experimental shift values. The NN regression line almost coincides with the target line $Y = X$.

The difference between the regression plots calculated for the both new and old assignments can be visually evaluated from Figures 13.16 and 13.17. For the old assignment both methods of chemical shift prediction show significant scattering of the calculated chemical shifts and a significant deflection of both plots from the target line. The application of both methods of chemical shift prediction in combination with the data stored in a database containing structures accompanied by their assigned NMR spectra allowed us to confidently determine full chemical shift assignment, which was previously impossible.

## 13.5   Elucidating "Undecipherable" Chemical Structures Using Computer-Assisted Structure Elucidation Approaches

A plethora of examples presented in this book show that it is frequently difficult to unambiguously elucidate a single molecular structure from the experimental spectra due to the ambiguity of the spectrum–structural information contained

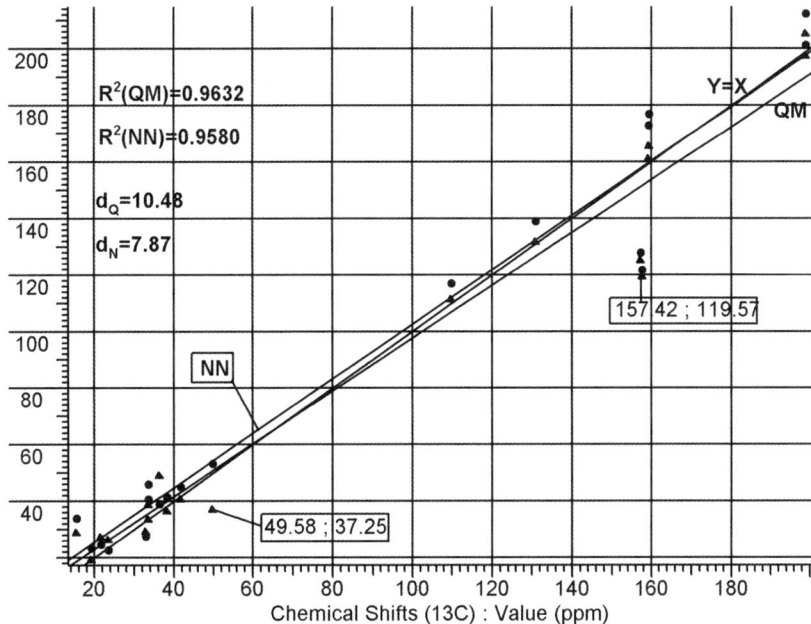

Database: Proposed Structures
• Chemical Shifts (13C) : QM Calc. (ppm) (Current Record) (19 pts)
▲ Chemical Shifts (13C) : NN Calc. (ppm) (Current Record) (19 pts)

**Figure 13.17** Correlation plots of $^{13}$C chemical shift values predicted for the old nimbosodione assignment by DFT (circles) and NN (triangles) methods *versus* experimental shift values. Both methods of chemical shift calculation show significant scattering of the calculated chemical shifts and significant deflection from the target line.

in 2D NMR data. Nevertheless, 2D NMR has continued to be extended with new experimental techniques on an ongoing basis: for instance, H2BC,[41] 1,1-ADEQUATE, 1,n-ADEQUATE and a combination of hyphenated approaches such as HSQC-TOCSY.[42] The application of these techniques can help in removing the uncertainties in the topological distances between some of correlating nuclei and directly influences the ability to elucidate the structure of an unknown compound.

The latest progress in the development of CASE expert systems allows us to declare that the latter should be considered as an integral part of a spectroscopist's armory for the quick and reliable structure elucidation of molecular structures. We will corroborate this position using two recent examples.

*Example 1*
Kummerlöwe *et al.*[43] investigated one of the products obtained by reacting an azide containing a 1,5-enyne group in the presence of electrophilic iodine sources. Initially, the researchers tried to elucidate the structure of this new

compound using classical methods commonly employed in such cases. High-resolution MS provided the molecular formula for the unknown: $C_{16}H_{18}IN$, $m/z = 351.0486$ [351.0484 calculated for C16H18NI (M+)]. The following spectroscopy data were acquired: IR spectrum, 1D $^1H$, and $^{13}C$ spectra in combination with 2D COSY, HSQC, $^1H$-$^{13}C$ HMBC, and $^1H$-$^{15}N$ HMBC experiments. Eleven fragments were identified from the data: a phenyl group, a methyl group, five methylene groups (three forming an isolated chain), a tertiary nitrogen atom, an iodine atom, and four quaternary carbon atoms. The $^1H$-$^{13}C$ HMBC spectrum revealed 63 long-range correlations and the $^1H$-$^{15}N$ HMBC spectrum exposed seven cross peaks, thereby correlating almost every fragment with every other fragment and indicating a very compact structure. A 2D 1,1-ADEQUATE spectrum[42] was also recorded on a Bruker Avance 900 MHz spectrometer equipped with a 5 mm cryogenically cooled TXI probehead optimized for proton detection. In the reported work,[43] this additional data did not help to elucidate the structure.

Since classical NMR analysis failed, the authors[43] decided to make an attempt to solve the problem in an unconventional way by using residual dipolar couplings (RDCs).[44] In accordance with the methodology associated with RDC, they assumed that as long as sufficient anisotropic parameters can be measured, and a large enough set of structural models can be constructed, it should be possible to identify the correct chemical structure.

In order to measure the RDCs the compound was aligned in a stretched polystyrene/chloroform gel. The corresponding scalar couplings were measured in a chloroform solution sample. Fourteen proposed structures, including several models that were unlikely (see Figure 13.18) were tested using the experimental data. The analysis suggested that structure **Ba** is the correct one.

To confirm the structure suggested by the RDC data almost 100 mg of the reaction product was synthesized and a 2D INADEQUATE spectrum[42] was acquired using 3 days of spectrometer time. In addition, labeling the starting material of the reaction with $^{15}N$-azide and measuring $^{13}C$-$^{15}N$ couplings for the $^{15}N$-labeled compound was performed. Both additional experiments clearly supported structure **Ba**.

Our further data analysis showed that the $^1H$-$^{13}C$ HMBC spectra contained nine NSCs (those having $^nJ_{HC}$, $n > 3$).[45] These NSCs produced seven

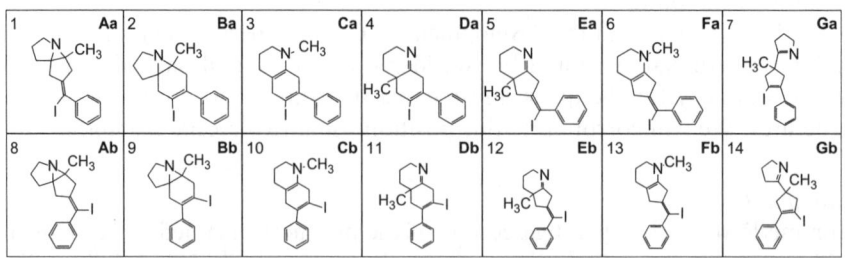

**Figure 13.18**   Potential structures of an unknown reaction product.

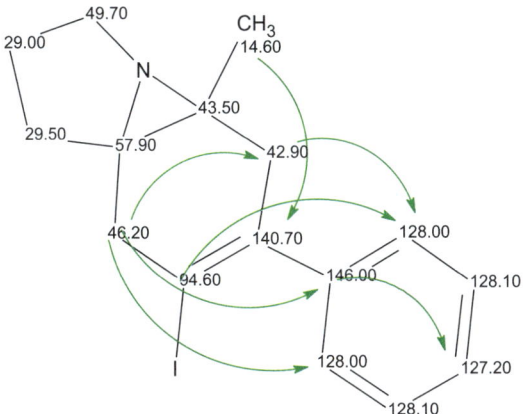

**Figure 13.19** Structure **Ba**. The arrows show the non-standard $^1$H-$^{13}$C HMBC correlations (NSCs). The connectivity 46.2 to 128.0 corresponds to $^5J_{CH}$.

non-standard C-to-C connectivities, which was probably the main reason that structure elucidation failed using the traditional approach for interpretation. Moreover, two unexpected *intense* $^5J_{CH}$ cross-peaks correlating two protons with the *ortho*-carbons of the phenyl group (see Figure 13.19) were identified in the $^1$H-$^{13}$C HMBC spectrum. We suggest that this can be explained as a result of hindered rotation of the phenyl group due to the large volume of the iodine atom. The corresponding part of the HMBC spectrum is presented in Figure 13.20.

At the same time, the authors[43] found that structure **Ba** was almost certainly excluded from the potential set of structures, because the $^{13}$C chemical shifts predicted by ChemDraw and presented in this work differed significantly from the experimental data. The corresponding regression plot is presented in Figure 13.21 for which the MAE was 4.65 ppm with linear regression described by $R^2 = 0.982$.

The highly complex nature of the 2D NMR data prompted the authors to conclude that the problem could not be solved by a classical approach. In making this decision, the NMR data were, however, considered in isolation from algorithmic assisted approaches such as those available in CASE software such as StrucEluc. Therefore the experimental data presented in the work[43] were analyzed using this software program with several modes of problem solving examined.

*Run 1.* The molecular formula, 1D-$^{13}$C, HSQC, $^1$H-$^{13}$C HMBC, and $^1$H-$^{15}$N HMBC spectra were input into the program. All five HMBC peaks marked in the reference[43] as very weak were ignored for the first run to reduce the possible number of NSCs. A MCD was automatically created as shown in Figure 13.22.

As a result of the logical analysis of the MCD, the program discovered the presence of NSCs in the HMBC spectrum, which suggested that FSG was

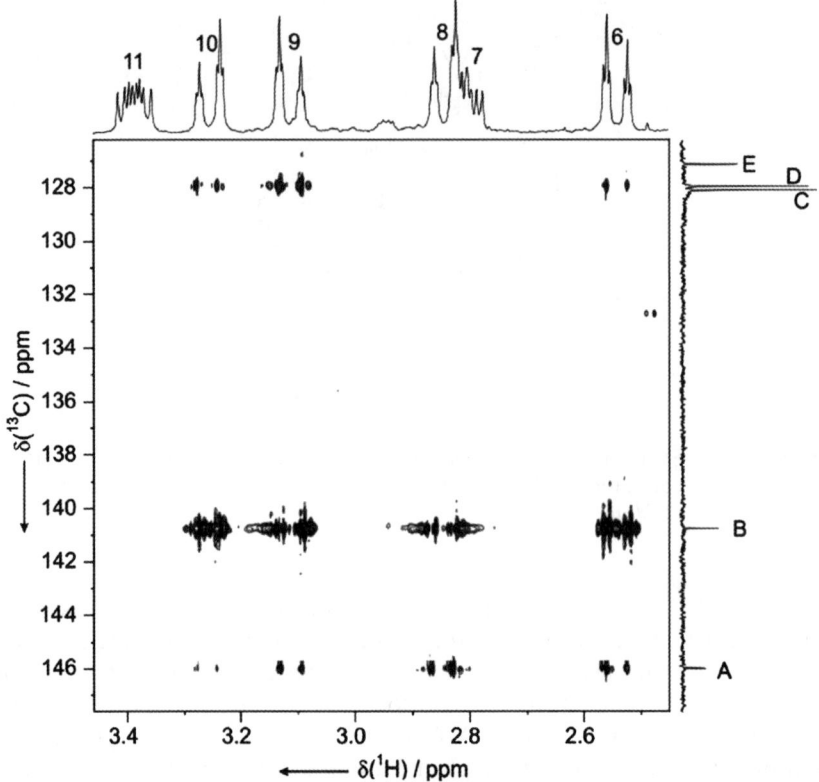

**Figure 13.20**    A selected region of the $^1$H-$^{13}$C HMBC spectrum showing the unusually intense signals between protons 9 (3.12 ppm) and 10 (3.26 ppm) of the methylene group and the carbon atoms D (128.0 ppm) in the *ortho* position of the phenyl group.

necessary. FSG was run assuming that the HMBC data contain an unknown number of NSCs, each of them being of unknown length. No assumptions or user interventions were used. As a result of structure generation accompanied by spectral and structural filtration, three possible structures were output in 13 min. $^{13}$C and $^1$H chemical shift predictions using the NN algorithms were performed and the structural file was then ranked in ascending order of the $^{13}$C chemical shift average deviation (Figure 13.23).

Figure 13.23 shows that the correct structure **Ba** was identified as the most probable structure and its $^{13}$C deviation is significantly (almost twice) smaller than that calculated with ChemDraw. The chemical shift assignment for structure #1 suggested by the prediction algorithms fully coincided with that suggested by the authors.[43] The proposed structure **Bb** (#2 in Figure 13.23) was also generated, but was declined based on the chemical shift predictions. Structure #3 results as a logical consequence from the experimental data but can be rejected due to the high chemical shift deviations.

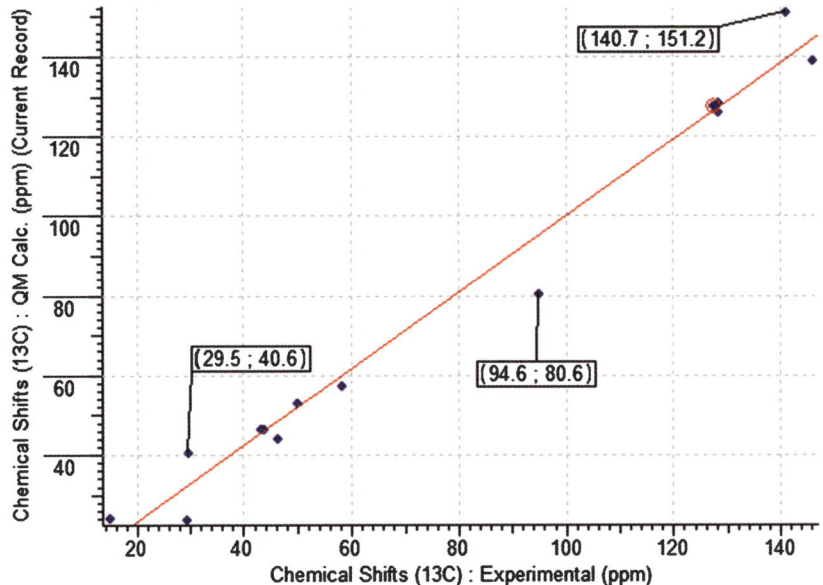

**Figure 13.21**    The regression plot of ChemDraw $^{13}$C chemical shift calculated *versus* experimental values. The first value shown in the boxes is the experimental shift and the second value is that calculated by the ChemDraw program.

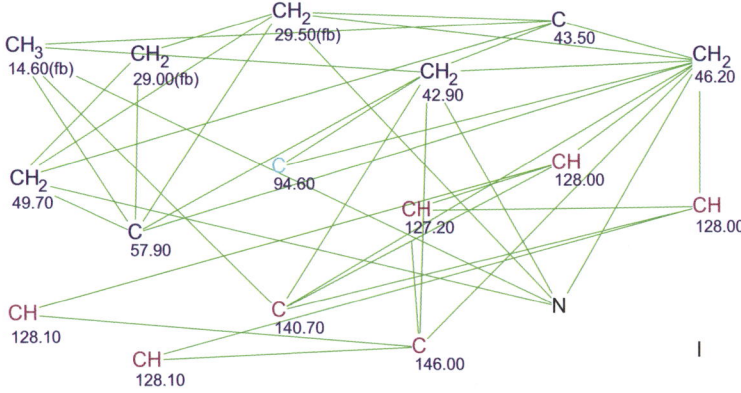

**Figure 13.22**    A MCD. The atom hybridization assigned by the program is marked by the following colors: blue, *sp³*; violet, *sp²*; pale blue, *not sp*. The label "fb" denotes the prohibiting of being adjacent to neighboring heteroatoms.

*Run 2.* All HMBC correlations, without exclusion and including the seven NSCs, were used and 1,1-ADEQUATE correlations were also added to the 2D NMR data (see Figure 13.24).

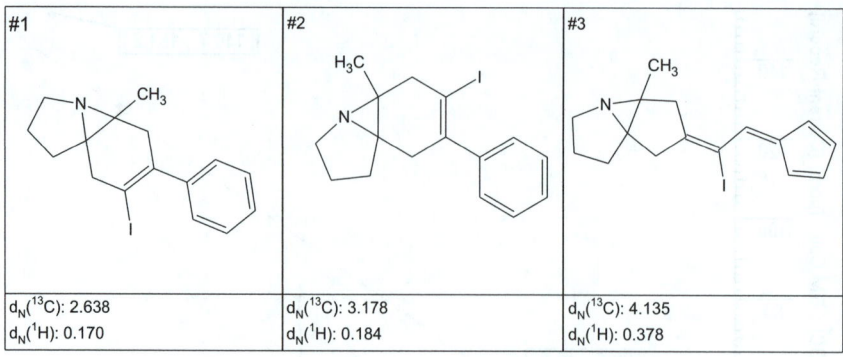

| #1 | #2 | #3 |
|---|---|---|
| $d_N(^{13}C)$: 2.638 | $d_N(^{13}C)$: 3.178 | $d_N(^{13}C)$: 4.135 |
| $d_N(^1H)$: 0.170 | $d_N(^1H)$: 0.184 | $d_N(^1H)$: 0.378 |

**Figure 13.23**   The output structural file ranked in ascending order of average $^{13}C$ chemical shift deviation.

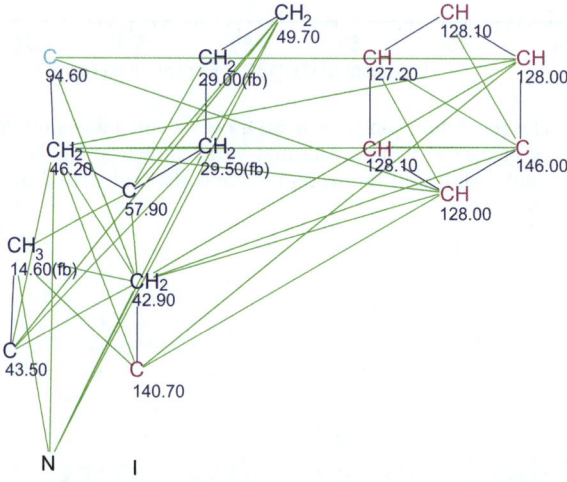

**Figure 13.24**   A MCD. Atom hybridization assigned by the program is marked by the following colors: blue, $sp^3$; violet, $sp^2$; pale blue, *not sp*. The label "fb" denotes prohibition of neighboring with heteroatoms. 1,1-ADE-QUATE correlations are denoted by blue lines.

FSG was run with the following result: only one correct structure #1 was generated in 0.7 s. The application of StrucEluc therefore allowed us to instantly and unambiguously find the correct structure from the HMBC and 1,1-ADE-QUATE data. As for the proposed structures 3–7 and 10–14 (shown in Figure 13.18), all of them can be easily rejected on the basis of visual comparison of experimental $^{13}C$ chemical shifts with the predicted values (see bar-based spectra in Figure 13.25). Figure 13.25 allows for the easy detection of significant outliers in the predicted spectra and avoids the need to check these structures using additional approaches such as the measurement of RDCs. In Section 13.3 we demonstrated

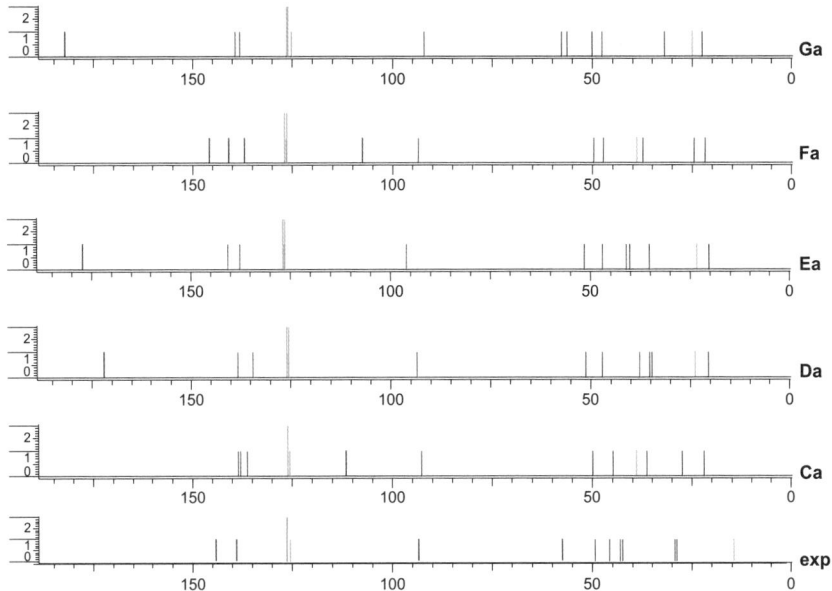

**Figure 13.25** A comparison of the $^{13}$C NMR experimental spectrum and calculated spectra for proposed structures **Ca–Ga**.

how fast NN chemical shift prediction accompanied with bar-graph-based spectrum comparison allowed users to avoid time consuming quantum mechanical chemical shift calculations to choose the corrected structure among a series of competing possibilities. Note that when bar-based spectra are compared, the correct structure is distinguished visually without utilizing any examination of the correctness of the chemical shift assignments in the calculated spectra.

*Example 2*
The second example reviewed in this section was inspired by an article published by Gross and co-workers.[46] They suggested a new method of determining the structures of small molecules based on atomic force microscopy (AFM).[47,48] To validate the usefulness of this approach as an adjunct to the other tools available for organic structure analysis they studied the natural product cephalandole A, (**13.27**), $C_{16}H_{10}N_2O_2$,

**13.27**

which had previously been misassigned by Wu *et al.*[49] and later corrected by Mason *et al.*[50] This example was already discussed in Chapter 12, in which it has been shown that incorrect structure could be immediately declined by

NMR chemical shift prediction using algorithms implemented in the StrucEluc system. However, both groups of authors[49,50] did not present 2D NMR data corresponding to the investigated compound. Gross and co-workers[46] explain that this compound was selected for testing the SPM method, because it meets all three criteria specified previously that render structure analysis especially challenging:[51] the ratio of heavy atoms to protons is approx. 2:1, and the O and N atoms at positions 1 and 4, respectively, interrupt the carbon skeleton completely, separating the two parts of the molecule. In addition, the carbonyl at C2 is distanced from the nearest proton by four bonds and is not expected to show correlations in an HMBC experiment. The molecular formula indicates that there were 13 degrees of unsaturation in the structure.

$^1$H-$^{13}$C HMBC and very sparse COSY data were used by the authors[46] to elucidate the structure. On the basis of NMR data analysis the authors suggested four structures consistent with the available data (see Figure 13.26). Structure #1 is the accepted structure of cephalandole A, and #2 is the previously misassigned structure of this compound. Structures #3 and #4 were also considered as possible.

Gross and co-workers[46] have demonstrated that the SPM approach is really capable of helping to select structure #1 as the most probable one using the analysis of molecular images, and this gives spectroscopists a new independent tool to distinguish molecules that may have similar structures.

The 1D and 2D NMR spectra acquired by the authors[46] to analyze this problem were input into StrucEluc and the MCD was created. No user intervention or data corrections were made. Checking of the MCD detected the presence of NSCs in the 2D NMR data and the fuzzy generation mode was therefore employed for structure generation. As a result the program produced an output file of 11 structures in 1 m 50 s, and structure #1 in the Figure was selected as the most ideal candidate using $^{13}$C NMR chemical shift prediction (see Figure 13.27).

The $^{13}$C deviations were $d_A(^{13}C) = 1.03$ and $d_N(^{13}C) = 1.25$ ppm for the HOSE code-based and NN prediction modes respectively. The second ranked structure gave corresponding values of 2.92 and 2.75 ppm. Structure #2 (Figure 13.26), initially suggested by Wu and co-workers[49] as the correct structure, was also generated and ranked 9$^{th}$ in the file with deviations of $d_A(^{13}C) = 4.10$ and $d_N(^{13}C) = 3.70$ ppm; hence it should definitely be rejected. It is worthy to note that the proposed structures of #3 and #4 were not generated at all. A question arises from analysis of Figure 13.27: because the generated structure #4

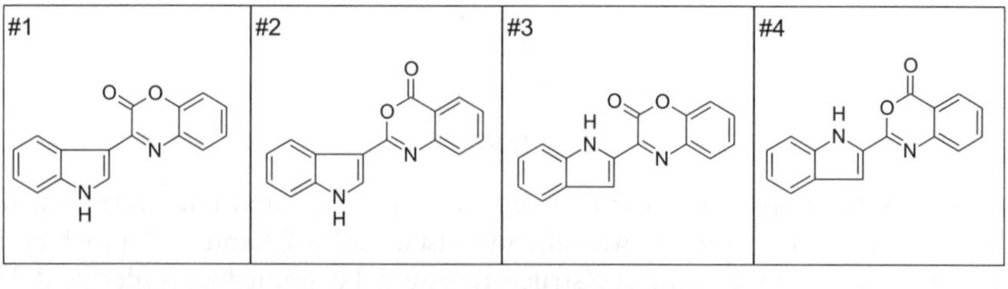

**Figure 13.26**   Proposed structures of cephalandole A.

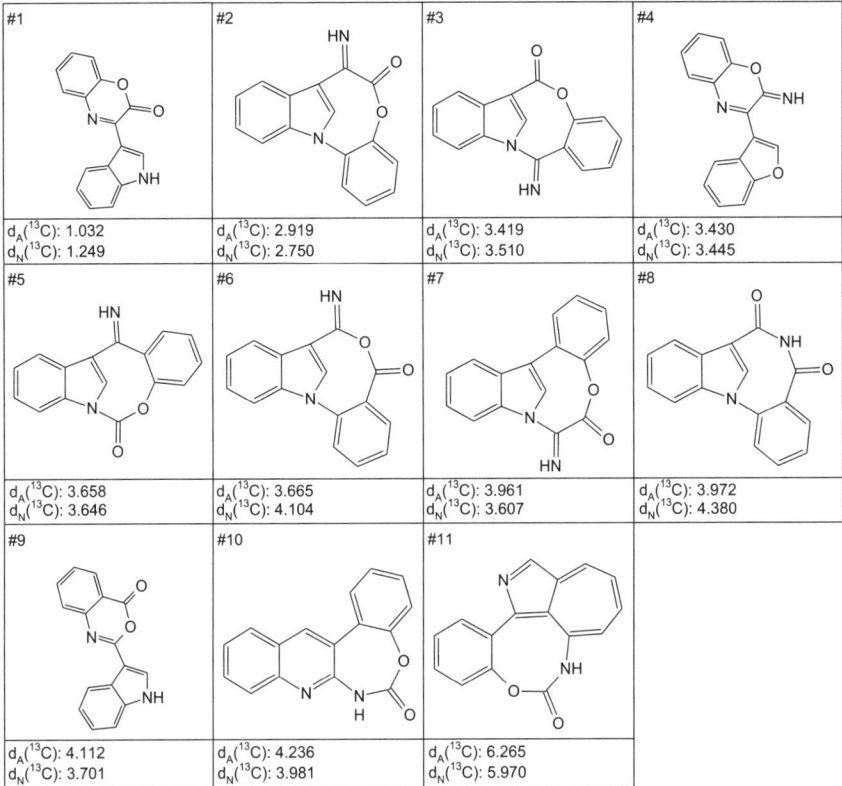

| #1 | #2 | #3 | #4 |
|---|---|---|---|
| $d_A(^{13}C)$: 1.032 $d_N(^{13}C)$: 1.249 | $d_A(^{13}C)$: 2.919 $d_N(^{13}C)$: 2.750 | $d_A(^{13}C)$: 3.419 $d_N(^{13}C)$: 3.510 | $d_A(^{13}C)$: 3.430 $d_N(^{13}C)$: 3.445 |
| #5 | #6 | #7 | #8 |
| $d_A(^{13}C)$: 3.658 $d_N(^{13}C)$: 3.646 | $d_A(^{13}C)$: 3.665 $d_N(^{13}C)$: 4.104 | $d_A(^{13}C)$: 3.961 $d_N(^{13}C)$: 3.607 | $d_A(^{13}C)$: 3.972 $d_N(^{13}C)$: 4.380 |
| #9 | #10 | #11 | |
| $d_A(^{13}C)$: 4.112 $d_N(^{13}C)$: 3.701 | $d_A(^{13}C)$: 4.236 $d_N(^{13}C)$: 3.981 | $d_A(^{13}C)$: 6.265 $d_N(^{13}C)$: 5.970 | |

**Figure 13.27**  Structural output file.

(Figure 13.27) is more preferable than structure #9 (suggested by Wu *et al.*[49]) and has the configuration similar to correct structure, it would be interesting to know whether the AFS method is capable of distinguish between structures #1 and #4 shown in Figure 13.27.

In summary, we have considered two examples of structures that were deemed too difficult to elucidate using traditional methods of 1D and 2D NMR spectra structural interpretation. In both cases the researchers[43,46] used new, more challenging techniques[44,47,48] to perform small molecule structure elucidation. We have demonstrated that the application of a CASE approach is a viable alternative to such new unusual methods and, in these cases, at least, could solve both problems quickly and reliably.

The examples discussed show that a modern CASE expert system should be considered as an integral part of a spectroscopists' armory for reliable structure elucidation. We suggest that now the possibility of solving a structural problem from NMR data should be evaluated by taking into account the growing capabilities of contemporary expert systems. We believe that in future CASE software will become a common tool for NMR spectroscopists to apply to the elucidation of not only challenging structures but also for solving routine everyday problems.

# 13.6   Discussion

The examples considered in this Chapter have shown that a systematic approach based on CASE principles allowed one to determine all structures which do not contradict the experimental data: the 1D and 2D NMR spectra and the molecular formula. Genuine structures are distinguished using a procedure of ranking structures in an ascending order of deviation between the experimental and calculated NMR spectra. Erroneous structures suggested by other researchers are also generated and are ranked lower in the ordered output file if the structures do not contradict the initial set of constraints. The program automatically rejects structures suggested by researchers that cannot be considered as consistent with the initial data.

When QM methods are used for structure elucidation it is desirable to reduce the set of structures as much as possible to prevent superfluous human labor and computational expenses. The minimization of a set of candidate structures and justification of structural hypotheses is achieved by the application of StrucEluc. We conclude that the optimal way to solve the spectrum–structural problem is the creation of structural hypotheses using a 2D NMR-based expert system and revealing the most probable structure(s) by empirical methods of chemical shift prediction. For additional verification of the preferred structures, the application of QM prediction methods is very useful. Moreover, non-empirical methods can play a decisive role if the verified structures contain fragments which are absent from the database utilized for training the empirical methods of prediction. If 2D NMR data are not available then the application of fast but accurate incremental and NN-based algorithms[26,31] for the preliminary probability-based estimation of each structure suggested by a researcher can be extremely helpful. We suggest that a systematic approach based on the application of an expert system for structure elucidation from NMR data is the most practical and rational way to approach the problem. The most convincing evidence of the effectiveness of the QM methods as a tool for molecular structure elucidation could be the application of these methods to the 2 or 3 top structures of output files produced by an expert system such as StrucEluc. This approach would allow one to compare conclusions based on empirical and QM calculations.

# References

1. M. E. Elyashberg, A. J. Williams and G. E. Martin, *Prog. NMR Spectrosc.*, 2008, **53**, 1.
2. K. C. Nicolaou and S. A. Snyder, *Angew. Chem., Int. Ed.*, 2005, **44**, 1012.
3. M. E. Elyashberg, A. J. Williams and K. A. Blinov, *Nat. Prod. Rep.*, 2010, **27**, 1296.
4. K. A. Blinov, M. E. Elyashberg, S. G. Molodtsov, A. J. Williams and E. R. Martirosian, *Fresenius' J. Anal. Chem.*, 2001, **369**, 709.
5. T. Lindel, J. Junker and M. Kock, *J. Mol. Model.*, 1997, **3**, 364.
6. J.-M. Nuzillard, *Chin. J. Chem.*, 2003, **21**, 1263.

7. K. P. Schulz, A. Korytko and M. E. Munk, *J. Chem. Inf. Comput. Sci.*, 2003, **43**, 1447.

8. A. Bagno, F. Rastrelli and G. Saielli, *Chem. Eur. J.*, 2006, **12**, 5514.

9. A. Bagno and G. Saielli, *Theor. Chem. Acc.*, 2007, **117**, 603.

10. A. Balandina, D. Saifina, V. Mamedov and S. Latypov, *J. Mol. Struc.*, 2006, **791**, 77.

11. G. Barone, L. Gomez-Paloma, D. Duca, A. Silvestri, R. Riccio and G. Bifulco, *Chem. Eur. J.*, 2002, **8**, 3233.

12. V. Barone, P. Cimino, O. Crescenzi and M. Pavone, *J. Mol. Struc.*, 2007, **811**, 323.

13. P. Cimino, L. Gomez-Paloma, D. Duca, R. Riccio and G. Bifulco, *Magn. Reson. Chem.*, 2004, **42**, S26.

14. S. D. Rychnovsky, *Org. Lett.*, 2006, **8**, 2895.

15. B. Schlegel, A. Hartl, H.-M. Dahse, F. A. Gollmick, U. Gräfe, H. Dörfelt and B. Kappes, *J. Antibiot.*, 2002, **55**, 814.

16. J. A. J. Porco, S. Su, X. Lei, S. Bardhan and S. D. Rychnovsky, *Angew. Chem. Int. Ed.*, 2006, **45**, 1.

17. A. J. Williams, M. E. Elyashberg, K. A. Blinov, D. C. Lankin, G. E. Martin, W. F. Reynolds, J. A. Porco, C. A. Singleton and S. Su, *J. Nat. Prod.*, 2008, **71**, 581.

18. M. E. Elyashberg, K. A. Blinov, A. J. Williams, S. G. Molodtsov and G. E. Martin, *J. Chem. Inf. Model.*, 2006, **46**, 1643.

19. D. Sanz, R. M. Claramunt, A. Saini, V. Kumar, R. Aggarwal, S. P. Singh, I. Alkorta and J. Elguero, *Magn. Reson. Chem.*, 2007, **45**, 513.

20. A. Balandina, V. Mamedov, F. Xavier, F. Bruno and S. Latypov, *Tetrahedron Lett.*, 2004, **45**, 4003.

21. A. A. Balandina, V. A. Mamedov, E. A. Khafizova and S. K. Latypov, *Russ. Chem. Bull.*, 2006, **55**, 2256.

22. CambridgeSoft Corporation, http://www.cambridgesoft.com/.

23. Advanced Chemistry Development. ACD/NMR Predictors. Prediction suite includes $^1$H, $^{15}$N, $^{19}$F, $^{31}$P NMR prediction, http://www.acdlabs.com.

24. Modgraph, http://www.modgraph.co.uk/product_nmr.htm.

25. W. Robien, http://felix.orc.univie.ac.at/~wr/csearch_server_ info.html.

26. Y. D. Smurnyy, K. A. Blinov, T. S. Churanova, M. E. Elyashberg and A. J. Williams, *J. Chem. Inf. Model.*, 2008, **48**, 128.

27. Upstream Solutions GMBH, *NMR Prediction Products (SpecTool):* http://www.upstream.ch/products/nmr.html.

28. M. E. Elyashberg, K. A. Blinov and A. J. Williams, *Magn. Reson. Chem.*, 2009, **47**, 371.

29. K. A. Blinov, D. Carlson, M. E. Elyashberg, G. E. Martin, E. R. Martirosian, S. G. Molodtsov and A. J. Williams, *J. Magn. Reson. Chem.*, 2003, **41**, 359.

30. M. E. Elyashberg, K. A. Blinov, S. G. Molodtsov, A. J. Williams and G. E. Martin, *J. Chem. Inf. Comput. Sci.*, 2004, **44**, 771.

31. K. A. Blinov, Y. D. Smurnyy, T. S. Churanova, M. E. Elyashberg and A. J. Williams, *Chemom. Intell. Lab. Syst.*, 2009, **97**, 91.

32. W.-G. Kim, J.-W. Kim, I.-J. Ryoo, J.-P. Kim, Y.-H. Kim and I.-D. Yoo, *Org. Lett.*, 2004, **6**, 823.
33. W. Steglich and V. Hellwig, *Org. Lett.*, 2004, **6**, 3175.
34. M. E. Elyashberg, K. A. Blinov, S. G. Molodtsov, Y. D. Smurnyy, A. J. Williams and T. S. Churanova, *J. Cheminform.*, 2009, http://www.jcheminf.com/content/1/1/3.
35. Atta-Ur-Rahman, *Nuclear Magnetic Resonance: Basic Principles*, Springer-Verlag, Berlin, 1985.
36. B. Blümich, *Essential NMR for Scientists and Engineers*, Springer, Berlin, 2005.
37. H. Friebolin, *Basic One- and Two-Dimensional Spectroscopy*, Wiley-VCH, Weinheim, 2005.
38. J. Meiler, R. Meusinger and M. Will, *J. Chem. Inf. Comput. Sci.*, 2000, **40**, 1169.
39. I. Ara, B. S. Siddiqui, S. Faizi and S. Siddiqui, *J. Nat. Prod.*, 1990, **53**, 816.
40. A. P. Li, P. Y. Bie, X. S. Peng, T. X. Wu, X. F. Pan, A. S. C. Chan and T. K. Yang, *Synth. Commun.*, 2002, **32**, 605.
41. N. T. Nyberg, J. Ø. Duus and O. W. Sørensen, *J. Am. Chem. Soc.*, 2005, **127**, 6154.
42. S. Berger and S. Braun, *200 and More NMR Experiments*, Wiley-VCH, Weinheim, 2004.
43. G. Kummerlowe, B. Crone, M. Kretschmer, S. F. Kirsch and B. Luy, *Angew. Chem., Int. Ed. Engl.*, 2011, **50**, 2643.
44. G. Kummerlowe, S. Schmidt and B. Luy, *Open Spectrosc. J.*, 2010, **4**, 16.
45. S. G. Molodtsov, M. E. Elyashberg, K. A. Blinov, A. J. Williams, E. E. Martirosian, G. E. Martin and B. Lefebvre, *J. Chem. Inf. Comput. Sci.*, 2004, **44**, 1737.
46. L. Gross, F. Mohn, N. Moll, G. Meyer, R. Ebel, W. M. Abdel-Mageed and M. Jaspars, *Nat. Chem.*, 2010, **2**, 821.
47. F. J. Giessibl, *Rev. Mod. Phys.*, 2003, **75**, 949.
48. L. Gross, F. Mohn, N. Moll, P. Liljeroth and G. Meyer, *Science*, 2009, **325**, 1110.
49. P.-L. Wu, Y.-L. Hsu and C.-W. Jao, *Nat. Prod.*, 2006, **69**, 1467.
50. J. J. Mason, J. Bergman and T. Janosik, *J. Nat. Prod.*, 2008, **71**, 1447.
51. P. Crews, J. Rodriguez and M. Jaspars, *Organic Structure Analysis*, Oxford University Press, Oxford, 2010.

# CHAPTER 14

# An Evaluation of the Performance of the Structure Elucidator System

In this Chapter we will discuss some of the performance characteristics of the Structure Elucidator (StrucEluc) system that were obtained in 2005 as a result of the many investigations utilizing the software algorithms contained within the software.[1] The investigations were performed on 250 problems. The distribution of structures relative to the number of skeletal atoms is shown in Figure 14.1. Comparing Figure 7.3 with Figure 14.1 shows that the overwhelming majority of problems contain a number of skeletal atoms that corresponds to the right wing of the curve mapping the distribution of molecules existing in the StrucEluc database with a specific number of skeletal atoms. From the histogram in Figure 14.1 it is seen that the most typical structures are those where the number of skeletal atoms varies from 20 to 30. At the same time more than 100 molecules contain from 30 to 90 skeletal atoms, which immediately invalidates the limit declared by others for deterministic expert systems (see Chapter 6). The performance criteria are segregated into a series of topics that will be self-explanatory based on the topic headings.

## 14.1   Time Required to Solve Problem

The total time necessary to solve a problem, $T_{tot}$, can be represented as the sum of the following components:

$$T_{tot} = t_{MCD} + t_g + t_{flt} + t_{spr}$$

New Developments in NMR No. 1
Contemporary Computer-Assisted Approaches to Molecular Structure Elucidation
By Mikhail Elyashberg, Antony Williams and Kirill Blinov
© Royal Society of Chemistry 2012
Published by the Royal Society of Chemistry, www.rsc.org

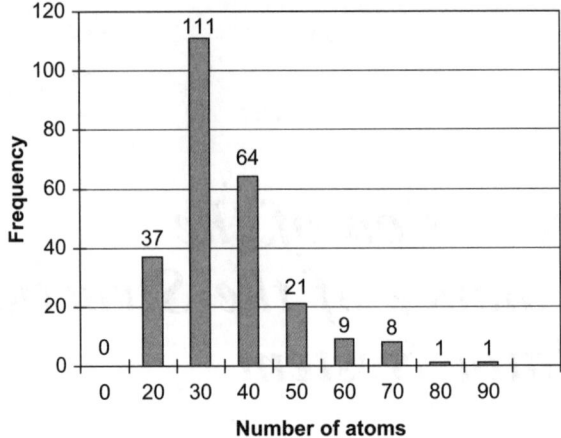

**Figure 14.1**   The distribution of structures relative to the number of skeletal atoms
across the 250 example problem set. The bars correspond to the number
of problems in a given range. The first bar comprised of 37 problems,
involved molecules with from 1–20 atoms. The second bar with 111
problems involved from 21–30 atoms, *etc.*

where $t_{MCD}$ is the time required to create the MCD (note: when the problem is
solved in the common mode this value is practically zero); $t_g$ is the time of
structure generation; $t_{flt}$ is the time consumed to performing spectral filtering on
the generated structural file (a process applied either during or just after
structures are generated simply to save disk storage) and the removal of
duplicates; and $t_{spr}$ is the time required to perform spectrum prediction on the
resulting output file and the ranking of the structures in ascending order of the
$d_A$ deviation. In reality, all procedures necessary for selection of the most
probable structure are performed automatically if the command "find best
structure" is initiated. The histogram presented in Figure 14.2 shows that
the majority of problems studied in this representative work were solved in
less than 100 s and that, as a rule, this time does not exceed 20–25 min
(1000–1500 s).

Structure generation times are shown as a histogram in Figure 14.3. These
data demonstrate that $t_g$ comprises only a small part of the $T_{tot}$ value. In
particular, for 200 out of 250 problems $t_g$ is less than 1 min.

## 14.2   Efficiency of Filtering Generated Structures

Figure 14.4 shows a histogram plot illustrating the number of structures output
into the resultant file prior to filtering and removal of duplicates.

Analysis of Figure 14.4 allows us to conclude that for 75% of the problems
the output file of generated structures contains less than 1000 structures and
only in 2% of the cases does the output file grow to around half a million
structures. Recalling that the number of potential isomers for a molecular

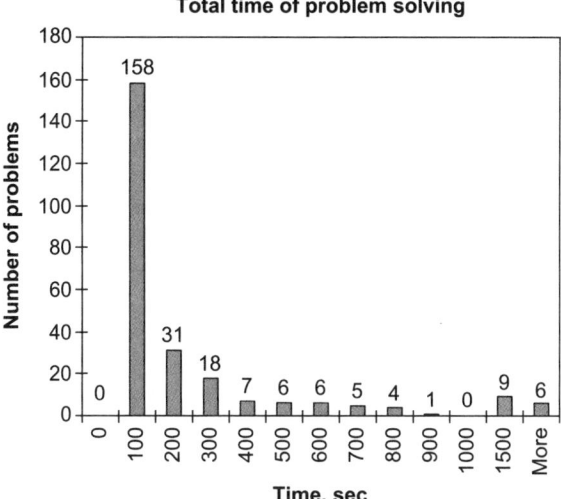

**Figure 14.2** A histogram showing the total time required to elucidate the structures within the test problem set.

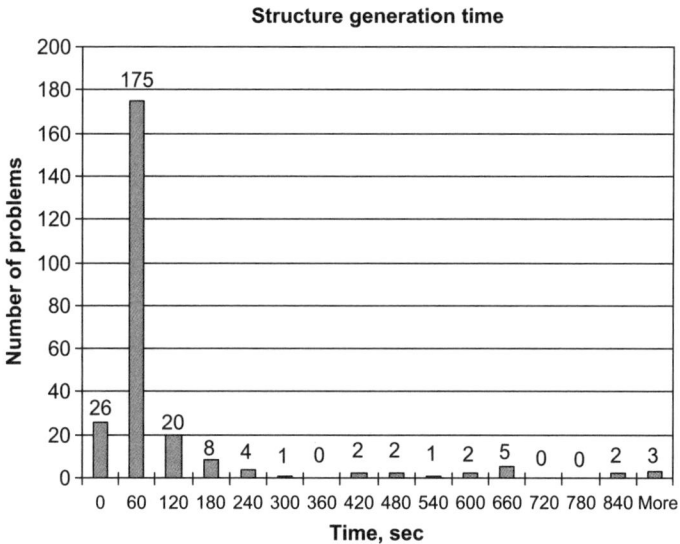

**Figure 14.3** The distribution of structure generation times across the entire problem set.

formula comprising a natural product can be astronomical (see Chapter 1), the results obtained can be considered as particularly successful. The results provide evidence that the statement made by other researchers[2,3] regarding the

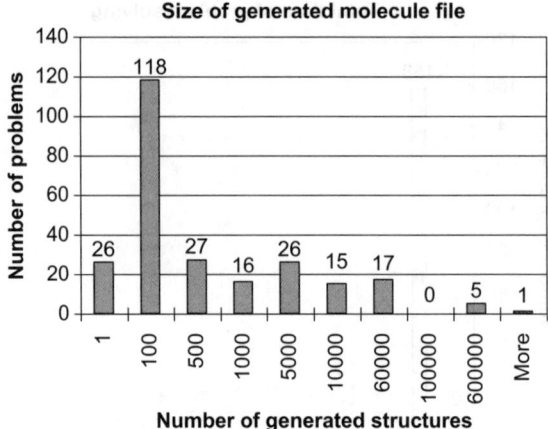

**Figure 14.4**   A histogram of the number of structures in the output file prior to filtering and removal of duplicates.

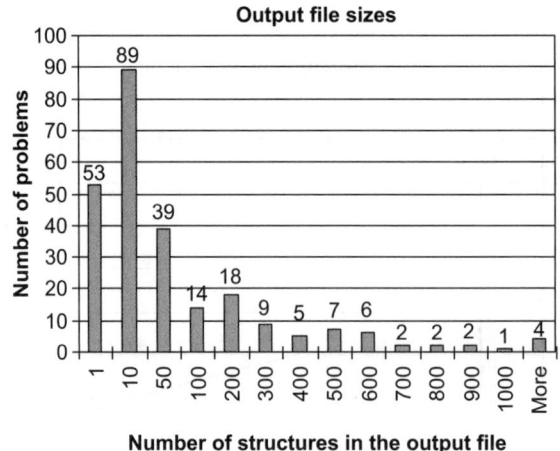

**Figure 14.5**   The distribution of structures in the reduced output files resulting from filtering and removal of duplicates.

limited possibilities of deterministic expert systems due to the potential threat of a combinatorial explosion is in fact unjustified.

Figure 14.5 illustrates the distribution of the size of the output file obtained following both spectral and structural filtering of the initial file of generated structures followed by removal of duplicates. A comparison of the distributions presented in Figures 14.4 and 14.5 illustrates the high efficiency of the filtering process, since the number of generated structures is reduced

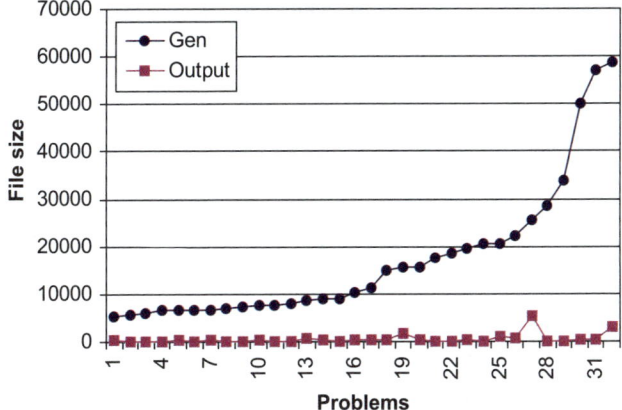

**Figure 14.6** The reduction in the number of generated structures after spectral and structural filtering. The data are provided for a subset of problems with output files containing between 5000 and 50 000 structures.

dramatically as a result of the application of the filtering procedures. The filtering efficiency is obvious in Figure 14.6 where the initial and final output files are shown for problems with output files containing between 5000 and 50 000 structures.

## 14.3 Evaluation of Methods for Selection of the Most Probable Structure

A step-by-step method for selection of the most probable structure contained within the output file obtained using StrucEluc was fully described in Chapter 9. The $d_A$ values represent the average deviations of the "accurately" calculated $^{13}C$ NMR chemical shifts relative to the experimental shifts. These values play a decisive role in the process of selecting the correct structure. It is generally assumed that the first ranked structure with the minimum $d_A$ value is the most probable structure match. For the problem sets examined, for approx. 99% of the problems $d_A$ ranking placed the correct structure in the first position. The distribution of problems with first order rankings is presented in Figure 14.7 where $d_A(1)$ values, the deviations corresponding to the first ranked structures, are presented. In approx. 60% of the cases the $d_A(1)$ value is less than 2 ppm and the average value is 2.09 ppm. To reveal the general trend of $d_A$ values within a ranked output file the average magnitudes corresponding to the ten first structures in the ranked output file were calculated (see Figure 14.8).

The Figure shows that $d_A$ deviations increase for the first four structures by an increment of approx. 1 ppm from each structure to the next. Average similarity coefficients were also considered for the ten structures as represented in Figure 14.9. The similarity coefficients were calculated by comparing all structures with that ranked first. Recall that the first ranked structure was

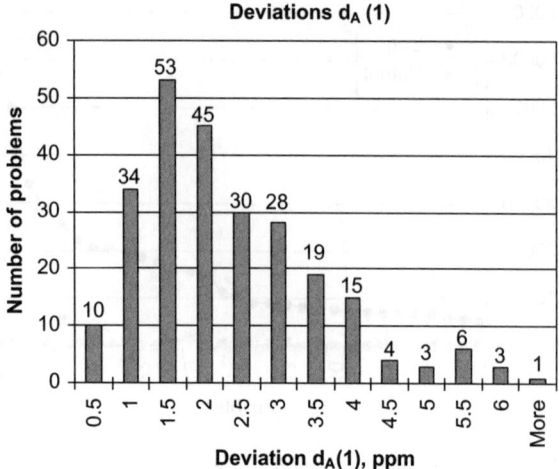

**Figure 14.7**   A histogram illustrating the distribution of problems with deviations corresponding to the first ranked structures, $d_A(1)$ values.

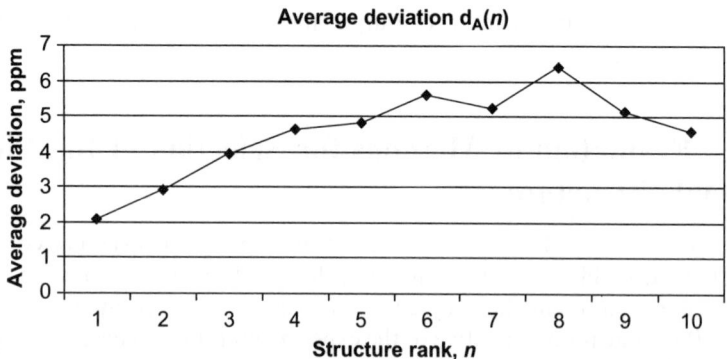

**Figure 14.8**   The average $d_A$ magnitudes corresponding to the 10 first structures of the ranked output file.

correct in 99% of the cases. Figure 14.9 shows that the average similarity coefficients drop slowly starting from the second structure whose average similarity coefficient is equal to 0.67. The observed dependence confirms that, in general, the main assumption common to molecular spectroscopy is that similar structures have similar spectra.

Experience has shown that the probability of coincidence of the first ranked structure with the correct one depends on the difference $\Delta_{2-1}$ between $d_A(2)$ and $d_A(1)$: the greater the difference then the greater the probability that the first ranked structure is the target structure. The distribution of $\Delta_{2-1}$ values for output files containing more than one structure across the problem set is

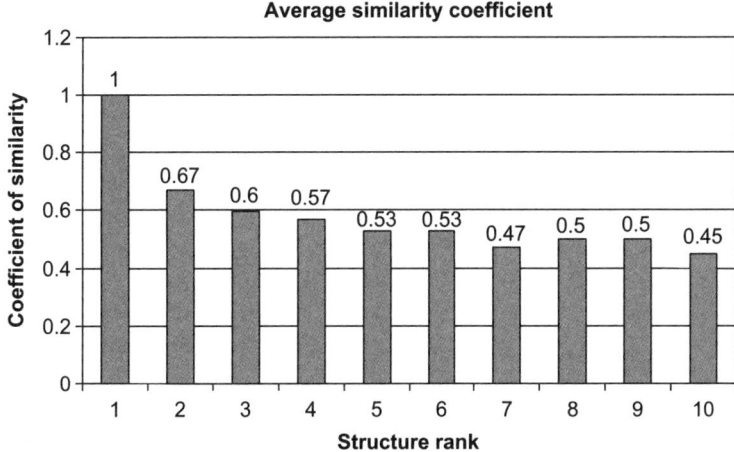

**Figure 14.9**    The average similarity coefficients of the first ten ranked structures. The first structure was the reference structure for similarity.

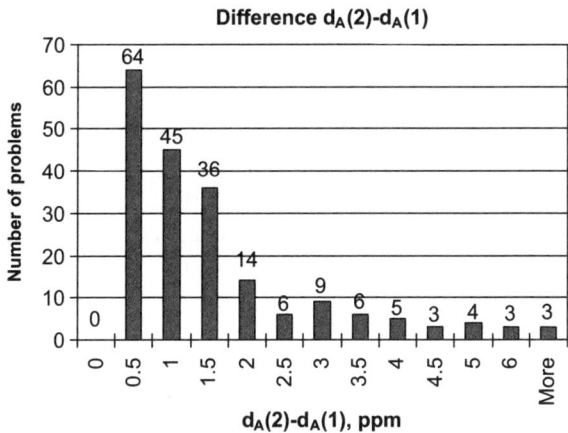

**Figure 14.10**    The distribution of problems as a function of the difference in deviations between the first and second ranked structures, $d_A(2) - d_A(1)$ values.

presented in Figure 14.10. Investigations performed on a large number of problems confirmed the following empirical fact which was established earlier:[4] if the value $\Delta_{2-1} \geq 1$ ppm, then the structure ranked first has a significant probability of being the correct solution to the problem. At the same time, it was also found that the correct structures were distinguished by structural ranking in many cases even when the difference $\Delta_{2-1}$ was very small. It turned out that only in several problems was the correct structure not placed in the first position.

Analysis of the improperly ranked solutions establishes that the cause of the incorrect structure ranking is most frequently the absence of appropriate structures from the database that allow precise prediction of the chemical shifts of certain carbon atoms characterized by unusual environments. In other cases, the carbon atom environment was not poorly represented *per se*, but the chemical shift of the given atom was unusual due to some particular spatial effects. As demonstrated in Sections 9.6.2 and 11.2, an improper ranking can be accounted for in terms of the large number of isomer molecules for which several of the top ranked structures are very similar. For instance, when we input into the StrucEluc software the 1D and 2D NMR (HSQC, HMBC, COSY) data for the recently published[5] molecule, belizeanolide ($C_{81}H_{32}O_{20}$), the following solution was obtained: $k = 938044 \rightarrow 7845 \rightarrow 3926$, $t_g = 3$ h 9 m. The three best structures identified by the program from nearly 4000 hypothetical molecules are shown in Figure 14.11. The correct structure was placed in third position. The difference in deviations $d(3) - d(1)$ is very small $-0.08$ ppm. Here the QM $^{13}C$ chemical shift calculation is unlikely to be helpful due to the large size of the molecule. In such a situation only additional experimental data, chemical knowledge and chemical common sense can help solve the problem. The following empirical observation was deduced: for the majority of *improperly ranked* structures, the difference $\Delta_{2-1}$ is less than 1. A large $d_A(1)$ value in combination with a small $\Delta_{2-1}$ value is suggestive of either improper structural ranking or of incorrect $^{13}C$ chemical shift assignment, the latter due to misassignment by the investigator.

For example, consider one problem[6] solved with StrucEluc and using fuzzy structure generation (FSG) with the parameters $m = 3$, $a = 2$. The correct structure, **14.1**, with the carbon assignments suggested by the authors[6] was placed in the second ranked position by $^{13}C$ NMR prediction. The solution was characterized with the following parameters: $d_A(1) = 5.06$ ppm, $d_A(2) = 5.20$ ppm, and $\Delta_{2-1} = 0.14$ ppm. The deviations are obviously large, whereas their difference is small.

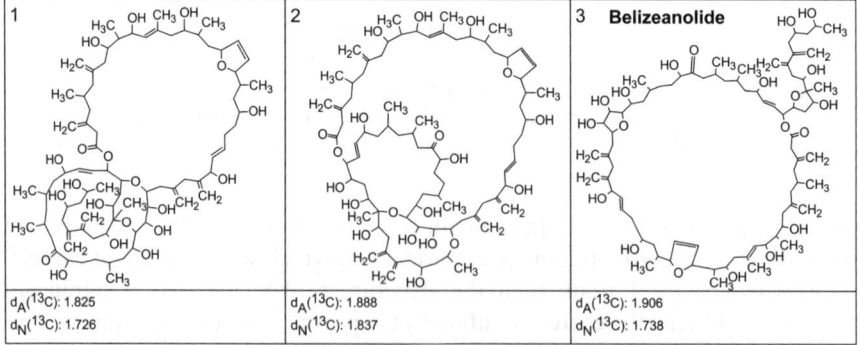

**Figure 14.11**   The first three structures of the ordered output file resulting from the structure elucidation of belizeanolide molecule.

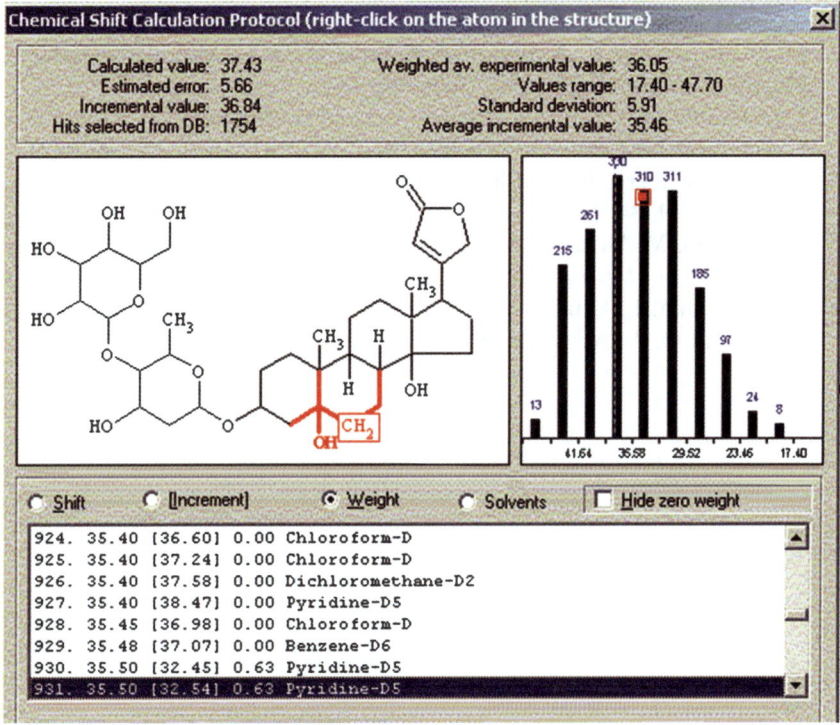

**14.1**

A comparison of the chemical shifts calculated by ACD/CNMR Predictor with the experimental values assigned to structure **14.1** showed that all values were close except for two shifts. The predicted values for atoms C(15.60) and C(29.80) were 35.46 ppm and 19.00 ppm, respectively. To explain the discrepancies in the chemical shifts, the chemical shift calculation protocol feature of the ACD/CNMR predictor was used (see Section 3.1.3.3). The protocol related to the methylene group at 15.60 ppm is displayed in Figure 14.12.

**Chemical Shift Calculation Protocol (right-click on the atom in the structure)**

Calculated value: 37.43    Weighted av. experimental value: 36.05
Estimated error: 5.66    Values range: 17.40 - 47.70
Incremental value: 36.84    Standard deviation: 5.91
Hits selected from DB: 1754    Average incremental value: 35.46

| | | | | |
|---|---|---|---|---|
| 924. | 35.40 | [36.60] | 0.00 | Chloroform-D |
| 925. | 35.40 | [37.24] | 0.00 | Chloroform-D |
| 926. | 35.40 | [37.58] | 0.00 | Dichloromethane-D2 |
| 927. | 35.40 | [38.47] | 0.00 | Pyridine-D5 |
| 928. | 35.45 | [36.98] | 0.00 | Chloroform-D |
| 929. | 35.48 | [37.07] | 0.00 | Benzene-D6 |
| 930. | 35.50 | [32.45] | 0.63 | Pyridine-D5 |
| 931. | 35.50 | [32.54] | 0.63 | Pyridine-D5 |

**Figure 14.12** The ACD/CNMR chemical shift calculation protocol. The methylene group with its environment is colored red in one of the reference structures displayed from the database.

Figure 14.12 indicates that 1754 structural hits were selected from the database and that the majority of reference structures support a chemical shift prediction for the $CH_2$ group of approx. 36 ppm. Taking this into account with the large value of $d_A(1)$ suggests that either the deduced structure is incorrect or, more likely, the carbon assignment for the carbon atoms associated with the shifts of 15.6 and 29.8 should be interchanged. Assuming that the number of non-standard correlations (NSCs) in the HMBC data exceeded 3, FSG was repeated with parameters of $m = 4$, $a = 2$. 4884 structures were generated and the correct structure with the properly assigned chemical shifts was selected using the usual ranking procedure ($d_A = 3.34$ ppm). This approach suggests that the authors[5] had, indeed, made an incorrect spectral assignment in the case of **14.1**.

Traditionally, the developers of expert systems aspire to minimize the structure generation time and the size of the output file. However, the results presented above indicate that an output file containing many thousands of structures cannot seriously hamper the selection of the most probable structure, since the methods described push the actual structure to the top of the rank ordered file.

# References

1. M. E. Elyashberg, K. A. Blinov, A. J. Williams, S. G. Molodtsov and G. E. Martin, *J. Chem. Inf. Model.*, 2006, **46**, 1643.
2. J.-L. Faulon, *J. Chem. Inf. Comput. Sci.*, 1996, **36**, 731.
3. C. Steinbeck, *J. Chem. Inf. Comput. Sci.*, 2001, **41**, 1500.
4. M. E. Elyashberg, K. A. Blinov and E. R. Martirosian, *Lab. Autom. Inf. Manag.*, 1999, **34**, 15.
5. Napolitano, M. Norte, J. M. Padron, J. J. Fernandez and A. Z. Daranas, *Angew. Chem., Int. Ed.*, 2009, **48**, 796.
6. N. Uchiyama, F. Kiuchi, M. Ito, G. Honda, Y. Takeda, O. K. Khodzhimatov and O. A. Ashurmetov, *J. Nat. Prod.*, 2003, **66**, 128.

# Conclusions

This book has hopefully communicated the history, challenges, developments and capabilities of CASE systems. In this section we will summarize and review the observations we have made through this volume.

The development of CASE expert systems (ES) has progressed dramatically from the first prototypes, resembling rudimentary toys from a contemporary point of view, to systems which are now practical analytical tools that may be applied to problems in organic and analytical chemistry.

In spite of the great interest and practical expectations of CASE systems the process by which these rudimentary systems were converted into modern powerful applications consumed decades of effort and innovation due to the enormous complexity of the problem. The primary challenge is that a computer, while capable of performing calculations very quickly, "thinks" slowly. It was therefore very difficult to teach a software program to mimic a spectroscopist's reasoning during the process of deciphering structural information encoded into spectra. The CASE approach is based on the principles of artificial intelligence and it took many years to seek optimal approaches to implement these principles into an expert system architecture.

During the first 25 years of CASE development a general workflow for the CASE systems was tested on many models of expert systems using MS, IR, $^{13}$C and $^{1}$H NMR spectra. It was eventually revealed that the amount of structural information which could be extracted from these spectra was limited by their nature. These forms of spectra did not allow for the elucidation of structures of large complex organic molecules, for example, natural products. Meanwhile the automated structure elucidation of newly isolated chemicals was certainly in strong demand by chemists. Nowadays millions of new compounds are added to the CAS registry in a single year and the determination of chemical structures remains one of the primary challenges for the chemist.

A second era of CASE system development was initiated when homonuclear and heteronuclear 2D NMR spectroscopic techniques became available to

New Developments in NMR No. 1
Contemporary Computer-Assisted Approaches to Molecular Structure Elucidation
By Mikhail Elyashberg, Antony Williams and Kirill Blinov
© Royal Society of Chemistry 2012
Published by the Royal Society of Chemistry, www.rsc.org

spectroscopists. The algorithms and strategies that had proven successful in 1D NMR CASE systems were extended to include 2D NMR spectra and new algorithms were developed as necessary. A series of 2D NMR based expert systems then appeared which demonstrated the possibility to elucidate the chemical structures of organic molecules containing over 100 skeletal atoms.

The success of these systems resulted from the development of a general theoretical approach to structure elucidation using "axiomatic" knowledge, the interrelations between structures and spectra, and factual knowledge, large databases containing hundreds of thousands of structures and millions of molecular fragments associated with their assigned NMR spectra. According to this approach, first a *partial axiomatic theory* applicable to the solution of a given problem is created using the intersection of processing spectral data and a system knowledgebase. A solution to the problem is then sought as a *finite* set of all (without any exclusion) possible corollaries (structures) deduced from the axiomatic theory. The set is finite because the number of isomers corresponding to a given molecular formula is finite. A CASE system ultimately acts like a logical machine inferring a full set of consequences following from a set of axioms and hypotheses. The imposition of structural constraints during the inference of structures leads, as a rule, to a manageable number of structures which can be generated using fast and effective algorithms.

The selection of the most probable structure from all conceivable candidates is performed using NMR spectrum prediction. The most reliable selection is possible using $^{13}$C NMR spectrum calculations, though the prediction of $^{1}$H, $^{15}$N, $^{31}$P or $^{19}$F chemical shifts are also very helpful for supporting the identification of the "best" structure. The best structure is that for which the average chemical shift deviation between the experimental and calculated spectra is minimal. Two kinds of approaches for chemical shift calculation are presently available—both empirical and quantum-mechanical prediction.

The empirical approach is realized by employing algorithms based on additive rules (incremental methods), artificial neural nets (ANN) and structural databases encoded by HOSE codes (fragmental approach). These methods are (a) rather fast, (b) applicable to molecules of any size, (c) amenable to full automation and (d) provide sufficient accuracy for the reliable determination of the best structure. They can be applied to large structural files containing tens of thousands of structures while the speed of chemical shift prediction by both incremental and ANN approaches is of 6000–10 000 chemical shifts per second at an accuracy of 1.6–1.8 ppm. A drawback exists in that the empirical methods fail to accurately calculate chemical shifts of substructures that were absent from a database used for training the corresponding prediction program.

A quantum-mechanical approach commonly uses the GIAO approximation of the DFT theory. GIAO calculations require the participation of a qualified specialist and are not likely to be automated. The calculations are very slow (in comparison with the speed of empirical methods) and cannot be applied to the shift prediction of large molecules and certainly not for large structural files. If the appropriate calculation protocol (using DFT functional and basis sets) is successfully chosen and the calculated chemical shifts are scaled, then QM methods can provide accuracy similar to that common for empirical methods. In contrast to empirical methods the possibility of quantum-mechanical chemical shift prediction, in principle, does not depend on chemical composition and the exotic nature of the structures under examination.

The most rational strategy for the prediction of NMR spectra can be formulated as follows: if the empirical approaches struggle to predict the NMR spectrum for a part of a structure then a model molecule including this challenging substructure should be constructed and QM calculations should be performed. Since the size of a model molecule is likely rather modest the calculation will not be too time consuming. Ultimately the predicted spectrum will be a sum of the spectra calculated using both methods—empirical and quantum-mechanical.

Nowadays chemical structure elucidation is commonly considered to be complete when the relative or hopefully the absolute stereochemistry of a new compound is also determined. CASE methods, in combination with QM chemical shift calculations, can also be used for this goal. In practice several stereoisomers, a total of $2^N$ where $N$ is the number of stereogenic centers, can be selected by the researcher as potential structures and then the best stereoisomer is selected on the basis of QM NMR chemical shift prediction using either $^{13}$C and/or $^1$H prediction. A question that can be posed is whether it is possible to make selection of the isomer set to be tested less subjective? Experiments show that the calculation of $^{13}$C chemical shifts for a full set of $2^N$ stereoisomers using a fragmental approach, whose algorithms are sensitive to stereochemistry, allows the perspective stereoisomers to be characterized by the minimal deviations between the experimental and predicted spectra and moving the most likely structure to the top of the list. The fragmental approach can therefore be used as a preliminary filter for selecting the stereoconfigurations that can then be checked using QM calculations.

More precise methods for the determination of stereochemistry in organic molecules developed in the framework of CASE systems is based on the application of geometrical constraints imposed by NOESY/ROESY 2D NMR spectra. The constraints are introduced into a molecular mechanics program and utilize a penalty function, depending on the constraints, to distinguish the most probable stereoisomer among the list of $2^N$ possible isomers. For molecules containing too many stereocenters ($N > 7$–8) the application of a genetic algorithm leads to an enormous reduction in computational costs. For instance, the correct stereochemistry of the brevetoxin B molecule (where $N = 23$) was determined in about 3 h using a typical modern PC. The limitation of this method is in its applicability only to rigid molecules which is, fortunately, typical for many natural products. A modern CASE expert system is capable therefore not only of establishing the structure of a new compound but also of determining its stereochemistry.

The success or failure of any expert system depends on the degree to which the system has to be adjusted to the properties of the information that needs to be processed. The structural information carried within 2D NMR spectra is *fuzzy by nature* (since the lengths of the HMBC correlations are not strictly defined), *incomplete* (not all possible correlations are observed), frequently *contradictory* (the presence of COSY and HMBC "nonstandard" correlations (NSCs) with $^nJ_{HH,CH}$ for which $n = 4,5,6$) and *uncertain* (the number of NSCs and their lengths are unknown). Signal overlap makes the correlation lengths *ambiguous*, and if there is a deficit of hydrogen atoms then the 2D NMR data are sparse and the data is *insufficient* to allow structure elucidation. In addition, due to the *subjective* misinterpretation of some 2D NMR peaks the information can become *false*.

These challenging properties of 2D NMR data clearly indicate that many challenges exist for either the researcher or the computer program modeling the path from the spectra to the structure of an unknown. An expert system, ACD/Structure Elucidator (StrucEluc), has been developed with the goal of taking into account *all* of these challenges during the treatment of the data starting from the processing of the raw 2D NMR spectra. Among contemporary 2D NMR based expert systems, ACD/Structure Elucidator is likely the most advanced CASE software system available. This flexible expert system provides a series of operating regimes and selection of the most appropriate one can be conducted in an interactive mode. The Common Mode delivers an *ab initio* solution: that is all plausible structures are generated from atoms under constraints derived from the 2D NMR data and axiomatic knowledge built into the system. If the initial information is good then the output file will contain the correct structure and can be detected by spectrum prediction. It is worthy to note that the possibility of obtaining a solution does not depend on the size of the molecule: it depends only on the quality of the initial information and especially on the number of homonuclear and heteronuclear 2D correlations. The number of correlations is higher with the higher degree of saturation.

The StrucEluc system contains an algorithm that is capable of revealing the presence of non-standard correlations (NSCs) in 2D NMR data. If the initial logical analysis of the data detects the presence of NSCs then the system tries to remove the contradictions and solves the problem. The most efficient and powerful tool to overcome the contradictions in spectral data is Fuzzy Structure Generation (FSG). This mode allows a correct solution to be found even in those cases when 2D NMR data contain an *unknown number* of nonstandard correlations of *unknown lengths*. Particularly, problems have been solved for which the number of NSCs was up to 15, while the *n* value describing the long-range nature of the correlations varied between 4 and 6 for $^nJ_{HH,CH}$.

To overcome the lack of structural information (the high degree of unsaturation, low spectra quality, *etc.*), the so-called Fragment Mode turned out to be very efficient. StrucEluc performs a fragment search in a database containing *ca.* 2.3 million fragments using a $^{13}C$ NMR subspectrum-based search. It determines the most perspective fragments (if found) which are encoded into the 2D NMR data and then utilizes them for subsequent structure generation. The Fragment Mode can, however, fail: appropriate fragments may not be found because the molecule of the unknown may be built of new structural moieties that were previously unknown. To overcome this difficulty the following practical approach is offered. Most frequently, the origin of an analyzed compound is known, as it is typical for natural products, so a set of related compounds with assigned NMR spectra may be available. The StrucEluc program automatically creates a User Fragment Library from the related structures and then can use it to perform a standard fragment search. A successful User Fragment library can often be produced even from a single related structure. This approach has solved numerous problems that could not be solved using human interpretation or standard CASE approaches.

The initial information fed to the CASE program can be false, that is it contains incorrect hypotheses (structural constraints) introduced by the researcher. This is revealed during the process of problem solving. Either an empty output structural file or a large deviation value is calculated for the best structure and serves as an identifier of false information. The detection of incorrect "axioms" is attained iteratively using trial methods. StrucEluc has fast algorithms for both structure generation and NMR spectrum prediction and allows quick evaluation of the quality of the initial information and thereby can save the researcher significant time.

The system is capable of elucidating the structure of a new compound and determining a set of the most probable stereoisomers or the preferable stereoisomer and its 3D model. A large number of physicochemical parameters of the generated molecules can be predicted by prediction algorithms implemented into StrucEluc and this simplifies the subsequent investigations of the new compound properties.

In earlier versions of CASE expert systems the developers were pushed to reduce the output file as much as possible to alleviate the selection process to identify the correct structure. This goal can only be achieved by utilizing as many different structural constraints as possible which unfortunately increases the risk of involving incorrect hypotheses: at least one wrong hypothesis results in failure to get the correct solution! Implementing fast and accurate empirical methods for NMR shift prediction into StrucEluc removed the challenge of handling large output files and the system is capable of selecting the correct structure from among tens of thousands of candidates. As a result all questionable hypotheses can be omitted from the set of "axioms" composing the partial axiomatic theory associated with a given problem and therefore the reliability of the obtained solution markedly increases.

The practicality and efficiency of StrucEluc was corroborated by elucidating hundreds of complex natural product structures as well as by the successful application of the system in numerous companies and scientific institutions around the world.

A comparison of the problem solutions obtained in parallel using both traditional and CASE methods has clearly demonstrated the indisputable advantage of computer assisted structure elucidation approaches. A human expert cannot possibly enumerate all possible structures that satisfy all experimental data and, moreover, it turns out that the suggested structures are frequently not the logical consequences of the analysis of the data provided. This circumstance, in combination with the quantum-mechanical calculation of the NMR chemical shifts of all structures (as customized by some researchers) for selecting the preferred isomer allows one to conclude that this methodology is extremely unproductive. Using a systematic approach to problem solving, StrucEluc most frequently finds the correct structure in several seconds or minutes.

Due to the complexity and difficulty associated with the manual interpretation of 2D NMR data, it is not surprising that incorrectly elucidated structures appear rather frequently in peer-reviewed journals. The analysis of experimental data published in a large series of articles with the aids of StrucEluc has convincingly shown that if researchers used a CASE system in their research then errors in spectral interpretation would not occur and the correct structure would be obtained. In the worst case, the program would give either a hint or exact indication to the chemist that the elucidated structure is wrong. A large value of the deviation between the experimental and predicted spectra for the best structure suggests that both the spectral data and the set of assumptions introduced during the process of structure elucidation should be rechecked.

In spite of the impressive progress in CASE development it should be understood that expert systems will never fully exclude the human expert from the process of molecular spectral analysis. The level of automation will only asymptotically approach the potential of full automation. Structure elucidation implies a synergistic interaction between the spectroscopist and computer, which is a direct consequence from the principle of complementarity originally established by Niels Bohr. An expert system is nothing more than a powerful amplifier of the human intellect. We may expect that as CASE algorithms improve, and computers become faster, that more complex problems will be solvable.

Recently it was shown that StrucEluc can successfully and quickly elucidate chemical structures that were considered by experienced spectroscopists as unsolvable. This evidence unambiguously demonstrates that a modern CASE expert system should be considered as an integral part of a spectroscopists' armory for structure elucidation. The capabilities of NMR experimental techniques are best proven in conjunction with the mathematical aids developed for spectral data analysis and the logical inference of all structures consistent with the experimental data and other available information. We believe that in future CASE software will become a common tool for NMR spectroscopists much like the software that today is an integral part of X-ray crystallography.

We expect that in the near future the further development of expert systems will make such software applications versatile analytical tools that will ultimately become indispensable, not only for structure elucidation but also for the determination of the most probable relative stereochemistry of a newly isolated or synthesized natural product. We also believe that the teaching of CASE methods in universities will help a new generation of chemists to work more efficiently. It will eventually lead to such expert systems becoming routine tools available in the majority of organic and analytical chemistry laboratories.

# Subject Index

References to figures are given in *italic* type. References to tables are given in **bold** type.